Communications
in Computer and Information Science   **2229**

Rationale

The CCIS series is devoted to the publication of proceedings of computer science conferences. Its aim is to efficiently disseminate original research results in informatics in printed and electronic form. While the focus is on publication of peer-reviewed full papers presenting mature work, inclusion of reviewed short papers reporting on work in progress is welcome, too. Besides globally relevant meetings with internationally representative program committees guaranteeing a strict peer-reviewing and paper selection process, conferences run by societies or of high regional or national relevance are also considered for publication.

Topics

The topical scope of CCIS spans the entire spectrum of informatics ranging from foundational topics in the theory of computing to information and communications science and technology and a broad variety of interdisciplinary application fields.

Information for Volume Editors and Authors

Publication in CCIS is free of charge. No royalties are paid, however, we offer registered conference participants temporary free access to the online version of the conference proceedings on SpringerLink (http://link.springer.com) by means of an http referrer from the conference website and/or a number of complimentary printed copies, as specified in the official acceptance email of the event.

CCIS proceedings can be published in time for distribution at conferences or as postproceedings, and delivered in the form of printed books and/or electronically as USBs and/or e-content licenses for accessing proceedings at SpringerLink. Furthermore, CCIS proceedings are included in the CCIS electronic book series hosted in the SpringerLink digital library at http://link.springer.com/bookseries/7899. Conferences publishing in CCIS are allowed to use Online Conference Service (OCS) for managing the whole proceedings lifecycle (from submission and reviewing to preparing for publication) free of charge.

Publication process

The language of publication is exclusively English. Authors publishing in CCIS have to sign the Springer CCIS copyright transfer form, however, they are free to use their material published in CCIS for substantially changed, more elaborate subsequent publications elsewhere. For the preparation of the camera-ready papers/files, authors have to strictly adhere to the Springer CCIS Authors' Instructions and are strongly encouraged to use the CCIS LaTeX style files or templates.

Abstracting/Indexing

CCIS is abstracted/indexed in DBLP, Google Scholar, EI-Compendex, Mathematical Reviews, SCImago, Scopus. CCIS volumes are also submitted for the inclusion in ISI Proceedings.

How to start

To start the evaluation of your proposal for inclusion in the CCIS series, please send an e-mail to ccis@springer.com.

Bin Xu · Bing Qin · Yannis Tzitzikas ·
Zhichun Wang · Xian-Ling Mao · Ming Liu ·
Pavlos Fafalios · Yongbin Liu · Zhiliang Tian ·
Michalis Mountantonakis
Editors

China Conference on Knowledge Graph and Semantic Computing and International Joint Conference on Knowledge Graphs

International Joint Conference, CCKS-IJCKG 2024
Chongqing, China, September 20–22, 2024
Proceedings

Springer

Editors
Bin Xu
Tsinghua University
Beijing, China

Yannis Tzitzikas
FORTH
University of Crete
Heraklion, Greece

Xian-Ling Mao
Beijing Institute of Technology
Beijing, China

Pavlos Fafalios
Technical University of Crete
Heraklion, Greece

Zhiliang Tian
National University of Defense Technology
Changsha, Hunan, China

Bing Qin
Harbin Institute of Technology
Harbin, China

Zhichun Wang
Beijing Normal University
Beijing, China

Ming Liu
Harbin Institute of Technology
Harbin, China

Yongbin Liu
University of South China
Hengyang, Hunan, China

Michalis Mountantonakis
Institute of Computer Science (ICS),
the Foundation for Research and Technology
Heraklion, Greece

ISSN 1865-0929   ISSN 1865-0937 (electronic)
Communications in Computer and Information Science
ISBN 978-981-96-1808-8   ISBN 978-981-96-1809-5 (eBook)
https://doi.org/10.1007/978-981-96-1809-5

This Springer imprint is published by the registered company Springer Nature Singapore Pte Ltd.
The registered company address is: 152 Beach Road, #21-01/04 Gateway East, Singapore 189721, Singapore

If disposing of this product, please recycle the paper.

Preface

This volume contains the papers presented at CCKS-IJCKG 2024: the China Conference on Knowledge Graph and Semantic Computing and International Joint Conference on Knowledge Graphs held during September 20–22, 2024, in Chongqing.

CCKS is organized by the Technical Committee on Language and Knowledge Computing of the Chinese Information Processing Society, and was previously held in Beijing (2016), Chengdu (2017), Tianjin (2018), Hangzhou (2019), Nanchang (2020), Guangzhou (2021), Qinhuangdao (2022), and Shenyang (2023).

CCKS is the merger of two previously held relevant forums, i.e., the Chinese Knowledge Graph Symposium (CKGS) and the Chinese Semantic Web and Web Science Conference (CSWS). CKGS was previously held in Beijing (2013), Nanjing (2014), and Yichang (2015). CSWS was first held in Beijing in 2006 and has been the main forum for research on Semantic (Web) technologies in China for a decade. Since 2016, CCKS has brought together researchers from both forums and covers wider fields, including knowledge graphs, the Semantic Web, linked data, natural language processing, knowledge representation, graph databases, information retrieval, and knowledge aware machine learning. It aims to become the top forum on knowledge graph and semantic technologies for Chinese researchers and practitioners from academia, industry, and government.

The 13th International Joint Conference on Knowledge Graphs (IJCKG 2024) was a premium academic forum on Knowledge Graphs. The mission of IJCKG 2024 was to bring together international researchers in the Knowledge Graph community and other related areas to present their innovative research results or novel applications of Knowledge Graphs. IJCKG evolved from the Joint International Semantic Technology Conference (JIST): a joint event for disseminating research results regarding the Semantic Web, Knowledge Graphs, Linked Data, and AI on the Web.

The 18th China Conference on Knowledge Graph and Semantic Computing and The 13th International Joint Conference on Knowledge Graphs (CCKS-IJCKG 2024) was jointly undertaken by Chongqing University and held in Chongqing from September 20th to 22th, 2024.

The theme of this conference was "Knowledge Graphs and Large Language Models", and the conference aimed to explore the intersection of these two cutting-edge fields and innovative opportunities for their integration, explore the role and application of knowledge graphs in cross-platform, cross-field, and other AI tasks, and study knowledge representation, knowledge storage, knowledge mining, knowledge fusion, knowledge reasoning, and other key technologies of knowledge graph development in the context of Artificial General Intelligence, guide the transformation of knowledge graph-related technologies, and lay the foundation for the ultimate realization of Artificial General Intelligence. The conference program included workshops, keynotes, a frontiers and trends forum, industry forums, young scholars forums, evaluations and competitions, paper presentations, posters, and demos. We invited well-known scholars to introduce the latest developments in related fields and development trends, and well-known R&D experts in industry to share practical experience, and promote industry-university-research cooperation.

As for peer-reviewed papers, 168 submissions were received in the following eight areas,

– Knowledge Graph-Enhanced Large Language Models
– Knowledge Representation and Reasoning
– Knowledge Graph Construction and Knowledge Integration
– Graph Database and Knowledge Management
– Machine Learning on Graphs
– Knowledge Retrieval and Information Retrieval
– Knowledge Graph and Large Language Model Applications
– Knowledge Graph Open Resources

During the reviewing process, each submission was assigned to at least three Program Committee members. The committee decided to accept 30 full papers. This CCIS volume contains revised versions of 30 English full papers.

Additionally, the evaluation track this year set up 11 tasks, which attracted more than 2400 teams to participate, forming a highly influential event. Besides the bonuses and certificates issued to the top three teams in each task, the committee also encouraged them to submit evaluation papers. After peer review by experienced researchers and task organizers, 11 papers were accepted (after revision) for inclusion in this volume of proceedings.

The hard work and close collaboration of a number of people contributed to the success of this conference. We would like to thank the Organizing Committee and Program Committee members for their support, and the authors and participants who are the primary reason for the success of this conference. We also thank Springer for their trust and for publishing the proceedings of CCKS-IJCKG 2024.

Finally, we appreciate the sponsorships from Meituan as Diamond Sponsor, Zhipu AI as platinum sponsor, Baidu, Ant Group, CISDI GROUP, and K2Data as gold sponsors, and SHOUHENG SOFTWARE, Top AI, BaYou, Global Tone Communication Technology, and HPCMaster as silver sponsors.

September 2024

Bin Xu
Bing Qin
Yannis Tzitzikas
Zhichun Wang
Xian-Ling Mao
Ming Liu
Pavlos Fafalios
Yongbin Liu
Zhiliang Tian
Michalis Mountantonakis

Organization

General Chairs

Bin Xu — Tsinghua University, China
Bing Qin — Harbin Institute of Technology, China
Yannis Tzitzikas — University of Crete, Greece

Program Committee Chairs

Zhichun Wang — Beijing Normal University, China
Xian-Ling Mao — Beijing Institute of Technology, China
Ming Liu — Harbin Institute of Technology, China
Pavlos Fafalios — Technical University of Crete, Greece

Local Chairs

Jiang Zhong — Chongqing University, China
Dongsheng Zou — Chongqing University, China
Yongpan Sheng — Southwest University, China

Publicity Chairs

Linmei Hu — Beijing Institute of Technology, China
Wenjie Li — Hunan First Normal University, China
Beatriz Esteves — Ghent University, Belgium

Publication Chairs

Yongbin Liu — University of South China, China
Zhiliang Tian — National University of Defense Technology, China
Michalis Mountantonakis — FORTH-ICS, Greece

Tutorial Chairs

Shasha Li   National University of Defense Technology,
                                         China
Dimitris Plexousakis   FORTH-ICS and University of Crete, Greece

Evaluation Chairs

Tianxing Wu   Southeast University, China
Yuanzhe Zhang   Institute of Automation, Chinese Academy of
                                         Sciences, China

Frontiers and Trends Forum Chair

Ningyu Zhang   Zhejiang University, China

Young Scholars Forum Chair

Yuxiao Dong   Tsinghua University, China

Knowledge Graph Education Forum Chairs

Xin Wang   Tianjin University, China
Meng Wang   Tongji University, China

Poster/Demo Chairs

Wen Zhang   Zhejiang University, China
Manolis Koubarakis   University of Athens, Greece

Sponsorship Chairs

Changliang Xu   Hangzhou Institute for Advanced Study, UCAS,
                                         China
Lei Hou   Tsinghua University, China

Industry Track Chairs

Feiyu Xiong   Shanghai Institute of Algorithm Innovation, China
Xiaowang Zhang   Tianjin University, China
Mehdi Rezagholizadeh   Noah's Ark Lab, Huawei Technologies, China

Registration Chair

Rongzhen Li   Chongqing University, China

Website Chairs

Pengfei Cao   Institute of Automation, Chinese Academy of
                                 Sciences, China
Yannis Marketakis   FORTH-ICS, Greece

Area Chairs

Knowledge Representation and Reasoning

Qiannan Zhu   Beijing Normal University, China
Giorgos Flouris   ICS-FORTH, Greece

KG Construction and Integration

Yuting Wu   Beijing Jiaotong University, China
Konstantin Todorov   University of Montpellier, France

Graph Database and Knowledge Management

Qingfu Zhu   Harbin Institute of Technology, China
Haridimos Kondylakis   ICS-FORTH, Greece

Knowledge and Information Retrieval

Ai Qingyao   Tsinghua University, China
Carlos Bobed   University of Zaragoza, Spain

Machine Learning on Graphs

Yumin Shang	Beijing University of Posts and Telecommunications, China
Xiang Zhao	National University of Defense Technology, China
Danilo Dessi	GESIS, Germany

Knowledge Graph, NLP and LLMs

Zhen Wu	Nanjing University, China
Damien Graux	Huawei Technologies Ltd., China

Knowledge Graphs and Applications

Daifeng Li	Sun Yat-sen University, China
Grigoris Antoniou	Leeds Beckett University, UK

Knowledge Graph Open Resources

Tianxing Wu	Southeast University, China
Theodore Patkos	FORTH ICS, Greece

Program Committee

Alessandro Faraotti	IBM, Italy
Ana Iglesias-Molina	Universidad Politécnica de Madrid, Spain
Antonis Bikakis	University College London, UK
Bi Guanqun	Institute of Information Engineering, Chinese Academy of Sciences, China
Binxia Xu	University College London, UK
Carlos Bobed	University of Zaragoza, Spain
Chen Zhang	Peking University, China
Cheng Ling	Shanghai Jiao Tong University, China
Chengzhi Zhang	Nanjing University of Science and Technology, China
Da Li	ICT, China
Da Li	UCAS, China
Daifeng Li	Baidu, China
Damien Graux	Huawei Technologies R&D (UK) Ltd., UK
Danilo Dessì	GESIS – Leibniz Institute for the Social Sciences, Germany
Denghao Ma	Beijing Information Science and Technology University, China

Despina-Athanasia Pantazi	National and Kapodistrian University of Athens, Greece
Dong Zhang	Sun Yat-sen University, China
Dongfang Li	Harbin Institute of Technology, Shenzhen, China
Dongsheng Guo	Quan Cheng Laboratory, China
Edelweis Rohrer	Universidad de la República, Uruguay
Elena Montiel-Ponsoda	Universidad Politécnica de Madrid, Spain
Fernando Bobillo	University of Zaragoza, Spain
Floriano Scioscia	Polytechnic University of Bari, Italy
Gang Wu	Northeastern University, China
George Baryannis	University of Huddersfield, UK
George Stamoulis	National and Kapodistrian University of Athens, Greece
Georgios Meditskos	Aristotle University of Thessaloniki, Greece
Grigoris Antoniou	Leeds Beckett University, UK
Giorgos Flouris	ICS-FORTH, Greece
Giorgos Savathrakis	Institute of Computer Science - FORTH, Greece
Gong Cheng	Nanjing University, China
Guozheng Rao	Tianjin University, China
Hanlei Zhang	Tsinghua University, China
Hao Wang	Beijing Institute of Technology, China
Haridimos Kondylakis	Institute of Computer Science, FORTH, Greece
Heng Yu	Beijing Normal University, China
Heng-Da Xu	Beijing Institute of Technology, China
Hideaki Takeda	National Institute of Informatics, Japan
Hu Nan	Southeast University, China
Huaiyu Wan	Beijing Jiaotong University, China
Huang Qi	Jiangxi Normal University, China
Hui Huang	Harbin Institute of Technology, China
Huiying Li	Southeast University, China
Ilias Tachmazidis	University of Huddersfield, UK
Irini Fundulaki	ICS-FORTH, Greece
Jia Li	Peking University, China
Jiang Wenbin	Institute of Computing Technology, CAS, China
Jiaxin Ding	Shanghai Jiao Tong University, China
Jiazhan Feng	Peking University, China
Jihao Shi	HIT, China
Jingchang Chen	Harbin Institute of Technology, China
Jinglong Gao	Harbin Institute of Technology, China
Jinliang Lu	Institute of Automation, Chinese Academy of Sciences, China
Jinpeng Li	Peking University, China

Jinyu Guo	University of Electronic Science and Technology of China, China
Jose L. Martinez-Rodriguez	Autonomous University of Tamaulipas, Mexico
Juan Li	Zhejiang University, China
Jun Pang	Wuhan University of Science and Technology, China
Kai Xiong	Harbin Institute of Technology, China
Kai Zhang	Zhejiang University City College, China
Kang Xu	Nanjing University of Posts and Telecommunications, China
Katerina Papantoniou	ICS-Forth, Greece
Ken Fukuda	National Institute of Advanced Industrial Science and Technology, Japan
Liang Hong	Wuhan University, China
Liang Wen	Beijing Normal University, China
Liwen Zheng	Beijing University of Posts and Telecommunications, China
Long Bai	Institute of Computing Technology, Chinese Academy of Sciences, China
Ludovico Boratto	University of Cagliari, Italy
Manolis Koubarakis	National and Kapodistrian University of Athens, Greece
Mao Hongli	Beijing Institute of Technology, China
Meng Wang	Tongji University, China
Naoki Fukuta	Shizuoka University, Japan
Ningyu Zhang	Zhejiang University, China
Peng Peng	Hunan University, China
Qiyang Wang	Shanghai Jiao Tong University, China
Quzhe Huang	Peking University, China
Raúl García-Castro	UPM, Spain
Ruifang He	Tianjin University, China
Ruihan Hu	Beijing University of Posts and Telecommunication, China
Ruijie Wang	University of Zurich, Switzerland
Ruitao Feng	HTWK, China
Ryandhimas E. Zezario	Academia Sinica, Taiwan
Salim Hafid	University of Montpellier, France
Sebastien Montella	Huawei Research, UK
Sheng Bi	Southeast University, China
Shenghao Yang	Tsinghua University, China
Shiyao Cui	Institute of Information Engineering, Chinese Academy of Sciences, China
Shuang Zeng	ByteDance, China

Shusaku Egami   National Institute of Advanced Industrial Science
                             and Technology, Japan
Shuyi Wang   Tianjin Normal University, China
Shuyuan Zhao   Beijing Jiaotong University, China
Shuzheng Si   Peking University, China
Sotiris Batsakis   University of Huddersfield, UK
Sven Hertling   University of Mannheim, Germany
Takahiro Kawamura   National Agriculture and Food Research
                             Organization, Japan
Tan Hexiang   ICT, China
Theodore Patkos   Institute of Computer Science, FORTH, Greece
Theofilos Mailis   Greece
Vasilis Efthymiou   Harokopio University of Athens, Greece
Wei Hu   Nanjing University, China
Wei Huang   Institute of Computing Technology, Chinese
                             Acadamy of Sciences; University of Chinese
                             Academy of Sciences, China
Wei Huang   Chongqing Institute of Green and Intelligent
                             Technology, Chinese Academy of Sciences,
                             China
Wei Shen   Nankai University, China
Weijie Wu   Shanghai Jiao Tong University, China
Weixin Zeng   National University of Defense Technology,
                             China
Wen Zhang   Zhejiang University, China
Wenliang Chen   Soochow University, China
Wenpeng Lu   Qilu University of Technology, China
Wenqiang Liu   Tencent, China
Xiao Liu   Peking University, China
Xiaodong Feng   Sun Yat-sen University, China
Xiaowang Zhang   Tianjin University, China
Xu Lei   Wuhan University, China
Xue-Feng Xi   Suzhou University of Science and Technology,
                             China
Yang Li   Northeast Forestry University, China
Yao Li   Shanghai Jiao Tong University, China
Yaojia Lv   Harbin Institute of Technology, China
Yaojie Lu   Institute of Software, Chinese Academy of
                             Sciences, China
Yinglong Ma   NCEPU, China
Yizheng Zhao   Nanjing University, China
Yongbin Liu   Tsinghua University, China
Yongpan Sheng   Southwest University, China

Yongrui Chen	Southeast University, China
Youhuan Li	Hunan University, China
Yuhang Tian	Beijing Institute of Technology, China
Yu-Ming Shang	Beijing University of Posts and Telecommunications, China
Yuxia Geng	Zhejiang University, China
Yuxuan Lai	Peking University, China
Zechang Xiong	Beijing Jiaotong University, China
Zeming Xing	Nanjing University, China
Zequn Sun	Nanjing University, China
Zhen Wu	Nanjing University, China
Zheng Zihao	Harbin Institute of Technology, China
Zhixu Li	Fudan University, China
Zirui Wu	Peking University, China
Zixuan Li	Institute of Computing Technology, Chinese Academy of Sciences, China
Ziyang Chen	National University of Defense Technology, China

Sponsors

Diamond Sponsor

Platinum Sponsor

Gold Sponsors

Silver Sponsors

BaYou
八友科技

((•)) GTCOM

HP C Master

Contents

Graph Database and Knowledge Management

Machine Learning on Graphs

Knowledge Retrieval and Information Retrieval

Knowledge Graph and Large Language Model Applications

Knowledge Graph Open Resource

Poster and Demo

Evaluations

Knowledge Representation
and Reasoning

KG-Diffusion: An Improved Knowledge Graph Completion with Diffusion

Jiawei Meng$^{(\boxtimes)}$ and Wen Zhang

Zhejiang University, Hangzhou, China
{mjw.cs,zhang.wen}@zju.edu.cn

Abstract. Knowledge graphs (KGs) often suffer from missing links, hindering downstream tasks that rely on their completeness. In order to complete the KGs, existing knowledge graph completion (KGC) methods primarily leverage semantic or structural information within the KG. However, relying solely on this information often leads to suboptimal prediction performance. To overcome this limitation, we propose KG-diffusion, a novel KGC method designed to generate missing links and thereby enhance KG completion. KG-diffusion employs an encoder-decoder architecture operating in a continuous vector space. The encoder was used to learn KG embeddings, while Gaussian noise is incrementally introduced into the entity embeddings. The decoder's objective is to reconstruct the original embeddings from these noise representations. Recognizing the importance of conditioning in the generative process, we introduce an aggregator module that enhances conditioning by integrating information from neighboring entities. This enriched conditioning process guides the decoder towards more accurate generation. Extensive evaluations on two publicly available benchmark datasets demonstrate that KG-diffusion achieves state-of-the-art performance in KGC tasks.

Keywords: Knowledge Graph · Knowledge Graph Completion · Neural Network · Diffusion Model · Generation

1 Introduction

Knowledge graphs (KGs) store factual knowledge as triples (head entity, relation, tail entity). These rich factual content makes KGs indispensable in many fields, such as semantic retrieval [27], knowledge question answering [28], recommendation system [10,25]. However, the inherent incompleteness of KGs hinders their practical utility. Knowledge graph completion (KGC) tackles this challenge by predicting the missing tail (head) entity based on the given head (tail) entity and relation, forming complete triple, has recently received a lot of attention.

Two primary KGC paradigms exist: embedding-based (KGE) and GNN-based approaches. KGE methods, lauded for their simplicity and efficiency, project KG entities and relations into low-dimensional vector spaces [1,3,17,19]. Scoring functions then assess the plausibility of candidate triples. However, KGE

B. Xu et al. (Eds.): CCKS-IJCKG 2024, CCIS 2229, pp. 3–15, 2025.
https://doi.org/10.1007/978-981-96-1809-5_1

methods struggle to capture the structural information inherent in KGs and exhibit a heavy reliance on the chosen scoring function. To address these limitations, Graph Neural Networks (GNNs) were incorporated into the KGC task [2,9,21]. GNNs effectively model KG structure, enabling the learning of implicit entity representations through the aggregation of information from neighboring entities. Despite their success, most existing GNN-based KGC methods focus exclusively on the local topology of the KG, neglecting the rich information encoded in the global distributions [26], which limits prediction accuracy.

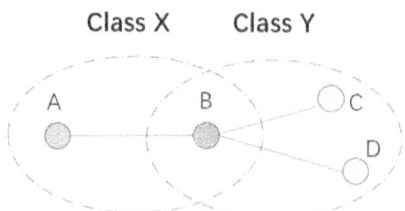

Fig. 1. Example of uncertainty in KGs. Class X and class Y are classes from two different domains. Thus, entity B has different meanings in different classes. For example, "Apple" can refer to either a kind of fruit or a company.

The inherent uncertainty associated with entities and relations in KGs arises from their participation in multiple triples, as shown in Fig. 1. This necessitates a representation that captures both their location and uncertainty. Multi-dimensional Gaussian distributions $\mathcal{N}(\mu, \sigma)$ offer a suitable solution [5]. This probability distribution representation of entities and relations constitutes a crucial piece of global structural information, which are difficult to be modeled by previous methods. Therefore, effectively learning and integrating the local structure representation embodied by the KG with the global probability distribution representation is paramount for achieving superior KGC accuracy.

Diffusion model, a type of generative model [8], have achieved remarkable success in recent years, particularly in image synthesis [7,16,30], text generation [6,11] and molecule generation [23]. Notably, these models demonstrate the capability to learn the implicit probability distribution within complex datasets.

Inspired by this, we propose leveraging diffusion models to capture the global probability distribution inherent in knowledge graphs and transform the link prediction task into a conditional entity generation problem. In this work, we introduce KG-diffusion, a novel knowledge graph completion method based on diffusion models. KG-diffusion captures the global probability distribution of a KG and guides the generation of missing links. We build upon the KGDM [26] diffusion model with several key improvements. KG-diffusion is designed as a network with the encoder-decoder architecture. The encoder projects KG entities and relations into a low-dimensional vector space, generating embeddings, while Gaussian noise is injected into the entity embeddings. The decoder uses an enhanced denoiser module based on the CE-Denoiser to recover original entity embeddings from corrupted representations, conditioned on provided information (relations and neighboring entities). To improve the quality of the denoiser's output in conditional generation model, we introduce an aggregation module. This module aggregates information from neighboring entities and relations, enriching the guiding conditions fed to the denoiser. We explore summation, mean, and multiplication as candidate aggregation functions and sample a fixed number of neighbors during the aggregation process to enhance training and reasoning efficiency.

In summary, the following are our key contributions:

- We introduce KG-diffusion, a novel knowledge graph completion method leveraging diffusion models. KG-diffusion uses an encoder-decoder architecture to generate high-quality entity embeddings from noise, conditioned on existing entities and relations, to complete missing triples within the KG. Unlike prior approaches, KG-diffusion explicitly models the probability distribution inherent in KGs, leading to a new entity-generation approach for KGC tasks.
- To enhance entity generation quality, KG-diffusion incorporates an aggregator module. This module aggregates embeddings of sampled neighboring entities and relations, enriching the conditioning information and strengthening constraints during decoding, leading to more accurate entity generation.
- We verify KG-diffusion on two benchmark datasets, FB15k-237 and WN18RR. KG-diffusion achieved Hits@10 scores of 0.684 on WN18RR and 0.619 on FB15k-237, which are 18.3% and 1.8% higher than the state-of-the-art methods, respectively.

2   Preliminaries

2.1   Problem Formulation

Knowledge graph is usually defined as $\mathcal{G} = (\mathcal{E}, \mathcal{R}, \mathcal{T})$, in which $\mathcal{E}$ is a collection of entities, $\mathcal{R}$ is a collection of relations, $\mathcal{T} = \{(h, r, t) \mid h, t \in \mathcal{E}; r \in \mathcal{R}\}$ is the set of fact triples. h, r, t are the head entity, relation and tail entity of the triple, respectively. Knowledge graph completion usually refers to the task of given an incomplete triple $(h, r, ?)$ of a missing entity, or $(?, r, t)$, the KGC model will predict the missing entities in it.

2.2   Diffusion Models

Diffusion models have been shown to be effective generative models [8], have showed success in image synthesis [7,16,30], text generation [6,11] and molecule generation [23] recently, which transforms data into Gaussian noise by forward and recovers data from pure noise by backward process. Specifically, for a given time step $T = \{0, 1, ..., T\}$, z_0 and original data distribution, and the forward process gradually has disturbance for the gaussian noise $z_T \sim \mathcal{N}(0, \mathbf{I})$:

$$q(z_t \mid z_{t-1}) = \mathcal{N}(z_t; \sqrt{1 - \beta_t} z_{t-1}, \beta_t \mathbf{I}) \tag{1}$$

where $\beta_t \in (0, 1)$, controls the disturbance noise at time step t. Based on reparameterization trick, let $\alpha_t = 1 - \beta_t$ and $\bar{\alpha}_t = \sum_{s=1}^{t} \alpha_s$, we can directly get any intermediate variable z_t from z_0:

$$q(z_t \mid z_0) = \mathcal{N}(z_t; \sqrt{\bar{\alpha}_t} z_0, \sqrt{1 - \bar{\alpha}_t} \mathbf{I}) \tag{2}$$

The reverse process is an approximate posterior $q(z_{t-1} \mid z_t)$, to predict the noise of the current time step t and denoise to restore to the next state of reverse z_{t-1}:

$$p_\theta(z_{t-1} \mid z_t) = N(z_{t-1}; \mu_\theta(z_t, t), \textstyle\sum_\theta(z_t, t)) \tag{3}$$

where $\mu_\theta(\cdot)$ and $\sum_\theta(\cdot)$ to be parameterized by the denoising neural networks e.g., Transformer [12,22] and UNet [4,14]:

$$\mu_\theta(z_t, t) = \frac{1}{\sqrt{\alpha_t}} (z_t - \frac{\beta_t}{\sqrt{1 - \bar{\alpha}_t}} f_\theta(z_t, t)) \tag{4}$$

3 Method

In this section, we introduce the modules of KG-diffusion. Figure 2 depicts the architecture of KG-diffusion, a novel knowledge graph completion method that leverages a conditional diffusion model within an encoder-decoder framework. The KG-diffusion training process consists of two distinct stages: forward and reverse. KG-diffusion comprises three core modules: **Encoder** utilizes a combination of embedding layers trained jointly. These layers project entities and relations within the knowledge graph (KG) into a lower-dimensional vector space. **Aggregator** focuses on conditioning information crucial for the decoding process (refer to Fig. 2). It incorporates two key steps, Neighbor Sampling and Embedding Aggregation. **Decoder** also referred to as the link generator, leverages a denoising architecture (the CE-Denoiser) to reconstruct the original entity embedding.

3.1 Encoder

In order to apply the continuous diffusion model and map the discrete knowledge graph information into the continuous space, we jointly learned three different embeddings emb_e, emb_r and emb_T, which map the entity $e \in \mathcal{E}$, relation $r \in \mathcal{R}$ and timestep $t \in T$ to the continuous vector space $emb_e(e)$, $emb_r(r)$ and $emb_T(t)$ respectively.

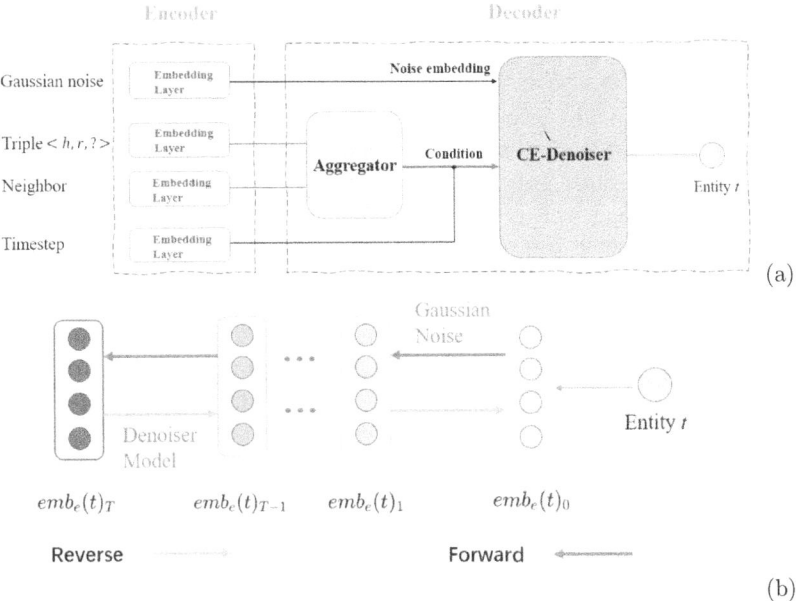

Fig. 2. Architecture of KG-diffusion. (a) Architecture of Encoder, Decoder and Aggregator. (b) Forward process and reverse process.

3.2   Aggregator

Our analysis revealed that triplet-based features within the knowledge graph are insufficient for capturing the intricate control necessary for precise generation outcomes. This limitation will be further corroborated through the ablation study. To address this challenge, we propose a condition Aggregator that leverages an aggregation function to incorporate both triplet features and local subgraph features as conditioning elements for refinement. Specifically, given a head entity denoted by h and a relation denoted by r, we integrate information from their one-hop neighbors into the guidance conditions. For a head entity h, its one-hop neighbors encompass entities directly connected to it via edges. Similarly, for a relation r, one-hop neighbors are defined as relations that share common connected entities. The condition will be augmented aggregated neighbor embeddings:

$$h_{agg} = f_{agg}(emb_e(h), \bigoplus_{(h,r,x) \in \mathcal{T}} emb_e(x)) \tag{5}$$

$$r_{agg} = f_{agg}(emb_r(r), \bigoplus_{Nei(r,x)=True} emb_r(x)) \tag{6}$$

where f_{agg} is a aggregation function and operator $\bigoplus$ is 'sum' in our experiment, $Nei(r, x)$ judge whether relations r and x have a common entity. Finally, the denoising conditions were obtained:

$$c = (emb_e(h), emb_r(r), h_{agg}, r_{agg}) \tag{7}$$

The same procedure is used for predicting head entities. However, incorporating all neighbors can lead to significant computational overhead. Observations indicate an uneven distribution of one-hop neighbors, with most entities having only a few dozen, as shown in Fig. 3. To address this, we use a fixed-size neighbor sampling strategy, which has proven effective.

An ablation experiment will validate that relying solely on entity and relation conditioning for denoising generation is insufficient. This approach degrades output quality. Conversely, our proposed condition Aggregator effectively guides entity generation while maintaining efficiency. KG-diffusion addresses data leakage by using neighborhood samples exclusively from the prior knowledge graph, ensuring a clear separation between training and testing datasets.

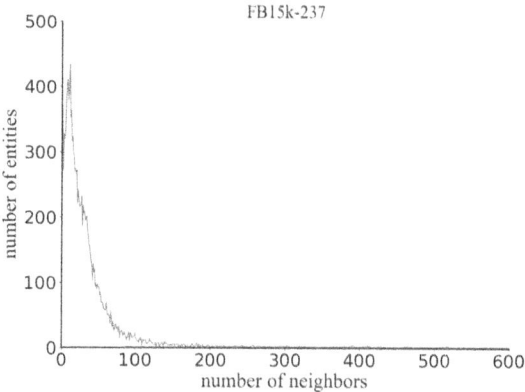

Fig. 3. Distribution of one-hop neighbors in FB15k-237. The horizontal axis represents the number of neighbors of the entity, and the vertical axis represents the total number of entities with the same number of neighbors.

3.3   Decoder

We input the Gaussian noise $emb_e(t)_k$ and the aggregated condition c into the link generator, and decode it to obtain the target entity. Specifically, most of the existing denoisers are used in the field of image or text generation, such as UNet and Transformer, and cannot be directly applied to knowledge graphs where there is a large amount of structured and unstructured information. Inspired by the MLP-based CE-Denoiser proposed by the KGDM [26], we regard this MLP with residual blocks as the denoiser of KG-diffusion.

CE-Denoiser. A key of CE-Denoiser is CE-Denoiser Block. Different from UNet of image Denoiser and Transformer of text Denoiser, CE-Denoiser Block consists of 6 alternating layers of MLP. Layer norm (Layernorm, LN) is applied before each layer, and uses residual connections around each sub-layer [24]. Considering that the triple form of the knowledge graph is simple enough, CE-Denoiser Block does not use the self-attention mechanism.

In CE-Denoiser, given the head entity $emb_e(h)$, relation $emb_r(r)$, time embedding $emb_T(k)$ and the noise embedding $emb_e(t)_k$ of the tail entity at time k, after CE-Denoiser Block (CE), the intermediate features are obtained:

$$E = CE(k, emb_e(t)_k, f_r(emb_e(h), emb_r(r))) \qquad (8)$$

where $f_r(.)$ is the scoring function. In our method, $f_r(h) = emb_e(h) + emb_r(r)$. Next, the intermediate feature E passed by a linear layer to obtain the predicted noise episode:

$$\epsilon = LinearLayer(LN(E)) \qquad (9)$$

3.4   Forward and Reverse

This section describes our Forward and Reverse process for implementing DDPM [7].

Forward. In the Forward stage of KG-diffusion, given the triples (h, r, t) used for training, the respective embeddings of entities and relations are first obtained through the Encoder, and then for the entity embedding $emb_e(t)$ that needs to be predicted, according to Eqs. 1 and 2, we projects the embedding of the tail entity t to the noise embedding $emb_e(t)_k$ by gradually adding Gaussian noise at each time step $k \in T$. The same procedure is used for predicting head entities.

Reverse. In the Reverse stage, KG-diffusion recovers the original entity embedding $emb_e(t)_0$ from the Gaussian noise embedding according to the given denoising conditions according to the reverse steps defined by Eqs. 3 and 4. Specifically, the posterior probability distribution of the Reverse stage approximation is expressed as:

$$p_\theta(emb_e(t_{k-1}) \mid emb_e(t_k), c) = \mathcal{N}(emb_e(t_{k-1}); \mu_\theta(emb_e(t_k), k, c), \sigma_k^2 \mathbf{I}) \qquad (10)$$

where $\sigma_k^2 \mathbf{I}$ is a constant. To predict entities, through the heavy parameter technique, μ_θ can be represented as:

$$\mu_\theta(emb_e(t_k), k, c) = \frac{1}{\sqrt{\alpha_k}}(emb_e(t) - \frac{\beta_k}{\sqrt{1 - \bar{\alpha}_k}}, f_\theta(emb_e(t_k), k, c)) \qquad (11)$$

3.5   Training and Inference

Based on the negative sampling technique proposed by [13,26], the final loss function we need to optimize is:

$$Loss = -\log\sigma(\gamma - D(emb_e(h), emb_e(t)_0)) - \sum_{i=1}^{n}\frac{1}{k}\log\sigma(D(emb_e(h), emb_e(t_i')_0) - \gamma) \qquad (12)$$

where t_i' is a tail entity of the negative sample triplet corresponding to the groundtruth triplet (h, r, t), $emb_e(t)_0$ and $emb_e(t_i')_0$ are the entity embeddings predicted by the denoising model. γ is a fixed margin, σ is the sigmoid function, D is the L1 distance. The same procedure is used for predicting head entities. During the inference stage, given an entity and relation, KG-diffusion samples neighbors from the graph of prior knowledge, then uses conditions to guide the Decoder to generate missing entity.

4   Experiment

In this section, we will demonstrate the effectiveness of our methods with comprehensive experiments. We first detail our experimental settings in Sect. 4.1. Subsequently, we present the results to address the following research questions (RQs):

– **RQ1**: Does KG-diffusion outperform existing baseline methods in knowledge graph completion tasks?
– **RQ2**: Does each module within KG-diffusion (Aggregator, Decoder) contribute significantly to the model's performance?

4.1   Experiment Setting

Datasets. To evaluate our proposed KG-diffusion method,we selected the two most commonly used dataset benchmarks for KGC tasks, FB15k-237 [18] and WN18RR [3], following the standard training/testing split.

Baselines. We compared with the following three types of KGC methods. KGE methods: TransE [1], DistMult [29] and RotatE [17]. GNN-based methods: ComplEX [19] COMPGCN [20], RGCN [15]. Diffusion methods: KGDM [26].

Evaluation Metrics. For the evaluation metrics, we calculated the Mean Rank (MR), the Mean Reciprocal Rank (MRR) and the hit rate with rank less than or equal to 1, 3 and 10 (H@1, H@3 and H@10). Lower MR, higher MRR and higher hit rate indicate better performance of the model.

Implementation Details. We use Adam as optimizer and apply grid search to tune the hyper-parameters, including learning rate lr, batch size b, embedding dimension d and neighbor sampled number of neighbor n. For each triple, we sample one-hop subgraphs, and use mean as aggregation function. We conduct all the experiments on Nvidia GeForce 3090 GPUs with 24 GB RAM.

Table 1. Results of knowledg graph completion on FB15k-237 and WN18RR. The best value in each column is marked in bold. The second best is underlined. Some baseline results are from the corresponding papers.

Model	FB15K-237					WN18RR				
	MR	MRR	H@1	H@3	H@10	MR	MRR	H@1	H@3	H@10
KGE Methods										
TransE	173	0.330	0.231	0.369	0.528	3384	0.226	0.014	0.401	0.529
DisMult	254	0.241	0.155	0.263	0.419	5110	0.43	0.39	0.44	0.49
RotatE	177	0.338	0.241	0.375	0.553	3340	0.476	0.428	0.492	0.571
GNN Methods										
ComplEx	333	0.241	0.189	0.301	0.452	5541	0.476	0.419	0.475	0.532
CompGCN	197	0.355	0.264	0.390	0.535	3533	0.479	<u>0.443</u>	0.494	0.546
RGCN	221	0.273	0.182	0.303	0.456	2719	0.402	0.345	0.437	0.494
GraiL	2053	–	–	–	–	2539	–	–	–	–
Diffusion Methods										
KGDM	<u>87</u>	**0.416**	**0.366**	**0.492**	<u>0.608</u>	<u>2843</u>	<u>0.502</u>	**0.449**	<u>0.510</u>	<u>0.578</u>
KG-diffusion (Ours)	**86**	<u>0.405</u>	<u>0.322</u>	<u>0.442</u>	**0.619**	**2274**	**0.507**	0.415	**0.560**	**0.684**

4.2 Main Results (RQ1)

Link Prediction Performance on Benchmark Datasets. Table 1 summarizes the results of our primary experiment on link prediction, conducted using two established benchmark datasets. As evident from the table, KG-diffusion achieves superior performance compared to most baseline approaches. On the FB15k-237 dataset, KG-diffusion surpasses the best baseline (KGDM) by 1.1% in Mean Rank (MR) and 1.8% in Hits@10 (H@10). However, it secures the second-best position in other evaluation metrics, performing slightly lower than KGDM. On the WN18RR dataset, KG-diffusion shows an 18.3% improvement in H@10 over the best baseline (KGDM) and leads in most other metrics.

Our proposed diffusion model-based approach demonstrates competitive performance against state-of-the-art (SOTA) baselines. Notably, it achieves superior effectiveness on the WN18RR dataset, even if it doesn't secure the top position in all metrics. These results convincingly validate the efficacy of our method for knowledge graph completion tasks, successfully addressing Research Question 1 (RQ1).

4.3 Ablation Study (RQ2)

There are three primary modules in KG-diffusion, Encoder, Aggregator and Decoder. We evaluate the effect of different modules on FB15K237 and WN18RR. 'Transformer' and 'RNN' means replacing the CE-Denoiser as denoiser with these two models, respectively. 'w/o Residual' indicates that the residual block in the CE-Block is not used. 'w/o Aggregator' means that no condition Aggregator is used, meaning only the embeddings of the given triples are used as condition. As can be seen from Table 2, the performance of KG-diffusion decreases significantly after the removal of different modules.

The results demonstrate a substantial performance decline across all metrics on both datasets after removing any of these modules. Notably, substituting the CE-Denoiser with Transformers or RNNs led to a significant drop exceeding 35.2% and 44.0% in Hits@10 performance on WN18RR. This observation suggests that the CE-Denoiser is particularly adept at capturing triplet features within this context compared to Transformer and RNN-based approaches. Furthermore, on WN18RR, removing the condition Aggregator resulted in a particularly pronounced decrease of 23.7% in Hits@10 performance. This finding underscores the critical role of the aggregation step in enhancing control over the entity generation process by incorporating contextual information.

This ablation study effectively addresses Research Question 2 (RQ2) by highlighting the indispensability of each module within KG-diffusion for achieving optimal performance.

Table 2. Results of ablation study on FB15k-237 and WN18RR. The best value in each column is marked in bold. We performed the following ablation experiments: (1) replacing CE-Denoiser with other models as denoiser, and 2. No residual blocks or aggregators are enabled.

Model	FB15K-237					WN18RR				
	MR	MRR	H@1	H@3	H@10	MR	MRR	H@1	H@3	H@10
Transformer	357	0.202	0.142	0.224	0.358	3496	0.268	0.172	0.289	0.443
RNN	675	0.134	0.083	0.129	0.221	4330	0.210	0.123	0.303	0.383
w/o Residual	342	0.200	0.172	0.247	0.353	3229	0.301	0.174	0.371	0.509
w/o Aggregator	302	0.292	0.266	0.323	0.440	3018	0.339	0.261	0.378	0.522
KG-diffusion (Ours)	**86**	**0.405**	**0.332**	**0.442**	**0.619**	**2274**	**0.507**	**0.415**	**0.560**	**0.684**

5   Conclusion and Limitation

We introduce KG-diffusion, a novel and efficient knowledge graph completion method based on diffusion models. KG-diffusion leverages an encoder-decoder framework to bridge the gap between the discrete nature of knowledge graphs and the continuous space used by diffusion models, and provides a promising solution for incorporating control conditions within diffusion models applied to knowledge graphs, effectively constraining generation quality. Experiments demonstrate KG-diffusion's effectiveness, achieving competitive performance against state-of-the-art methods. While KG-diffusion shows promising results, some limitations warrant further exploration. Firstly, the evaluation primarily focused on the quality of generated entities. Future work will delve deeper into the distribution of generated entities, particularly investigating whether the distribution of the generated entities is significantly different from baseline methods. Secondly, the reliance on diffusion models introduces a computational bottleneck due to their high time complexity and training difficulty compared to baseline methods. We plan to investigate more efficient models that better balance performance and computational efficiency.

Acknowledgment. This work is founded by National Natural Science Foundation of China (NSFC62306276), Zhejiang Provincial Natural Science Foundation of China (No. LQ23F020017), Yongjiang Talent Introduction Programme (2022A-238-G), Ningbo Natural Science Foundation (2023J291), and Fundamental Research Funds for the Central Universities (226-2023-00138).

References

1. Bordes, A., Usunier, N., Garcia-Durán, A., Weston, J., Yakhnenko, O.: Translating embeddings for modeling multi-relational data. In: Proceedings of the 26th International Conference on Neural Information Processing Systems, NIPS 2013, vol. 2
2. Defferrard, M., Bresson, X., Vandergheynst, P.: Convolutional neural networks on graphs with fast localized spectral filtering. In: Advances in Neural Information Processing Systems 29: Annual Conference on Neural Information Processing Systems (2016)
3. Dettmers, T., Minervini, P., Stenetorp, P., Riedel, S.: Convolutional 2d knowledge graph embeddings. In: Proceedings of the Thirty-Second AAAI Conference on Artificial Intelligence, (AAAI 2018), the 30th innovative Applications of Artificial Intelligence (IAAI 2018), and the 8th AAAI Symposium on Educational Advances in Artificial Intelligence (EAAI 2018)
4. Dhariwal, P., Nichol, A.Q.: Diffusion models beat gans on image synthesis. In: Advances in Neural Information Processing Systems 34: Annual Conference on Neural Information Processing Systems 2021, NeurIPS 2021
5. Feng, W., Zha, D., Guo, X., Dong, Y., He, Y.: Representing knowledge graphs with gaussian mixture embedding. In: Knowledge Science, Engineering and Management - 14th International Conference, KSEM 2021
6. Gong, S., Li, M., Feng, J., Wu, Z., Kong, L.: Diffuseq: sequence to sequence text generation with diffusion models. In: The Eleventh International Conference on Learning Representations, ICLR 2023
7. Ho, J., Jain, A., Abbeel, P.: Denoising diffusion probabilistic models. In: Advances in Neural Information Processing Systems 33: Annual Conference on Neural Information Processing Systems 2020, NeurIPS 2020
8. Kingma, D.P., Welling, M.: Auto-encoding variational bayes. arXiv preprint arXiv: 1312.6114
9. Kipf, T.N., Welling, M.: Semi-supervised classification with graph convolutional networks. In: 5th International Conference on Learning Representations, ICLR 2017
10. Koren, Y., Bell, R.M., Volinsky, C.: Matrix factorization techniques for recommender systems

11. Li, X., Thickstun, J., Gulrajani, I., Liang, P., Hashimoto, T.B.: Diffusion-lm improves controllable text generation. In: Advances in Neural Information Processing Systems 35: Annual Conference on Neural Information Processing Systems 2022, NeurIPS 2022

12. Lin, Z., et al.: Text generation with diffusion language models: a pre-training approach with continuous paragraph denoise. In: International Conference on Machine Learning, ICML 2023

13. Mikolov, T., Sutskever, I., Chen, K., Corrado, G.S., Dean, J.: Distributed representations of words and phrases and their compositionality. In: Advances in Neural Information Processing Systems 26: 27th Annual Conference on Neural Information Processing Systems (2013)

14. Rombach, R., Blattmann, A., Lorenz, D., Esser, P., Ommer, B.: High-resolution image synthesis with latent diffusion models. In: Proceedings of the IEEE/CVF Conference on Computer Vision and Pattern Recognition (2022)

15. Schlichtkrull, M.S., Kipf, T.N., Bloem, P., van den Berg, R., Titov, I., Welling, M.: Modeling relational data with graph convolutional networks. In: The Semantic Web - 15th International Conference, ESWC 2018

16. Song, J., Meng, C., Ermon, S.: Denoising diffusion implicit models. In: 9th International Conference on Learning Representations, ICLR 2021

17. Sun, Z., Deng, Z., Nie, J., Tang, J.: Rotate: knowledge graph embedding by relational rotation in complex space. In: 7th International Conference on Learning Representations, ICLR 2019

18. Toutanova, K., Chen, D.: Observed versus latent features for knowledge base and text inference. In: Proceedings of the 3rd Workshop on Continuous Vector Space Models and their Compositionality, CVSC 2015

19. Trouillon, T., Welbl, J., Riedel, S., Gaussier, É., Bouchard, G.: Complex embeddings for simple link prediction. In: Proceedings of the 33nd International Conference on Machine Learning, ICML 2016

20. Vashishth, S., Sanyal, S., Nitin, V., Talukdar, P.: Composition-based multi-relational graph convolutional networks. arXiv preprint arXiv: 1911.03082

21. Vashishth, S., Sanyal, S., Nitin, V., Talukdar, P.P.: Composition-based multi-relational graph convolutional networks. In: 8th International Conference on Learning Representations, ICLR 2020

22. Vaswani, A., et al.: Attention is all you need. In: Advances in Neural Information Processing Systems 30: Annual Conference on Neural Information Processing Systems 2017

23. Vignac, C., Krawczuk, I., Siraudin, A., Wang, B., Cevher, V., Frossard, P.: Digress: discrete denoising diffusion for graph generation. In: The Eleventh International Conference on Learning Representations, ICLR 2023

24. Wang, Q., et al.: Learning deep transformer models for machine translation. In: Proceedings of the 57th Conference of the Association for Computational Linguistics, ACL 2019

25. Wong, C., et al.: Improving conversational recommender system by pretraining billion-scale knowledge graph. In: ICDE 2021

26. Long, X., Zhuang, L., Li, A., Wei, J., Li, H., Wang, S.: Kgdm: a diffusion model to capture multiple relation semantics for knowledge graph embedding. In: The 38th Annual AAAI Conference on Artificial Intelligence, AAAI 2024

27. Xiong, C., Power, R., Callan, J.: Explicit semantic ranking for academic search via knowledge graph embedding. In: Proceedings of the 26th International Conference on World Wide Web, WWW 2017

28. Xu, Z., Zhang, W., Ye, P., Chen, H., Chen, H.: Neural-symbolic entangled framework for complex query answering. In: Advances in Neural Information Processing Systems 35: Annual Conference on Neural Information Processing Systems 2022, NeurIPS 2022
29. Yang, B., Yih, W., He, X., Gao, J., Deng, L.: Embedding entities and relations for learning and inference in knowledge bases. In: 3rd International Conference on Learning Representations, ICLR 2015
30. Zhao, S., et al.: Uni-controlnet: all-in-one control to text-to-image diffusion models. arXiv preprint, arXiv: 2305.16322

Cardiovascular Disease Knowledge Graph Reasoning Method Based on ConvKB Link Predication

Yu Song, Yongqi Zhu, Kunli Zhang$^{(\boxtimes)}$, Yingjie Han, Chenkang Zhu, and Bohan Yu

School of Computer and Artificial Intelligence, Zhengzhou University, Zhengzhou 450001, HN, China
ieklzhang@zzu.edu.cn

Abstract. Intelligent question answering systems based on knowledge graphs exhibit robust question understanding, high accuracy, and scalability. This paper explores the complexities of multi-hop queries in the context of knowledge graph question answering, with cardiovascular disease serving as a case study to examine reasoning methodologies for these intricacies. We propose a knowledge reasoning model, designated as Convolutional Neural Network Link Prediction (LPCNN), which is designed to address these challenges. LPCNN employs link prediction methods to achieve knowledge graph reasoning tasks. The core word and the question are mapped to the head entity and relation, then encode it. A convolutional neural network-based knowledge graph embedding model is employed for knowledge graph representation and candidate triple scoring. The highest-scoring tail entity is selected as the inference chain, thereby obtaining answers for corresponding hop counts. Experiments conducted on medical cardiovascular knowledge graph and public domain knowledge graph demonstrate that LPCNN can perform intelligent multi-hop reasoning for question answering. LPCNN exhibits certain advantages over other methods in reasoning for complex questions in medical cardiovascular knowledge graphs.

Keywords: Knowledge reasoning · Link prediction · Cardiovascular disease

1 Introduction

Intelligent question answering refers to systems based on natural language processing techniques, which analyze and understand user queries, extract relevant information from vast datasets, and ultimately provide responses in natural language form [1]. Early medical intelligent question answering systems employed rule-based and pattern-matching techniques, such as MedQA [2], AskHERMES [3], enquireMe [4]. With limitations in answer extraction quality. With the development of deep learning and natural language processing technologies, medical intelligent question answering has evolved to analyze questions and answers from a semantic perspective in addition to rule-based and pattern-matching methods, resulting in improved answer quality, as seen in systems like MEANS [5], AskCuebee [6]. There are three main approaches to intelligent question answering: intelligent question answering based on question-answer pairs, machine reading comprehension-based intelligent question answering, and knowledge graph-based intelligent question answering [7]. The knowledge graph-based approach involves identifying the core words and intentions in queries, then retrieving and reasoning answers from the knowledge graph. In the medical domain, a wealth of medical resources can be used to construct medical domain knowledge graphs. Building intelligent question answering systems based on this foundation enables semantic modeling of medical entities and relationships, thereby better understanding the meaning of queries and improving answer accuracy.

Knowledge reasoning is the core of knowledge graph question answering. One-hop reasoning refers to a reasoning method where only one step of reasoning between the question and existing knowledge is needed to obtain the answer [8]. In one-hop questions, identifying the core entities and relationships in the question is sufficient to determine the correct answer. However, in practical application scenarios, many questions involve multiple domains and knowledge points, requiring multiple steps of reasoning to obtain the answer. Multi-hop reasoning involves analyzing multiple entities and relationships [9], and the selection of relationships is crucial to the reasoning results. In medical intelligent question answering, different knowledge receives varying levels of attention. For example, relationships like "clinical symptoms" and "drug treatment" receive higher attention, with more descriptions, and they are densely distributed in the knowledge graph. Conversely, relationships like "incidence rate" and "mortality rate" receive relatively lower attention, with most descriptions being accurate numerical values, and they are sparse in the knowledge graph. Based on this characteristic, different knowledge reasoning methods can be used to obtain accurate knowledge chains and answers.

We propose a knowledge reasoning model based on convolutional neural network link prediction for complex multi-hop questions in cardiovascular medicine knowledge graphs. LPCNN employs a convolutional neural network-based knowledge graph embedding model, namely ConvKB, for knowledge graph embedding representation. The RoBERTa model is employed to encode the question and core words. By treating the question and core words as relations and head entities, respectively, LPCNN combines the triple scoring function of the ConvKB model to score the tail entities during each hop of reasoning. The entity with the highest score is selected as the answer or chain link. The main contributions of this paper are as follows:

- We constructed the CvdKGQA dataset for medical cardiovascular knowledge graph question answering and reasoning, which can be utilized for researching techniques for complex multi-hop question reasoning in medical knowledge graphs.
- We proposed the LPCNN model, consisting of knowledge graph embedding, question embedding, and answer selection modules, for conducting research on knowledge reasoning techniques using link prediction methods.
- We combining LPCNN with the Large Language Model (LLM) [10], through the experiments and analyses on the CvdKGQA and WebQSP datasets [11] demonstrate the superiority of the LPCNN model.

2 Related Work

In Multi-hop Knowledge Graph Question Answering (Multi-hop KGQA), it is necessary to analyze various entities and relationships in the knowledge graph to obtain relevant information. In this process, knowledge reasoning can assist the system in automatically inferring the associations between entities and relationships involved in user queries, thereby providing more accurate answers. Knowledge reasoning in Multi-hop KGQA mainly consists of logic-based, path-based, and embedding-based methods. The logical-based method transforms the question into logical expressions and utilizes logical reasoning techniques to search for answers in the knowledge graph that match the expressions. The path-based method starts from the topic entity in the user query and searches along multiple triple paths in the knowledge graph to find answer entities or relationships. The embedding-based method involves transforming the question and candidate answers into semantic vector representations in a shared vector space for processing.

Based on embedding methods, multi-hop question answering knowledge reasoning can better acquire semantic correlations, enhance model reasoning capabilities, and adapt to diverse application scenarios. According to the processing methods of embedding vector representations, it can be divided into three approaches: semantic matching, memory networks, and graph neural networks.

Semantic matching methods involve calculating the semantic matching between the embeddings of questions and candidate answers, and then obtaining the final answer by ranking the candidate answers. Bordes et al. [12] expressed the semantics of questions and knowledge graphs using embedding methods and applied them to respond to simple questions in KGQA. To achieve multi-hop question answering reasoning and address the challenges of multi-hop knowledge reasoning, Bordes et al. [13] encoded subgraphs of questions and answers in the knowledge graph to obtain richer semantic representations and infer answers to complex questions. Dong et al. [14] proposed methods to learn embedded representations of answer chains, question contexts, and types for complex multi-hop knowledge reasoning in the absence of annotated data.

Methods based on memory networks involve the long-term preservation of knowledge from graphs transformed into vectors, including features vector representation, memory updates, output feature mapping, and transformation output. Weston et al. [15] proposed a readable and writable external memory module used to preserve question information in multi-hop knowledge reasoning to achieve long-term memory goals. Miller et al. [16] introduced a key-value memory network model that completes multi-hop reasoning tasks by iteratively updating stored key-value pairs of memories. Xu et al. [17] proposed a mechanism for updating user question representations to enhance multi-hop knowledge reasoning capabilities. Zhou et al. [18] proposed an interpretable reasoning network, a memory network reasoning model with interpretability, using an interpretable step-by-step reasoning process to answer questions.

Methods based on graph neural networks have stronger expressive power when processing knowledge graph data, fully considering the entity and relationship information of thematic entities in the graph. Schlichtkrull et al. [19] proposed the Relation Graph Convolutional Network (R-GCN), which extends GCN by aggregating specific relationships, making it suitable for encoding multi-relationship graphs to predict answers to multi-hop questions. Teru et al. [20] proposed the GraIL framework, which utilizes a multi-relationship graph model based on attention mechanisms to perform inductive relation prediction on knowledge graphs. The attention mechanism focuses on two adjacent entities, their relationships, the target relationship to be predicted, and scores are assigned based on the embedding representations of the two adjacent entities, subgraphs, and the predicted relationship, with the highest scored entity being selected as the answer.

Methods based on graph neural networks can effectively capture the complex structures and semantic correlations between entities and relationships in knowledge graphs, possessing stronger expressive power. However, they come with high computational complexity and poor interpretability. On the other hand, methods based on memory networks can preserve long-term path knowledge, utilizing enhanced long-path information to improve reasoning capabilities. In contrast, methods based on semantic matching can reflect the semantic similarity between user queries and knowledge graphs. They are simple, intuitive, highly interpretable, and widely applicable, suitable not only for simple questions but also for complex multi-hop question-answering scenarios. Through rich semantic representations and reasoning capabilities, they can handle more complex knowledge reasoning tasks.

3   Methodology

The LPCNN model treats knowledge reasoning in knowledge graph question-answering as a link prediction task. It takes the core entity in the query as the starting point for knowledge reasoning. The corresponding entity in the knowledge graph is denoted as h, and encoding is performed after omitting the core entity from the query. This encoding is then treated as the relationship in the knowledge graph denoted as r. Thus, the knowledge reasoning task is transformed into a link prediction task in the knowledge graph. The task of predicting o given known h and r in (h, r, o). By leveraging the transitivity among multi-hop triplets in the ConvKB knowledge graph embedding model, complex multi-hop question reasoning is accomplished. The architecture of the LPCNN model, as shown in Fig. 1, mainly consists of three modules:

The knowledge graph embedding module utilizes the ConvKB model, based on convolutional neural networks, to embed all entities in the knowledge graph.

The question embedding module replaces the core words in the query and encodes the words in the query using the pre-trained RoBERTa model.

The answer selection module scores each candidate triplet using the question encoding and the ConvKB model, and determines the answer for the current hop based on the scoring results.

3.1   Knowledge Graph Embedding Module

In ConvKB, each triplet (S, P, O) is represented as a 3-column matrix $A = [V_h, V_r, V_t] \in R^{k \times 3}$, where each column vector represents an element of the triplet. Using a filter $\omega \in R^{1 \times 3}$ operating on the convolutional layer, it focuses on the global relationships between same-dimensional entities embedded in triplet (V_h, V_r, V_t) and also generalizes

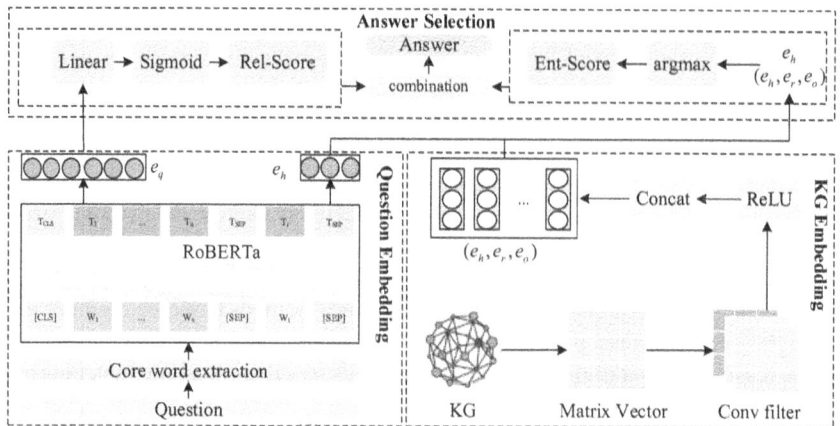

Fig. 1. Structure diagram of LPCNN model

excessive features in the model. This ω operation is repeated on each row of A, ultimately generating a feature map $V = [v_1, v_2, ..., v_k] \in R^k$, as shown in Eq. (1):

$$v_i = g(\omega \cdot A_i + b) \tag{1}$$

where $b \in R$ is a bias term, g is an activation function.

Different filters ω are used for knowledge graph embedding representations to generate different features. Let Ω and τ denote the sets of filters and the number of filters, respectively, $\tau = |\Omega|$. This yields τ-dimensional feature maps, which we concatenate them into a single vector. Finally, we scored the triplet (h, r, t) through dot product calculation, where the weight vector $W \in R^{\tau k}$; Fig. 2 illustrates the computational process in ConvKB.

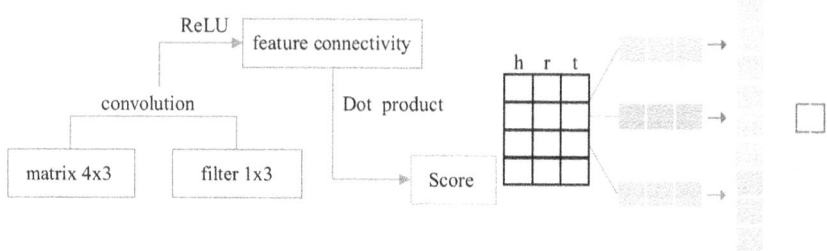

Fig. 2. Graph for computing triplet scores in ConvKB

The scoring function f is shown in Eq. (2):

$$f(h, r, t) = concat(g[v_h, v_r, v_t] * \Omega) * W \tag{2}$$

where Ω and W are shared parameters, $*$ represents the convolution operator, and *concat* represents the concatenation operation.

LPCNN utilizes the Adam optimizer to train the knowledge graph embedding representation model by minimizing the loss function ℓ. The L_2 regularization of the model's weight vectors is performed, with the loss function depicted in Eqs. (3) and (4):

$$\ell = \sum (h, r, t) \in \{GUG'\} \log\big(1 + \exp(1_{h,r,t} * f(h, r, t))\big) + \frac{\lambda}{2}\|w\|_2^2 \tag{3}$$

$$l_{(h,r,t)} = \begin{cases} 1, & if \ (h, r, t) \in G \\ -1, & if \ (h, r, t) \in G' \end{cases} \tag{4}$$

G' is a set of invalid triplets generated by corrupting valid triplets from G.

3.2 Question Embedding Module

The question embedding module embeds a natural language question q into a fixed-dimensional vector $e_q \in C^d$ using a feedforward neural network. The RoBERTa model is chosen to embed the question q into a 768-dimensional vector. This vector is then passed through four fully connected linear layers with ReLU activation functions, and finally mapped to a complex space.

Given an input text $T = \{t_1, t_2, ..., t_n\}$, the working principle of the RoBERTa layer is to utilize a Transformer structure to construct a multi-layer bidirectional encoder network, capable of reading the entire text sequence. Each layer can integrate contextual information, where T represents the word vectors corresponding to $H = \{h_1, h_2, ..., h_n\}$. H contains language information representations obtained during the model's pretraining process. Learning the embedding representation of the question q with encoded representation e^q, topic entity set $h \in E$ and the candidate answer entity set $a \in E$ is illustrated in Eqs. (5) and (6):

$$\varphi(e_h, e_q, e_a) > 0 \, \forall a \in A \tag{5}$$

$$\varphi(e_h, e_q, e_{\overline{a}}) > 0 \, \forall \overline{a} \notin A \tag{6}$$

φ is the scoring function learned in the knowledge graph embedding module. For each question, the final score $\varphi(\cdot)$ is computed over all candidate answer entities $a\prime \in E$. Learning is achieved by minimizing the binary cross-entropy loss between the scores and the target labels, where the target label is assigned 1 for correct answers and 0 otherwise.

3.3 Answer Selection Module

The answer selection module primarily reasons based on the knowledge graph embedding representations and question encoding results. During reasoning, the model scores the input $(head, question)$ against all candidate entities $a\prime \in E$ that could potentially be answers. The scoring process employs the ConvKB scoring method described in the knowledge graph embedding module, as shown in Eq. (7):

$$R_{a\prime} = \underset{a\prime \in \varepsilon}{argmax}\varphi(e_h, e_r, e_{a\prime}) \tag{7}$$

The final collection of scored candidate entity answers is denoted as $R_{a'}$. LPCNN learns a scoring function $S(r, q)$, which ranks each relationship $r \in R$ for a given question q. Here, h_r represents the encoding of the relationship, $q' = (< s >, w_1, w_2, ..., w_{|q|}, < /s >)$ represents the word sequence input to RoBERTa for question, and the scoring function is defined as the dot product between the output of RoBERTa's last layer hidden state $RoBERTa(h_q)$ and the embedding representation of the relationship $r(h_r)$. The calculation formula is shown in Eqs. (8) and (9):

$$h_q = RoBERTa(q') \tag{8}$$

$$S(r, q) = sigmoid(h_q^T h_r) \tag{9}$$

Among all relationships, we select the relationships with scores greater than a threshold as the candidate relationship set, denoted as R_a. For each candidate entity a', we sort them based on their scores. And we calculate the paths between the candidate answer entities and the core entity based on their scores. The highest scoring path between relations h and a' is determined, and its tail entity is selected as the final answer. Thus, the relationship scores for each candidate answer entity is the intersection between them, as shown in Eq. (10):

$$RelScore_{a'} = |R_a \cap R_{a'}| \tag{10}$$

The final answer entity is determined by the linear combination of relationship scores and ConvKB scores, as shown in Eq. (11):

$$e_{ans} = R_{a'} + \gamma RelScore_{a'} \tag{11}$$

γ is a learnable hyperparameter. When computing the final answer score, the scoring mechanism of the ConvKB model is adopted. The ConvKB model possesses transitivity, jointly representing multi-hop relationships during embedding representation. Therefore, when calculating the score, if there are multi-hop relationships in the knowledge graph, the score tends to be higher. This allows for greater improvement in scoring for some complex multi-hop combinations of relationships. By combining the relationship scores of candidate answer entities $RelScore_{a'}$, the multi-hop combination most relevant to the question can be selected, and the entities along that path are chosen as the answer. For a 2-hop question like 'What medical devices are used in surgery for acute coronary syndrome?', LPCNN selects the 1-hop relation 'surgical treatment' and the 2-hop relation 'surgical consumables', and combines the candidate entity scoring results to obtain the answer 'stent'.

4 Experiment

4.1 Dataset

CvdKG Dataset. We conducted an analysis of the most frequently asked questions related to cardiovascular diseases, referenced existing annotation standards, combined the characteristics of cardiovascular diseases, established a knowledge description system, and formulated annotation standards. The extraction of knowledge is conducted through the utilisation of manual, rule-based, and deep learning annotation techniques, with text similarity employed for the integration of knowledge derived from multiple sources. The constructed cardiovascular disease knowledge graph (CvdKG) comprises 15 entities and 52 relationships, resulting in a total of 36,378 entities and 68,316 triples. The CvdKG is capable of providing data support for intelligent question-answering about cardiovascular diseases. The CvdKGQA dataset was constructed based on the CvdKG, with data collected from the cardiovascular disease knowledge graph and cardiovascular disease online Q&A platforms, manually rewritten and templates extracted. Firstly, template questions are generated based on the typical one-hop, two-hop and three-hop combinations present within the knowledge graph. These generated template questions are then manually evaluated. Based on the evaluation results, the templates are adjusted in order to align the questions as closely as possible with human natural language questioning habits. The statistics of the CvdKGQA dataset are presented in Table 1.

Table 1. Statistics of the CvdKGQA dataset

hops	train	validation	test
1-hop	12,218	1,526	1,526
2-hop	14,312	1,845	1,845
3-hop	36,319	4,539	4,539

WebQSP Dataset.   The target database is Freebase. The defining feature of the WebQSP dataset is its relatively limited question scope, which is compensated for by the extensive knowledge graph it draws upon. The WebQSP dataset comprises natural language questions at one and two hops. Given the considerable scale of the Freebase knowledge base, we employ the GraftNet model to prune the knowledge graph, ensuring that only the specified predicates and triples are retained. The WebQSP dataset comprises a training set, a validation set, and a test set, with 2,998, 100, and 1,639 data points, respectively.

To validate the effectiveness of the LPCNN model in complex multi-hop question reasoning, experiments were conducted on the processed CvdKGQA dataset and WebQSP datasets.

4.2   Experimental Results and Analysis

In order to evaluate the effectiveness of the proposed knowledge reasoning model LPCNN, comparative experiments will be conducted using the inference models Graft-Net [21] of graph neural networks and the knowledge graph embedding models [22] RESCAL, TuckER, and ComplEX in EmbedKGQA [23] as benchmarks. Furthermore, an attempt will be made to utilise the large language model, Quenmax [24], to score the triples for the purpose of completing the inference chain selection, in contrast with the LPCNN answer selection module. In the experiments, the evaluation metric employed is Hits@1, with the results presented in Table 2. The WebQSP dataset comprises natural language questions of 1-hop and 2-hop complexities. The evaluation metrics for both 1-hop and 2-hop questions have been integrated.

Table 2. The experimental results of LPCNN on the CvdKGQA and WebQSP datasets.

model	CvdKGQA			WebQSP	
	1-hop	2-hop	3-hop	1-hop	2-hop
GraftNet	38.6	77.3	87.7	66.4	
RESCAL	39.3	80.4	**92.4**	65.4	
TuckER	40.7	77.2	90.9	63.7	
ComplEX	41.4	78.4	91.6	66.6	
LP-quenmax	40.2	80.3	88.6	-	
LPCNN	**42.3**	**83.3**	91.2	**67.5**	

Experimental Results. The experimental results demonstrate that ConvKB exhibits superior performance in the 1-hop and 2-hop datasets relative to the baseline models in CvdKGQA. In the more complex 3-hop dataset, the ConvKB model exhibits a 0.3% superior performance compared to the TuckER model, while its performance declines in comparison to the RESCAL and ComplEX models. The ConvKB, RESCAL, and ComplEX models are all knowledge graph embedding models with transitivity. They outperform TuckER, which lacks transitivity, in reasoning over complex questions, thereby demonstrating the feasibility of using knowledge graph embedding models with transitivity for link prediction inference in complex question reasoning. In the answer selection module, LPCNN also demonstrates an outstanding ability in selecting the correct answer chain, when compared to the large language model Queen-max.

The WebQSP dataset presents a more challenging environment than the CvdKGQA dataset, with a greater number of relationships and triples. The training samples are insufficient, and there are also issues with complex relationship distributions. The LPCNN model achieves an accuracy of 67.5% on the WebQSP dataset, indicating that the link prediction knowledge reasoning method based on ConvKB has more accurate reasoning capabilities when reasoning on large knowledge graphs.

Experimental Analysis. To further verify the performance of the link prediction knowledge reasoning question answering model, the performance of 2-hop data from the Cvd-KGQA dataset in ConvKB, RESCAL, and ComplEX with different training batches is analysed. The results demonstrate that the ConvKB model exhibits the highest accuracy, with a consistently increasing trend throughout the training process. The decline in accuracy during training for the RESCAL and ComplEX models is more pronounced than that for ConvKB, indicating that ConvKB exhibits superior performance on the 2-hop dataset. However, in the initial training batch, the accuracy of the RESCAL model is considerably higher than that of ConvKB and ComplEX, suggesting that it captures more nuanced textual feature information during the initial learning phase.

Case study Analysis. A comparative analysis of the models used in the experiments revealed discrepancies in performance during the knowledge reasoning phase. To illustrate, in the two-hop question "What medical devices are used in the surgery of acute coronary syndrome?", the inference processes of the ConvKB, RESCAL, and ComplEX models are presented in Table 3. The results yielded by both the ComplEX and RESCAL models were ultimately erroneous. In the case of the ComplEX model, an error occurred during the encoding of the question as relations, resulting in the inversion of relations between 1-hop and 2-hop. This ultimately led to the generation of no final answer output. In the RESCAL model, while a reference answer was provided in the form of a medical device, the selected relationships did not correspond consistently with the question. Specifically, the encoded part of the question in the 1-hop phase represented the relationship as "surgical consumables." In the chain prediction knowledge reasoning model, the selection of relationship representations and their order in question representation play a decisive role in determining whether the final answer selection is correct. In conclusion, ConvKB offers valuable contributions to the field of multi-hop knowledge reasoning question answering models.

Table 3. Results of 2-hop inference in the CvdKGQA dataset

Model	1hop	2hop	Answer
ComplEX	surgical consumables	surgical treatment	NULL
RESCAL	surgical consumables	surgical consumables	NULL
LPCNN	surgical treatment	surgical consumables	instruments

5   Conclusion

In this paper, we propose a knowledge reasoning model, LPCNN, which is based on convolutional neural network link prediction. The model is designed for use in answering complex multi-hop questions in the field of medical knowledge graph question answering. LPCNN employs link prediction for knowledge reasoning, wherein core words and questions are treated as head entities and relations, respectively, and encoded using the RoBERTa model. A convolutional neural network-based knowledge graph embedding model is employed for knowledge graph representation and triple scoring, with the tail entity with the highest score selected as the inference chain or answer. The accuracy rates on the CvdKGQA and WebQSP datasets are 72.3% and 67.5%, respectively, representing an average improvement of 1.6% and 2.1% compared to the RESCAL model. In the CvdKGQA dataset, the accuracy for two-hop questions is improved by 2.9%. This evidence supports the feasibility of employing a link prediction knowledge reasoning model based on ConvKB for reasoning over complex multi-hop questions in medical knowledge graphs.

We demonstrate the effectiveness of the LPCNN method in the context of cardiovascular medical knowledge graph question-answering. In future work, we will attempt to combine the LPCNN method with large language models and apply it to other medical knowledge graphs or knowledge graphs in other domains, with a view to further evaluating the performance of the proposed method.

Acknowledgments. The Science and Technology Innovation 2030- "New Generation of Artificial Intelligence" Major Project [No.2021ZD0111000], and Henan Provincial Science and Technology Research Project [No. 232102211039].

References

1. Shuifa, S., et al.: Review of graph neural networks applied to knowledge graph reasoning. J. Front. Comput. Sci. Technol. **17**(1), 27 (2023)
2. Al-Smadi, B.S.: Deberta-bilstm: a multi-label classification model of arabic medical questions using pre-trained models and deep learning. Comput. Biol. Med. **170**, 107921 (2024)
3. Cao, Y., et al.: Askhermes: an online question answering system for complex clinical questions. J. Biomed. Inform. **44**(2), 277–288 (2011)
4. Milne-Ives, M., et al.: The effectiveness of artificial intelligence conversational agents in health care: systematic review. J. Med. Internet Res. **22**(10), e20346 (2020)
5. Abacha, A.B., Zweigenbaum, P.: Means: a medical question-answering system combining nlp techniques and semantic web technologies. Inf. Process. Manage. **51**(5), 570–594 (2015)
6. Abdi, A., Idris, N., Ahmad, Z.: Qapd: an ontology-based question answering system in the physics domain. Soft. Comput. **22**, 213–230 (2018)
7. Chen, X., Jia, S., Xiang, Y.: A review: knowledge reasoning over knowledge graph. Expert Syst. Appl. **141**, 112948 (2020)
8. Jiang, Y., Chang, S., Wang, Z.: Transgan: Two transformers can make one stronggan. arXiv preprint arXiv:2102.07074 1(3) (2021)
9. Jiang, C., Ma, N., Wan, F.: Research on machine reading comprehension model based on bidirectional attention. In: 2023 IEEE International Conference on Sensors, Electronics and Computer Engineering (ICSECE), pp. 1157–1161. IEEE (2023)
10. Zhao, W.X., et al.: A survey of large language models. arXiv preprint arXiv:2303.18223 (2023)
11. He, G., Lan, Y., Jiang, J., Zhao, W.X., Wen, J.R.: Improving multi-hop knowledge base question answering by learning intermediate supervision signals. In: Proceedings of the 14th ACM International Conference on Web Search and Data Mining, pp.553–561 (2021)
12. Wang, Q., Mao, Z., Wang, B., Guo, L.: Knowledge graph embedding: A survey of approaches and applications. IEEE Trans. Knowl. Data Eng. **29**(12), 2724–2743 (2017)
13. Wang, C., Tan, X.P., Tor, S.B., Lim, C.: Machine learning in additive manufacturing: State-of-the-art and perspectives. Addit. Manuf. **36**, 101538 (2020)

14. Dong, L., Wei, F., Zhou, M., Xu, K.: Question answering over freebase with multicolumn convolutional neural networks. In: Proceedings of the 53rd Annual Meeting of the Association for Computational Linguistics and the 7th International Joint Conference on Natural Language Processing (Volume 1: Long Papers), pp. 260–269 (2015)

15. Minaee, S., Kalchbrenner, N., Cambria, E., Nikzad, N., Chenaghlu, M., Gao, J.: Deeplearning–based text classification: a comprehensive review. ACM Comput. Surv. (CSUR) **54**(3), 1–40 (2021)

16. Miller, A., Fisch, A., Dodge, J., Karimi, A.H., Bordes, A., Weston, J.: Key-value memory networks for directly reading documents. arXiv preprint arXiv:1606.03126 (2016)

17. Xu, K., Lai, Y., Feng, Y., Wang, Z.: Enhancing key-value memory neural networks for knowledge-based question answering. In: Proceedings of the 2019 Conference of the North American Chapter of the Association for Computational Linguistics: Human Language Technologies, Volume 1 (Long and Short Papers), pp. 2937–2947(2019)

18. Zhou, M., Huang, M., Zhu, X.: An interpretable reasoning network for multi-relation question answering. arXiv preprint arXiv:1801.04726 (2018)

19. Schlichtkrull, M., Kipf, T.N., Bloem, P., Van Den Berg, R., Titov, I., Welling, M.: Modeling relational data with graph convolutional networks. In: The semantic web: 15th international conference, ESWC 2018, Heraklion, Crete, Greece, 3–7 June 2018, proceedings 15, pp. 593–607. Springer (2018). https://doi.org/10.1007/978-3-319-93417-4_38

20. Wu, L., et al.: Graph neural networks for natural language processing: a survey. Foundat. Trends® Mach. Learn. **16**(2), 119–328 (2023)

21. Sun, H., Dhingra, B., Zaheer, M., Mazaitis, K., Salakhutdinov, R., Cohen, W.W.: Open domain question answering using early fusion of knowledge bases and text. arXiv preprint arXiv:1809.00782 (2018)

22. Sardana, A.: Embedkgqa: improving multi-hop question answering over knowledge graphs using knowledge base embeddings. In: ML Reproducibility Challenge 2020 (2020)

23. Saxena, A., Tripathi, A., Talukdar, P.: Improving multi-hop question answering over knowledge graphs using knowledge base embeddings. In: Proceedings of the 58th Annual Meeting of the Association for Computational Linguistics, pp. 4498–4507(2020)

24. Achiam, J.,et al.: Gpt-4 technical report. arXiv preprint arXiv:2303.08774 (2023)

25. Akdemir, E., Barışçı, N.: A review on deep learning applications with semantics. Expert Syst. Appli., 124029 (2024)

26. Zeng, D., Huang, T., Zhang, Z., Jiang, L.: Entity neighborhood awareness and hierarchical message aggregation for inductive relation prediction. Inf. Process. Manage. **61**(4), 103737 (2024)

Evolutionary Graph Network with Time-Aware Attention for Temporal Knowledge Graph Reasoning

Lijie Li[1], Yongyi Wang[1], Lingfu Wang[2], Jiahang Li[1], Ye Wang[1(✉)], Qilong Han[1], and Tao Ren[3]

[1] Harbin Engineering University, Harbin, China
{lilijie,wyywyy,lizhi01,wangye2020,hanqilong}@hrbeu.edu.cn
[2] University of Electronic Science and Technology, Chengdu, China
wanglf@std.uestc.edu.cn
[3] State Key Laboratory of Intelligent Game, Institute of Software Chinese Academy of Sciences, Beijing, China
rentao22@iscas.ac.cn

Abstract. Temporal Knowledge Graph Reasoning is an essential research area in the field of Temporal Knowledge Graphs, which make predictions about the future based on historical events. However, existing models have some drawbacks. (1) Current methods ignore influence of time information in historical snapshots. (2) Only the final prediction utilizes feature mining of temporal attributes. (3) The global information captured along the timeline is not complete enough. To address the aforementioned problems, we propose a temporal knowledge graph reasoning model named TIA-Net in this paper. TIA-Net is a convolution graph network model that integrates time-aware information using a multi-head attention mechanism. Firstly, the model incorporates time information encoding at each historical timestamp, temporal dependencies are captured from multiple perspectives using a multi-head attention mechanism. Secondly, the model utilizes a relational convolution graph network with evolutionary representation to capture structural dependencies among events in each snapshot. Lastly, the frequency characteristics of events, obtained by a global frequency adjustment mechanism, are used to improve prediction accuracy. Extensive experiments on five benchmark datasets show that TIA-Net outperforms the state-of-the-art temporal knowledge graph reasoning methods in most cases. Our source code is available at https://github.com/Cur-pro/TIA-Net.

Keywords: Temporal knowledge graph · Knowledge representation and reasoning · Time awareness · Attention mechanism

1 Introduction

Currently, rapidly growing data in the real world often exhibits complex temporal dynamics. To address this challenge, Temporal Knowledge Graphs (TKGs) are

B. Xu et al. (Eds.): CCKS-IJCKG 2024, CCIS 2229, pp. 28–40, 2025.
https://doi.org/10.1007/978-981-96-1809-5_3

proposed, where events are represented by a quadruple (subject, relation, object, time) [10]. Temporal Knowledge Graph Reasoning (TKGR) aims to infer answers to a specific query at a future moment based on static historical subgraphs from various time points, as shown in Fig. 1.

Fig. 1. Examples of different historical facts related to the query in ICEWS14.

Typically, human reasoning patterns are usually divided into three aspects. (1) People can consider the multidimensional impact of recent events on speculating potential future scenarios. (2) People can infer the cyclical nature of certain events. (3) People can promote reasoning by considering the frequency of events.

In this work, we propose a TKGR model named **TIA-Net**, which applying an evolutionary graph network with a time-aware information multi-head attention mechanism. In summary, the main contributions of this paper are as follows:

- TIA-Net simultaneously considers temporal dependencies, structural dependencies and the frequency characteristics of events.
- TIA-Net learns the representations of temporal information, entities, and relations at different historical timestamps by a time-aware attention method. Meanwhile, the fused representations are input to the model to capture temporal dependencies by a multi-head attention mechanism.
- TIA-Net effectively captures the frequency characteristics of events. The model calculates the frequency of events across various time spans and incorporates a global frequency adjustment mechanism for the final adjustment.
- Experiments on five TKG public datasets show that TIA-Net outperforms the state-of-the-art TKGR methods in most cases.

2   Related Works

2.1   Interpolation Reasoning

The historical subgraphs from moments 0 to t are known. However, each graph may generate new entities or relationships as time progresses. The purpose of interpolation reasoning is to complete the graph at the moment $tim \in [0, t]$ by inferring new entities and relationships to improve the information. There are many existing methods of interpolation reasoning. For example, TTransE [3] improved TransE [1] by adding temporal information to embedding relations to perform inference. HyTE [2] modeled temporal information as a hyperplane. DE-SimplE [4] characterized temporal information by learning embedding with different timestamps. TNTComplEx [9] modeled facts with temporal information as a fourth-order tensor. However, these methods only focus on known times-tamps and cannot capture the evolutionary patterns of sequence subgraphs, so they cannot be applied to the task of predicting future moments.

2.2 Extrapolation Reasoning

The purpose of extrapolation reasoning is to predict an event that occurs at the moment $t + 1$ based on a known subgraph of the history from moments 0 to t. This type of task is closer to real-life applications and presents more significant challenges, e.g., stock trend prediction. Know-Evolve [21] was the first work that utilized temporal point processes for multi-relational category inference. RE-GCN [14] captured the evolutionary patterns of historical subgraphs by looping R-GCN [17] and temporal gate cells, yet it failed to capture long-term temporal dependencies. CENET [22] improved on CyGNet [7] by focusing on emerging entities through contrast learning and classifiers. On the other hand, TiRGN [11] advanced entities and relationships in historical subgraphs and encodes temporal information. DHUNET [15] assigned different attention to historical subgraphs depending on the query. However, it failed to notice the characteristics of events by cyclical timing. Some additional models, such as DREAM [24] and CluSTeR [13], used reinforcement learning for retrieval inference. L^2TKG [23], GHT [20], and CEN [12] modeled by graph neural networks and recurrent neural networks. Nevertheless, none of the above models simultaneously captured the temporal dependence of historical snapshots, evolutionary representation structure dependence, and global frequency information.

3   Method

3.1   Problem Definition

A temporal knowledge graph is a collection of knowledge graphs with temporal information. According to the sequential nature of time, a static knowledge graph of events at each moment, (*i.e.*,timestamped snapshots) can be defined as $G = (G_1, G_2, \ldots, G_t)$. The timestamped snapshot G_t, denoted as $G_t = (E, R, F_t)$, represents the static knowledge graph of events for the moment where $t \in T$. Furthermore, each fact is a quadruple $F_t = (s, r, o, t)$. All notations presented in this paper are detailed in Table 1.

The main task in this paper is the entity prediction task. The query $(s, r, ?, t+1)$ denotes a known entity s and relation r speculating about the unknown entity at the moment $t + 1$. Query $(?, r, o, t + 1)$ is the same.

Table 1. The Summary of important notations.

Notations	Descriptions
$G = \{E, R, F_t, T\}$	Temporal knowledge graph
E, R, T	Entity set, Relationship set, Timestamp set
F_t	Fact set in TKG at timestamp t
H, R	Embedding matrices of entity and relation
$\mathbf{v^c}, v^{nc}$	Periodic and non-periodic time embeddings
s, r, t	Entity, Relation, Timestamp

3.2 Model Overview

The framework of TIA-Net is shown in Fig. 2, which includes the Evolutionary Representation Learning Module [11], Time-aware Attention Module, and Frequency Adjustment Module. TIA-Net simultaneously considers temporal dependencies, structural dependencies and the frequency characteristics of events from these three modules.

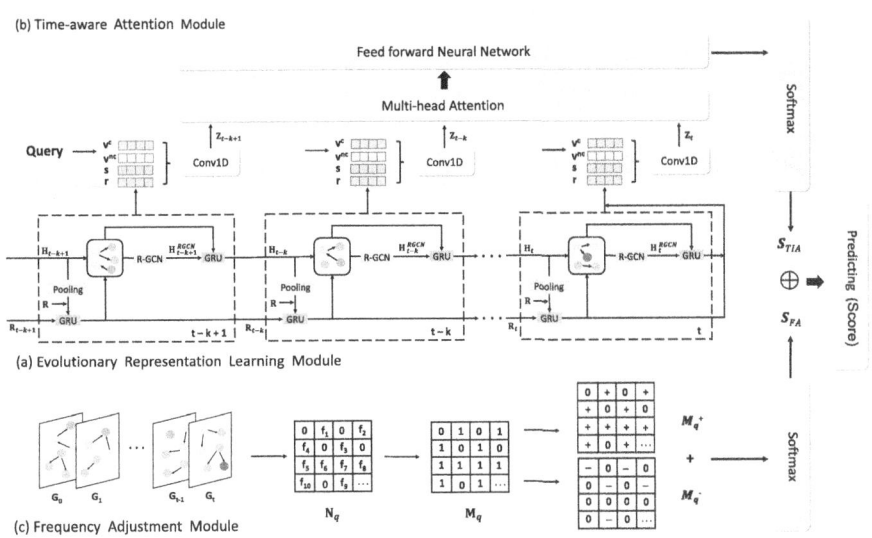

Fig. 2. The framework of TIA-Net.

3.3 Evolutionary Representation Learning Module

In this module, the historical subgraphs of the k timestamps before the timestamp $t + 1$ is defined as $G_{t-k+1:t}$. Meanwhile, $G_{t-k+1:t}$ are modeled along the timeline, as shown in part (a) of Fig. 2.

Representation Learning. TIA-Net utilizes an n-layer Relational Graph Convolutional Network (R-GCN) for information aggregation to comprehensively capture the semantics of all entities in a historical subgraph. Specifically, the entity embedding representation is captured through the l layers message passing architecture, where $l \in [0, n-1]$. The aggregation operation above is defined as:

$$h_o^l = \sigma \left(\frac{1}{|N_e(o)|} \sum_{(s,r,o) \in G_{t-k+1:t}} W_r^{l-1} \left(h_s^{l-1} + r^{l-1}\right) + W_{loop}^{l-1} h_o^{l-1} \right) \quad (1)$$

where h_o^l and h_o^{l-1} denote the embedding representations of the entities in layer l and layer $l-1$, respectively. $\sigma(\cdot)$ adopts the RReLU activation function. $N_e(o)$ symbolizes the set of entities in the neighborhood of entity o. W_r^{l-1} and W_{loop}^{l-1} denotes the trainable parameter.

Evolutionary Process. In order to capture the sequential dependencies of subgraphs at the previous timestamps, TIA-Net uses a dual GRU gating unit to update the representations of entities and relations. The entity GRU unit is as follows:

$$\mathbf{H}_{t+1} = \mathrm{GRU} \left(\mathbf{H}_t^{RGCN}, \mathbf{H}_t \right) \quad (2)$$

where $\mathbf{H}_{t+1}$ and $\mathbf{H}_t \in \mathbb{R}^{|E| \times d}$ are the entity embedding matrices at t and $t+1$, respectively. $\mathbf{H}_t^{RGCN}$ is the entity embedding matrix after the aggregation of representation learning at timestamp t. For relations, the module first uses the average pooling operation before updating relations representations.

$$\mathbf{r}_{t+1}^{\mathrm{Mean}} = \left[MP \left(\mathbf{H}_t, H_{t+1}^{\mathbf{r}} \right); \mathbf{r} \right] \quad (3)$$

where $H_{t+1}^{\mathbf{r}}$ is the representation of all entities connected to relation $\mathbf{r}$ at $t+1$. $\mathbf{r}_{t+1}^{\mathrm{Mean}}$ is the result of mean pooling operation at t+1. Afterward, the relation GRU is used to update the representation of relations.

$$\mathbf{R}_{t+1} = \mathrm{GRU} \left(\mathbf{R}_{t+1}^{\mathrm{Mean}}, \mathbf{R}_t \right) \quad (4)$$

where $\mathbf{R}_{t+1}, \mathbf{R}_t \in \mathbb{R}^{|R| \times d}$ denotes the relational embedding matrix at $t+1$ and t. $\mathbf{R}_{t+1}^{\mathrm{Mean}}$ is composed of $\mathbf{r}_{t+1}^{\mathrm{Mean}}$ of all relations.

3.4   Time-Aware Attention Module

Time Awareness. Define $F_1(s, r, o, t_1)$ and $F_2(s, r, o, t_2)$ as periodic events. TIA-Net captures the periodic character of the temporal information by encoding periodic and non-periodic encoding of t_1 and t_2. Periodic encoding satisfies the properties of periodic functions. Therefore, the periodic encoding of both events is the same. The similarity of the two events is higher than that without considering the periodicity feature obviously, so the prediction is more accurate.

The time information at $t+1$ is encoded using the time information encoder as follows:

$$\mathbf{v}^c = \sin\left(\boldsymbol{w}_c(t+1) + \boldsymbol{b}_c\right) \tag{5}$$

$$\mathbf{v}^{nc} = \boldsymbol{w}_{nc}(t+1) + \boldsymbol{b}_{nc} \tag{6}$$

where $sin\,(\cdot)$ is the periodic activation function. v^c is the periodic time vector, and v^{nc} is the non-periodic time vector. w_c and w_{nc} is the trainable weight parameter. b_c and b_{nc} is the trainable bias parameter.

For query $(s, r, ?, t+1)$, we focus on the timestamps from $t-k+1$ to t. First, we obtain the embedding representation $\bar{\mathbf{s}}$ and $\bar{\mathbf{r}}$ from the embedding matrices $\mathbf{H}$ and $\mathbf{R}$ at a specific timestamp $\tau \in [t-k+1, t]$ by Conv-TransE [19]. Then, we define a time-aware vector which fuses $\mathbf{v}^c$, $\mathbf{v}^{nc}$, $\bar{\mathbf{s}}$, and $\bar{\mathbf{r}}$ with the one-dimensional convolution through a convolution operator at timestamp τ. The formula is as follows:

$$m_e^n = \sum_{i=0}^{I-1} \boldsymbol{w}_m(i,0)\hat{s}(n+i) + \boldsymbol{w}_m(i,1)\hat{r}(n+i)$$
$$+ \boldsymbol{w}_m(i,2)\hat{\mathbf{v}}^c(n+i) + \boldsymbol{w}_m(i,3)\hat{\mathbf{v}}^{nc}(n+i) \tag{7}$$

$$\mathbf{p}_\tau = \text{Conv1d}\left([\bar{\mathbf{s}}; \bar{\mathbf{r}}; \mathbf{v}^c; \mathbf{v}^{nc}]\right) \tag{8}$$

where e and I are the number and width of convolution kernels. $n \in [0, d]$ is the entries in the output vector, and i denotes the offset. The symbol $\hat{\ }$ denotes padding of the current vector. We set the number of channels to $ch = 50$. $\mathbf{p}_\tau \in \mathbb{R}^{ch \times d}$ denotes the time-aware vector at timestamp $\tau \in [t-k+1, t]$. Then, we use a linear function to change the dimensionality.

$$\mathbf{Z}_\tau = \boldsymbol{w}_1 \mathbf{p}_\tau + \boldsymbol{b}_1 \tag{9}$$

where $\mathbf{Z}_\tau \in \mathbb{R}^d$ is the output after the transformation at timestamp τ. Hence, the final set of time-aware vectors is $\mathbf{Z}_{t-k+1:t} = \{\mathbf{Z}_{t-k+1}, \mathbf{Z}_{t-k+2}, ..., \mathbf{Z}_t\}$.

Multiple Attention Mechanisms. This module employs different attention to each history subgraph according to query. Considering that humans tend to prioritize the most recent events in their reasoning process, the effect of $\mathbf{Z}_t$ is paramount. Therefore, $\mathbf{Z}_t$ is used as the benchmark to assign attention to the set of time-aware vectors $\mathbf{Z}_{t-k+1:t)}$.

$$\boldsymbol{Q} = \mathbf{W}_q \mathbf{Z}_t, \quad \boldsymbol{K} = \mathbf{W}_k \mathbf{Z}_{t-k+1:t}, \quad \boldsymbol{V} = \mathbf{W}_v \mathbf{Z}_{t-k+1:t}, \tag{10}$$

where $\boldsymbol{Q}$ is the query matrix, $\boldsymbol{K}$ is the key matrix and $\boldsymbol{V}$ is the value matrix. $\mathbf{W}_q, \mathbf{W}_k, \mathbf{W}_v \in \mathbb{R}^{64 \times d}$ are the trainable parameters. In the module, the number of heads is set to 8. Attention mechanisms is as follows:

$$\text{Self\_Attention}(\boldsymbol{Q}, \boldsymbol{K}, \boldsymbol{V}) = \text{Softmax}\left(\frac{\mathbf{W}_q \mathbf{z}_t \left(\mathbf{W}_k \mathbf{Z}_{t-k+1:t}\right)^T}{\sqrt{d_k}}\right) \mathbf{W}_v \mathbf{Z}_{t-k+1:t} \tag{11}$$

where $\sqrt{d_k}$ is used to prevent gradient explosion and bias in attention scores as the scaling factor. This module uses a feed-forward neural network (FFN) containing $2^{\wedge}11$ hidden units.

$$FFN(\boldsymbol{a}) = \mathbf{W}_{fir}\left(\text{ReLU}\left(\mathbf{W}_{sec}\boldsymbol{a}\right)\right) \tag{12}$$

where $a \in \mathbb{R}^d$ is the output of the multi-head attention, and $\mathbf{W}_{fir}$, $\mathbf{W}_{sec} \in \mathbb{R}^{2^{\wedge}11 \times d}$ are the trainable parameters. $\boldsymbol{x}$ is the final output. Then, we use residual connections and layer normalization on the multi-head attention mechanism and the FFN. The final scores of the time-aware attention module are as follows:

$$\boldsymbol{S}_{\text{TIA}} = \text{Softmax}\left(\text{mm}\left(\boldsymbol{x}, \mathbf{Z}_t\right)\right) \tag{13}$$

where $\boldsymbol{x} \in \mathbb{R}^d$, $\mathbf{Z}_t \in \mathbb{R}^{|E| \times d}$. The final output $\mathbf{x}$ and the entity embedding $\mathbf{Z}_t$ at timestamp t are multiplied using matrix multiplication from Conv-TransE [19].

3.5   Frequency Adjustment Module

Frequency Adjustment Module incorporates frequency information into the final score through a global frequency adjustment mechanism.

We define a series of sparse matrices to react to the repeating state of the entity at timestamp $\tau \in [0, t]$. $v_\tau^{s,r}$ as each row in the matrices is a multi-hot vector. If the event (s, r, o_i) occurs at timestamp τ, the value in the ith dimension of $v_\tau^{s,r}$ is set to 1. Each dimension of $|E|$-dimensional vector $\mathbf{N}_{t+1}$ represents the historical frequency of all entities. The formula is as follows:

$$\mathbf{N}_{t+1} = v_0^{s,r} + v_1^{s,r} + \cdots + v_t^{s,r} \tag{14}$$

where $v_\tau^{s,r} \in \mathbb{R}^{|E|}$ and the size of sparse matrices is $|E| \times |R| \times |E|$. For query $(s, r, ?, t+1)$, the purpose of this module is to reward frequent recurring events by a positive score M_q^+ and penalize for never occurring events by a negative score M_q^-.

$$M_q^+ = \text{Softmax}\left(\mathbf{N}_{t+1}\right) \cdot \delta \tag{15}$$

where $\delta = 1/2$. M_q^- is set to -1e9 where the dimensions with a value of 0 in M_q. Ultimately, the score of Frequency Adjustment Module is calculated as:

$$\boldsymbol{S}_{FA} = M_q^+ + M_q^- \tag{16}$$

3.6   Training Strategy

For a prediction task query $(s, r, ?, t + 1)$, the model integrates the scores of the two modules. The formula is as follows:

$$\mathbf{S}(o \mid s, r, t + 1) = \mathbf{TIA\_Net}(s, r, t + 1) = \boldsymbol{S}_{FA} + \boldsymbol{S}_{TIA} \qquad (17)$$

The entity prediction task of a query is considered as a multicategorization task. Therefore, the task uses cross-entropy loss as a loss function. The formula is as follows:

$$\mathcal{L} = -\sum_{t \in T} \sum_{i \in E} \sum_{j \in E} o_i^t \cdot \ln \mathbf{S}(y_i^j \mid s, r, t + 1) \qquad (18)$$

where o_i^t is the i-th ground-truth object entity at the timestamp t. $\mathbf{S}(y_i^j \mid s, r, t + 1)$ is the probability of predicting the j-th entity as the object at timestamp t.

4   Experiments

4.1   Experimental Setup

Datasets. We evaluate TIA-Net on five public datasets to demonstrate its efficiency. These five datasets are YAGO [16], WIKI [10], ICEWS14 [5], ICEWS05-15 [3], and ICEWS18 [22]. We categorize each dataset into training, validation, and testing sets according to 80%, 10%, and 10%, and the other details are shown in Table 2.

Table 2. Public dataset details.

Data	Entities	Relation	Training	Validation	Test	Time gap
YAGO	10,623	10	161,540	19,523	20,026	1 year
WIKI	12,554	24	539,286	67,538	63,110	1 year
ICEWS14	6,869	230	74,845	8,514	7,371	24 h
ICEWS05-15	10,094	251	368,868	46,302	46,159	24 h
ICEWS18	23,033	256	373,018	45,995	49,545	24 h

Baseline Methods. We compare TIA-Net with some TKGR models in our experiments. The interpolation reasoning models are TransE [1] and DE-SimplE [4]. The extrapolation reasoning models are CENET [22], L2TKG [23], DREAM [24], TiRGN [11], DHUNET [15], GHT [20], CEN [12], CyGNet [7], TANGO [6], and RE-GCN [14], RE-NET [8], RGCRN [18]. Among them, TiRGN and DHUNET are the most relevant to our work.

Evaluation Metrics. This section adopts Mean Reverse Ranking (MRR) and Hits@1/3/10, which are commonly used in TKGR. The experimental results are reported utilizing a time-aware filter.

Experimental Details. The embedding dimension for all the datasets is set to 200. The number of layers in R-GCN is set to 2, and the dropout rate is set to 0.2. The number of $Conv1D$ kernels incorporating the temporal information is set to 50, and the dropout rate is set to 0.2. The length k of history subgraph is set to 3. In addition, n_layers is set to 1. The learning rate of AMSGrad optimizer is set to 0.001. The batch size is set to the size of each timestamp for training and testing. In Table 3 and Table 4, the experimental results labeled 2, are derived from our work on reproducing its open source code and default parameters. The experimental results labeled 1 have data derived from their own papers. Others are taken from the papers of TiRGN and GHT.

4.2   Results of TKGR

Entity prediction results in TIA-Net are shown in Tables 3 and Table 4, where the bolded font indicates the best results and the underlined portion indicates the second best results. Overall, TIA-Net achieved the best results on the vast majority of metrics. TiRGN and DHUNET are the most similar models to our study, but TIA-Net outperforms them on most of the evaluation indicators. TiRGN considers the periodic nature of time in TKGR. However, it ignores the influence of different historical timestamps on the predicted timestamps. The DHUNET considers the influence of historical moments on the predicted moments. However, it does not take into account that time is characterized by periodicity. In contrast, TIA-Net defines a time-aware vector that considers both the historical influence and time-periodic characteristics. Meanwhile, the final inference is performed by frequency adjustment. Therefore, it performs better. The most significant improvement is on the YAGO dataset. However, TIA-Net is inferior to DHUNET under the Hits@1 metric. The reason is that the YAGO dataset does not have static graph data, while DHUNET extracts unseen entities from the global static knowledge graph for prediction, which leads to better results for DHUNET. TIA-Net is slightly lower than CENET on the MRR and Hits@1 metrics for ICEWS18. The reason is that ICEWS18 is a large-scale and complex dataset, and CENET uses comparative learning and binary classifiers to make the results more discriminative.

4.3   Ablation Study

Ablation experiments retain the evolutionary representation learning module to verify the validity of (b) and (c) modules, as shown in Table 5. Ablation experiments select YAGO and ICEWS14 datasets. The model cannot cope with time's development and events' evolution when the time-aware module is deleted. For the YAGO dataset, it has a lot of historical repetitive events. The model

Table 3. Performance in percentage for entity prediction task on ICESW14, ICEWS18 and ICEWS05-15.

Method	ICEWS14				ICEWS18				ICEWS105-15			
	MRR	H@1	H@3	H@10	MRR	H@1	H@3	H@10	MRR	H@1	H@3	H@10
TTransE	13.43	3.11	17.32	34.55	8.31	1.92	8.56	21.89	15.71	5	19.72	38.02
DE-SimplE	32.67	24.43	35.69	49.11	19.3	11.53	21.86	34.8	35.02	25.91	38.99	52.75
RGCRN	38.48	28.52	42.85	58.1	28.02	18.62	31.59	46.44	44.56	34.16	50.06	64.51
RE-NET	39.86	30.11	44.02	58.21	29.78	19.73	32.55	48.46	43.67	33.55	48.83	62.72
xERTE	40.79	32.7	45.67	57.3	29.31	21.03	33.51	46.48	46.62	37.84	52.31	63.92
RE-GCN	42	31.63	47.2	61.65	32.62	22.39	36.79	52.68	48.03	37.33	53.9	68.51
CyGNet	37.65	27.43	42.63	57.9	27.12	17.21	30.97	46.85	40.42	29.44	46.06	61.6
TANGO	-	-	-	-	28.97	19.51	32.61	47.51	42.86	32.72	48.14	62.34
GHT[1]	37.4	27.77	41.66	56.19	27.4	18.08	30.76	45.76	41.5	30.79	46.85	63.73
CEN[1]	42.2	32.08	47.46	61.31	31.5	21.7	35.44	50.59	-	-	-	-
CENET[1]	53.35	49.61	54.07	60.62	**51.06**	**47.1**	51.92	58.82	-	-	-	-
L2TKG[1]	47.4	35.36	-	71.05	33.36	22.15	-	55.04	57.43	41.86	-	80.69
DREAM[1]	51.7	42	56.4	72.4	39.1	28	45.2	62.7	56.80	47.30	65.10	78.60
TiRGN[2]	44.43	33.86	49.84	64.77	33.66	23.17	37.98	54.22	50.04	39.25	56.14	70.71
DHUNET[2]	63.54	49.98	72.54	88.41	50.03	36.47	56.99	77.19	57.69	44.63	65.37	82.63
TIA-Net	**63.68**	**50.23**	**72.96**	**88.43**	50.30	36.49	**57.62**	**77.73**	**57.73**	**44.86**	**65.55**	**82.64**

Table 4. Performance in percentage for entity prediction task on YAGO, WIKI.

Method	YAGO				WIKI			
	MRR	H@1	H@3	H@10	MRR	H@1	H@3	H@10
TTransE	-	-	-	-	-	-	-	-
DE-SimplE	-	-	-	-	-	-	-	-
RGCRN	65.76	62.25	67.56	71.69	65.79	61.66	68.17	72.99
RE-NET	66.93	58.59	71.48	86.84	58.32	50.01	61.23	73.57
xERTE	84.19	80.09	88.02	89.78	73.6	69.05	78.03	79.73
RE-GCN	82.3	78.83	84.27	88.58	78.53	74.5	81.59	84.7
CyGNet	68.98	58.97	76.8	86.98	58.78	47.89	66.44	78.7
TANGO	63.34	60.04	65.19	68.79	53.04	51.52	53.84	55.46
GHT[1]	-	-	-	-	-	-	-	-
CEN[1]	-	-	-	-	78.93	75.05	81.9	84.9
CENET[1]	84.13	**84.03**	84.23	-	68.39	68.33	68.36	-
L2TKG[1]	-	-	-	-	-	-	-	-
DREAM[1]	-	-	-	-	-	-	-	-
TiRGN[2]	85.75	81.04	90.06	92.7	81.69	77.73	85.32	87.27
DHUNET[2]	86.81	82.58	90.48	92.98	86.24	79.35	91.50	98.81
TIA-Net	**88.02**	81.73	**93.08**	**99.22**	**86.54**	**79.67**	**91.76**	**98.87**

ignores the information about the frequency of entity occurrences globally when the frequency adjustment module is deleted. In summary, the performance of the complete TIA-Net model outperforms any individual module's performance.

Table 5. Ablation Study on ICEWS14 and YAGO.

Method	YAGO				ICEWS14			
	MRR	H@1	H@3	H@10	MRR	H@1	H@3	H@10
w.o. Time-aware attention module	82.62	78.52	85.41	89.64	41.95	32.57	46.27	60.24
w.o. Frequency adjustment module	75.90	70.83	78.64	85.27	38.69	28.99	43.14	57.86
TIA-Net	**88.02**	**81.73**	**93.08**	**99.22**	**63.68**	**50.23**	**72.96**	**88.43**

4.4   Influence of Historical Timestamps

We analyze the impact of different timestamps on the query timestamps in Fig. 3. The horizontal axis of the graph represents the historical timestamps. The vertical axis represents the predicted timestamps. Define the numbers of ground-truth facts in different historical timestamps as the effects on a query. Obviously, the ground-truth emphasis of predicted timestamps on historical information decreases as the historical distance increases. Meanwhile, historical information closer to the predicted timestamps is more important.

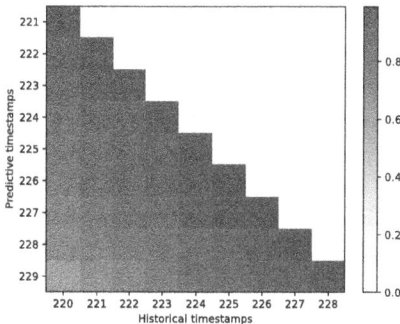

Fig. 3. Influence of different historical timestamps related to the query on WIKI.

5   Conclusions

In this paper, we propose a temporal knowledge graph extrapolation reasoning model named TIA-Net. Firstly, the model incorporates time information encoding at each historical timestamp and utilizes a multi-head attention mechanism to capture temporal dependencies. Secondly, the model utilizes a relational convolution graph network with evolutionary representation to capture structural dependencies among events in each snapshot. Lastly, the frequency characteristics of events, obtained by a global frequency adjustment mechanism, are used to improve prediction accuracy. Extensive experiments on five benchmark datasets demonstrate that TIA-Net outperforms the state-of-the-art TKG reasoning methods in most cases. We will focus on classifying different frequency entities by contrastive learning in the future.

References

1. Bordes, A., Usunier, N., Garcia-Durán, A., Weston, J., Yakhnenko, O.: Translating embeddings for modeling multi-relational data. In: Proceedings of the 26th International Conference on Neural Information Processing Systems, vol. 2. pp. 2787–2795 (2013)
2. Dasgupta, S.S., Ray, S.N., Talukdar, P.: Hyte: hyperplane-based temporally aware knowledge graph embedding. In: Proceedings of the 2018 Conference on Empirical Methods in Natural Language Processing, pp. 2001–2011 (2018)
3. García-Durán, A., Dumančić, S., Niepert, M.: Learning sequence encoders for temporal knowledge graph completion. In: Proceedings of the 2018 Conference on Empirical Methods in Natural Language Processing, pp. 4816–4821 (2018)
4. Goel, R., Kazemi, S.M., Brubaker, M., Poupart, P.: Diachronic embedding for temporal knowledge graph completion. In: Proceedings of the AAAI Conference on Artificial Intelligence, pp. 3988–3995 (2020)
5. Han, Z., Chen, P., Ma, Y., Tresp, V.: Explainable subgraph reasoning for forecasting on temporal knowledge graphs. In: International Conference on Learning Representations (2020)
6. Han, Z., Ding, Z., Ma, Y., Gu, Y., Tresp, V.: Learning neural ordinary equations for forecasting future links on temporal knowledge graphs. In: Proceedings of the 2021 Conference on Empirical Methods in Natural Language Processing, pp. 8352–8364 (2021)
7. Ji, S., Pan, S., Cambria, E., Marttinen, P., Philip, S.Y.: A survey on knowledge graphs: representation, acquisition, and applications. IEEE Trans. Neural Netw. Learn. Syst., 494–514 (2021)
8. Jin, W., Qu, M., Jin, X., Ren, X.: Recurrent event network: autoregressive structure inferenceover temporal knowledge graphs. In: Proceedings of the 2020 Conference on Empirical Methods in Natural Language Processing (EMNLP), pp. 6669–6683 (2020)
9. Lacroix, T., Obozinski, G., Usunier, N.: Tensor decompositions for temporal knowledge base completion. Stat., 10 (2020)
10. Leblay, J., Chekol, M.W.: Deriving validity time in knowledge graph. In: Companion proceedings of the Web Conference 2018, pp. 1771–1776 (2018)
11. Li, Y., Sun, S., Zhao, J.: Tirgn: time-guided recurrent graph network with local-global historical patterns for temporal knowledge graph reasoning. In: International Joint Conferences on Artificial Intelligence, pp. 2152–2158 (2022)
12. Li, Z., et al.: Complex evolutional pattern learning for temporal knowledge graph reasoning. In: Proceedings of the 60th Annual Meeting of the Association for Computational Linguistics (Volume 2: Short Papers), pp. 290–296 (2022)

13. Li, Z., Jin, X., Guan, S., Li, W., Guo, J., Wang, Y., Cheng, X.: Search from history and reason for future: two-stage reasoning on temporal knowledge graphs. In: Proceedings of the 59th Annual Meeting of the Association for Computational Linguistics and the 11th International Joint Conference on Natural Language Processing (Volume 1: Long Papers), pp. 4732–4743 (2021)
14. Li, Z., et al.: Temporal knowledge graph reasoning based on evolutional representation learning. In: Proceedings of the 44th international ACM SIGIR conference on research and development in information retrieval, pp. 408–417 (2021)
15. Liu, K., Zhao, F., Xu, G., Wang, X., Jin, H.: Temporal knowledge graph reasoning via time-distributed representation learning. In: 2022 IEEE International Conference on Data Mining (ICDM), pp. 279–288 (2022)
16. Mahdisoltani, F., Biega, J., Suchanek, F.M.: Yago3: a knowledge base from multilingual wikipedias. In: The annual Conference on Innovative Data Systems Research (2013)
17. Schlichtkrull, M., Kipf, T.N., Bloem, P., van den Berg, R., Titov, I., Welling, M.: Modeling relational data with graph convolutional networks. In: Gangemi, A., et al. (eds.) ESWC 2018. LNCS, vol. 10843, pp. 593–607. Springer, Cham (2018). https://doi.org/10.1007/978-3-319-93417-4_38
18. Seo, Y., Defferrard, M., Vandergheynst, P., Bresson, X.: Structured sequence modeling with graph convolutional recurrent networks. In: Cheng, L., Leung, A.C.S., Ozawa, S. (eds.) ICONIP 2018. LNCS, vol. 11301, pp. 362–373. Springer, Cham (2018). https://doi.org/10.1007/978-3-030-04167-0_33
19. Shang, C., Tang, Y., Huang, J., Bi, J., He, X., Zhou, B.: End-to-end structure-aware convolutional networks for knowledge base completion. In: Proceedings of the AAAI Conference on Artificial Intelligence, pp. 3060–3067 (2019)
20. Sun, H., Geng, S., Zhong, J., Hu, H., He, K.: Graph hawkes transformer for extrapolated reasoning on temporal knowledge graphs. In: Proceedings of the 2022 Conference on Empirical Methods in Natural Language Processing, pp. 7481–7493 (2022)
21. Trivedi, R., Dai, H., Wang, Y., Song, L.: Know-evolve: deep temporal reasoning for dynamic knowledge graphs. In: International Conference on Machine Learning, pp. 3462–3471 (2017)
22. Xu, Y., Ou, J., Xu, H., Fu, L.: Temporal knowledge graph reasoning with historical contrastive learning. In: Proceedings of the AAAI Conference on Artificial Intelligence, pp. 4765–4773 (2023)
23. Zhang, M., Xia, Y., Liu, Q., Wu, S., Wang, L.: Learning latent relations for temporal knowledge graph reasoning. In: Proceedings of the 61st Annual Meeting of the Association for Computational Linguistics (Volume 1: Long Papers), pp. 12617–12631 (2023)
24. Zheng, S., Yin, H., Chen, T., Nguyen, Q.V.H., Chen, W., Zhao, L.: Dream: adaptive reinforcement learning based on attention mechanism for temporal knowledge graph reasoning. In: Proceedings of the 46th International ACM SIGIR Conference on Research and Development in Information Retrieval, pp. 1578–1588 (2023)

The Framework Design of a Semantic Role-Based Knowledge Graph for Natural Disaster Emergency Response

Yuexiang Yang, Yujie Chen[✉], and Yanqing Liu

School of Management, China University of Mining and Technology-Beijing,
Beijing 10085, China
lyq6581@126.com

Abstract. Addressing the characteristics of natural disaster emergency texts, which include a wide variety and complexity of entity relationships as well as inconsistent word order and sentence structure, this paper studies and constructs semantic role types in the field of natural disasters, and designs a unique knowledge graph framework based on semantic roles, starting from the perspective of semantic role labeling and utilizing plan texts in the field of natural disaster emergency response. Utilizing deep learning methods to validate the feasibility of knowledge extraction from this research perspective, the study implements the storage and visualization of natural disaster emergency knowledge based on the Neo4j graph database. It analyzes the feasibility of this knowledge graph in assisting emergency decision-making and supporting information inquiry.

Keywords: Natural disaster emergency response · Knowledge graph · Semantic role

1 Introduction

In the field of knowledge graph construction, the current mainstream approach is to build the schema layer from the top down based on the knowledge system, and then realize the data layer based on the mapping of the schema layer [1]. However, emergency texts related to natural disasters exhibit characteristics such as a wide variety and complexity of entity relationships as well as inconsistent word order and sentence structure. The construction of ontological models is heavily influenced by the subjectivity of the constructors [2]. For instance, the sentence "地震发生后，社区地震应急负责人和管理人以及志愿者应将震感强度、观察到的房屋倒塌 地面破坏、人员伤亡等情况，向上级部门紧急报告" involves a significant amount of semantic knowledge, with strong correlations existing between these knowledge points. Entities typically refer to specific objects or concepts, including people, locations, objects, events, and so on [3]. As seen from the given example sentence, it is difficult to construct and distinguish these entities. Moreover, constructing relations to describe the semantic relationships between entities poses an even greater challenge. To complement knowledge and maximize the restoration of semantic information during the process of knowledge graph construction, this paper approaches the task from the perspective of semantic role labeling to build a knowledge graph.

2   The Issues with Traditional Semantic Role Construction

In the task of semantic role labeling, in order to better suit the field of natural disaster emergency response, we refer to relevant semantic role classification systems and utilize the predicate-argument structure of PropBank [4–6] as the underlying logic. By considering the unique characteristics of the natural disaster response domain, we add, delete, modify, and adapt multiple semantic role labeling systems [7–10] to construct a semantic role type that is suitable for the natural disaster emergency response domain discussed in this paper.

In the process of core semantic role labeling, we follow the principle of inclusiveness, while for the adjunct semantic role labeling, we adhere to the principle of granularity [11]. Ultimately, we have constructed six categories of core semantic roles (including Agent, Patient, Goal, Experiencer, Leader and Source) and eleven categories of adjunct semantic roles (including Condition, purpose, Basis, Space, Scope, Manner, Requirements, Time, Frequency and Status). The relevant definitions and examples from the natural disaster emergency response plan texts are presented in Table 1 and Table 2.

2.1   Loss of Semantic Information Due to Traditional Triples Format

In a knowledge graph, knowledge is represented in the form of triples, which consist of (head entity, relationship, tail entity)(h, r, t) [14]. For example, in the triple (China, capital of, Beijing), it indicates that "Beijing" is the "capital" of "China". According to the traditional format of triples (h, r, t), the triples extracted based on semantic roles in this context should be (agent, predicate, patient), which are the most fundamental and important core triples representing the core disaster response actions/behaviors/event information in a single textual sentence. Adjunct semantic roles such as condition, basis, and time serve as supplements to the basic triple (agent, predicate, patient), providing information like when, where, or how. However, such semantic information cannot be fit into the traditional triple format, leading to a loss of semantic information.

For example, in the text statement regarding natural disaster response, "地震发生后, 社区地震应急负责人和管理人以及志愿者应将震感强度、 观察到的房屋倒塌、地面破坏、人员伤亡等情况,向上级部门紧急报告". After decomposing and annotating the semantic roles in the context of natural disaster emergency response, "地震发生后" is a condition, "社区地震应急负责人和管理人以及志愿者" is the agent, "震感强度、 观察到的房屋倒塌、地面破坏、人员伤亡等情况" is the patient, "上级部门" is the Goal, and "紧急报告" is the predicate. According to the traditional triple format, it should be (社区地震应急负责人和管理人以及志愿者, 紧急报告, 震感强度、 观察到的房屋倒塌、地面破坏、人员伤亡等情况). However, the condition "地震发生后" and the goal "向上级部门" cannot be appropriately placed in the traditional triple format, leading to a loss of semantic information (see Fig. 1).

Table 1. Definitions and Examples of Core Semantic Roles

Core Semantic Roles	Definition	Example
Agent [7–9, 12]	In the predicate-argument structure, it typically serves as the subject role, thus also being the subject in actions, behaviors, or events related to natural disaster response	社区应编制地震应急预案。
Patient [8, 9]	In the P-A Structure, it usually serves as the object role, and thus is also the object in actions, behaviors, or events related to natural disaster response	社区应配备应急通讯、广播、应急照明等器具。
Goal [8, 9]	It refers to the receiver of the information, transferred goods, or other items targeted by the core predicate in the domain of natural disasters	搜索人员确定受困人员位置后应向指挥员报告。
Experiencer [8, 9]	It refers to the subject participants who are of equal status but not the agent in the actions, behaviors, or events related to natural disaster response	搜索人员确定受困人员位置后应向指挥员报告，并与营救人员交接受困人员及工作场地相关信息。
Leader [8, 9]	It refers to another non-core subject that has a possessive relationship with the main subject in the actions, behaviors, or events related to natural disaster response	地震应急避难场所疏散安置指挥部在抗震救灾指挥部指挥下组织受灾群众向安全地带转移。
Source [8, 9]	It refers to the source of the information, transferred goods, or other items in the field of natural disasters that are targeted by the core predicate	国家海洋环境预报中心及时从中国海洋石油总公司等单位获取海冰信息。

2.2 Confusion in Semantic Information Due to Traditional Triple Format

Moreover, in the field of natural disaster emergency response, there exist numerous actions, behaviors, and events, with interactions and overlaps between semantic roles. The traditional triple modeling approach has structural limitations and cannot handle the widespread occurrence of conflicting actions, behaviors, or events.

For instance, when the same agent exhibits different actions/behaviors/events under different conditions, such as in the natural disaster response text sentences 特大沙尘暴灾害后,省级人民政府立即组成救灾指挥部" and "省级人民政府在接到强、特强沙尘暴预警后,应督促气象部门要加强对沙尘暴灾害的监测和预报. After extracting and fusing entity relations from these two texts, the semantic

Table 2. Definitions and Examples of Adjunct Semantic Roles

Adjunct Semantic Roles	Definition	Example
Condition [8–10]	It refers to the prerequisite for the occurrence of actions, behaviors, or events in the field of natural disaster response	搜索人员开展搜索行动时，应将人工搜索、仪器搜索等方式结合使用。
Purpose [8–10]	It refers to the purpose or intention behind actions, behaviors, or events related to natural disaster response	社区应迅速组织社区居民疏散。
Basis [8–10]	It refers to the basis or reference for the occurrence of actions, behaviors, or events related to natural disaster response	救援队应根据工作场地评估情况制定搜索方案。
Space [8–10]	It refers to the natural or social space where actions, behaviors, or events related to natural disaster response take place	社区工作人员进入本社区的各居住区，指导和帮助居民应对地震灾害。
Scope [8–10]	The scope or defined boundaries that actions, behaviors, or events involve or are limited to	各有关地方和部门对核电站等核工业生产科研重点设施，做好事故防范处置工作。
Manner [8–10]	It refers to the specific means or methods for the occurrence of actions, behaviors, or events related to natural disaster response	海区局应将赤潮监测预警信息等综合信息以《赤潮快报》形式及时报自然资源部海洋预警监测司。
Direction [8–10]	It refers to the direction, ending location, or destination of actions, behaviors, or events related to natural disaster response	受困人员移出后，应转移至医疗处置区。
Requirements [8–10]	It refers to the requirements for the occurrence of actions, behaviors, or events related to natural disaster response	社区地震应急预案应根据实际需要适时进行修订，修订时间不宜超过 5 年。
Time [8–10]	It refers to the time point or period when actions, behaviors, or events related to natural disaster response occur	海冰预报从每年 11 月始至翌年 3 月终冰为止。

(continued)

Table 2. (*continued*)

Adjunct Semantic Roles	Definition	Example
Frequency [8–10]	The frequency of actions, behaviors, or events related to natural disaster response	信息报送频次不低于每周2次。
Status [8–10, 12, 13]	It refers to the state maintained by the object due to the occurrence of actions, behaviors, or events related to natural disaster response	工业和信息化部优先保障抗震救灾指挥通信联络和信息传递畅通。

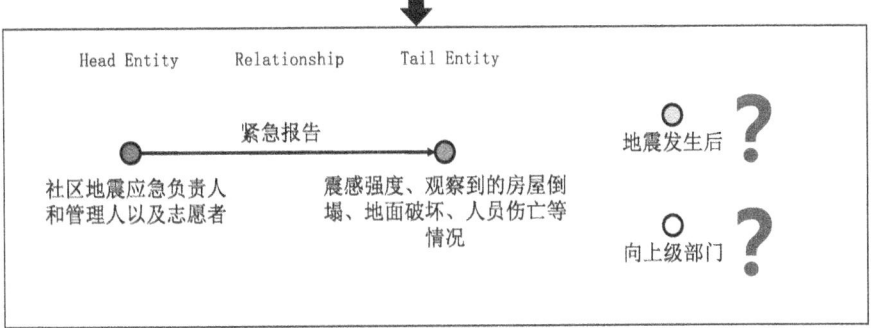

Fig. 1 Loss of Semantic Information

in-formation regarding "condition" is missing, and "省级人民政府" corresponds to different relations and tail entities. The traditional triple format can lead to confusion in semantic information. Assuming that when the condition "发生重、"发生重、特大沙尘暴灾害后"takes place, what the "省级人民政府" should do is unclear in the absence of the semantic information "condition." All entity relations connected to it would become potential answers to the question, which obviously does not align with standard knowledge(see Fig. 2).

3   Knowledge Graph Framework Design

This article differs from the mainstream approach of building knowledge graphs based on ontological models. Instead, it parses textual statements from the perspective of semantic roles, which is why it is not suitable for traditional triple forms. There is an urgent need to construct a new knowledge framework to incorporate textual knowledge such as natural disaster emergency response standard plans into triples. This framework should achieve the structuring of textual knowledge, the fusion of knowledge connections, and simultaneously support intelligent question answering and decision-making assistance.

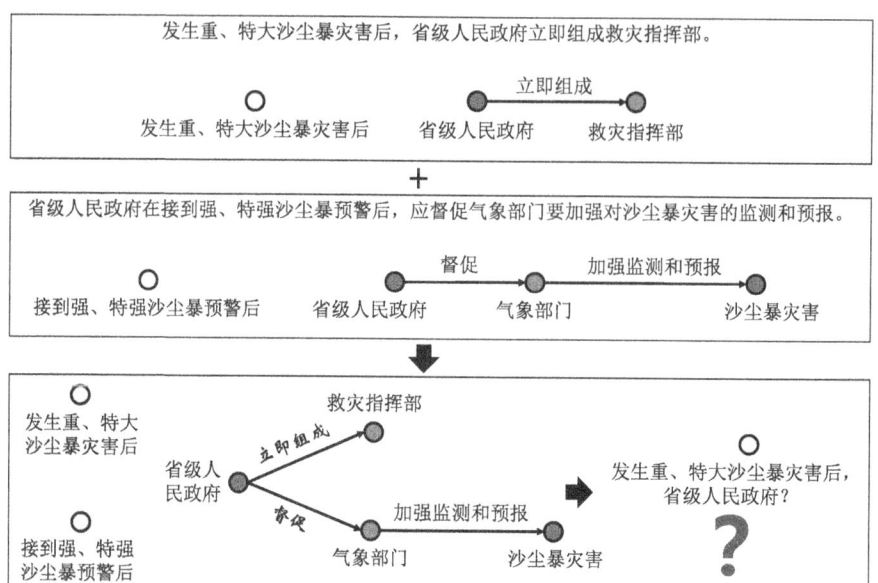

Fig. 2 Confusion in Semantic Information

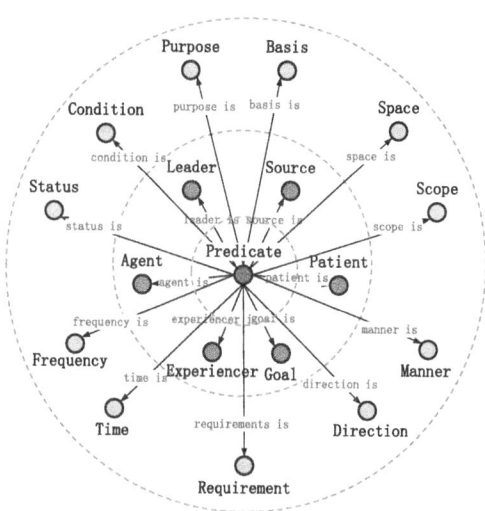

Fig. 3 Knowledge graph framework based on semantic role

Therefore, a knowledge framework is innovatively proposed based on semantic roles: on the basis of traditional triples, predicates and all semantic roles are taken as entities, and the relationships between predicates and semantic roles are taken as entity relationships, named as "semantic role  is". With predicates as the center, it points to semantic roles, with core semantic roles forming the core layer and adjunct semantic roles forming the adjunct layer. The knowledge graph framework diagram is shown in the Fig. 3.

In this framework, the core semantic roles play a crucial role. The core semantic roles are components directly related to the predicates, including subjects, objects, and so on. They help understand the subject, object, and third-party participants of natural disaster response actions /behaviors /events. By connecting the predicates with the core layer, we can obtain the core semantics of the entire natural disaster response action/behavior/event, namely, the subject- predicate-object structure semantics.

In addition to the core semantic roles, adjunct semantic roles provide additional semantic information. Adjunct semantic roles often involve more semantic relationships such as conditions, grounds, manners, and so on. These help enrich the semantic representation of the text and can more comprehensively present the context of natural disaster response actions/behaviors/events. By incorporating adjunct semantic roles into the framework, it is possible to better capture the details and background information in the text.

As can be seen from the predicate-argument structure, the predicate is the key point connecting all semantic roles and the core of the entire sentence. Merging predicates would lead to confusion in semantic information, making it inconvenient for subsequent knowledge queries. Therefore, predicate nodes will not be merged; instead, the semantic role section, which involves various organizational entities, information, resources, and other real-world entities, with a significant amount of duplicates, will undergo entity fusion. This approach not only avoids confusion in se-mantic information but also eliminates duplicate information within the knowledge graph, unifying entities and relations.

The example sentence can be dissected and analyzed according to the knowledge framework depicted in **Fig. 3**, "救援队[agent]进入工作场地前[condition],应确定[predicate]工作场地范围[patient],并设置[predicate]警示带[patient]。"Since there are two core predicates, the sentence can be decomposed into two sets of semantic in-formation,the triples information is respectively:

{(确定,agent is,救援队),(确定,patient is,工作场地范围),(确定,condition is,进入工作场地前)}.

{(设置,agent is,救援队),(设置,patient is,警示带),(设置,condition is,进入工作场地前)}.

From the triples information, it can be observed that there are duplicate entities. Therefore, an entity fusion process is applied to merge the semantic role entities with duplicates, as illustrated in Fig. 4.

Based on the structured knowledge parsed using the knowledge framework designed in this document, it is clearly visible that under the condition of "进入工作场地前","救援队" should "确定""工作范围" and "设置" 警示带",The core semantics and ancillary semantic information of this natural disaster response action/behavior/event can provide comprehensive information for decision support.

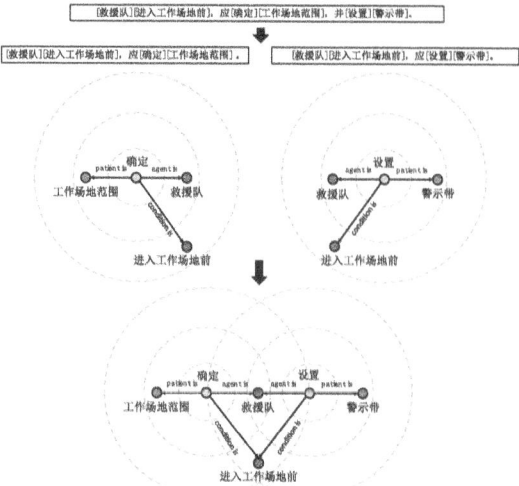

Fig. 4 Example diagram of knowledge framework triples

4 Related Experiments

This paper selects sentences of actions/behaviors/events from the text of natural disaster emergency response plans, which are usually in the subject- predicate-object structure and must contain core semantic roles such as predicates, agents, and patients, and may include additional semantic roles. These sentences are extracted as the experimental data for this paper. The text data is cleaned and pre-processed, and sequence labeling is performed based on predicates and semantic roles. Adjunct semantic roles focus on detailed semantic and contextual information, thus they should be annotated as comprehensively as possible. Ultimately, 844 predicates and 2,110 semantic roles were annotated as the experimental dataset.

To verify the feasibility of the aforementioned knowledge framework, this paper utilizes the classical sequence labeling model BERT-BiLSTM-CRF [15–17] for knowledge annotation and extraction. Based on BERT, RoBERTa is adopted [18], and whole word masking (wwm) is introduced, resulting in the RoBERTa a-wwm-BiLSTM-CRF model. The framework of the model is shown in the Fig. 5.

To validate the effectiveness of pre-trained language models in sequence labeling tasks, comparative experiments were conducted on a series of derivative models based on BERT-BiLSTM-CRF. A five-fold cross-validation method was adopted, and dropout was used for model optimization. Based on the performance of multiple training and validation runs of the models and the changes in the loss values, the model parameters were adjusted. The final parameter settings for each model are presented in the Table 3.

This paper adopts precision (P), recall (R), and F1 score as evaluation metrics to verify the accuracy of the model's recognition. The final evaluation results of the experiment are shown in Table 4. As can be seen from the experimental results, with the increase in the complexity of the model structure, the performance of sequence labeling also improves. The RoBERTa-wwm-BiLSTM-CRF model achieved the best results, with an

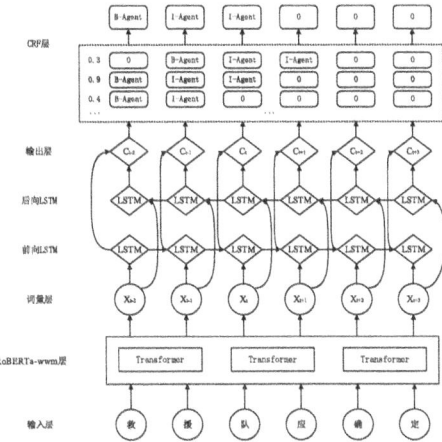

Fig. 5 RoBERTa-wwm-BiLSTM-CRF Model Framework

Table 3. Experimental model parameters set

Model	Batch_size	Lr	Maxseq_length	dropout	epoch
BiLSTM-CRF	20	4e-4	512	0.2	60
Bert-BiLSTM-CRF	16	1e-5	512	0.4	60
Bert-wwm-BiLSTM-CRF	16	1e-5	512	0.4	60
RoBERTa-wwm-BiLSTM-CRF	16	1e-5	512	0.4	60

F1 score of 0.77, demonstrating that knowledge graphs can be constructed from the perspective of semantic role labeling.

Table 4. Experimental model evaluation results

Model	Precision	Recall	F1
BiLSTM-CRF	0.40	0.50	0.44
Bert-BiLSTM-CRF	0.63	0.77	0.69
Bert-wwm-BiLSTM-CRF	0.66	0.79	0.71
RoBERTa-wwm-BiLSTM-CRF	0.73	0.80	0.77

Based on the knowledge graph framework designed in this paper, we construct triples from the semantic role labeling results. Using the Cypher language, these triples are imported into the Neo4j graph database in CSV format. The predicates and semantic roles are used as node labels, and the content is stored as node content. Nodes and rela-tionships are created to enable the storage and visualization of the natural disaster emergency knowledge graph. Knowledge graph display diagram is shown in Fig. 6.

Fig. 6 Knowledge graph display diagram

5 Emergency Decision-Making Information Inquiry

Constructing a natural disaster emergency knowledge graph based on semantic roles can help decision-makers better understand the responsibilities and relationships between various agencies while maximizing the restoration of semantic information, thereby assisting relevant emergency departments in making more rapid and accurate decisions during disaster responses.

(1) By utilizing the natural disaster emergency knowledge graph, relevant emergency departments and staff can query the responsibilities of different organizational departments under various disasters. For instance, using the query such as "MATCH (B)-[:施事是]- > (A) MATCH (B)-[:条件是]- > (C) WHERE A.name = '省级人民政府' AND C.name = '发生重、特大沙尘暴灾害后'MATCH (B)-[r]-(related) RETURN A,B,C,r,related". One can find out the responsibilities of "省级人民政府" as the agent when"发生重、特大沙尘暴灾害后. These responsibilities include "报送" 舆论导向to "国家林业局相关机构", "立即组成" "救灾指挥部", "做好" "舆论导向" and other measures (see Fig. 7).

(2) Assist various levels of emergency management departments in querying the reporting departments for relevant natural disaster information. When natural disasters occur, various levels of emergency management departments usually need to report disaster information to higher-level departments through an information re-porting mechanism, which facilitates the Party Central Committee and various levels of emergency management departments to obtain relevant information in time during natural disasters and carry out relevant emergency deployment. For example, using the query " MATCH (B)-[:施事是]- > (A) MATCH (B)-[:受事是]- > (C) MATCH

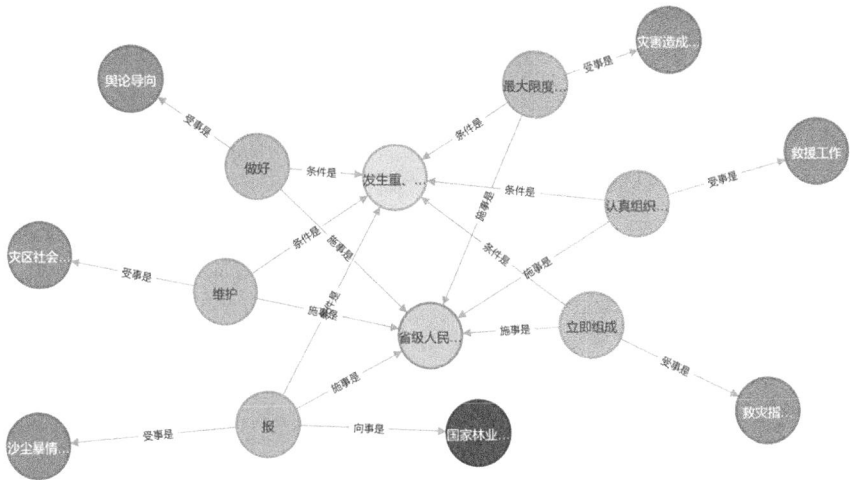

Fig. 7 Example Diagram for Agency Responsibility Query Using Conditional Statements

(B)-[:条件是]- > (D) WHERE A.name CONTAINS '海区局' AND C.name CON-TAINS '赤潮监测预警信息' AND D.name CONTAINS '赤潮灾害' MATCH (B)-[r]-(related) RETURN A, B, C, D, r, related;". The relevant information such as the reporting department, frequency, and form can be queried (see Fig. 8).

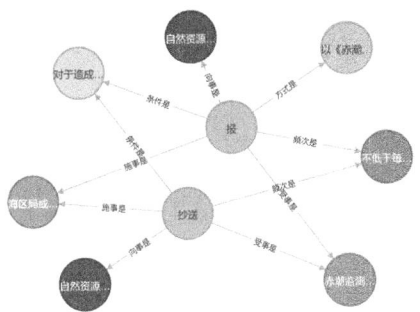

Fig. 8 Example Diagram for Information Reporting Department Inquiry

(3) Assist in querying the resource sharing and collaborative work between emergency management authorities at various levels and other related organizations during the emergency response process. For example, using the query" MATCH (B)-[:施事是]- > (A) MATCH (B)-[:受事是]- > (C) WHERE A.name = '省级重大生物灾害防治指挥机构' AND C.name CONTAINS '监测' MATCH (B)-[r]-(related) RETURN A, B, C, r, related;". One can find that the "省级重大生物灾害防治指挥机构" needs to accept the leadership of the "省级人民政府" when carrying out "监测"(see Fig. 9).

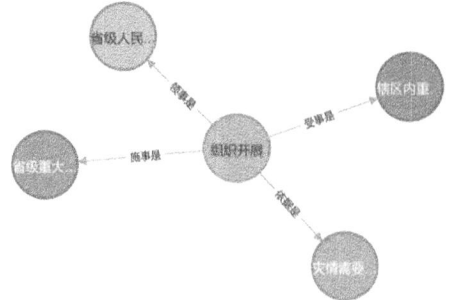

Fig. 9 Example Diagram for Multi-level Organizational Leadership Relationship Inquiry

The above-mentioned inquiries about relevant information can effectively facilitate the coordination and response of multiple organizational departments in the emergency management of natural disasters.

6 Conclusion

Focusing on the characteristics of emergency texts related to natural disasters, this paper studies and constructs semantic role types in the field of natural disasters from the perspective of semantic role labeling. It designs a knowledge graph framework with unique features based on semantic roles, and combines deep learning methods to verify the feasibility of this research approach for knowledge extraction. The paper realizes the storage and visualization of emergency knowledge related to natural disasters, and analyzes the feasibility of the knowledge graph assisting in emergency decision-making and supporting information inquiry. This paper also explores action/behavior/event statements in unstructured content. However, the research and construction method of knowledge graphs based on semantic roles is not universally applicable to multimodal information [19]. In the field of constructing emergency text knowledge graphs for natural disasters, it is urgent to combine different research methods to achieve the integration of cross-modal information in emergency texts related to natural disasters, ultimately completing the construction of a knowledge graph for natural disaster emergencies to facilitate precise decision-making.

Acknowledgments. This article was supported by National Key R&D Program of China(2021YFF0600401)and the central basic business research funding project "Research and application of risk classification technology for product quality and safety based on data driven "(552023Y-10371).

References

1. Zhang, J.,Zhang, X., Wu, C.: Survey of Knowledge Graph Construction Techniques. Comput. Eng. 1–16 (2021)
2. Liu, C., Zou, Z., Tong, J.: LSI-based semantic retrieval model for scientific data in solar-terrestrial space field. J. Univ. Chin. Acad. Sci. 33(05), 711–719 (2016)
3. Liu, W., Chen, L., Ren, Z.: Few-shot relation reasoning model based on graph neural networks and meta-learning. Comput. Eng. (2024)
4. Kingsbury, P., Palmer, M.: Propbank: the next level of treebank. In:Proceedings of Treebanks and lexical Theories, vol. 3 (2003)
5. Kingsbury, P., Palmer, M.: From TreeBank to PropBank. LREC, 1989–1993 (2002)
6. Bonial, C., Hwang, J., Bonn, J., Conger, K., Babko-Malaya, O., Palmer, M.: English propbank annotation guidelines. Center for Computational Language and Education Research Institute of Cognitive Science University of Colorado at Boulder, vol. 48 (2012)
7. Yuan, Y.: The hierarchical relationship and semantic characteristics of argumentative roles. Chin. Teaching World 03, 10–22 (2002)
8. Yang, M., Chang, B.: Semantic role classification based on peking university Chinese NetBank. J. Chin. Inform. Process. 25(02), 3–8 (2011)
9. LU, C.: Linguistics for knowledge engineering. Tsinghua University Press (2010)
10. Xue, N.: Annotation guidelines for the Chinese proposition bank. Brandeis University, Draft (2007)
11. Wang, C., Qian, Q., Xun, E.: Construction of semantic role bank for chinese verbs from the perspective of ternary collocation. J. Chin. Inform. Process. 34(09), 19–27 (2020)
12. Song, H., Cao, C., Wang, Y.: Construction of a finely-grained training dataset for Chinese semantic-role labeling. J. Chin. Inform. Process. 36(12), 52–66 (2022)
13. Song, H., Cao, C., Wang, Y.: A fine-grained annotated dataset for Chinese semantic-role labeling. J. Chin. Inform. Process 37(01), 16–32 (2023)
14. Huang, B., Wu, S., Wang, W.: KG-LLM-MCom: a survey on integration of knowledge graph and large language model. J. Wuhan Univ. (Nat. Sci. Edn.), (2024)
15. Zhang, Q., Fu, L., Wang, X.: Scholar home page information extraction based on BERT-BiLSTM-CRF. Appli. Res. Comput. 37(S1), 47–49 (2020)
16. Wang, Y., Zheng, Y., Yang, Q.: Hierarchical information extraction method based on joint sequence annotation. Comput. Appli. Softw. 38(08), 167–174 (2021)
17. Cheng, S., Li, Z., Wei, T.: Threat intelligence entity relation extraction method integrating bootstrapping and semantic role labeling. J. Computer Appli. 43(05), 1445–1453 (2023)
18. Liu, Y., OTT, M., Goyal, N.: Roberta: A robustly optimized bert pretraining approach. arxiv preprint arxiv, 1907.11692 (2019)
19. Chen, Y., Li, Y., Wen, M.: Research and comprehensive review on multi-modal knowledge graph fusion techniques. Comput. Eng. Appli. (2024)

Research on Automatic Extraction
of Emergency Response Standards Concept
Hierarchy Based on LDA

Wenling Liu[1,2][(✉)] [iD], Yuexiang Yang[1][(✉)], Xinyu Tu[1], and Wan Wang[3]

[1] China University of Mining and Technology-Beijing, Beijing 100083, China
xfwenling19@163.com, 201901@cumtb.edu.cn
[2] Taishan University, Taian 271400, China
[3] China National Institute of Standardization, Beijing 100191, China
wangwan@cnis.ac.cn

Abstract. In order to enhance the supporting role of emergency response standards in emergency management procedures and improve the utilization of normative documents such as emergency response standards, the Latent Dirichlet Allocation (LDA) model is used to implement automatic extraction of emergency response standards concept hierarchy. The construction of standard ontology model based on emergency procedures is realized. The emergency response standards are collected and sorted. And a dictionary of emergency response standard terms is built by extracting the standards terms. The standard titles are statistically analyzed by word cloud and the key points of standard formulation are found from the perspective of the field. The LDA model is used to extract the standard content topics. The standard concept hierarchies are extracted through summarizing the topic models and the standard ontology model is built. The role of the standard ontology is demonstrated through the rainstorm disaster scenario.

Keywords: Emergency response standards · Latent dirichlet allocation (LDA) · Concept hierarchy · Ontology · Topic extraction

1 Introduction

Emergencies are characterized by their sudden onset and the uncertainty regarding their timing, location, and manner of occurrence. The nature of such emergencies exhibits significant variability. Additionally, emergencies are often complex, resulting from a combination of multiple factors. Their impact is usually widespread, affecting normal social order and stability within a certain range, thereby posing a threat to public safety. In order to reduce and prevent the occurrence of emergencies, mitigate the harm to the nation and the public, and enhance the capability and efficiency of emergency response, China has established an emergency management system known as the "One Plan and Three Systems". The digitization of emergency response standards is a crucial component in improving this emergency management system. In the field of emergency response, relevant standard documents encompass a substantial amount of normative emergency

B. Xu et al. (Eds.): CCKS-IJCKG 2024, CCIS 2229, pp. 54–66, 2025.
https://doi.org/10.1007/978-981-96-1809-5_5

management information. This information spans multiple dimensions, including foundational, technical, management, and service aspects. Furthermore, it provides standardized support for all elements involved in the emergency response process, covering the entire life-cycle from preparedness, monitoring and warning, response and recovery.

Emergency response standards are crucial sources of knowledge, operational guidelines, and decision-making support for emergency management. They ensure the standardization and consistency of unified command and coordinated response efforts, as well as regulate the processes and phases of emergency management, providing beneficial supplements to the emergency management mechanisms. Therefore, improving the efficiency of utilizing emergency response standards is a significant aspect of emergency management. Based on this, this paper employs the Latent Dirichlet Allocation (LDA) model to automatically extract conceptual levels from emergency response standards and to construct an emergency standards ontology model. This approach aims to further understand the characteristics and patterns of the entire emergency response process, thereby laying the foundation for standards digitization.

2   Related Research

In recent years, emergencies have occurred frequently in China, such as the COVID-19 pandemic and the severe flooding in Zhengzhou. These emergencies have not only resulted in significant loss of life and property but also affected national security and social stability. Domain topic extraction is a crucial step in responding to and analyzing such emergencies. In emergency situations, rapidly and accurately acquiring and understanding the core content of emergencies is vital for decision-makers, the media, and the public. Domain topic extraction aims to identify and categorize key information related to a specific domain from large volumes of unstructured data. By extracting topics, corresponding hierarchical relationships of concepts can be obtained.

Scholars have already conducted research on topic extraction in the field of emergency. For instance, regarding online texts about emergency, Cao Shujin et al. [1] utilized the TF-IDF and LDA models to mine the latent topics in public opinion on Weibo during public health emergencies. Similarly, Zhao Rongying et al. [2] employed the LDA model to analyze the evolution of content and intensity in online discussions on the Zhihu platform following public health emergencies. Furthermore, Pourebrahim et al. [3] explored information posted by the public on Twitter during Hurricane Sandy's impact on the United States and, based on the analysis results, formulated corresponding rescue plans and disaster loss assessment schemes. Xiaohui Bian et al. [4] used the LDA model to analyze public sentiment expressed on social media during major public health emergencies, further revealing the emotional evolution under different topics. Han et al. [5] analyzed public opinion on social media during the Shouguang flood disaster in Shandong, discovering that the topics and sentiments of public opinion evolved along with the progression of the disaster. These studies demonstrate that the literature is closely related to significant events at the time [6]. In the context of accident report texts, Pengxiang Zhang [7] implemented rules to extract thematic paragraphs from railway equipment accident reports. Similarly, Zhanglu Tan et al. [8] utilized the LDA model to mine topics related to coal mine safety hazards. In the automatic extraction of domain conceptual hierarchy, Tang Xiaobo et al. [9] achieved the automatic extraction of conceptual hierarchical relationships in the field of academic paper evaluation. Gao Jinsong et al. [10] used the LDA model to construct concepts of Dunhuang murals consisting of six major categories. Zhang Wei et al. [11] utilized spectral clustering to generate a framework of content hierarchy relationships.

In summary, the application of the LDA model for topic extraction in the field of emergency has shown significant effectiveness, particularly in processing online incident texts and accident report texts. Furthermore, it enables the construction of conceptual models. However, its application to the study of emergency response standards remains in its early stages. The content of emergency response standards is rich. To effectively express standard knowledge in a standardized and semantic manner, it is necessary to clarify the conceptual hierarchy. The LDA model performs well in analyzing text semantics and can effectively achieve topic extraction from unstructured documents [12]. Therefore, this paper utilizes the LDA model to extract topics from emergency response standards, forming a conceptual hierarchy of these standards. Furthermore, an ontology model is constructed, which contributes to enhancing the understanding and capability in responding to emergency.

3 Research Approach and Technology

3.1 Research Approach

This paper aims to achieve the automatic extraction of conceptual hierarchies from emergency response standards. It conducts an in-depth analysis from both the titles and the content of the standards, and further constructs an ontology model that aligns with the emergency response process. Firstly, the emergency response standards were collected. Subsequently, the standard texts were organized and preprocessed, and a terminology dictionary was constructed. Following this, word vectorization was performed by integrating information from the terminology dictionary. Next, topic modeling was conducted from both the standard titles and standard content, extracting a conceptual hierarchy of the standards. Finally, an ontology model conforming to the emergency response process was constructed. As shown in Fig. 1.

3.2 Technology

LDA Model. This paper utilizes the LDA model to extract the conceptual hierarchy of emergency response standards.The LDA model was proposed by Blei et al. in 2003 [13]. The LDA topic model is a commonly used unsupervised topic generation model that includes three granularities: documents, topics, and words[14]. The core idea of the LDA topic model is to find the topic words for each input document by specifying the number of topics and to determine the probability distribution of each topic in new documents. The calculation formula is as follows.

$$P(\boldsymbol{W}, \boldsymbol{Z}, \boldsymbol{\theta}, \boldsymbol{\varphi}; \alpha, \beta) = \prod_{j=1}^{M} P(\theta_j; \alpha) \prod_{j=1}^{K} P(\varphi_i; \beta) \prod_{t=1}^{N} P(Z_{j,t}|\theta_j) P(W_{j,t}|\varphi_{Z_{j,t}}) \quad (1)$$

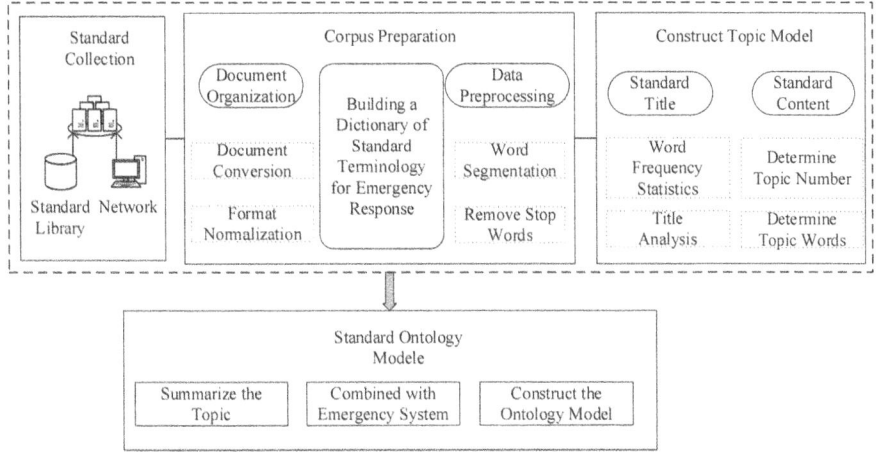

Fig. 1. Framework of the research

W, Z represents the topic words and topics generated by the model. θ, φ represents the distributions of document topics and the topic words under Each topic, randomly generated by a Dirichlet distribution with parameters α, β. $P(\theta_j; \alpha)$ Represents the probability of each topic in a document given the parameter α. $P(\varphi_i; \beta)$ represents the probability of topic words under each topic in a document given the parameter β. $P(Z_{j,t}|\theta_j)$ represents the probability of a topic given the document under parameter θ. $P(W_{j,t}|\varphi_{Z_{j,t}})$ represents the probability of a topic word given the topic under parameter φ_Z.

Perplexity. The number of topics in an LDA model needs to be predetermined and can vary depending on the content and size of the corpus. In this paper, the number of topics is determined by calculating the perplexity. Perplexity is often used to evaluate the performance of probabilistic language models. The calculation formula for perplexity is as follows.

$$Perplexity(D) = \exp(-\frac{\sum_{d=1}^{M} \log(p(W_d))}{\sum_{d=1}^{M} N_d}) \tag{2}$$

In the formula, D represents all words, M represents the number of documents, N_d represents the number of words in the d document, and $p(W_d)$ represents the probability of a word occurring in the d document. A lower perplexity value indicates a better generative capability of the model [15], making the extracted topics more interpretable. Overall, the perplexity value generally decreases as the number of topics increases.

4 Conceptual Hierarchy Extraction

4.1 Data Collection and Data Preprocessing

A total of 1,589 emergency response standards were collected for this paper, including 257 national standards, 651 industry standards, 574 local standards, and 127 group standards. These were categorized by the type of emergency as follows: 269 standards for natural disaster, 715 standards for accident disaster, 459 standards for public health emergency, and 146 standards for social security emergency.

To ensure effective topic extraction, the text preprocessing stage involved the normalization of both standard titles and standard content. In the standard titles, phrases like "Part X" were removed. In the standard content, information such as tables of contents and dates, which are not relevant for analysis, was excluded. The processed content then underwent word segmentation and stop-word removal. The standard content included a module of specialized terminology. This terminology module specifies vocabulary in the field of emergency response standards. Therefore, when extracting standard terminology, a standard terminology dictionary was constructed to enhance the accuracy of word segmentation and lay the foundation for topic extraction and determination of conceptual hierarchy [16].

4.2 Analysis of Emergency Response Standard Titles

The titles of emergency response standards include their respective fields and the standardized objects involved. Analyzing these titles can provide an overall summary of the standards topics. This paper compiles the titles of the four major categories of emergency response standards. The titles are segmented using Jieba, and word frequency statistics are performed. The word cloud representing the statistical analysis results of the titles for the four major categories of emergency response standards is shown in Fig. 2.

Fig. 2. Word cloud of emergency response standard titles

From Fig. 2, it can be seen that in the category of natural disaster response standards, there are many standards related to meteorological disasters and risk assessment. In the category of accident disasters, there are numerous standards related to firefighting. For public health emergency, many standards are aimed at addressing the COVID-19 pandemic. In the category of social security emergency, there are numerous anti-terrorism standards. The top 20 most frequent words in the titles of each category of standards are shown in Table 1.

Table 1 shows the word frequencies after segmenting the titles of the four major categories of emergency response standards. It highlights the key areas in the field of emergency response standards.

(1) **Natural Disasters Standards.** From the perspective of research objects, there are many standards related to meteorological disasters, earthquakes, forest disasters, and geological disasters. Natural disasters in China mainly include floods, earthquakes, droughts, snow disasters, low temperature freezing, forest and grassland fires, and

Table 1. Word frequency statistics of emergency response standard titles

Type	Word
Natural disaster	disaster, meteorology, assessment, specifications, risk, technical specifications, emergency, level, earthquake, lightning, early warning, forest, geological disaster, monitoring, guidelines, code of practice, investigation, technology, fire, heavy rain.
Accident disaster	code, firefighting, fire safety, specifications, information, emergency, safety, management, construction, unit, accident, classification, fire, capability, code of practice, method, production, handling, requirements, guidelines.
Public health emergency	Prevention and control, pandemic, technical specifications, guidelines, procedures, technology, COVID-19, specifications, port, monitoring, health, border, disinfection, animal, management area, handling, infectious disease, epidemic, quarantine.
Social security emergency	Prevention, anti-terrorism, management, specifications, information security, code, requirements, technology, information, finance, incident, security, protection, public security, criminal cases, industry, guidelines, classification, Wuhan, management system.

geological disasters, which is consistent with the results shown in Table 1. This indicates that China has studied frequently occurring natural disasters and developed standards to respond to their occurrence. The formulation of natural disaster standards enables more effective implementation of rescue actions. Rescue actions should be initiated immediately after a natural disaster, including the resettlement of affected people and the allocation of relief supplies.

(2) **Accident Disasters Standards.** From the perspective of standard forms, there are many standards related to codes, management, classification, guidelines, and methods. From the perspective of standard content, there are many standards related to fire disasters and production safety accidents. Firefighting and fire safety standards are designed for firefighting and rescue operations. Therefore, both are included in the category of fire disaster standards. Standards for fire accidents rank first among accident disaster standards. Fire accidents can occur in various locations and for various reasons, so the formulation of standards related to fires enables timely firefighting and fire suppression.

(3) **Public Health Emergency Standards.** The number of standards related to the novel coronavirus is relatively high. This is because since the outbreak of the COVID-19 pandemic in 2019, the development of the epidemic has been widely followed by the entire nation. The formulation of standards provides a systematic approach to combating the epidemic. In addition to standards for COVID-19 prevention and control, there are also numerous standards related to symptoms at border ports, diseases related to animals, and infectious diseases. For public health emergencies, the development of medical plans is a crucial step, which includes both clinical data and indicator-based data. Currently, these data are scattered and mostly exist in

non-structured forms such as text, which not only inconveniences users' extraction but also increases the difficulty of computer processing. Through the analysis of standards, resource integration and utilization can be achieved, providing decision-making support for the deployment of medical plans for public health emergencies.

(4) **Social Security Emergency Standards.** Anti-terrorism prevention standards are the most numerous. According to statistics, this category of standards accounts for 53% of all collected standards for social security emergencies. This indicates the importance attached to anti-terrorism work by various provinces in China in recent years, and targeted anti-terrorism standards can be formulated based on provincial conditions.

4.3 Conceptual Hierarchy Extraction in Emergent Event Response Standards

This study selected 100 emergency response standard documents, including both cross-disaster generic standards and the four major categories of standards. Since emergency plans contain measures for responding to various emergency, 72 emergency plans were included in the corpus. The LDA model was used to construct the conceptual hierarchy of emergency response standards. Due to the differences in the format of standard texts and online texts, this study followed the settings of alpha and beta values in governance policy texts in free trade zones [17], setting the value of alpha to 0.1 and beta to 0.01. Perplexity gradually decreases with the increase in the number of topics. However, after reaching a certain number of topics, it begins to increase. Therefore, the number of topics at the lowest point is selected as the optimal number of topics. This study calculated perplexity for 1–20 topics, and it reached the lowest point at the 16th topic, as shown in Fig. 3. Therefore, the number of topics was set to 16.

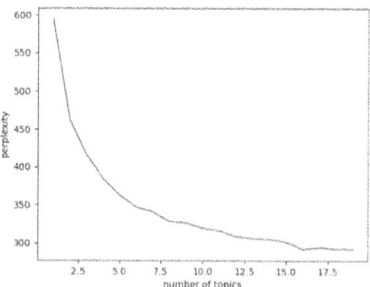

Fig. 3. Perplexity curve

Through the LDA model, the top frequent words for each topic were obtained, and the top 4 topic words were selected for display. By summarizing these frequent words, 16 topics were derived, as shown in Table 2.

Based on the thematic words in the table, the thematic content can be inferred. Combining with the emergency emergency process and emergency system, the 16 thematic contents can be classified into the following categories.

(1) Fine-grained classification of standard knowledge elements. Topic 2 reflects the fine-grained processing of standard content knowledge. It specifies fine-grained elements

Table 2. The theme mining results of emergency response standards.

Number	Word	Topic
TOPIC1	Supervisory Department, Office, Command Center, Leadership Group	Emergency Response Organization
TOPIC2	Classification, Symbol, Section, Name	Sign Classification
TOPIC3	Fire, Firefighting, Water Supply, Alert	Fire Incident Handling and Rescue
TOPIC4	Special, Overall, According to Law, Legal, Regulations	Response Basis
TOPIC5	Disinfection, Water Supply, Detection, Ventilation	Post-disaster Disposal Measures
TOPIC6	Community, Residents, Public, Volunteers	Non governmental organization
TOPIC7	Municipal Government, Command Center, Office, Special Equipment	Government Organization
TOPIC8	Geological Disaster, Early Warning, Indicators, Function	Natural Disaster Warning and Monitoring
TOPIC9	Center, Detection, Forecast, Prediction	Emergency Monitoring
TOPIC10	Storm Surge, Alarm, Planning, Experiment	Emergency Warning
TOPIC11	Emergency Scene, Leadership Group, Command, Notification	Disaster Emergency Scene Handling
TOPIC12	Risk, Plan, Risk Management, Risk Assessment	Prevention Preparation
TOPIC13	Construction, Equipment, Installation, Location	Hazard-affected carriers
TOPIC14	Railway, Earthquake, Flood Control, Transportation	Railway Traffic-related Emergencies
TOPIC15	Service, Evaluation, Project, Data, Method	Post-disaster Investigation and Evaluation
TOPIC16	Meteorology, Lightning, Flood Control, Drought Resistance	Meteorological Disaster Emergency Handling

through classification, symbol signs, etc. For example, the standard "GB 2893–2008 Safety Colors" specifies different colors, where red means prohibited and yellow means warning. By specifying fine-grained knowledge, it provides a basis for the subsequent application of knowledge and selection of indicators.

(2) Participants in Emergency Response. Topics 1, 6, and 7 illustrate participants in emergency response from three perspectives. Topic 6 represents non-governmental organizations, while topics 1 and 7 represent governmental organizations. These three topics highlight the main organizational participants in emergency response,

including both decision-making and implementation levels. Determining the participants topics can promptly identify response subjects during emergencies, thereby ensuring coherent information transmission.

(3) Emergency Response Process. The emergency process includes four aspects, including preparedness, monitoring and warning, response, and recovery. Nine out of the 16 topics involve these four aspects. This also reflects that the formulation of standards targets emergencies and aims to provide decision-making and indicator support for emergency response.

(4) Specific Disaster Analysis. Due to the varying numbers of standards for different types of disasters, with more standards for natural disasters, topic 16 appears to represent meteorological disasters.

Through the automatic extraction of emergency response standard topics, the following content can be summarized. From the perspective of participants, it includes governmental organizations, non-governmental organizations, and individuals. From the perspective of emergency processes, it includes disposal and rescue, post-disaster disposal measures, monitoring, warning, accident scene handling, preparation, post-disaster investigation, etc., which can be summarized as the four emergency response processes. The specific topics further decompose these four processes. Other topics include sign classification, response basis, hazard-affected carriers, railway-related accidents, meteorological accidents, etc. Due to their different aspects, they can be decomposed into the levels of emergency response basis, accident carrier, and emergency type. Through summarizing and combining with the emergency response standard system, the emergency response standard conceptual hierarchy model is obtained, as shown in Table 3.

4.4 Applications

Based on the conceptual hierarchy structure derived from the emergency response standards discussed above, an ontology model can be constructed in accordance with these emergency standards. Ontology can represent the conceptual hierarchy in emergency response standards in a structured form, making it suitable for computer systems and reducing knowledge redundancy. An ontology can conceptualize and organize situational and environmental awareness data (events, activities) related to any type of emergency [18]. Saad E et al. [19] constructed an ontology comprising four modules, including actors, environment, tasks, and others, based on discussions and interviews with firefighters and domain experts. Wei Xu and Sisi Zlatanova [20] proposed a method for developing an ontology for disaster management response, which provides appropriate information at the right time and place to support emergency response when the integration of geographic information is constrained. Deepika Shukla et al. [21] proposed the Disaster Management Ontology (DMO), which facilitates the allocation of tasks among the necessary authorities at various stages of a disaster and provides a knowledge-based decision support system for the victims. Therefore, following Perez's categorization of the five major ontology modeling meta-languages [22], this paper defines the Emergency Response Standard Ontology (ERSO) as a collection.

ERS0 = { ERS0-Classes, ERS0-Attributes, ERS0-Relations, ERS0-Axioms, ERS0-Instances}.

Table 3. Concept hierarchy of emergency response standards.

Primary concepts	Secondary concepts	Example
Participant	Government organization	Government departments, emergency command agencies, etc.
	Non governmental organization	Social groups, volunteer teams, etc.
	Unit	Managers, technicians, etc.
Emergency procedures	Preparedness	Risk assessment, hazard inspection, etc.
	Monitoring and warning	Forecasting, alarms, etc.
	Response	Information transmission, command management, etc.
	Recovery	Evaluation, statistical investigation, etc.
Response basis	Emergency response standard	《GB 2893–2008 Safety Colors 》, etc.
	Emergency plan	《National Food Safety Accident Emergency Plan》, etc.
Hazard-affected carriers		Building, equipment, etc.
Emergency type	Natural disaster	Floods, inundation, etc.
	Accident disaster	Explosions, gas leaks, etc.
	Public health emergency	COVID-19 pandemic, etc.
	Social security emergency	School stampede accidents, etc.

Among them, ERSO-Classes represent the classes in the ontology's conceptual hierarchy. ERSO-Attributes represent attributes in the ontology, such as 'National Standard' being the level attribute of the standard '《GB 2893–2008 Safety Colors》'. ERSO-Relations, ERSO-Axioms, and ERSO-Instances respectively represent the relationships, axioms, and instance collections in the ontology. Relationships can be expressed as relationships between concepts, relationships between concepts and instances, relationships between instances, etc. The axiom set represents the eternal truths in the ontology, such as 'Compensation' being a subclass of 'Recovery and Reconstruction', where the inheritance relationship between them is a subclass relationship axiom. The instance collection represents the lowest granularity elements in the ontology.

In order to demonstrate the role of standard ontology construction, this paper uses a scenario case of a rainstorm disaster, with standards as the guiding basis, to construct the ontology model. A partial display is shown in Fig. 4.

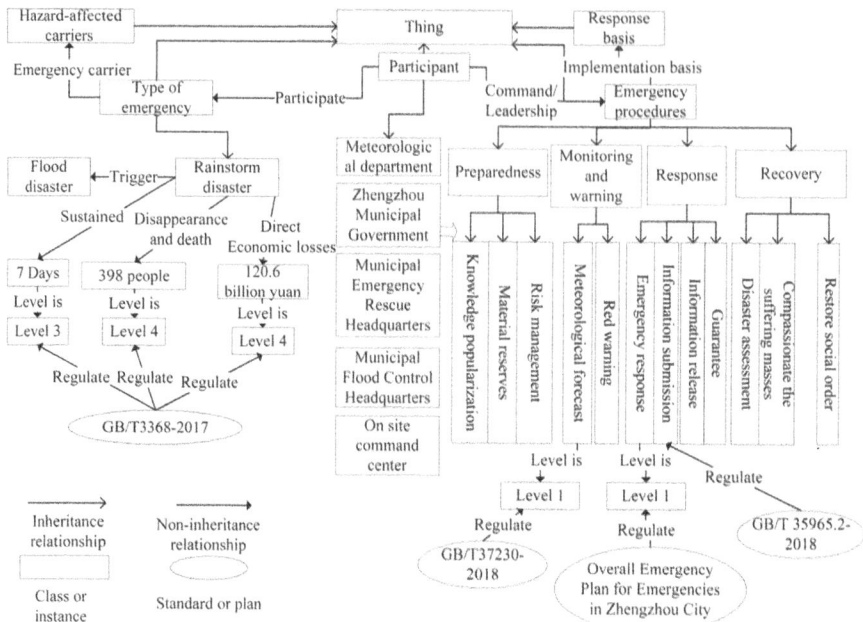

Fig. 4. Emergency model of rainstorm disaster based on standard ontology

5   Conclusion

Knowledge from emergent event response standards plays a decision-support role in handling emergencies. However, the current utilization of standard knowledge often involves searching entire documents, resulting in low knowledge utilization efficiency. To enhance the utility of standard knowledge, this paper employs the LDA model to achieve automatic extraction of the conceptual hierarchy of emergent event response standards and constructs an ontology model for these standards based on this hierarchy. Standard ontology is a crucial step for knowledge extraction and improving knowledge utilization. By using the topic extraction model to build the standard ontology, this approach reduces manual involvement, saves time, and enables in-depth analysis of standard content. This paper performs mining and analysis of emergent event response standard content, identifying key issues in standard formulation from a domain perspective. It also extracts themes from the standard content from a deeper level, achieving automatic extraction of the conceptual hierarchy. This process identifies five concepts, including emergency participants and emergency processes. Based on this framework, an ontology model for emergency events in the case of heavy rainfall was constructed. This lays the foundation for the utilization of standard knowledge and the selection and determination of indicators. Future work should focus on further enriching the ontology model to enhance its broader applicability.

Acknowledgments. This study was funded by the National Key R&D Program of China (Project Number: 2021YFF0600400) and Research and application of risk classification technology for product quality and safety based on data driven (Project Number: 552023Y-10371).

Disclosure of Interests. The authors have no competing interests.

References

1. Shujin, C., Wenyu, Y.: Topic mining and evolution analysis of public opinion on microblog of public health emergencies. J. Inform. Resou. Manag. **10**(06), 28–37 (2020)
2. Rongying, ZH. Ruru,CH. Zhan, CH., et al.: Research on thetheme evolution of public health emergencies based on Zhihu. J. Inform. Resour. Manag. **10**(06), 28–37(2020)
3. Nastaran, P., Selima, S., John, E., et al.: Understanding communication dynamics on Twitter during natural disasters: a case study of Hurricane Sandy. Inter. J. Disaster Risk Reduct. **37**(C), 101176–101176(2019)
4. Xiaohui, B., Tong, X.: Evolution of public sentiments during COVID-19 pandemic. Data Analy. Knowl. Dis. **6**(07), 128–140 (2022)
5. Han, X.,Wang, J.: Using social media to mine and analyze public sentiment during adisaster: a case study of the 2018 Shouguang City flood in China. ISPRS Inter. J. Geo-Inform. **8**(4), 185 (2019)
6. Xiaopeng, Y.: Research progress and prospect of emergency management based on CNKI and CiteSpace. China Saf. Sci. J. **32**(08), 185–193 (2022)
7. Pengxiang, Z.H.: Information extraction method for railway equipment accidents based on multi-dimensional character feature representation. China Saf. Sci. J. **32**(06), 109–114 (2022)
8. Zhanglu, T., Ze, W., Xiao, C.H.: Research on topic extraction for coal mine hidden danger based on LDA. China Saf. Sci. J **26**(06), 123–128 (2016)
9. Xiaobo, T., Qiongfu, W., Hao, M.: Automatic extraction of concept hierarchies based on word co-occurrence and word vector—taking the academic paper evaluation as an example. Inform. Sci.,1–10 (2022)
10. Jinsong, G., Ruiling, S.H., Jiawei, F.: The construction of the cultural relics information resources topic graph with literature evidence-based practice. Inform. Sci. **40**(03), 12–20 (2022)
11. Wei, Z.H., Hao, W., Sanhong, D., et al.: Research on hierarchy identification of Chinese terms in the field of E-government. J. China Soc. Sci. Tech. Inform. **40**(01), 62–76 (2021)
12. Tazibt, A., Aoughlis, F.: Latent Dirichlet allocation–based temporal summarization. Inter. J. Web Inform. Syst. **15**(1), 83–102 (2019)
13. Blei, D. M., Ng, A. Y., & Jordan, M. I. Latent dirichlet allocation. J. Mach. Learn. Res. **3**, 993–1022 (2003)
14. Blei, D., Griffiths, T., Jordan, M.: The nested chinese restaurant process and bayesian nonparametric inference of topic hierarchies. J. ACM (JACM) **57**(2), 130 (2010)
15. Peng, G., Yuefen, W.: Identifying optimal topic numbers from sci-tech information with LDA model. New Technol. Libr. Inform. Serv. **36**(9), 42–50 (2016)
16. Linyu, S.H., Ming, Y., Chang, Y., et al.: Research on text classification of railway safety incidents based on BLS. China Saf. Sci. J. **32**(06), 103–108 (2022)
17. Lei, L., Zige, L.: Cluster analysis of free trade zone governance policy text based on LDA: a case study of liaoning free trade zone. J. Jishou Univ. (Soc. Sci.) **42**(02), 23–34 (2021)
18. Gaur, M., Shekarpour, S., Gyrard, A., et al. empathi: an ontology for emergency managing and planning about hazard crisis. In: 2019 IEEE 13th International Conference on Semantic Computing (ICSC), pp. 396–403. IEEE (2019)

19. Saad, E., Hindriks, K.V., Neerincx, M.A.: Ontology design for task allocation and management in urban search and rescue missions. In: 10th International Conference on Agents and Artificial Intelligence, ICAART 2018, pp. 622–629. SciTePress (2018)
20. Xu, W., Zlatanova, S.. Ontologies for disaster management response. Geomatics Solutions for Disaster Management, pp. 185–200. Springer, Berlin Heidelberg (2007)
21. Shukla, D., Azad, H.K., Abhishek, K., et al. Disaster management ontology-an ontological approach to disaster management automation. Sci. Rep. 13(1), 8091 (2023)
22. Perez, A., Benjamins, V.: Overview of knowledge sharing and reuse components:ontologies and problem solving methods. In: Proceedings of the UCAI-99 Workshop on Ontologies and Problem-Solving Methods, pp.1–15. CEUR Publications, Stockholm (1999)

Knowledge Graph Construction and Knowledge Integration

Beyond Isolation: Multi-agent Synergy for Improving Knowledge Graph Construction

Hongbin Ye[1(✉)], Honghao Gui[2], Aijia Zhang[1], Tong Liu[1], and Weiqiang Jia[1]

[1] Zhejiang Lab, Hangzhou, China
yehongbin@zhejianglab.com
[2] Ant Group, Hangzhou, China

Abstract. This paper introduces CooperKGC, a novel framework challenging the conventional solitary approach of large language models (LLMs) in knowledge graph construction (KGC). CooperKGC establishes a collaborative processing network, assembling a team capable of concurrently addressing entity, relation, and event extraction tasks. Experimentation demonstrates that fostering collaboration within CooperKGC enhances knowledge selection, correction, and aggregation capabilities across multiple rounds of interactions.

Keywords: Knowledge graph construction · Information extraction · Agent cooperation

1 Introduction

In the era of information abundance, constructing comprehensive knowledge graphs [17,25,36] has emerged as a pivotal task. The advent of LLMs, such as GPT-3 [1] and ChatGLM [4], has revolutionized natural language processing by showcasing unparalleled proficiency in understanding and generating human-like text. However, the application of these models to KGC remains an intricate challenge, as this task necessitates not only language understanding but also precise extraction of elements within the confines of predefined schemas. Recent investigations [33] reveals that the raw textual data utilized to train large language models may lack task-specific schemas, resulting in a weakened semantic grasp and structural analysis of the underlying schema. Therefore we contends that a shift from traditional parameter-based paradigms to a more nuanced approach, like *Chain-of-Thought* (CoT) [32,37], can address the challenges posed by multi-step inference problems inherent in KGC. Embracing the profound insights from the *Society of Mind* (SOM) [16], which conceptualizes the mind as a complex system emerging from the interactions of simple components, our research explores the transformative potential of LLM-based agents in multi-agent systems. Taking inspiration from pioneering work of [13], we employ the multi-agent debate framework for collaborative self-reflection on challenging tasks. Collaboration is

B. Xu et al. (Eds.): CCKS-IJCKG 2024, CCIS 2229, pp. 69–81, 2025.
https://doi.org/10.1007/978-981-96-1809-5_6

Table 1. Comparison with previous methods. The upper half represents LLM-based KGC method, while the lower shows the emerging multi-agent approach.

	Has multiple agents involved?	Has personalized agents?	Has interactive rounds?	Involves chain of thought processes?	Accomplishes multiple tasks in parallel?
AutoKG [39]	✗	✗	✗	✗	✗
ChatIE [33]	✓ (2-agents)	✗	✓ (2-rounds)	✗	✗
GPT-NER [28]	✗	✓	✓ (>3-rounds)	✓	✗
GPT-RE [27]	✗	✓	✗	✓	✗
CoT-ER [15]	✓ (3-agents)	✗	✓ (3-rounds)	✓	✗
LM vs LM [2]	✓ (2-agents)	✓	✓ (3-stages)	✓	✗
Multiagent Debate [3]	✓ (2-agents)	✗	✓ (3-stages)	✓	✗
MAD [13]	✓ (3-agents)	✓	✓ (3-stages)	✓	✗
PRD [10]	✓ (2-agents)	✓	✓ (3-stages)	✓	✗
SPP [30]	✓ (>3-agents)	✓	✓ (4-stages)	✓	✗
Our CooperKGC	✓ (3-agents)	✓	✓ (3-stages)	✓	✓ (3-tasks)

defined as an iterative refinement process, wherein each round generates a new answer based on prior answers and self-reflection. This iterative feedback fosters continuous improvement, making our collaborative approach adept at tackling problems that elude single-agent solutions.

Specifically, our dedicated team of agents comprises experts proficient in various tasks, including named entity recognition, relation extraction, and event extraction. In our approach **CooperKGC**, we construct a collaborative team of agents, each specializing in distinct tasks to simulate the nuanced teamwork prevalent in human society. The integration of open interaction, expertise refinement, and adaptability to others' opinions mirrors the foundations of a cohesive society. Our exploration into diverse collaboration strategies reveals key insights: (1) Inclusion of agents with varied expertise enhances collaboration outcomes. (2) While model hallucinations [35] may arise, effective communication among team members mitigates these drawbacks. (3) Substantial team collaboration enhances extraction results on target tasks; however, an intriguing observation emerges that increasing cooperation rounds doesn't invariably yield superior results. In our collaborative mechanism, balancing interaction frequency ensures the expert agent's beliefs remain undisturbed by excessive external authoritative information, aligning with fundamental theories of sociology [5,6,24].

2 Related Work

2.1 LLM-Based Knowledge Graph Construction

Recent years have witnessed a surge of interest in leveraging the remarkable advancements achieved by LLMs within the realm of KG. Notably, [39] delves into the application of LLMs in KG construction and reasoning tasks. Building on this foundation, [38] integrates KG structural information into LLMs, employing self-supervised structural embedding pre-training. [33], in a novel perspective, proposes a multi-turn question and answer architecture. Furthermore, [19] pinpoints the unspecified task description as a key factor hindering the performance of contextual information extraction. To address this, a guided

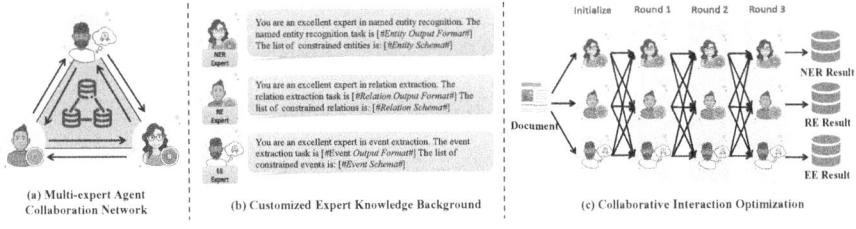

Fig. 1. The overview of our CooperKGC.

learning framework is introduced to enhance the extraction model's alignment with specified guidelines. Departing from the conventional isolation of KGC as a singular task, our approach advocates a departure from such isolation by fostering collaboration among a group of expert model agents in a multi-round social environment.

2.2 Interactive Collaboration of Multiple Agents

Recent developments highlight the effectiveness of collaborative efforts among language model agents, offering potential to enhance individual LLM capabilities. Various interaction architectures have emerged, assigning agents to specific roles. For example, setups like that in [3] engage two agents in debates, enhancing factuality and reasoning albeit with increased computational costs. Similarly, [2] introduces an examiner LLM to validate claims, uncovering factual errors through division of labor. Contributions such as the ChatLLM network [7] foster dialogue and collective problem-solving among language models. Others introduce judges to summarize debates and provide conclusions [13,34], or frameworks incorporating peer review and discussion to address self-enhancement bias [10]. Our study presents an interactive architecture tailored for a knowledge graph construction team, focusing on multiple information extractions and feedback compatibility.

3 Methodology

Illustrated in Fig. 1, we introduces a collaborative framework CooperKGC, aimed at advancing knowledge graph construction by concurrently extracting component elements such as entities, relations, and events. Notably, our method could extend beyond the confines of the selected three tasks, offering flexibility through a dynamically formulated team collaboration network tailored to specific task requirements.

3.1 Construction of Multi-expert Agent Collaboration Network

Traditional methods treat expert agents, each equipped with distinct back-ends, as isolated nodes within the collaborative network. These nodes independently contribute to task-solving through separate thinking chains, and a central adjudication node amalgamates and rectifies their responses. However, this conventional solution reveals two flaws: (1) The adjudication node, functioning as the central hub, exhibits low fault tolerance and demands substantial reasoning ability to assimilate opinions from nodes spread across diverse collaborative networks; (2) The team heavily relies on the ruling node as the sole consensus mechanism, hindering effective interactions between participants in the KGC task. In response to these limitations, we advocate a decentralized collaborative network communication scheme. Here, each expert agent backend, responsible for handling a specific task, establishes a bidirectional communication channel with any other expert agent backend. Despite the asynchronous nature of message production during practical operations, we adopt rounds as the fundamental unit of interaction to accomplish designated tasks and facilitate replica communication among expert agents. It is noteworthy that, although our approach draws inspiration from the Byzantine Fault Tolerance [8] to form a distributed network, the message records held by each agent node differ. In the process of replica communication, we implement message simplification, whereby extraction results complying with schema constraints are distilled. The formalization of the abstract collaboration network comprises three fundamental components:

Expert Nodes. Expert nodes embody agents proficient in specific sub-tasks within KGC. They assimilate context from their peers at the preceding time step and formulate responses based on the input text $\mathcal{X}$. Notably, an Expert node can take various forms, including a vanilla LLM guided by explicit instructions, a self-reflective agent with a chain of thinking, or an agent explicitly leveraging domain knowledge through external knowledge bases or tool libraries. With this foundation, our focus shifts to the collaborative functions between agents. Formally, the response r_i^t of the i-th agent at the t-th round is expressed as a function $\mathcal{F}_t^i$, mapping from the base input text $\mathcal{X}$, prompt p_t^i, and predecessor expert agent's replicas $\mathcal{R}_{t-1}$: $r_t^i = \mathcal{F}_t^i(\mathcal{X}, p_t^i, \mathcal{R}_{t-1})$, where $\mathcal{R}_{t-1} = \{r_{t-1,j} | j = 1, 2, ...\}$. Let $\mathcal{A}$ be the set of all expert nodes and T be the maximum round.

Communication Edges. A two-way communication channel facilitates the exchange of insights among expert nodes $\mathcal{A}$ in the KGC collaboration network. In this context, we define $\mathcal{E}$ as the set encompassing all edges within the system. Recognizing the nuanced distinctions in information dissemination, we establish directional edges, represented by $e_{m,n} = (a_{t-1}^m, a_t^n) \in \mathcal{E}$, where a_{t-1}^m and a_t^n signify the adjacent agent responsible for transmitting replica. It was, a_t^n can perceive the replica passed from a_{t-1}^m as its contextual input. Thus, the expert nodes are intricately linked through these communication edges, constituting the interactive communication units $\mathcal{C} = (\mathcal{A}, \mathcal{E})$.

Replicas Delivery. In the interactive communication unit $\mathcal{C}$, replica delivery serves as the conduit guiding the flow of information from an agent in $(t - 1)$-th round to the input message queue of another agent in t-th round. To streamline this intricate exchange, we designate a specific simplification function $\mathcal{S}$ to simple the information: $d_{t-1} = \mathcal{S}(r_{t-1})$, where $\mathcal{S}$ predigest the complex CoT

Algorithm 1: The Optimization Process of **CooperKGC**

Input: Input Text $\mathcal{X}$, Expert Nodes $\mathcal{A}$, Communication Edges $\mathcal{E}$,
       Communication Unit $\mathcal{C}$, Round $\mathcal{N}$

Output: KGC result $\mathcal{Y}^i$ for each $a_i \in \mathcal{A}$

for $a_i \in \mathcal{A}$ **do**
    | /* Initial extraction results   */
    | $r_0^i = \mathcal{F}_0^i\left(\mathcal{X}\|\mathcal{P}_o, \mathcal{P}_t, \mathcal{P}_c\right); d_0^i = \mathcal{S}(r_0^i);$
end

for $t = 1; \mathcal{N}$ **do**
    | **for** $a_i \in \mathcal{A}$ **do**
    | | /* Replicas delivery by edges   */
    | | $\mathcal{D}_{t-1}^i \leftarrow Transfer(e_{m,i}), \forall i, e_{m,i} = (a_{t-1}^m, a_t^i) \in \mathcal{E};$
    | | /* Refine results by referring others   */
    | | $r_i^t = \mathcal{F}_t^i\left(\mathcal{X}\|\mathcal{D}_{t-1}^i\|\mathcal{P}_v\|\mathcal{P}_o, \mathcal{P}_t, \mathcal{P}_c\right); d_t^i = \mathcal{S}(r_t^i);$
    | **end**
end

/* Extract final answer, filter d_t^i whose format does not comply with
   the constraints   */
$\mathcal{Y}^i \leftarrow \text{filter\_ans}(d_t^i | a_t^i \in \mathcal{A}, \mathcal{C}, \mathcal{X});$

---

reasoning process. Therefore, the replicas queue collected by the i-th expert node is expressed as $\mathcal{D}^i = \left\{d_{t-1}^j | j \neq i\right\}$.

3.2 Customized Expert Knowledge Background

In order to unleash the ability of different expert agents to collaborate on complex extraction problems, we introduce customized expert knowledge background. This context comprises three key components: (1) Opening statement $\mathcal{P}_o$, where each expert agents is presented with a directive elucidating how it can contribute its unique expertise to address a KGC task; (2) Task definition $\mathcal{P}_t$, which outlines the specifics of the knowledge graph extraction, including the targeted elements and the guiding schema; and (3) In-context demonstration $\mathcal{P}_c$, involving the selection of a limited set of $\mathcal{M}$ instances. The overarching objective of this in-context demonstration is to furnish LLMs with illustrative examples.

Opening Statement. As first part of the prompt, $\mathcal{P}_o$ contains a high-level instruction: *"You are a knowledge graph constructor, need to synthesise relation extraction agent, named entity recognition agent, and event extraction agent to constitute an extraction collaborative team, which guides the agents to refine their results by referring to the extraction answers of others."*

Task Definition. The task description $\mathcal{P}_t$ can be further decomposed into three components, as exemplified by the RE agent:(1) The first sentence of the task description, *"You are an excellent expert in relation extraction."* is a constant that tells the LLM that it needs to focus on the relation extraction task; (2) The second sentence defines the output format of the task: *"Each result is returned as a tuple, e.g. [(head entity 1, relation type 1, tail entity 1), ...]"*. (3) The third sentence points to a specific list of relation types : *"The list of constrained relations is: [#Relation 1: [#Head Entity Type 1, #Tail Entity Type 2]...]"*.

In-context Demonstration. Some studies [14,15,27] show improvements in contextual learning by selecting few-shot demonstrations based on similarity. Our contextual prompts $\mathcal{P}_c$ are introduced as N-way K-shot sampling of the demonstration samples $\mathcal{M} = N \times K$, providing direct evidence about the task and references to predictions. However, limited by the input tokens of the LLMs, a single prompt may not contain all supported instances, so we use a sentence embedding similarity-based approach to select the $\mathcal{M}$ examples with the closest Euclidean distance as contexts.

3.3 Collaborative Interaction Optimization

In the context of team collaboration optimization, the need for meticulous decomposition design diminishes, thus we reach to the periphery of the age-old adage, "Two heads are better than one." As shown in Algorithm 1, after collecting replicas by other expert agents, we further provide collaboration prompts $\mathcal{P}_v$: *The relation extraction answer you gave in the last round of collaboration was "##LAST_ROUND_RESULT##". The answer given by the NER expert agent was "##NER_RESULT##", The EE expert agent was "##EE_RESULT##". You should refer to other members to revise your answer."*

4 Experiments

We conduct comprehensive experiments to evaluate the performance by answering the following research questions:

- **RQ1**: How does our CooperKGC perform through teamwork when competing against SOTA?
- **RQ2**: What is the impact of the expert agents and the communication rounds in multi-round interactions in teamwork?
- **RQ3**: How effective is the proposed CooperKGC in extracting different types of entities, relations and events?

4.1 Experiment Settings

Dataset. As to the NER task, we conduct experiments on the following popular benchmark: **Conllpp** [31], **OntoNotes5.0** [20] and **MSRA** [9]. For RE task, we conduct experiments on the following popular benchmark: **NYT11-HRL** [23], **Re-TACRED** [21], and **DuIE2.0** [11]. For EE task, there are two standard datasets: **ACE05** [26] and **DuEE1.0** [12].

Table 2. F1-score results for 3 KGC tasks (NER, RE, EE) on the 8 datasets.

Model	NER			RE			EE	
	Conllp	OntoNotes5.0	MSRA	NYT11-HRL	RE-TACRED	DUIE2.0	ACE05	DUEE1.0
AutoKG (0-shot)	50.6	40.4	56.8	12.5	17.2	26.9	20.7	68.7
ChatIE (0-shot)	58.4	47.5	57.7	37.5	43.9	68.4	29.7	72.0
CoT-ER (0-shot)	60.1	52.6	57.3	45.3	44.2	68.7	43.1	73.1
AutoKG (1-shot)	55.3	40.9	56.8	26.5	22.5	43.6	26.9	71.2
ChatIE (1-shot)	61.3	49.2	59.2	44.7	47.5	70.2	31.2	74.2
CoT-ER (1-shot)	61.1	53.7	58.7	47.4	48.3	71.5	45.3	74.1
CooperKGC (0-shot)	**61.3** (10.7)	**53.8** (13.4)	**60.2** (3.4)	**45.7** (33.2)	**47.1** (29.9)	**72.2** (45.3)	**47.2** (26.5)	**79.5** (10.8)
CooperKGC (1-shot)	**61.5** (6.2)	**55.4** (14.5)	**60.9** (4.1)	**49.2** (22.7)	**51.2** (28.7)	**73.6** (30.0)	**47.5** (20.6)	**81.3** (10.1)

Baselines. In our experimental framework, we opt for **AutoKG** [39] as the implementation of Vanilla LLMs for KGC realm, which defines an end-to-end extraction workflow through the manual templates. Expanding on this foundation, **ChatIE** [33] refines the extraction process using a two-round method. Taking RE as an example, this method entails the initial extraction of the relation, followed by the output of the associated entity span. This sequential approach mirrors a cognitive model's thought process, explicitly delineating the steps of task decomposition. Further, **CoT-ER** [15] introduces an explicit evidence reasoning method, characterized by three rounds of processing. In the first and second rounds, the LLM is required to output concept-level entities corresponding to head and tail entities. Subsequently, in the third round, the extraction of relevant entity spans occurs, establishing a specific relationship between these two entities with explicit evidence.

4.2 Performance Comparison with SOTA (RQ1)

Our study conducts comprehensive experiments on 0-shot and 1-shot settings across 8 datasets, each with 100 samples from test/valid sets, evaluating results using micro F1. We use the "gpt-3.5-turbo" API for both baseline models and proposed methods, with a temperature parameter set to 0.0 and average results reported over three runs. Our method sets a maximum of 4 rounds and 3 KGC team members. For the English dataset, the default customized expert knowledge background for the NER task is based on Conllpp, the RE task is NYT11-HRL, and the EE task is ACE05. Similarly, for the Chinese dataset, the NER task is based on MSRA, the RE task is DUIE2.0, and the EE task is DUIE1.0. Table 2 presents F1-score results for 3 KGC tasks across datasets, revealing:

(1) CooperKGC enhances overall performance across diverse tasks. Compared to the vanilla method, CooperKGC shows significant improvements in both 0-shot and 1-shot settings. In contrast, for a simple extraction approach like AutoKG with a single round of LLM calls, on the one hand, the overly heavy information input for task comprehension and rule constraints poses a challenge for a single model. On the other hand there is a lack of sufficient inference steps for a self-debugging process. Our approach, multiple rounds of interactions

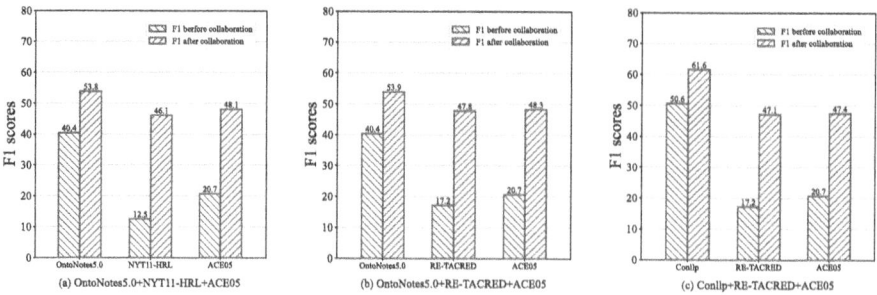

Fig. 2. Equipping KGC agents with different expert knowledge backgrounds.

alleviate this anxiety of requiring "hit-and-miss" reasoning, making it easier to explicitly identify erroneous intermediate feedbacks during the interactions.

(2) **Teamwork is an effective implicit reasoning chain.** Taking the NYT11-HRL dataset as an example, although ChatIE improved by 25.0 over the baseline in the 0-shot setting while achieving an improvement of 18.2 over the baseline in the 1-shot setting, we believe that the gain stems from decomposing the extraction process into two phases. Among the first stage is determining the types of relations involved in a given sentence, which often involves multiple relations in a single sentence. The second stage then designs triple extraction templates for each relation, which clearly indicates the sub-tasks to be accomplished in each stage. CoT-ER uses head-to-tail mapping to induce LLM to generate explicit evidence of reasoning, resulting in an improvement of 32.8 over the baseline in the 0-shot setting. CooperKGC outperforms both, with a 33.2 improvement in the 0-shot and a 22.7 improvement in the 1-shot setting. We believe that building collaborative teams contributes to "*Brain Storming*" [18], where each round of the brainstorming process is performed by the members of team. By collecting evidence from other members in each round of interactions, agent's responses is fine-tuned from the previous round. Although there is no reasoning path, this proactive optimisation shows more encouraging prospects than passive methods.

4.3   Analysis of Team Members and Interaction Rounds (RQ2)

To further investigate the impact brought by the combination of intelligences with different expert knowledge backgrounds on team collaboration, we introduce an experiment to analyse the diverse combination of team members. Specifically, We experiment on 0-shot setting and the number of team members is fixed to 3. By replacing the expert knowledge backgrounds representing NER agent, RE agent, and EE agent, we analyse which kind of expert knowledge backgrounds (mainly the schema constraints in the task description $\mathcal{P}_t$) could produce better benefits for the team construction goals. Figure 2 shows the results of equipping KGC agents with different expert knowledge backgrounds, and we observe that `combination b (OntoNotes5.0+RE-TACRED+ACE05)` allows EE expert agents to achieve the best extraction performance, and the richer variety of relation types guided by RE-TACRED allows EE agents to discover more potential arguments compared with `combination a`. In addition, `combination b` achieves a more comprehensive improvement compared to `combination c`. We analyse the schema of OntoNotes5.0 versus Conllp and find that three of the entity categories are the same (*"PER"*, *"LOC"*, *"ORG"*), while the remaining 15 more specialised entity categories refine the *"MISC"* category in Conllp, which results in benefits in extraction performance for the RE-TACRED and ACE05 datasets. We therefore conclude that more specialised expert agents, i.e., equipped with fine-grained schema constraints, can bring more insightful information to guide teamwork.

Table 3. Micro-F1 Performance under different member assignments.

Team Members	Conllp	NYT11-HRL	ACE05
3-Agent	61.3	45.7	47.2
3-Agent + ONTONOTES	58.4	**46.3**	48.3
3-Agent + RE-TACRED	**62.2**	38.4	47.4
3-Agent + BOTH	58.6	38.9	**48.4**
3-Agent (ALL CONLLP)	60.8	-	-
3-Agent (ALL NYT-HRL)	-	44.9	-
3-Agent (ALL ACE05)	-	-	29.1

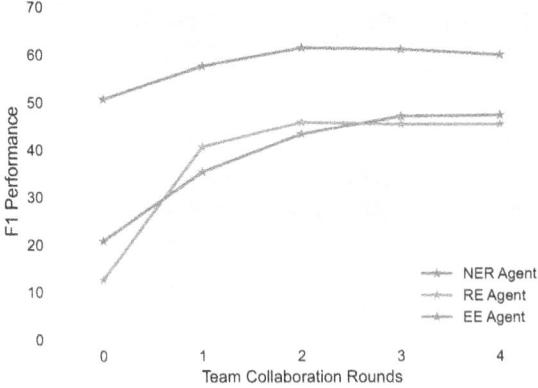

Fig. 3. Micro-F1 Performance with Increased Rounds.

Another question is whether it is possible to equip with more agents to make more gains for our team. Table 3 shows the results of both kinds of experiments, the upper one is to add additional agents to the original team,

Table 4. A qualitative example of CooperKGC on expert agent collaboration.

R	NER Agent	RE Agent	EE Agent
0	(LOC, Palestinian section of the border crossing)	(Palestinian section of the border crossing, location-located_in, Israeli troops) (Israeli troops, person-nationality, Palestinians)	[{Trigger Type: Conflict-Attack, Trigger Word: taken over, Arguments: (Attacker, Israeli troops),(Target: Palestinian section of the border crossing)}, {Trigger Type: Movement:Transport, Trigger Word: return, Arguments: (Destination, the Palestinian section of the border crossing)}]
1	(PER, Israeli troops), (LOC, border), (ORG, police)	(Israeli troops, location-located_in, the Palestinian section of the border crossing), (Israeli troops, person-nationality, Israeli), (Six Palestinian police officers, person-nationality, Palestinians)	[{Trigger Type: Conflict-Attack, Trigger Word: uprising, Arguments: (Attacker, Israeli troops), (Place, the Palestinian section of the border crossing)}, {Trigger Type: Movement:Transport, Trigger Word: return, Arguments: (Destination, the Palestinian section of the border crossing)}]
2	(PER, Israeli troops), (LOC, border), (ORG, police), (PER, Six Palestinian police officers)	(Israeli troops, person-place_lived, the Palestinian section of the border crossing), (Six Palestinian police officers, person-nationality, Palestinians)	[{Trigger Type: Conflict-Attack, Trigger Word: uprising, Arguments: (Attacker, Israeli troops), (Place, Israeli)}, {Trigger Type: Movement:Transport, Trigger Word: return, Arguments: (Destination, border), (Artifact, Israeli troops)}]
3	(PER, Israeli troops), (LOC, border), (ORG, police), (PER, Six Palestinian police officers)	(Israeli troops, person-place_lived, the Palestinian section of the border crossing), (Six Palestinian police officers, person-nationality, Palestinians)	[{Trigger Type: Conflict-Attack, Trigger Word: uprising, Arguments: (Attacker, Israeli troops), (Place, Israeli)}, {Trigger Type: Movement:Transport, Trigger Word: return, Arguments: (Destination, border), (Artifact, Six Palestinian police officers)}]
4	(PER, Israeli troops), (LOC, border), (ORG, police), (PER, Six Palestinian police officers)	(Israeli troops, person-place_lived, the Palestinian section of the border crossing), (Six Palestinian police officers, person-nationality, Palestinians)	[{Trigger Type: Conflict-Attack, Trigger Word: uprising, Arguments: (Attacker, Israeli troops), (Place, Israeli)}, {Trigger Type: Movement:Transport, Trigger Word: return, Arguments: (Destination, the Palestinian section of the border crossing), (Artifact, Six Palestinian police officers)}]

and the results show that `Team (3-Agent+BOTH)` makes the extraction results of ACE05 improved by adding a NER Agent and a RE Agent. However, another risk is also demonstrated, in both `Team (3-Agent+OntoNotes)` and `Team (3-Agent+RE-TACRED)` it is observed that when more authoritative expert agents are introduced, it leads to a decrease in the extraction results of the agent for the same task, and this kind of unconscious opinion conformity is consistent with the concept of *"Presentation of Self"* [6] in sociology. In addition, inspired by *"Self-consistency"* [29], in the bottom of Table 3 we explore the difference between the performance of the self-consistent voting method and CooperKGC on a single task. Although the consistency method to some degree mitigates the randomness of the single agent producing the hallucinatory fact, it is nevertheless weaker than our results on all 3 representative datasets. We argue that a single perspective is unable to access the interactive information provided by other experts, and thus suffers from *"Information Cocoons"* [22].

Next, we provide an analysis of the impact of the number of collaboration rounds on multi-agent teams. In Fig. 3, we increase the number of rounds for interaction between agents while fixing the number of agents to 3. We find that the performance of the algorithm also increases with the number of collaboration rounds in the first 2 rounds on all three types of tasks. However, the NER agent performance achieves its best in round 2, the RE agent in round 3, and additional collaboration by the EE agent over 3 rounds leads to a final performance similar to 3 rounds collaboration. Therefore, we believe that for tasks with simple extraction structures, too many interactions may lead to the introduction of undesirable hallucinations, hence a balance between performance and collaboration costs needs to be achieved on a task-specific basis.

4.4   Case Study of Collaboration Process (RQ3)

To illustrate the effectiveness of our proposed CooperKGC for collaborative interactions in KGC teams, Table 4 provides a qualitative example demonstrating the intermediate process. Note the CoT reflection process such as "*After considering the extraction results of other agents...*" is skipped, and the input sentence is an example of EE task "*Six Palestinian police officers were allowed to return to the Palestinian section of the border crossing, which had been taken over by Israeli troops shortly after the start of the uprising.*" we compare the results of the EE agent with the groundtruth, while the results of the NER agent and the RE agent are only for reference since there is no groundtruth. The observations are as follows: (1) **Knowledge Selection.** In Round 2, the EE agent borrows the *LOC* entity "*border*" newly discovered by the NER agent in the previous round and adds an argument *(Destination, border)* to the original answer; (2) **Knowledge Correction.** In the 1st round of interactions, the EE agent corrects the wrong trigger word "taken over", which indicates that the team members have the ability to provide self-feedback; (3) **Knowledge Aggregation.** Although the EE agent puts a wrong argument *(Destination, border)* in round 3, it rectifies the hallucination facts generated in the interim by eliciting LLM semantic comprehension during the interaction.

5   Conclusion and Future Work

In this study, we initiated the formation of a KGC team by aggregating agents with diverse expertise. Our results highlight the collaborative potential of LLM agents, showcasing how agent networks can enhance task performance collectively. The emergence of human-like behaviors in collaboration aligns with sociological theories, leading to improvements in factuality, knowledge integration, and intellectual reasoning. Future research could draw insights from sociologically derived architectures, expanding the application of CooperKGC variants to solve diverse collaborative tasks.

References

1. Brown, T.B., Mann, B., Ryder, N., et al.: Language models are few-shot learners. In: NeurIPS (2020)
2. Cohen, R., Hamri, M., Geva, M., Globerson, A.: LM vs LM: detecting factual errors via cross examination. CoRR abs/ arXiv: 2305.13281 (2023)
3. Du, Y., Li, S., Torralba, A., et al.: Improving factuality and reasoning in language models through multiagent debate. CoRR abs/ arXiv: 2305.14325 (2023)
4. Du, Z., Qian, Y., Liu, X., et al.: GLM: general language model pretraining with autoregressive blank infilling. In: ACL, pp. 320–335. ACL (2022)
5. Durkheim, E.: The division of labor in society. In: Social stratification, pp. 217–222. Routledge (2018)
6. Goffman, E., et al.: The presentation of self in everyday life. 1959. Garden City, NY **259** (2002)
7. Hao, R., Hu, L., Qi, W., et al.: Chatllm network: more brains, more intelligence. CoRR abs/ arXiv: 2304.12998 (2023)
8. Lamport, L., Shostak, R.E., Pease, M.C.: The byzantine generals problem. ACM Trans. Program. Lang. Syst. **4**(3), 382–401 (1982)

9. Levow, G.: The third international Chinese language processing bakeoff: word segmentation and named entity recognition. In: SIGHAN@COLING/ACL, pp. 108–117. ACL (2006)

10. Li, R., Patel, T., Du, X.: PRD: peer rank and discussion improve large language model based evaluations. CoRR abs/arXiv: 2307.02762 (2023)

11. Li, S., et al.: DuIE: a large-scale Chinese dataset for information extraction. In: Tang, J., Kan, M.-Y., Zhao, D., Li, S., Zan, H. (eds.) NLPCC 2019. LNCS (LNAI), vol. 11839, pp. 791–800. Springer, Cham (2019). https://doi.org/10.1007/978-3-030-32236-6_72

12. Li, X., et al.: DuEE: a large-scale dataset for Chinese event extraction in real-world scenarios. In: Zhu, X., Zhang, M., Hong, Yu., He, R. (eds.) NLPCC 2020. LNCS (LNAI), vol. 12431, pp. 534–545. Springer, Cham (2020). https://doi.org/10.1007/978-3-030-60457-8_44

13. Liang, T., He, Z., Jiao, W., et al.: Encouraging divergent thinking in large language models through multi-agent debate. CoRR abs/ arxiv: 2305.19118 (2023)

14. Liu, J., Shen, D., Zhang, Y., et al.: What makes good in-context examples for gpt-3? In: DeeLIO@ACL, pp. 100–114. ACL (2022)

15. Ma, X., Li, J., Zhang, M.: Chain of thought with explicit evidence reasoning for few-shot relation extraction. CoRR abs/ arXiv: 2311.05922 (2023)

16. Minsky, M.: Society of mind. Simon and Schuster (1988)

17. Mondal, I., Hou, Y., Jochim, C.: End-to-end construction of NLP knowledge graph. In: ACL/IJCNLP, vol. ACL/IJCNLP 2021, pp. 1885–1895. ACL (2021)

18. Osborn, A.F.: Applied imagination (1953)

19. Pang, C., Cao, Y., Ding, Q., Luo, P.: Guideline learning for in-context information extraction. CoRR abs/ arXiv: 2310.05066 (2023)

20. Pradhan, S., Moschitti, A., Xue, N., et al.: Towards robust linguistic analysis using ontonotes. In: CoNLL, pp. 143–152. ACL (2013)

21. Stoica, G., Platanios, E.A., Póczos, B.: Re-tacred: addressing shortcomings of the TACRED dataset. In: AAAI, pp. 13843–13850. AAAI Press (2021)

22. Sunstein, C.R.: Infotopia: how many minds produce knowledge. Oxford University Press (2006)

23. Takanobu, R., Zhang, T., Liu, J., Huang, M.: A hierarchical framework for relation extraction with reinforcement learning. In: AAAI, pp. 7072–7079. AAAI Press (2019)

24. Tuckman, B.W.: Developmental sequence in small groups. Psychol. Bull. **63**(6), 384 (1965)

25. Vakaj, E., Tiwari, S., Mihindukulasooriya, N., et al.: NLP4KGC: natural language processing for knowledge graph construction. In: WWW, pp. 1111. ACM (2023)

26. Walker, C., Strassel, S., Medero, J., Maeda, K.: Ace 2005 multilingual training corpus (2006)

27. Wan, Z., Cheng, F., Mao, Z., et al.: GPT-RE: in-context learning for relation extraction using large language models. CoRR abs/ arXiv: 2305.02105 (2023)

28. Wang, S., Sun, X., Li, X., et al.: GPT-NER: named entity recognition via large language models. CoRR abs/ arXiv: 2304.10428 (2023)

29. Wang, X., Wei, J., Schuurmans, D., et al.: Self-consistency improves chain of thought reasoning in language models. In: ICLR. OpenReview.net (2023)

30. Wang, Z., Mao, S., Wu, W., et al.: Unleashing cognitive synergy in large language models: a task-solving agent through multi-persona self-collaboration. CoRR abs/ arXiv: 2307.05300

31. Wang, Z., Shang, J., Liu, L., et al.: Crossweigh: training named entity tagger from imperfect annotations. In: EMNLP-IJCNLP, pp. 5153–5162. ACL (2019)

32. Wei, J., Wang, X., Schuurmans, D., et al.: Chain-of-thought prompting elicits reasoning in large language models. In: NeurIPS (2022)
33. Wei, X., Cui, X., Cheng, N., et al.: Zero-shot information extraction via chatting with chatgpt. CoRR abs/ arXiv: 2302.10205 (2023)
34. Xiong, K., Ding, X., Cao, Y., et al.: Examining the inter-consistency of large language models: an in-depth analysis via debate. CoRR abs/ arXiv: 2305.11595 (2023)
35. Ye, H., Liu, T., Zhang, A., et al.: Cognitive mirage: a review of hallucinations in large language models. CoRR abs/ arXiv: 2309.06794 (2023)
36. Ye, H., Zhang, N., Chen, H., Chen, H.: Generative knowledge graph construction: A review. In: EMNLP, pp. 1–17. ACL (2022)
37. Yu, Z., He, L., Wu, Z., et al.: Towards better chain-of-thought prompting strategies: a survey. CoRR abs/ arXiv: 2310.04959 (2023)
38. Zhang, Y., Chen, Z., Zhang, W., Chen, H.: Making large language models perform better in knowledge graph completion. CoRR abs/ arXiv: 2310.06671 (2023)
39. Zhu, Y., Wang, X., Chen, J., et al.: Llms for knowledge graph construction and reasoning: Recent capabilities and future opportunities. CoRR abs/ arXiv: 2305.13168 (2023)

SCM-Net: Semantic-Contrastive Multimodal Framework for Enhanced Chinese NER

Zhuang Wang[1], Yijia Zhang[1(✉)], Jianyuan Yuan[1], Songtao Li[1], Mingyu Lu[2], and Hongfei Lin[3]

[1] College of Information Science and Technology, Dalian Maritime University Dalian, Dalian, China
zhangyijia@dlmu.edu.cn
[2] College of Artificial Intelligence, Dalian Maritime University Dalian, Dalian, China
[3] School of Computer Science and Technology, Dalian University of Technology Dalian, Dalian, China

Abstract. Current approaches to Chinese Named Entity Recognition (CNER) often struggle with the accurate identification of entities due to the inherent complexity of the Chinese language, including its script and phonetics. This study introduces three innovative solutions to address specific challenges in CNER. Firstly, the prevalent issue of irrelevant vocabulary inclusion is tackled by implementing a semantic similarity-based filtering method. This method employs cosine similarity calculations between matched words and the entire sentence, selecting only those words that exceed a predefined similarity threshold for integration into the character sequence. Secondly, the challenge of distinguishing relevant from irrelevant vocabulary is addressed through contrastive learning. By minimizing the representational distance between characters and matching words, and maximizing the distance from non-matching segmented words, the model's discriminatory power is significantly enhanced. Lastly, the complexity of Chinese characters, both in terms of their structure and phonetics, is addressed by incorporating multimodal features. Phonetic features are extracted by converting characters into pinyin, followed by word2vec embeddings, while morphological features are obtained through character images processed by a Vision Transformer (VIT) model. These multimodal features provide a comprehensive understanding of Chinese characters, thereby improving the accuracy and robustness of the CNER model. Collectively, these solutions present a significant advancement in addressing the unique challenges of CNER, offering a more nuanced and effective approach to entity recognition in Chinese texts.

Keywords: Named Entity Recognition · contrastive learning · multimodal features · cosine similarity calculations

B. Xu et al. (Eds.): CCKS-IJCKG 2024, CCIS 2229, pp. 82–97, 2025.
https://doi.org/10.1007/978-981-96-1809-5_7

1 Introduction

Chinese Named Entity Recognition (CNER) has evolved significantly over the years, transitioning from rule-based approaches to advanced machine learning techniques. The advent of statistical methods, such as Hidden Markov Models (HMM) and Conditional Random Fields (CRF), marked a significant shift, offering more flexibility and better performance [1]. Recently, the application of pre-trained language models like BERT and its adaptations for Chinese (e.g., Chinese BERT-wwm) has set new benchmarks in CNER. These models leverage vast amounts of data to learn rich contextual representations, significantly improving entity recognition accuracy [2]. Despite these advancements, current CNER models still face challenges. They often struggle with the inclusion of irrelevant vocabulary, hindering their ability to distinguish between relevant and irrelevant terms effectively. Additionally, the unique phonetic and morphological features of Chinese characters are not fully utilized, which could provide valuable clues for entity recognition. In current Chinese Named Entity Recognition (CNER) research, a prevalent issue is the accurate identification and differentiation of long phrases and the shorter terms they encompass. Traditional CNER methods often struggle with complex phrases containing multiple potential entities, especially when a long phrase embeds several possible short entity terms. As shown in Fig. 1, when dealing with a long phrase like "北京大学国际关系学院" (Peking University School of International Relations), traditional approaches might erroneously identify "北京" (Beijing), "大学" (University), "国际" (International), "关系" (Relations), and "学院" (School) as independent entities, overlooking their collective entity significance.

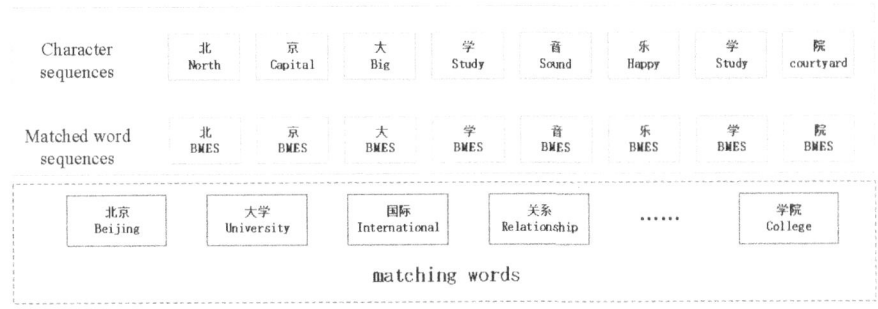

Fig. 1. Some examples demonstrating that there are some noises in the words provided by lexicon. These vocabulary words are not helpful for the task.

The contributions of this paper are significant. It introduces a more effective vocabulary filtering method for CNER, enhances the discriminatory power of CNER models through contrastive learning, integrates multimodal features for a comprehensive understanding of Chinese characters, and achieves state-of-the-art results on four public Chinese NER datasets: OntoNote, MSRA, Weibo, and Resume. The main contributions of our work are as follows:

1. Introduction of a semantic similarity-based method for more effective vocabulary filtering in CNER.
2. Implementation of contrastive learning to enhance the discriminatory power of CNER models.
3. Integrating phonetic and morphological features into CNER.
4. Achievement of state-of-the-art (SOTA) results on four public Chinese NER datasets: OntoNote, MSRA, Weibo, and Resume, surpassing previous benchmarks.

2　Related Work

Current research on Chinese named entity recognition (CNER) mainly involves two approaches: feature extraction based on deep neural networks and character-level information enhancement.

The feature extraction approach based on deep neural networks can efficiently and quickly obtain text sequence features, bypassing the complex and time-consuming feature engineering process. This method is replacing the traditional rule-based and machine learning-based named entity recognition approaches. A comprehensive survey by Li et al. [3] emphasizes the impact of deep learning on NER and highlights its significant advantages over traditional methods. One of the earliest and most influential methods in this field is the application of convolutional neural networks (CNNs) and recurrent neural networks (RNNs), including long short-term memory (LSTM) networks. Huang et al. [4] demonstrated the effectiveness of combining LSTM with conditional random fields (CRF) for sequence labeling, setting a new benchmark for NER performance. The introduction of pre-trained language models like BERT and GPT has further revolutionized this field. These models, pre-trained on large corpora, provide a deeper understanding of language context and semantics. Liu et al. [5] showed that fine-tuning BERT leads to state-of-the-art performance on NER tasks, surpassing previous deep learning models. Another important advancement is the integration of active learning with deep learning models. Shen et al. [6]. explored this in their work, demonstrating how active learning algorithms can improve the efficiency of NER models, especially in cases of limited labeled data. Deep learning has also made significant progress in domain-specific NER. For example, in biomedical NER, Habibi et al. [7] showed that deep learning models with word embeddings significantly improve the recognition of biomedical entities. Similarly, in the context of Chinese clinical literature, a deep neural network-based minimal feature engineering approach has been found to be effective.

The character-level information enhancement approach primarily focuses on incorporating more character features and contextual information into Chinese character embeddings to improve CNER performance. Liu et al. [8] compared experiments and found that character-level embeddings outperform word-level embeddings for CNER. However, using character vectors alone can lead to the loss of intrinsic character information, so enhancing character embeddings is necessary. Lu et al. [9] integrated pinyin information into Chinese character embeddings using a BiLSTM network to improve CNER performance. Zhang et al. [10] proposed a Lattice-LSTM method that incorporates dictionary information to enrich character representations and enhance entity recognition. Sui et al. [11] proposed a graph attention network-based method that integrates lexical knowledge into character representations by designing three word-character interaction graph networks in the graph layer. Zhang et al. [12] used local attention convolution algorithms to enrich the semantic information in character embeddings.

Although deep learning and pre-trained language models have significantly advanced CNER, they still face challenges in fully capturing the complexity of the Chinese language. One notable issue is the models' limited ability to capture subtle phonetic and visual differences, as well as the context-dependent nature of Chinese characters. Additionally, many models often include a large number of irrelevant or low-information words during the recognition process, leading to decreased accuracy. These challenges emphasize the need for a more comprehensive and context-sensitive approach to CNER.

3   Methods

3.1   Character Feature Extraction

In our Chinese Named Entity Recognition (CNER) model, the process of character feature extraction involves transforming the input Chinese character sequence $Y = (y_1, y_2, \ldots, y_n)$ into a vectorized representation. Each character is represented as:

$$l_i = e^c(y_i), l_i \in \mathbb{R}^d \qquad (1)$$

3.2   Semantic-Based Vocabulary Selection

In the realm of Chinese Named Entity Recognition (CNER), the vocabulary selection based on semantic similarity is a fundamental and crucial step, pivotal for the accurate identification and classification of entities. The process commences with an input Chinese sentence, represented as $Y = (y_1, y_2, \ldots, y_n)$, where each y_i is an individual character. This sequence undergoes a matching process against a pre-trained vocabulary list G, which is a comprehensive collection of words relevant to the CNER task. The outcome of this matching is a set of words M, found within the input sequence and serving as potential candidates for entity recognition.

As shown in Fig. 2, to encapsulate the contextual essence of the sentence, the sequence Y is fed into a pre-trained BERT model. This model, renowned for its proficiency in understanding language nuances and context, transforms the input characters into a rich, contextualized sentence representation, denoted as X. This transformation is mathematically represented as:

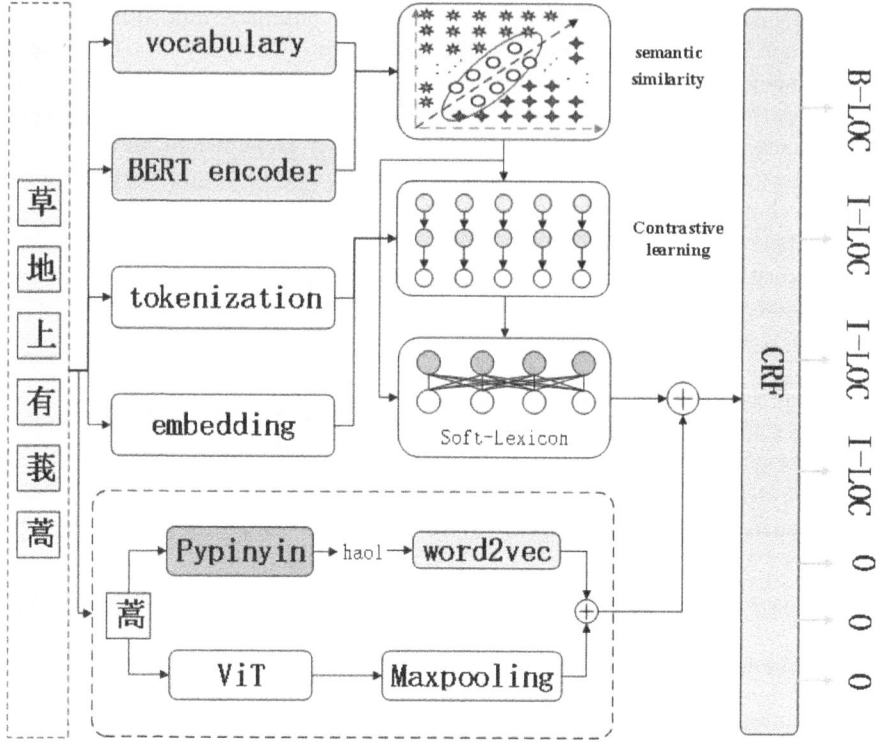

Fig. 2. An overview of our SCM-Net.

$$X = BERT(Y) \tag{2}$$

where $BERT$ signifies the function applied by the model.

The subsequent step involves calculating the semantic similarity between each word in the set M and the sentence representation X. This step is crucial for pinpointing words that are not only present in the sentence but also hold significant contextual relevance. Employing the cosine similarity metric, defined as:

$$\text{sim}\,(m_i, X) = \frac{m_i \cdot X}{\|m_i\| \, \|X\|}. \tag{3}$$

where m_i is the vector representation of a word and $\cdot$ denotes the dot product, this process identifies contextually relevant words. A predefined threshold a is then used to filter these words based on their similarity scores. Words with a similarity score exceeding a are selected, culminating in a refined matched word set $R = \{m_i \in M \mid \text{sim}\,(m_i, X) > a\}$. This set R comprises words that are not just present in the input sequence but are also contextually pertinent, thereby enhancing the precision of the entity recognition process that follows.

The soft-lexicon strategy allows for a more delicate integration of contextual information as it considers the semantic relevance of vocabulary to the entire sentence. This method can capture subtle nuances that rigid, context-ignorant approaches may overlook. In order to integrate vocabulary information into the character sequence representation, we refer to the soft-lexicon strategy for fusion operations. Specifically, each character in the input Chinese character sequence is associated with four corresponding word sets, which are words filtered through the semantic similarity module. According to the soft-lexicon strategy, we use the four segmentation tags "BMES" to label these words. Each character in the matched words is classified into $\{B, M, E, S\}$. When the character's corresponding word set is empty, we denote it with "none". For each character, it is comprised of the following four sets:

$$b\left(y_i\right) = \{r_{i,k}, \forall r_{i,k} \in R, i < k \leq n\} \tag{4}$$

$$m\left(y_i\right) = \{r_{j,k}, \forall r_{j,k} \in R, 1 \leq j < i < k \leq n\} \tag{5}$$

$$e\left(y_i\right) = \{r_{j,i}, \forall r_{j,i} \in R, 1 < j \leq i\} \tag{6}$$

$$s\left(y_i\right) = \{y_i, \exists y_i \in R\} \tag{7}$$

each character can be represented as:

$$h_i^{\text{bmes}} = [z\left(b\left(y_i\right)\right); z\left(m\left(y_i\right)\right); z\left(e\left(y_i\right)\right); z\left(s\left(y_i\right)\right)] \tag{8}$$

where z is used to obtain the feature representation of the word set, and the specific implementation details are as follows:

where $D(r)$ refers to the frequency of a word's occurrence in the dataset, which is used to obtain the embedding representation of vocabulary through an embedding lookup table. b refers to values that appear less than b times in the dataset and account for 10% of the total. The formula for calculating N is as follows:

$$N = \sum_{r \in (b \cup m \cup e \cup s)} N(r) + b' \tag{9}$$

through the above soft-lexicon strategy, we have obtained the representation of each character after integrating lexical information. The final representation of word features is as follows:

$$H^{\text{bmes}} = \left[h_1^{\text{bmes}}, h_2^{\text{bmes}}, \ldots, h_n^{\text{bmes}}\right] \tag{10}$$

3.3   Implementation of Contrastive Learning

In the previous section, we integrated the filtered vocabulary into the character sequence, obtaining the final word feature representation. Alongside this, the initial character feature representation of the input sequence, denoted as

$L = (l_1, l_2, \ldots, l_n)$, was established. To further enhance our model, we introduce contrastive learning, a technique that has shown significant promise in various domains of machine learning.

In the realm of Named Entity Recognition (NER), traditional models often grapple with the challenge of accurately distinguishing between contextually similar entities, a task made difficult due to the subtle linguistic nuances present in natural language. This limitation becomes particularly pronounced in dealing with ambiguous or polysemous words, leading to a decrease in the overall efficacy of the entity recognition process. To mitigate these issues, the field has seen a growing interest in the adoption of contrastive learning, a technique that has demonstrated considerable success in enhancing representation learning. Contrastive learning, as highlighted in recent studies, excels in amplifying the differences between distinct entities while maintaining similarities where relevant, thereby enabling models to develop more robust and discriminative features.

As shown in Fig. 2, in our Chinese Named Entity Recognition (CNER) model, the implementation of contrastive learning is a crucial step that follows the integration of semantically filtered vocabulary into the character sequence. After obtaining the final word feature representation and the initial character feature representation $L = (l_1, l_2, \ldots, l_n)$, we proceed to segment the original input sequence into a set of words, denoted as F. This segmentation forms the basis for creating a collection of negative examples, represented as $W = (w_1, w_2, \ldots, w_n)$, which are words in F that do not match the selected vocabulary. These words are then processed through a BERT model to extract their features, forming the segmented word feature set W.

The essence of contrastive learning in our model, inspired by recent advancements in the field [13] is to enhance the model's ability to distinguish between similar entities. This is achieved by maximizing the representational distance between the character features L and the negative examples W, while minimizing the distance between L and the positive examples H^{bmes}, where H^{bmes} represents the character features integrated with the matched word representations. Our ultimate goal is to minimize the loss of contrastive learning, which can be expressed as:

$$\mathcal{L}_c = -\log \frac{e^{\text{sim}\left(l_i, h_i^{\text{bmes}}\right)/\gamma}}{\sum_{j \in m} e^{\text{sim}(l_i, w_j)/\gamma}} \tag{11}$$

where $sim()$ is used to calculate the similarity between characters and matching words, as well as between characters and segmented words. refers to the number of vocabularies in the segmentation set, and γ is the temperature.

The contrastive learning loss is formulated to capture these objectives, combining components that penalize close distances between L and W, and reward close distances between L and H^{bmes}. This approach significantly improves the discriminative power of the feature representations, leading to more accurate entity recognition in the CNER model.

3.4   Integration of Multimodal Features

In the realm of Chinese Named Entity Recognition (CNER), traditional approaches predominantly focus on semantic and contextual analysis at the character level, often utilizing advanced models like BERT and LSTM. However, these methods tend to overlook the rich multimodal nature of Chinese characters, encompassing both orthographic and phonetic information, leading to challenges in accurately identifying entities represented by less common or complex characters.

To address these limitations, as shown in Fig. 2, our CNER model integrates the orthographic features of Chinese characters using the PIL library's Image-Draw to generate images, which are then resized to fit the input dimensions of the Vision Transformer (ViT) [1024x1024]. The ViT, known for its capability to capture intricate visual patterns, processes these images and outputs features that are further refined through a Maxpooling layer, extracting the essential structural features of each character.

$$T_i = \mathrm{Ma}\left(\mathrm{Vi}\left(\mathrm{Im}\left(y_i\right)\right)\right) \tag{12}$$

where $Im(\cdot)$ refers to obtaining the image representation of Chinese characters through ImageDraw. $Vi(\cdot)$ is the process of extracting character graphic features using ViT (Vision Transformer). $Ma(\cdot)$ refers to the Maxpooling layer.

Concurrently, the phonetic features are addressed by transcribing characters into their pinyin forms using the Pypinyin library [14], as exemplified by the character "翎" (líng) transcribed as "ling2", where "ling" represents the phonetic component, and "2" indicates the tone. These phonetic representations are input into a Word2Vec model, chosen for its effectiveness in capturing semantic relationships within phonetic structures, and the resulting phonetic feature vectors are upsampled to align with the orthographic features. The specific process is as follows:

$$Z_{i,m,n}^{phn} = e^{phn}\left(P\left(y_{m,n}^i\right)\right) \tag{13}$$

where $Z_{i,m,n}^{phn}$ refers to the final phonetic feature representation of the Chinese character $y_{m,n}^i$ obtained. i refers to the ith character in the input Chinese sequence $y_{m,n}$. $P(\cdot)$ denotes the process of obtaining the phonetic representation of Chinese characters using Pypinyin. $e^{phn}(\cdot)$ refers to the final phonetic embedding representation of Chinese characters obtained using the word2vec tool.

The final step involves concatenating these orthographic and phonetic feature vectors to form a comprehensive multimodal representation for each character, as follows:

$$a_i = t_i + z_i \tag{14}$$

where t_i refers to the graphic features of Chinese characters, and z_i refers to the phonetic features of Chinese characters. Through the above steps, we have obtained the glyph feature representation and the phonetic feature representation of the characters in the input sequence. The final multimodal feature representation of Chinese is as follows:

$$A = [a_1, a_2, \ldots, a_n] \tag{15}$$

This fusion of visual and phonetic aspects allows our model to overcome the limitations of traditional methods by providing a more holistic understanding of each character, significantly enhancing the model's ability to recognize and interpret entities, especially those represented by less common or ambiguous characters.

3.5 Output Layer

After obtaining character sequence representations enriched with lexical information and multimodal character features, these two sets of feature vectors are concatenated to form a comprehensive character feature representation as $F = [f_1, f_2, \ldots, f_n]$. This integrated representation not only merges the semantic and multimodal information of the characters but also enhances the model's capability to discern entity boundaries. Consequently, these concatenated feature vectors are fed into the CRF layer for sequence labeling. The CRF layer effectively decodes the most probable sequence of labels for the input sequence, thereby accurately identifying and classifying named entities within the input text, such as person names, locations, organizations, and more. The specific process is as follows:

$$p(b \mid F) = \frac{\prod_{i=1}^{N} \eth_i \left(l_{i-1}, F\right)}{\sum_{y' \in L} \prod_{i=1}^{N} \eth_i \left(y'_{i-1}, y'_i, F\right)} \tag{16}$$

where $\eth$ refers to the potential function, F is the input sequence with a length of N. During the model training process, we use maximum likelihood estimation, and the formula is as follows:

$$L(p(b \mid F)) = \sum_i \log p(b \mid F) \tag{17}$$

4   Experiments

To validate the effectiveness of our proposed SCM-Net model in Chinese Named Entity Recognition (CNER), we conducted a comprehensive series of experiments, benchmarking our model against current state-of-the-art approaches in the field. These comparative experiments were meticulously designed to not only demonstrate the efficacy of SCM-Net but also to provide a clear understanding of how our model stands in relation to existing prominent methods in Chinese NER.

4.1   Experiment Settings

To comprehensively validate the effectiveness of our proposed SCM-Net model in Chinese Named Entity Recognition (CNER), we conducted experiments on four publicly available datasets, each offering unique characteristics and challenges. The datasets include OntoNotes 4.0, MSRA, Weibo, and Resume. These datasets collectively enable a thorough assessment of our model across various text genres and entity types. **Hyperparameter**: In the experimental setup for our SCM-Net model, training was conducted on a single GPU for efficiency. Key parameters were set as follows: a learning rate of 0.01, batch size of 32, and a dropout rate of 0.1, with the Adam optimizer facilitating the training process. The model underwent 12 epochs to ensure adequate learning. For performance evaluation, we focused on F1 score, precision, and recall, which are critical metrics for assessing the effectiveness of our model in Chinese Named Entity Recognition tasks.

4.2   Baseline

In this section, we compare our model with several current Chinese entity recognition models to assess its effectiveness. Lattice-LSTM [8] integrates character and word information using a unique lattice structure. LR-CNN [15] enhances performance through innovative lexicon integration. Soft-Lexicon [16] merges dictionary data into character-level models for improved efficiency. FLAT [17] is a flat-lattice transformer model that speeds up processing while maintaining accuracy. MECT [18] combines multiple metadata embeddings in a cross-transformer architecture. NFLAT [19] extends FLAT to handle ambiguous word boundaries using non-flat lattice structures. ZEN [20] enhances Chinese text encoding with n-gram representations. MCL [21] employs multi-granularity contrastive learning for effective dictionary utilization. SSMI [22] optimizes character representations with semantic similarity and mutual information maximization.

4.3   Datasets

We used 4 Chinese named entity recognition standard datasets: Ontonotes 4.0 [23], MSRA [24], Resume [25,26], and Weibo [8]. The statistics of each dataset are shown in Table 1. Among them, Sent represents the sentence count, Char represents the character count, Entity represents the entity count.

(1) Ontonotes 4.0: The corpus is from news domain text, with a total of 28.05k entities in the training, validation, and test sets. There are 4 entity label types: PER, LOC, GPE, and ORG. The word segmentation labels are BMES, such as B-PER, M-PER, E-PER, S-PER, and the non-entity label is O.

(2) MSRA: The corpus is from news domain text, without a validation set. There are a total of 81k entities in the training and test sets. There are 3 entity label types: NS, NT, and NR. The word segmentation labels are BMES, and there is a non-entity label O.

Table 1. Dataset statistics table

Dataset	Type	Training set	Validation set	Test set
Ontonotes 4.0	Sent	15.7K	4.3K	4.3K
	Char	491.9K	200.5K	208.1K
	Entity	13.4K	6.95K	7.7K
MSRA	Sent	46.4K	–	4.4K
	Char	2169.9K	–	172.6K
	Entity	74.8K	–	6.2K
Weibo	Sent	1.4K	0.27K	0.27K
	Char	73.8K	14.5K	14.8K
	Entity	1.89K	0.39K	0.42K
Resume	Sent	3.8K	0.46k	0.48K
	Char	124.1K	13.9K	15.1K
	Entity	1.34K	0.16K	0.15K

(3) Weibo: The corpus is from social media platform text, with a total of 2.7k entities in the training, validation, and test sets. There are 4 entity label types: PER, LOC, GPE, and ORG, including the suffix labels NOM (Nominal Entity) and NAM (Named Entity). The word segmentation labels are BMES, such as B-PER.NOM, B-PER.NAM, and there is a non-entity label O.

(4) Resume: The corpus is from Sina finance resume text data, with a total of 1.65k entities in the training, validation, and test sets. There are 8 entity label types: NAME, CONT, RACE, TITLE, EDU, ORG, PRO, and LOC. The word segmentation labels are BMES, and there is a non-entity label O.

4.4 Main Result

As shown in Table 2. In the comprehensive evaluation of Chinese Named Entity Recognition (NER) models across OntoNote4, MSRA, Weibo, and Resume datasets, nuanced differences in model performances emerge. SCM-Net, our proposed model, demonstrates robust results, particularly excelling in the MSRA and Resume datasets with its innovative multimodal feature integration and contrastive learning approach. However, it's essential to acknowledge the specific strengths of other models in certain contexts. For instance, SSMI shows a remarkable ability to handle semantic complexities on the Weibo dataset, likely due to its focus on semantic similarity and mutual information maximization. This is in contrast to MCL, which excels in leveraging dictionary information through its multi-granularity contrastive learning, particularly evident in its performance on the same dataset.

Resume dataset, suggesting their effectiveness in lexicon integration and simplification. Meanwhile, FLAT and NFLAT, known for their lattice structures, offer advantages in processing speed, though they might not fully capture the intricate semantic relationships as effectively as SCM-Net or MCL.

Table 2. Results on OntoNote 4.0, MSRA, Weibo, and Resume datasets

Models	OntoNote4			MSRA			Weibo			Resume		
	P	R	F	P	R	F	P	R	F	P	R	F
Lattice-LSTM	76.35	71.56	73.88	93.57	92.79	93.18	53.04	62.25	58.79	94.18	94.11	94.46
LR-CNN	76.40	72.60	74.45	94.50	92.93	93.71	57.14	66.67	59.92	95.37	94.84	95.11
Soft-Lexicon	77.13	75.22	76.16	94.73	93.40	94.06	59.08	62.22	61.42	95.71	95.77	95.74
FLAT	–	–	76.45	–	–	94.35	–	–	63.42	–	–	94.93
MECT	77.57	76.27	76.92	94.55	94.09	94.32	61.91	62.51	63.30	96.40	95.39	95.89
NFLAT	75.17	79.37	77.21	94.92	94.19	94.55	59.10	63.76	61.94	95.63	95.52	95.58
ZEN	80.52	78.97	79.03	95.90	95.06	95.20	–	–	66.71	95.48	95.43	95.40
MCL	–	–	82.96	–	–	96.11	–	–	73.08	–	–	96.46
SSMI	82.46	84.61	83.52	96.15	96.49	96.32	71.53	73.18	72.83	97.48	97.18	97.33
SCM-Net	82.56	84.35	84.13	96.35	96.58	96.93	71.87	74.26	73.47	97.32	97.46	98.03

In direct comparisons, while SCM-Net shows an overall high performance, it's crucial to consider the context-specific advantages of other models. For example, the high F1-scores of SSMI and MCL on the Weibo dataset underscore their specialized approaches in semantic analysis and dictionary utilization, respectively. Similarly, ZEN's success on the OntoNote4 dataset highlights the importance of n-gram representations in certain linguistic contexts, a strategy different from SCM-Net's broader multimodal approach.

In comparing Chinese NER models on Weibo, SCM-Net's performance is notably distinct. For instance, while SSMI achieves an F1-score of 72.83, SCM-Net slightly improves upon this with a score of 73.47. This enhancement is likely due to SCM-Net's effective use of contrastive learning, which aids in distinguishing subtle differences in entity types, a common challenge in the Weibo dataset. In the context of the MSRA dataset, SCM-Net records an F1-score of 96.93, surpassing ZEN's 95.20. This indicates SCM-Net's adeptness in integrating and analyzing multimodal linguistic information, a step beyond the n-gram representations employed by ZEN. The model's ability to combine various features like character-level and syntactic information, allows for a more comprehensive linguistic understanding, leading to improved entity recognition. Additionally, when compared to MCL on the Resume dataset, SCM-Net's F1-score of 98.03 versus MCL's 96.46 suggests its enhanced capability in managing complex entity relationships. This can be largely attributed to SCM-Net's multimodal approach, which not only considers textual content but also incorporates additional linguistic features, providing a more contextually aware entity recognition process.

Text	The conference was held at [Jing'an LOC] District, [Shanghai LOC].	He moved to [Apple ORG] for his new project.	She studied at [Tsinghua University ORG] before joining [Google ORG].
Lattice-LSTM	(Jing'an LOC) x (Shanghai LOC) √	(Apple ORG) x	(Tsinghua University ORG) x (Google ORG) √
MECT	(Jing'an LOC) x (Shanghai LOC) x	(Apple ORG) √	(Tsinghua University ORG) x (Google ORG) x
SCM-Net	(Jing'an LOC) √ (Shanghai LOC) √	(Apple ORG) √	(Tsinghua University ORG) √ (Google ORG) √

Fig. 3. Three cases of the predictions by Lattice-LSTM, MECT and SCM-Net.

4.5   Ablation Result

The ablation study was conducted to evaluate the contribution of key components in the SCM-Net model. As show in Table 3, four experiments were performed, including the main SCM-Net model and three ablated versions: SCM-Net-NoMM (without multimodal features), SCM-Net-NoCL (without contrastive learning), and SCM-Net-AllLex (using all lexicon words without semantic similarity filtering). The performance was assessed on four datasets: OntoNote4, MSRA, Weibo, and Resume, using precision (P), recall (R), and F1-score (F) as metrics.

Table 3. Results of ablation experiments on four datasets

Models	OntoNote4			MSRA			Weibo			Resume		
	P	R	F	P	R	F	P	R	F	P	R	F
SCM-Net-NoMM	52.08	83.89	83.91	96.07	96.17	96.61	71.54	74.05	72.95	97.11	97.23	97.78
SCM-Net-NoCL	82.16	83.94	83.82	95.82	96.22	96.46	71.33	73.83	73.14	97.03	97.12	97.85
SCM-Net-AllLex	82.34	84.02	83.84	96.13	96.23	96.54	71.23	73.91	73.05	96.74	97.02	97.83
SCM-Net	82.56	84.35	84.13	96.35	96.58	96.93	71.87	74.26	73.47	97.32	97.46	98.03

SCM-Net-NoMM showed a significant drop in performance on the OntoNote4 dataset (F1-score: 83.91) compared to the main SCM-Net model (F1-score: 84.13), indicating the importance of multimodal features in capturing complex linguistic patterns. However, on the MSRA dataset, the performance remained relatively high (F1-score: 96.61), suggesting that multimodal features play a more crucial role in datasets with diverse linguistic contexts.

SCM-Net-NoCL exhibited a slight decrease in performance across all datasets, with the most notable difference on the MSRA dataset (F1-score: 96.46 for SCM-Net-NoCL vs. 96.93 for SCM-Net). This underscores the effectiveness of contrastive learning in enhancing the model's ability to differentiate between entity types, especially in datasets with a wide variety of entities.

SCM-Net-AllLex showed a marginal decrease in performance, particularly on the Resume dataset (F1-score: 97.83 for SCM-Net-AllLex vs. 98.03 for SCM-Net). This suggests that the semantic similarity-based lexicon selection contributes to the model's precision in entity recognition, although its impact is less pronounced than that of the other components. The main SCM-Net model consistently outperformed the ablated versions, demonstrating the combined effectiveness of multimodal features, contrastive learning, and semantic similarity-based lexicon selection. The highest performance was observed on the Resume dataset (F1-score: 98.03), highlighting the model's strength in datasets with complex entity structures.

4.6   Case Analysis

As show in Fig. 3 the first case, involving a complex entity "Jing'an District, Shanghai;; demonstrates SCM-Net's superior ability to recognize multi-word entities. SCM-Net correctly identifies the entire entity, while Lattice-LSTM only recognizes "Shanghai," and MECT fails to identify it. This case underscores the advantage of SCM-Net's multimodal feature integration, enabling it to capture more comprehensive entity structures compared to the other models.

In the second case, which presents an ambiguous entity context "Apple" SCM-Net and MECT correctly interpret "Apple" as an organization, whereas Lattice-LSTM misinterprets it as a fruit. This scenario highlights SCM-Net's effective use of contextual information, likely benefiting from its contrastive learning component, which aids in distinguishing between different entity interpretations.

The third case features a rare entity "Tsinghua University." SCM-Net successfully identifies both "Tsinghua University" and "Google" as organizations, while Lattice-LSTM only recognizes "Google," and MECT fails to identify either. This case illustrates SCM-Net's capability in identifying rare entities, possibly a result of its semantic similarity-based lexicon selection, which enhances the model's ability to process less common entities.

The case studies presented in this analysis clearly demonstrate the effectiveness of our proposed SCM-Net model. By accurately identifying complex, ambiguous, and rare entities, SCM-Net consistently outperforms other models like Lattice-LSTM and MECT. The integration of innovative features such as multimodal feature integration, contrastive learning, and semantic similarity-based lexicon selection distinctly contributes to its superior performance.

5   Conclusion

In this paper, we have introduced SCM-Net, a cutting-edge model tailored for Chinese Named Entity Recognition (NER), addressing several key challenges in this domain. To tackle the issue of complex entity structures, we introduced multimodal feature integration, enhancing the model's ability to interpret intricate linguistic patterns. For the challenge of distinguishing similar entities, we employed contrastive learning, which significantly improved the model's discriminative capabilities. Additionally, to better handle rare and nuanced entities, we incorporated a semantic similarity-based lexicon selection method, refining the model's focus on relevant lexical information. Our extensive experiments on datasets such as OntoNote4, MSRA, Weibo, and Resume demonstrate that SCM-Net achieves superior performance compared to existing models. This success is largely due to the effective synergy of these innovative approaches, each contributing to a more accurate and contextually sensitive NER process.In addition, we also consider incorporating other external information useful for NER tasks, such as Chinese strokes, to enhance the accuracy of entity recognition.

Acknowledgment. This work is supported by grant from the Natural Science Foundation of China (No. 62072070)

References

1. Zhou, G., Su, J.: Named entity recognition using an hmm-based chunk tagger. In: Proceedings of the 40th Annual Meeting of the Association for Computational Linguistics, pp. 473–480 (2002)
2. Devlin, J., Chang, M.-W., Lee, K., Toutanova, K.: Bert: Pre-training of deep bidirectional transformers for language understanding, arXiv preprint arXiv:1810.04805 (2018)
3. Li, J., Sun, A., Han, J., Li, C.: A survey on deep learning for named entity recognition. IEEE Trans. Knowl. Data Eng. **34**(1), 50–70 (2020)
4. Huang, Z., Xu, W., Yu, K.: Bidirectional lstm-crf models for sequence tagging, arXiv preprint arXiv:1508.01991 (2015)
5. Tsai, H., Riesa, J., Johnson, M., Arivazhagan, N., Li, X., Archer, A.: Small and practical bert models for sequence labeling, arXiv preprint arXiv:1909.00100 (2019)
6. Liu, Z., Zhu, C., Zhao, T.: Chinese named entity recognition with a sequence labeling approach: based on characters, or based on words? In: International Conference on Intelligent Computing, pp. 634–640 (2010)
7. Lu, Y., Zhang, Y., and Ji, D.: Multi-prototype chinese character embedding. In: Proceedings of the tenth international conference on language resources and evaluation (LREC 2016), pp. 855–859 (2016)
8. Zhang, Y., Yang, J.: Chinese ner using lattice lstm, arXiv preprint arXiv:1805.02023 (2018)
9. Sui, D., Chen, Y., Liu, K., Zhao, J., Liu, S.: Leverage lexical knowledge for chinese named entity recognition via collaborative graph network. In: Proceedings of the 2019 Conference on Empirical Methods in Natural Language Processing and the 9th International Joint Conference on Natural Language Processing (EMNLP-IJCNLP), pp. 3830–3840 (2019)
10. Feng, Z., Jian, H., Zhongjie, Z.: Lac-dglu: named entity recognition model based on cnn and attention mechanism. Comput. Sci. **47**(11), 212–219 (2020)
11. Shen, Y., Yun, H., Lipton, Z.C., Kronrod, Y., Anandkumar, A.: Deep active learning for named entity recognition, arXiv preprint arXiv:1707.05928 (2017)
12. Habibi, M., Weber, L., Neves, M., Wiegandt, D.L., Leser, U.: Deep learning with word embeddings improves biomedical named entity recognition. Bioinformatics **33**(14), i37–i48 (2017)

13. Zhou, R., Hu, Q., Wan, J., Zhang, J., Liu, Q., Hu, T., Li, J.: Wcl-bbcd: a contrastive learning and knowledge graph approach to named entity recognition, arXiv preprint arXiv:2203.06925 (2022)
14. Mai, C., et al.: Pronounce differently, mean differently: a multi-tagging-scheme learning method for Chinese ner integrated with lexicon and phonetic features. Inform. Process. Manag. **59**(5), 103041 (2022)
15. Gui, T., Ma, R., Zhang, Q., Zhao, L., Jiang, Y.-G., Huang, X.: Cnn-based chinese ner with lexicon rethinking. IJCAI **2019** (2019)
16. Ma, R., Peng, M., Zhang, Q., Huang, X.: Simplify the usage of lexicon in Chinese ner, arXiv preprint arXiv:1908.05969 (2019)
17. Li, X., Yan, H., Qiu, X., Huang, X.: Flat: Chinese ner using flat-lattice transformer, arXiv preprint arXiv:2004.11795 (2020)
18. Wu, S., Song, X., Feng, Z.: Mect: multi-metadata embedding based cross-transformer for Chinese named entity recognition, arXiv preprint arXiv:2107.05418 (2021)
19. Wu, S., Song, X., Feng, Z., Wu, X.-J.: Nflat: non-flat-lattice transformer for Chinese named entity recognition, arXiv preprint arXiv:2205.05832 (2022)
20. Diao, S., Bai, J., Song, Y., Zhang, T., Wang, Y.: Zen: pre-training chinese text encoder enhanced by n-gram representations," arXiv preprint arXiv:1911.00720 (2019)
21. Zhao, S., Wang, C., Hu, M., Yan, T., Wang, M.: Mcl: multi-granularity contrastive learning framework for Chinese ner. In: Proceedings of the AAAI Conference on Artificial Intelligence, vol. 37(11), pp. 14011–14019 (2023)
22. Qi, P., Qin, B.: Ssmi: semantic similarity and mutual information maximization based enhancement for chinese ner. In: Proceedings of the AAAI Conference on Artificial Intelligence, vol. 37(11), pp. 13 474–13 482 (2023)
23. Weischedel, R., et al.: Ontonotes release 4.0. LDC2011T03, Philadelphia, Penn.: Linguistic Data Consortium, vol. 17 (2011)
24. Levow, G.-A.: The third international chinese language processing bakeoff: word segmentation and named entity recognition. In: Proceedings of the Fifth SIGHAN Workshop on Chinese Language Processing, pp. 108–117 (2006)
25. Peng, N., Dredze, M.: Named entity recognition for chinese social media with jointly trained embeddings. In: Proceedings of the 2015 Conference on Empirical Methods in Natural Language Processing, pp. 548–554 (2015)
26. ——, Improving named entity recognition for chinese social media with word segmentation representation learning, arXiv preprint arXiv:1603.00786 (2016)

Biomedical Document Relation Extraction via Mention-Entity Double Fusion and Contrast Enhanced Inference

Huixian Cai, Yijia Zhang[(⊠)], Jianyuan Yuan, and Hongfei Lin

School of Information Science and Technology, Dalian Maritime University, Liaoning 116026, Dalian, China
{caihuixian,zhangyijia,jianyuany}@dlmu.edu.cn, hflin@dlut.edu.cn

Abstract. With the continuous advancement of research and technological progress, biomedical information is experiencing explosive growth. Relation extraction in the biomedical field holds significant research significance and practical value. However, in real-world scenarios, such as in healthcare, many relational facts often need to span multiple sentences to be fully expressed. Existing mainstream methods primarily employ sequential or graph models to represent entity relationships. However, these approaches often underutilize contextual information, leading to biases in entity relationship identification. To address this limitation, we introduce the MED-CDA model, namely, the **M**ention-**E**ntity **D**ouble Fusion and **C**ontrast **D**ata **A**ugmentation. This model comprises the Mention-Entity Double Fusion module and the Contrast data augmentation module. The Mention-Entity Double Fusion module utilizes a relation-specific mention attention network and a U-shaped document graph to model mention and entity information. Subsequently, by fusing the obtained features, it leverages contextual semantic information to fully capture the implicit logical relationships between entity pairs. The Contrast data augmentation module is a novel data augmentation method aimed at mitigating class imbalance issues in biomedical literature. This enables the model to focus more on text with fewer relation types, thereby enhancing its inference ability. Our experimental results exhibit substantial performance enhancements across three widely-used biomedical datasets—BIORED, CDR and GDA when compared to baseline models, underscoring its competitive edge.

Keywords: Biomedical information · Data augmentation · Class imbalance

1 Introduction

In the biomedical field, the task of relation extraction involves detecting and classifying relationships between different biomedical concepts mentioned in text. However, the diverse and specialized nature of entities and the variability in the form of natural language sentences make this task more complex than in general domains. Sentence-level relation extraction and document-level relation extraction refer to different levels of granularity in extracting relationships between entities from text. However, it relies only on

B. Xu et al. (Eds.): CCKS-IJCKG 2024, CCIS 2229, pp. 98–110, 2025.
https://doi.org/10.1007/978-981-96-1809-5_8

the local context within a sentence and cannot fully leverage cross-sentence contextual information, which may lead to ambiguity and reference issues. The same entity may have different meanings in different contexts, and sentence-level methods may struggle to accurately distinguish between them. In the past two years, with the application and development of large-language model technologies, scholars (Rajpurkar et al., Evans et al., Jumper et al., Lee et al.) have found that large models can analyze and process biomedical data on a larger scale more quickly. However, the experimental results show that the performance of GPT-3 still significantly lags after fine-tuning smaller pretrained language models.

On the other hand, document-level relation extraction involves extracting relationships between entities across multiple sentences or the entire document. It aims to capture relationships that span beyond individual sentences, considering the context provided by the entire document. For example, as shown in Fig. 1, in document-level biomedical relation extraction (Bio-DocuRE), relationship extraction is performed on medical biological entities throughout the entire document.

Each entity may appear multiple times in the text and may take different forms of representation. For example, "*iodine transport defect (ITD)*" is a type of disease, where "*ITD*", its abbreviation, represents the same disease with consistent coding. However, having the same expression does not necessarily mean referring to the same entity. For instance, "*Sodium/Iodide Symporter (NIS)*" denotes a gene or its product, while "*iodide*", despite sharing common expression, represents a chemical entity, presenting a stark contrast between the two. This necessitates the model's ability to accurately identify the category to which biomedical entities belong. In real medical scenarios, many relational facts often need to span multiple sentences to be fully expressed. Through examples, it is evident that inter-sentential relationships exist only in key sentences, while entity relationships across sentences span the entire document, with the number of inter-sentence relationships also being smaller than those across sentences. Therefore, there is an urgent need to advance research on biomedical relation extraction from the sentence level to the document level.

There has been limited focus on all mention pairs in the text, which can be problematic because coreferential mention pairs contribute unequally to specific relationships. These mention pairs associated with specific relationships have considerable contextual influence, aiding the model in understanding the associations between entities. Therefore, to obtain effective and comprehensive reasoning information, we propose a Mention-Entity Double Fusion module. Specifically, for reasoning between mentions, we utilize a novel relation-specific mention enhancement network. First, we encode the basic semantics of each relation into prototype representations. Then, we dynamically merge mentions referring to the same entity using a weighted sum, calculating the relevance weights between the prototype of a specific candidate relation and each mention's representation of the given entity. We use fully connected mention pairs to capture dependencies across sentences. For reasoning between entity pairs, we utilize a Document U-shaped Network, treating features between entity pairs as images. The model treats each relation type as a pixel-level mask, capturing global interdependencies. These two sets of features are then fused to obtain fused features, with refined mention and entity information re-extracted and improved. By leveraging multidimensional information, we can

[1]OBJECTIVE: **Iodide transport defect (ITD)** is a rare disorder characterised by an inability of the thyroid to maintain an **iodide** gradient across the basolateral membrane of thyroid follicular cells, that often results in **congenital hypothyroidism**.

[2] ... it has been shown to arise from abnormalities of the **sodium/iodide symporter (NIS)**.

[4]The diagnosis of **ITD** was suspected ... if any **iodide** uptake by the thyroid and salivary glands.

[8] ... **deletion of the coding sequence (nt 1314 through nt 1328)** ... of the adjacent intron.

[11] ... a vector expressing the mutant del-(439-443) NIS failed to concentrate **iodide** ... cause of the **ITD** in this patient.

[12] ... case of **congenital hypothyroidism** due to a new deletion in the NIS gene.

Intra-sentence entity pairs relations:
<Iodide transport defect (ITD), iodide> R[0]:Association
<ITD, iodide> R[0]:Association
<del-(439-443), ITD> R[1]:Positive
<del-(439-443), iodide> R[2]:Negative

Inter-sentence entity pairs relations:
<ITD, sodium/iodide symporter (NIS)> R[3]:Negative
<iodide, sodium/iodide symporter (NIS)> R[4]:Association
<congenital hypothyroidism, del-(439-443)> R[5]:Positive
<congenital hypothyroidism, **deletion of the coding sequence (nt 1314 through nt 1328)**> R[6]:Positive
<congenital hypothyroidism, sodium/iodide symporter> R[7]:Association

Fig. 1. An example illustration of Bio-DocuRE from the BioRED dataset.

fully utilize document information to help the model accurately identify relationships between biomedical entities.

In this paper, we propose a novel model for relation extraction tasks in biomedical texts, called the **MED-CDA** Model. We validate our approach on three open datasets in the biomedical domain. Our contributions are summarized as follows:

- We propose a novel module: the Mention-Entity Fusion (MED) module, for integrating mention and entity information. We fuse the mention and entity information to infer relationships between biomedical entities from a multigranularity perspective.
- To address the imbalance issue in biomedical text data, we employ a contrast data augmentation (CDA) approach to alleviate this phenomenon. Through this approach, we enhance the dataset size and leverage augmented data.
- We validate our model on three open datasets in the biomedical domain: BioRED and CDR. The results demonstrate that our model achieves promising F1 scores, which are higher than those of our baseline models.

2 Methodology

2.1 Problem Definition

For a given annotated document D, the task of extracting relationships between all entity pairs (e_h, e_t) mentioned in the document is to represent the head entity and tail entity, where each entity pair is also referred to as a relation instance. An entity may appear multiple times in the document, and its representation can vary, resulting in multiple corresponding mentions $\{m_j^i\}_{j=1}^{N_{e_i}}$ for each entity $\{e_i\}_{i=1}^n$.

2.2 Model Architecture

In this section, we will provide a detailed description of the overall framework of the DFCEI model proposed in Fig. 2. We will cover the following aspects: Encoder Layer; Information Exchange Layer; Information Fusion Layer; Contrast Enhanced Inference Layer and Output Layer

Encoder Layer

For a given document D of length l, we have $D = [w_t]_{t=1}^{l}$, where w_t is the word at location t. Building upon our previous efforts in relationship classification, we employ special tags * to denote the beginning and end positions of mentions. Subsequently, contextual embeddings of the document are derived using a pretrained language model (PrLM):

$$X = PrLM\left([w_1, ..., w_l]\right) = [x_1, ..., x_l] \tag{1}$$

where $X \in R^{l \times d}$ and d is the hidden dimension of the PrLM.

Information Exchange Layer

The same entity may appear multiple times in various sentences throughout an article. We employ a labeled graph convolutional neural network to generate mention and entity representations for each entity pair. In essence, we first extract mention information from the context using an attention matrix for mentions. Then, we identify the relevant head and tail entities to aid the model in effectively integrating mentions.

We adopt the context pooling method proposed by Zhou et al. (2021). For each mention pair, we initially aggregate the attention output for its mentions through mean pooling.

$$A_x = \sum_{j=1}^{N_{e_i}} (a_{m_x}) \tag{2}$$

where $a_{m_x} \in R^{H \times L}$ represents the self-attention weight at the position of mention m_x, H denotes the number of attention heads, and L is the document length.

Then the context query can be calculated as:

$$Q^{(h,t)} = \sum_{i=1}^{H} (A_h^i \cdot A_t^i) \tag{3}$$

$$C^{(h,t)} = H^T Q^{(h,t)} \tag{4}$$

$$G_h = tanh(W_s h_{e_h} + W_c c^{(h,t)}) \tag{5}$$

where $A_h^i \in R^{H \times L}$ is the aggregated attention output for head entity h, likewise for t. $Q^{(h,t)} \in R^L$ is the mean-pooled attention weight for entity pair (e_h, e_t) and $H \in R^{l \times d}$ is the contextual embedding of the whole document. Then the context vector $C^{(h,t)} \in R^d$ is fused with the entity representations. Where $G_h \in R^d$ is the context-enhanced representation of head entity h for entity pair (e_h, e_t). We obtain the tail entity representation G_t using the same computational approach.

Information Fusion Layer

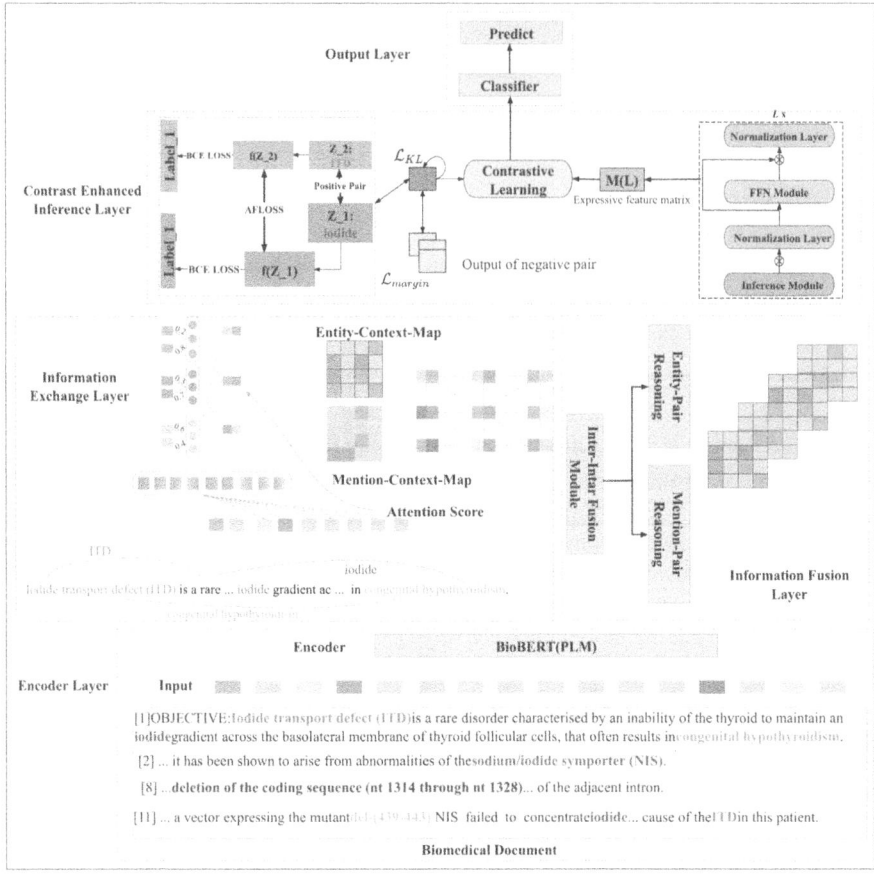

Fig. 2. The overall framework of our model.

After obtaining representations for the head and tail entities, we use them as queries and perform cross-attention operations.

Specifically, given an entity pair (e_h, e_t) and a sequence of mention embeddings $h_{m1}, h_{m2}..., h_{mn}$ for the head or tail entity, $h_{m_i} \in \mathbf{R}^d$ and n is the number of mentions. In accordance with prior research on contextual features $Z_{(h,t)} \in \mathbf{R}^d$ for this pair, the head entity e_h is computed as follows:

$$a^i_{(h,t)} = \frac{W_Q G^T_{(h,t)} W_K h_{m_i}}{\sqrt{d}} \tag{6}$$

$$\alpha^i_{(h,t)} = \frac{exp(a^i_{(h,t)})}{\sum_{j=1}^{p} exp(a^i_{(h,t)})} \tag{7}$$

$$Z^h(h, t) = \sum_{i=1}^{p} \alpha^i_{(h,t)} h_{m_i} \tag{8}$$

where $W_Q \in \mathbf{R}^{d*d}$ and $W_K \in \mathbf{R}^{d*d}$ are the query and key context matrixes, respectively, and d is the dimension of the hidden states. We obtain the tail entity representation $Z^t(h, t)$ in the same manner. By examining the formula, we can intuitively observe that the representations of the head and tail entities are not fixed but rather dynamically adjusted by the varying weights of the context on entity pairs.

We use grouped bilinear functions for feature fusion. The mention embedding G_s will be split into k equal-sized groups, such as $G_h = \left[G_h^1, G_h^2, G_h^3, ...G_h^k\right]$, and we obtain the following formula:

$$G_t = \left[G_t^1, G_t^2, G_t^3, ...G_t^k\right] \tag{9}$$

$G_h = \left[G_h^1, G_h^2, G_h^3, ...G_h^k\right]$ is referred to as the representation of the selected header, where $|k|$ is the number of head entity mentions. By analogy, G_t is the reference representation of the selected tail entity.

We utilize bidirectional attention to model the interaction between the query and context. Furthermore, we introduce an information fusion module to individually obtain the contextual representations of the head and tail mentions.

$$Head = Multi\,Head\,Attention\,Layer[G_h, G_t, G_t] \tag{10}$$

$$Tail = Multi\,Head\,Attention\,Layer[G_t, G_h, G_h \tag{11}$$

The information fusion module facilitates attentive interactions between head and tail entity mentions, resulting in contextual representations. Therefore, the multiheaded attention mechanisms are essentially identical. We use a shared multiheaded attention layer to implement this aspect. Additionally, this configuration reduces the number of parameters, thereby lowering computational costs.

$$P = Multi\,Head\,Attention\,Layer\,(Head, Tail) \tag{12}$$

Contrast Enhanced Inference Layer and Output Layer.
We concatenate the representations of entity pairs and mention pairs into an eigenvector $M^{(0)}, M^{(0)} = \left[P_{h,t}\right]_{N*N}$ where each row $M_{h,*}^{(0)}$ represents a head entity e_h and each column $M_{*,t}^{(0)}$ represents a tail entity e_t.

Our contrast-enhanced inference module delves into implicit relationships between entities by acquiring more expressive entity pair representations. Since all inference head self-attention modules follow the same calculation process, let us illustrate using the first entity pair as an example:

$$F_i^{(l,1)} = W_d\left[M_{h,i}^{(l)}; M_{i,t}^{(l)}\right] + b_d, i = 1,2, \ldots, N \tag{13}$$

where W_d and b_d are the training parameters, [;] represents the join operation, and then we obtain the output matrix $M_{h,t}^{(l,1)}$ for the first entity pair (e_h, e_t):

$$M_{h,t}^{(l,1)} = Attention(Q, K, V), where\,Q = M_{h,t}^{(l)}, K = V = \left[M_{h,t}^{(l)}; F_1^{(l,1)}; \ldots; F_N^{(l,1)}\right] \tag{14}$$

It should be noted that the upper corners of $M_{h,t}^{(l,1)}$ and $F_i^{(l,1)}$ represent the index values of the inference layer, while the lower corner represents the entity index. Subsequently, we sum all the outputs of the bull self-attention module to obtain its final output:

$$\widetilde{M}_{h,t}^{(l)} = LN(M^{(l)} + W_O[M_{h,t}^{(l,1)}; ...; M_{h,t}^{(l,4)}] + b_O \qquad (15)$$

where W_O and b_O are model parameters, and $LN(\cdot)$ are layer normalized functions. Finally, the output of the inference layer $l + 1$ is calculated:

$$M^{(l+1)} = LN\left(\widetilde{M}^{(l)} + FNN\left(\left(\widetilde{M}^{(l)}\right)\right)\right), \text{ where} \widetilde{M}^{(l)} = \left[\widetilde{M}^{(l)}\right]N * N \qquad (16)$$

After repeating the above process L times, we obtain a more expressive eigenmatrix $M^{(L)}$.

Negative Sample Generation.
The dataset $\{xi, yi\}_{i=1}^N$ is assumed to contain N training samples. We randomly removed a portion of words from each sample, with a deletion rate of $p \in [0, 1]$, and repeated this process N_{neg} times to generate N_{neg} negative samples. Subsequently, we obtain an augmented dataset $\{xi, yi, X_{neg}^i\}_{i=1}^N$, where $X_{neg}^i = \{x_i^j\}_{j=1}^{N_{neg}}$ are derived from xi. Further augmentation of the training set can be achieved by repeating the entire process naug times, resulting in $\{xi, yi, X_{neg}^i\}_{i=1}^{N*N_{neg}}$, thereby increasing the dataset size by N_{aug}.

Positive Sample Generation.
Inspired by Gao et al. (2021), we leverage the randomness of dropout to create positive samples. The core concept of R-Drop involves feeding each sample into the model twice during training, resulting in two prediction outcomes. More specifically, the fused feature matrix $M^{(l+1)}$ is input into the model twice at each step, yielding two corresponding predictions denoted as $\hat{y}_i^1$ and $\hat{y}_i^2$. Our objective is then to minimize the bidirectional Kullback-Leibler (KL) divergence between them, denoted as $\mathcal{L}_{KL}$:

$$\mathcal{L}_{KL} = \sum_{i=1}^N \frac{1}{2}\left[mathcalD_{KL}\left(\hat{y}_i^1 \| \hat{y}_i^2\right) + \mathcal{D}_{KL}\left(\hat{y}_i^1 \| \hat{y}_i^2\right)\right] \qquad (17)$$

Contrastive Loss
Given the diminished semantic coherence and integrity, it is reasonable to assume that negative samples are more likely to contain AD. To confirm this, we enforce that their disparities exceed a margin, denoted as m. Particularly, the encoder receives xi and X_{neg}^i as input and outputs their corresponding embedding representations hi and $\mathcal{H}_{neg}^i$. Then, these representations are fed into the classifier to obtain final scores $\tilde{y}_i$ and $\tilde{y}_i^j$ for $\tilde{x}_i$ and $\tilde{x}_i^j$, respectively.

$$\mathcal{L}_{margin} = \sum_{i=1}^N \max\left(0, m - \hat{y}_i + \frac{\sum_{j=1}^{N_{neg}} \tilde{y}_i^j}{N_{neg}}\right) \qquad (18)$$

where m is the margin between positive and negative samples. The final loss is a combination of the above three loss terms $\mathcal{L}_1$, $\alpha\mathcal{L}_{margin}$, $\mu\mathcal{L}_{KL}$.

Given that each target entity pair may have multiple mention pairs in a document, we utilize a classification scheme based on multiple instance learning to aggregate the predictions of all target mention pairs.

$$P(r_i) = \frac{\exp\left(l_{r_i}^{(head,tail)}\right) + \exp(g(head, tail))}{\exp(l_{r_i}^{(head,tail)}) + \exp(l_{TH}^{(head,tail)}) + \exp(g(head, tail)} \tag{19}$$

$$P(r_{TH}) = \frac{exp(l_{r_{TH}}^{(head,tail)})}{\sum_{r_j \in N_T \cup \{TH\}} exp(l_{r_j}^{(head,tail)})} \tag{20}$$

$$\mathcal{L}_1 = \sum (1 - P(r_i))^\gamma log(P(r_i)) log \sum_{1 \le i,j \le k} exp\left(g\left(head_i, tail_j\right)\right) + log(P(r_{TH})) \tag{21}$$

Our $\mathcal{L}_1$ loss consists of two parts: the first part is for positive classes, and the second part is for negative classes.

$$\mathcal{L} = \mathcal{L}_1 + \alpha \mathcal{L}_{\text{margin}} + \mu \mathcal{L}_{KL} \tag{22}$$

where α and μ are hyperparameters that govern the influence of positive and negative samples, respectively. In our model, we set $\alpha = 0.5$ and $\mu = 0.5$

3 Experiment

3.1 Datasets

We evaluated our model using document-level relational extraction datasets from two benchmark biomedical domains. Further details can be found in Table 1. The BIORED dataset is a relationship extraction dataset designed for the biomedical domain. It was created by Luo et al. in 2022 and comprises 600 PubMed abstracts. It includes 400 articles in the training set, 100 articles in the development set and 100 articles in the test set. It serves as a valuable resource for researchers to develop and evaluate biomedical relationship extraction algorithms.

Table 1. Details of the datasets.

Datasets	Docts	Relations	Avg # entities doc	Avg # mentions entity	Avg # relations doc	Avg # sentences doc
BIORED	600	9	3.8	20.4	10.8	11.9
CDR	1500	2	6.8	19.2	11.9	9.7

The CDR dataset is commonly used in the biomedical field for relation extraction tasks. This dataset contains information on the relationships between chemicals and diseases. A total of 1500 articles are evenly divided into training, development, and test sets.

3.2 Data Preprocessing

Training was conducted on an NVIDIA 3080 GPU with 16 GB of memory. Consistent with the CGM2IR model, we used the F1 score as the evaluation metric for document-level relation extraction performance. We set Batch Size is 4, Learning Rate is 2e-5, Epoch is 40 and Seed is 66.

3.3 Main Results

The Results of the BIORED Dataset: Table 2 details the main results of our model in the comprehensive experiment on the BIORED dataset. The LSR model conducts relation extraction by constructing a document-level graph and dynamically modeling various potential relationship types. However, this approach fails to effectively leverage coreference and semantic rules, potentially leading to suboptimal structure generation. On the other hand, the GAIN model constructs a heterogeneous graph based on mentions to capture the contextual rules of entities. Nevertheless, it overlooks the issue of oversmoothing in GCNs, where features of nodes in the graph converge to similar values after deep iterations. Sequential models such as ATLOP and SSAN prioritize global dependencies. However, these approaches fail to fully exploit local information, as potential references of the same entity contain valuable entity-related details essential for identifying relationships between entities.

Table 2. The main results for the BIORED dataset.

Method	P	R	F1
ATLOP (Zhou et al. 2020)	71.7	72.3	71.8
LSR (Nan et al. 2020)	63.8	65.4	66.3
GAIN (Zeng et al. 2020)	68.5	69.7	70.2
SSAN (Xu et al., 2021)	69.4	70.8	69.6
ATLOP (Zhou et al. 2020)	71.7	72.3	71.8
DHGCN (Sun et al. 2022)	-	-	77.1
CGM2IR (Zhao et al. 2022)	74.9	75.6	77.8
DGI (Wang et al. 2023)	75.4	76.1	76.9
Ours	**77.8**	**76.4**	**79.2**

The Results of the CDR Dataset: Table 3 offers comprehensive insight into the primary outcomes of our model compared to other models on the CDR dataset, showcasing its superiority over previous works. Our model attained an F1 score 2.1 percentage points higher than that of the baseline model (CGM2IR) and notably outperformed other representative models. Evidently, in contrast to sequential models, graph models exhibit enhanced utilization of context information in biomedical documents. Moreover, our model accentuated the interaction between mentions and entities, thereby facilitating more effective learning of CID relationship extraction.

Table 3. The main results for the CDR dataset.

Method	P	R F1	
EncAttAgg (Jiang et al. 2020)	59.9	70.9	64.9
GCNN (Sahu et al. 2019)	52.8	66.0	58.6
EoG (Christopoulou et al. 2019)	62.1	65.2	63.6
SSAN (Xu et al. 2021)	-	-	65.8
MGSN (Liu et al. 2021)	69.0	66.7	67.8
GLRE (Wang et al. 2020)	65.1	72.2	68.5
ATLOP (Zhou et al. 2020)	-	-	69.4
HANN (Zhao et al. 2022)	68.0	69.5	68.8
DHGCN (Sun et al. 2022)	-	-	73.1
SAIS (Xiao et al. 2022)	72.8	73.1	74.5
CGM2IR (Zhao et al. 2022)	-	-	73.8
DGI (Wang et al. 2023)	70.6	70.9	72.9
REGREx (Dao et al. 2023)	68.8	65.2	66.8
RDDCP (Dong et al. 2023)	70.4	70.9	71.6
Ours	**74.8**	**75.5**	**75.9**

3.4 Ablation Study

In this section, to further analyze the MED-CDA model, we also conducted ablation studies on the BIORED and CDR datasets to demonstrate the effectiveness of different modules and mechanisms within DFCEI. The results of the ablation studies are presented in Table 4. First, we investigated the impact of the inter-intar fusion module on the MED-CDA model. When this module was removed from the MED-CDA model, the F1 score decreased by 2.1.

Subsequently, we systematically analyzed the various components within the inter-intar module, carefully evaluating their effects on the experiments. As our research progressed, we assessed the influence of the entity pair inference module, which led to a significant decrease of 0.6 in the F1 score when removed from the model. Then, we eliminated the mention pair inference module, resulting in a decrease of 0.9 in the F1 score. Further exploration revealed that removing the contrast-enhanced inference module led to a significant decrease of 1.5 in the F1 score. Finally, eliminating the contrastive learning module resulted in a decrease of 0.4 percentage points in the F1 score. With the methodical removal of specific modules from the model, we observed differing degrees of decline in F1 scores. The comprehensive data from our meticulous ablation experiments on the CDR dataset are presented in detail in Table 5.

However, we found that the effect of the contrastive enhanced inference module is greater than that of the inter-intar fusion module on the CDR dataset. This may be attributed to the relatively simple relationships within the CDR dataset. Our inference

Table 4. Ablation Study of the MED-CDA model in BIORED dataset.

Model	F1
Full Model	79.2
o- Inter-Intar Fusion module	77.1
o- Entity-Pair Reasoning	76.5
o- Mention-Pair Reasoning	76.2
o-Contrast Enhanced Inference module	77.7
o-Enhanced Inference	76.4
o-Contrastive learning	77.3

Table 5. Ablation Study of the MED-CDA model on the CDR dataset.

Model	F1
Full Model	75.9
o- Inter-Intar Fusion module	73.8
o- Entity-Pair Reasoning	72.4
o- Mention-Pair Reasoning	72.9
o-Contrast Enhanced Inference module	74.5
o-Enhanced Inference	74.0
o-Contrastive learning	73.7

model encounters challenges in inferring underlying relationships, resulting in less effective information being obtained. Additionally, despite the large number of articles in the CDR dataset, the data are unbalanced. Our proposed data augmentation module can effectively improve this phenomenon.

4   Related Work

The objective of biomedical document-level relation extraction is to automatically extract relationships between various entities from biomedical texts and classify or annotate them, thereby extracting semantic associations through the identification of relationships between entities.

4.1   Biomedical Document-Level Relation Extraction (BioDLRE)

In the biomedical domain, sequence-based document-level relation extraction methods typically utilize recurrent neural networks (RNNs) or their variants to represent entities. This is done to capture contextual information within the text sequences and classify all candidate entity pairs. However, this method often fails to adequately exploit the mentioned interactions, overlooking relationships expressed across sentence boundaries. To address this challenge, Verga et al. introduced Bi-affine Relation Attention Networks (BRANs). BRANs integrate network architecture, multi-instance learning, and multi-task learning methods to extract entity relationships. However, this method relies solely on local context information and fails to consider semantic relationships between distant contexts, potentially resulting in information loss. To overcome this limitation, Li et al. propose a novel model based on multi-attention mechanisms: GCA is utilized to capture global semantic information within the document, enabling the model to account for semantic relationships between all sentences in the document.

4.2 Contrastive Learning

Contrastive learning has previously been applied to learn high-quality representations of images in computer vision. However, due to the lack of a universal method for data augmentation in text data, contrastive learning has not been widely utilized in natural language processing. In this work, Su et al. explored leveraging contrastive learning to enhance the text representation of the BERT model for relation extraction. The key aspect of this approach lies in seamlessly integrating linguistic knowledge into data augmentation, thereby offering a tailored contrastive pretraining step specifically designed for relation extraction tasks. Dong et al. proposed a relation extraction model named Relational Distance and Document-level Contrastive Pretraining (RDDCP). This model achieves coreference resolution through a straightforward and efficient mention replacement approach.

5   Conclusion and Future Work

In this paper, we introduce a document-level relation extraction method specifically tailored for the biomedical domain. We propose the MED-CDA model, which aims to enhance global and local information between mentions and entities. For mention information, we employ a Mention Enhancement Network to model it, while for entity information, we utilize a Document U-Net to treat features as images, capturing global dependency information. Finally, we employ a novel contrastive data augmentation method to mitigate the issue of relationship imbalance in the biomedical domain, allowing the model to focus more on texts with fewer relationship categories.

There is ample room for improvement in our future work. We can expand this method to include more specialized and extensive datasets to validate its effectiveness in handling relation extraction tasks. Additionally, when tackling class imbalance issues, we must address the challenge of achieving the right balance between recall and precision. Finally, we will delve into exploring the impact of context on evidence sentence pairs, which will be the primary focus of our future research endeavors.

Acknowledgments. This work is supported by grant from the Natural Science Foundation of China (No. 62072070).

References

1. Verga, P., Strubell, E., McCallum, A.: Simultaneously self-attending to all mentions for full-abstract biological relation extraction. In: Proceedings of the 2018 Conference of the North American Chapter of the Association for Computational Linguistics: Human Language Technologies, Volume 1 (Long Papers), pp. 872–884. Association for Computational Linguistics, New Orleans, Louisiana (2018)
2. Bhasuran, B., Natarajan, J.: Automatic extraction of gene-disease associations from literature using joint ensemble learning. PLoS ONE **13**(7), e0200699 (2018). https://doi.org/10.1371/journal.pone.0200699
3. Xu, Y., et al.: Star-BiLSTM-LAN for document-level mutation-disease relation extraction from biomedical literature. In: 2020 IEEE International Conference on Bioinformatics and Biomedicine (BIBM), Seoul, Korea (South), pp. 357–362 (2020). https://doi.org/10.1109/BIBM49941.2020.9313250
4. Quirk, C., Poon, H.: Distant supervision for relation extraction beyond the sentence boundary. arXiv preprint arXiv:1609.04873 (2016)
5. Peng, N., Poon, H., Quirk, C., et al.: Cross-sentence n-ary relation extraction with graph lstms. Trans. Associat. Comput. Linguist. **5**, 101–115 (2017)
6. Panyam, N.C., Verspoor, K., Cohn, T., et al.: Exploiting graph kernels for high performance biomedical relation extraction. J. Biomed. Seman. **9**, 1–11 (2018)
7. Gupta, P., Rajaram, S., Schütze, H., Runkler, T.: Neural relation extraction within and across sentence boundaries. In: Proceedings of the AAAI Conference on Artificial Intelligence, vol. 33(01), pp. 6513–6520 (July 2019).
8. Liu, X., Fan, J., Dong, S.: Document-level biomedical relation extraction leveraging pretrained self-attention structure and entity replacement: Algorithm and pretreatment method validation study. JMIR Med. Inform. **8**(5), e17644 (2020)
9. Kanjirangat, V., Rinaldi, F.: Enhancing biomedical relation extraction with transformer models using shortest dependency path features and triplet information. J. Biomed. Inform. **122**, 103893 (2021)
10. Rajpurkar, P., et al.: Chexnet: Radiologist-level pneumonia detection on chest x-rays with deep learning. arXiv preprint (2017). arXiv:1711.05225
11. Senior, A.W., et al.: Improved protein structure prediction using potentials from deep learning. Nature **577**(7792), 706–710 (2020)
12. Lee, J., et al.: BioBERT: a pre-trained biomedical language representation model for biomedical text mining. Bioinformatics **36**(4), 1234–1240 (2020)
13. Jumper, J., et al.: Highly accurate protein structure prediction with AlphaFold. Nature **596**(7873), 583–589 (2021)
14. Zhang, Z., et al.: Document-level relation extraction with dual-tier heterogeneous graph. In: Proceedings of the 28th International Conference on Computational Linguistics, pp. 1630–1641, Barcelona, Spain (Online). International Committee on Computational Linguistics (2020)

GVDExtractor: Document-Level Ternary Relation Extraction of Gene-Variant-Disease from Medical Literature

Na Li[1(✉)], Jiaxin Hu[1], Sen Ai[1], and Xiang Zhang[1,2(✉)]

[1] Southeast University, Nanjing 211189, China
{n.li,jiaxinhu,sen.ai,x.zhang}@seu.edu.cn
[2] Judicial Big Data Research Centre, School of Law, Southeast University,
Nanjing, China

Abstract. The automatic extraction of relations among genes, variants, and diseases is currently an urgent problem in medical research. Previous studies on relation extraction have mainly relied on short texts from general domain corpora, and have only been able to handle binary relations within a single sentence, making it difficult to address challenges such as relation diversity, cross-sentence relations, and sparse entity distribution. To address these issues, this paper proposes a Double Graph-Based Relation Extraction method based on graph neural networks, which aims to extract gene-variant-disease ternary relations from full-text medical literature (abbreviated *GVDExtractor*). The proposed method effectively integrates the discourse-level information and dependency structure information of documents through the document structure graph. Then the entity relation graph provides a natural expression structure for cross-sentence relations by directly connecting edges between entities and reducing their physical distance. Results from experiments on different datasets demonstrate the effectiveness and generalization of the proposed model, providing strong support for medical research.

Keywords: N-ary Relation Extraction · Document-level · Medical Literature Mining · Graph Neural Network

1 Introduction

The exploration of the ternary relation between genes, variants, and diseases can enhance our comprehension of the mechanisms underlying disease onset and progression, thereby enabling the development of more efficacious strategies for disease prevention and treatment [12]. Figure 1 illustrates a specific example of a ternary relation in medical literature, wherein two sentences denote the association between the gene IL23R, the disease CD (Crohn Disease), and the variant rs11209026 and rs11465804 located on the gene IL23R. Notably, the inference of the variant-disease relation in the second sentence requires contextual analysis and must be combined with the first sentence to deduce the ternary relation

© The Author(s), under exclusive license to Springer Nature Singapore Pte Ltd. 2025
B. Xu et al. (Eds.): CCKS-IJCKG 2024, CCIS 2229, pp. 111–123, 2025.
https://doi.org/10.1007/978-981-96-1809-5_9

<IL23R, rs11209026, CD> and <IL23R, rs11465804, CD>. Due to the professionalism and complexity of the medical field, there are several major challenges in effective relation extraction at the complex medical texts as follows:

Complex N-Ary Relations. Most complex relations cannot be fully captured by a single sentence and require reasoning across sentences. Especially, according to research [18], the number of entity relations in medical text is usually four to six times that of general field, among which a large number of relations are cross-linked to form n-ary relation knowledge, such as gene-variant-disease relation, gene-drug-disease relation, etc. Early medical n-ary relation extraction tasks depend on rules, co-occurrence or lexicography, which need lots of manual intervention. How to extract complex N-ary medical relations efficiently is often a weakness in current research.

> Two of these(**IL23R** on Chromosome 1 and CARD15 on Chromosome 16) correspond to genes previously reported to be associated with **CD**(Crohn Disease).
> ......4 paragraphs......
> In our data, two markers of the **IL23R** gene, **rs11209026** and **rs11465804**, gave the most significant association signals($p<10^{-9}$)
>
> GVD Relation: <**IL232R, rs11209026, CD**>,
>   <**IL232R, rs11465804, CD**>

Fig. 1. An instance of gene-variant-disease association. We have annotated the gene, variant, and disease entities in red, yellow, and blue, respectively, to indicate their association. (Color figure online)

Sparse Entity Distribution. Entity pairs are often spread across multiple sentences and paragraphs within a document. Traditional sentence-level relation extraction methods (Li et al., 2022 [16]) try to model long-distance dependency and perform multi-hop reasoning. In document-level relation extraction, dependency-based models (Gupta, Pankaj, et al., 2019 [7]) and methods that focus on entity and mention relations (Li, Jingye, et al., 2021 [15]; Xu, Benfeng, et al., 2021 [25]) have been proven effective. However, few studies have combined the strengths of both approaches, and there is a lack of sufficient use of document-level discourse structure in current research.

In response to the aforementioned challenges, we propose a gene-variant-disease relation extraction method called *GVDExtractor* (**G**ene-**V**ariant-**D**isease **E**xtractor) based on graph neural networks. The objective of *GVDExtractor* is to extract gene-variant-disease ternary relations from medical literature. *GVDExtractor* uses a double graph design: the **D**ocument **S**tructure **G**raph (DSG) and the **E**ntity **R**elation **G**raph (ERG). The DSG is constructed based on the discourse structure of the document and the dependency tree, encodes local context information and rich hierarchical structure information of entities. To focus on the most important information for relation extraction, a pruning strategy called KSL (**K**-distant **S**hortest dependency path for **L**owest common ancestor subtree) is proposed during the construction process. The rich semantic representation obtained from the DSG is then incorporated into the aggregation and propagation process of the ERG, which is composed of entities and their mentions. The model sequentially encodes the two graphs with different granularity mentioned above, and finally uses a classifier to obtain the relations in the document.

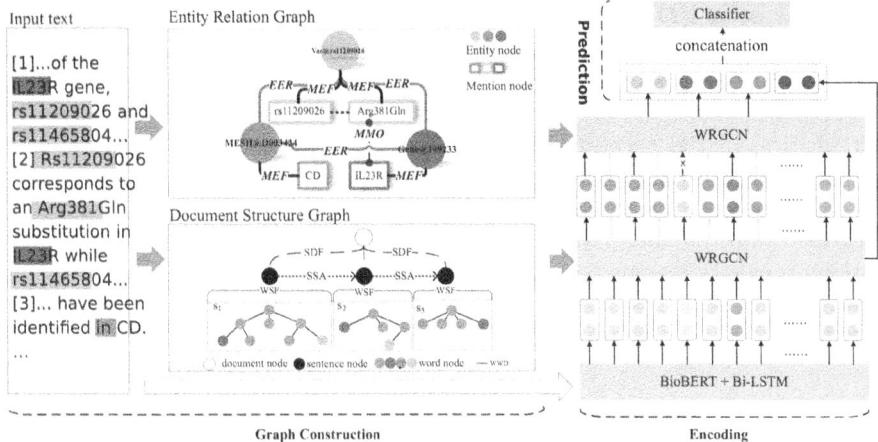

Fig. 2. The overall architecture. We employ the colors red, yellow, and blue to respectively denote information pertaining to genes, variants, and diseases. In the input text, the red segment signifies gene-related text, whereas during the inference, it indicates the vector representation of genes. (Color figure online)

In summary, this paper presents the following contributions:

(1) We propose a graph neural network-based method called *GVDExtractor* for extracting gene-variant-disease relations from medical literature. The double graph design makes full use of document structure and local context, to better solve the problems of document-level n-ary relation extraction.

(2) We design a pruning strategy called KSL, which expand the shortest dependency path (SDP) by including tokens that are up to distance K away from the lowest common ancestor (LCA) for entites. It significantly prunes redundant information that is irrelevant to relation extraction, thus enable efficient process of complex medical texts.

2   Method

In this section, we shall formulate gene-variant-disease relation extraction task and present our model *GVDExtractor* with comprehensive elaboration.

Notation. Given medical literature P and corresponding set of candidate entities $T = G \times M \times D = \{< e_g, e_m, e_d >| \; e_g \in G, e_m \in M, e_d \in D\}$, where G, M and D denote the sets of gene, mutation and disease entities. Each entity e_k^i is defined as a set of mentions, i.e., $e_k^i = m_1^{e_k^i}, m_2^{e_k^i}, \ldots, m_{N_m}^{e_k^i}$, where $k \in g, m, d$ and N_m represents the number of unique mentions for e_k^i. The task aim to predict the relation type r expressed in P for the candidate entity tuples in T, where $r \in R = \{a, na\}$. a and na denote the associated and non-associated relations.

Overview of *GVDExtractor*. Figure 2 illustrates the general architecture of *GVDExtractor*. The medical literature is first transformed into two graph representations: DSG and ERG. The DSG's nodes represent discourse units such as words, sentences and document, while the ERG's nodes represent mentions and entities. In the encoding stage, document is encoded using Bi-LSTM [8] and BioBERT [14] to obtain vector representations of words. Thereafter, the DSG is encoded to obtain word representations. These representations are then integrated into the node representations of the ERG, followed by further information propagation among nodes using graph neural networks. After undergoing two stages encoding, the resulting inference information is fed into a classifier to predict the type of gene-variant-disease ternary relation.

2.1 Graph Construction

Document Structure Graph. The nodes in a DSG can be categorized into three types: word, sentence and document. The sentence and document nodes establish the discourse structure of the document, with the document nodes serving as intermediaries that facilitate interactions among different sentences. Meanwhile, the edges in the document graph can be classified into the following types: (1) **W**ord-**W**ord **D**ependency (WWD): For each sentence $s_i = w_1, w_2, ..., w_n$, we employ the Standford CoreNLP[1] syntactic parsing tool to analyze its syntactic structure and obtain a syntactic dependency tree. The dependency relation between words in a sentence is then represented as a WWD edge; (2) **W**ord-**S**entence **AF**filiation (WSF): All word nodes w_k belonging to the sentence node s_i are directly connected to s_i with a dependency edge; (3) **S**entence-**S**entence **A**djacency (SSA): Sentence node s_i is directly connected to its adjacent sentence nodes with adjacency edges. (4) **S**entence-**D**ocument **AF**filiation (SDF): Sentence node s_i is also connected to the document node with a dependency edge.

Figure 3(a) illustrates an example of DSG. The dependency parsing trees in the DSG can be very complex and contain a lot of redundant information, such as prepositions, articles, conjunctions, and other vocabulary that is irrelevant to the target relation. Therefore, we adapted Zhang et al.'s pruning method [28], which incorporate **Of**f-**P**ath Information with **P**ath-**C**entric pruning (OP-PC) to develop a pruning strategy adapted to document-level multi-relational extraction, named K-distant SDP for LCA Pruning.

The core of KSL pruning is the LCA subtree formed by the paths between entities contains the shortest dependency path. By selecting nodes on the LCA tree that are K-distant away from the shortest dependency path, the robustness of pruning can be increased. We improved OP-PC to extract n-ary relation from medical literature. First, not only the intra-sentence pruning, the inter-sentence structural edges, i.e., SSA, SDF and WSF are preserved, ensuring the connectivity of the entire DSG. Second, for n-ary relation, KSL constructs the shortest dependency path for candidate entity triples within a sentence and preserves

---

[1] https://nlp.stanford.edu/.

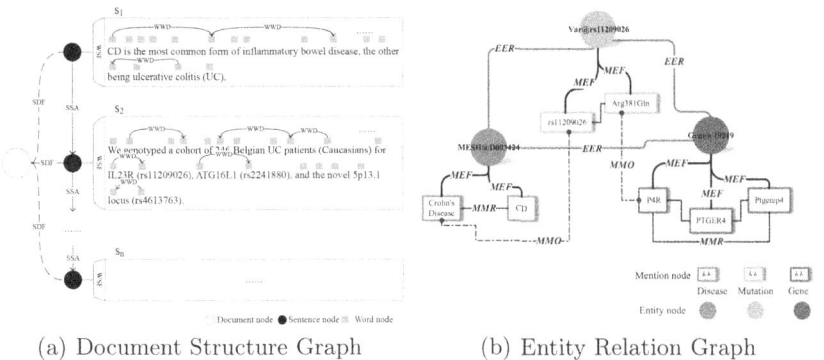

(a) Document Structure Graph   (b) Entity Relation Graph

Fig. 3. Examples of the Document Structure Graph (left) and Entity Relation Graph (right). Most of the WWD edges are omitted in the DSG.

the neighbor node along the path. For candidate entity triples across sentences, which are split into binary sub-relations for pruning. Third, to handle the multiple entity mentions in a sentence, only the shortest dependency paths and their neighbor nodes generated by binary sub-entity triples and ternary entity triples in the filtered candidate entity triple set T' are retained, avoiding the explosion of tuples and unnecessary retention of excessive shortest dependency paths.

Entity Relation Graph. The ERG expresses cross-sentence relations naturally through the direct connection between entities, effectively capturing long-distance dependency. There are two types of nodes in the ERG, namely mention nodes and entity nodes and four types of edges between them: (1) **M**ention-**M**ention co-**O**ccurrence edge (MMO): connects mentions belonging to the same sentence directly; (2) **M**ention-**M**ention co-**R**eference edge (MMR): connects mentions belonging to the same entity in order to capture non-local relations between mentions; (3) **M**ention-**E**ntity A**F**filiation edge (MEF): connects mentions belonging to the same entity to the entity node, which can transfer mention-level information to the entity level; (4) **E**ntity-**E**ntity **R**elation edge (EER): connects entities pairwise that belong to the same candidate entity tuple $< e_g, e_m, e_d >$. Figure 3(b) shows the candidate entity tuple $<$ Gene@19219, Var@rs1120926, MESH@D003424 $>$ in the ERG. The three entities have different mentions, and the co-reference and co-occurrence relations between mentions can be directly connected by MMR and MMO edges, explicitly constructing non-local dependency relations between mentions. The information between mentions and mentions is aggregated at the entity level, making the entity nodes fully integrate the contextual information of different mentions. Meanwhile, the entities are interconnected by EER edges for information exchange.

Relation Prediction. The prediction module first encodes the two graph, DSG and ERG, which can be divided into four layers: the word embedding layer encoded by BioBERT obtains the basic feature vector of words, the Bi-LSTM context encoding layer obtains the initial word vector representation, the DSG encoding layer obtains the final vector representation of word nodes, and the ERG encoding layer obtains the final vector representation of entities.

Word Embedding Layer. Given the $P'_{feature} = \{x_1, x_2, \ldots, x_n\}$ is obtained by encoding the document with BioBERT, where x_i is the basic feature vector representation of w_i. According to [1,28] and other studies, it is beneficial to use type embeddings and co-reference embeddings as additional features to enhance the input vector for relation extraction. Let t_i denote the type embedding of w_i and c_i denote the co-reference embedding of w_i. Finally, for each word w_i, its final word embedding representation X_i is denoted as $P'_{emb} = \{X_1, X_2, \ldots, X_n\}$.

$$X_i = [x_i, t_i, c_i] \tag{1}$$

Context Encoding Layer. To obtain contextual vector representations, we use Bi-LSTM to perform secondary modeling on the final output of the embedding layer, capturing contextual information provided by both directions.

$$\overrightarrow{h_i} = \overrightarrow{\text{LSTM}}(\overrightarrow{h_{i-1}}, X_i) \tag{2}$$

$$\overleftarrow{h_i} = \overleftarrow{\text{LSTM}}(\overleftarrow{h_{i+1}}, X_i) \tag{3}$$

$$H_i = [\overrightarrow{h_i}, \overleftarrow{h_i}] \tag{4}$$

$\overrightarrow{\text{LSTM}}$ and $\overleftarrow{\text{LSTM}}$ represent the LSTM in the forward and backward directions, respectively, and H_i is the concatenation of $\overrightarrow{h_i}$ and $\overleftarrow{h_i}$, i.e., the final context vector representation of the i-th word.

DSG and ERG Encoding Layer. According to studies [11,24], increasing the number of iterations can increase the risk of over-smoothing in graph neural networks. To improve the performance, we adopts the gating mechanism proposed by Gilmer et al. [3] in message passing neural networks to control the range of message propagation to the next layer, as shown in Eq. 5.

$$\mathbf{g}_i^{l+1} = \sigma \left(\sum_{t \in T} \sum_{j \in \mathcal{N}_i^t} \frac{\alpha_t^l}{|\mathcal{N}_i^t|} \mathbf{W}_t^l \mathbf{g}_j^l + \mathbf{W}_s^l \mathbf{g}_i^l \right) \tag{5}$$

$$gate_i^l = \text{sigmoid} \left(\mathcal{F}_g \left([\mathbf{u}_i^l; \mathbf{g}_i^l] \right) \right) \tag{6}$$

$$\mathbf{g}_i^{l+1} = gate_i^l \odot \tanh \left(\mathbf{u}_i^l \right) + \left(1 - gate_i^l \right) \odot \mathbf{g}_i^l \tag{7}$$

Where $\mathcal{F}_g$ is a linear transformation function used to control the gate level. The final hidden vector $\mathbf{g}_i^{l+1}$ is a combination of the non-linear transformation of the previous features and the update message obtained after gating. We refer to the R-GCN with added weight function and gate mechanism as *WRGCN*.

Relation Classification. Through the above three steps, we obtain the final representation vector e of the entity and the document node representation vector N_D in the DSG. Therefore, we utilize them to obtain comprehensive relation reasoning information, as shown in the Eq. 8.

$$I = [e_g, e_m, e_d, N_D] \tag{8}$$

Then sigmoid function is used as shown in Eq. 9, where W_b, W_a, b_a, and b_b are trainable parameters.

$$P(r \mid e_g, e_m, e_d) = sigmoid(W_b \sigma(W_a I + b_a) + b_b) \tag{9}$$

Cross-entropy loss function is employed for training, as shown in Eq. 10, where S represents the entire corpus.

$$\mathcal{L} = -\sum_{D \in S} \sum_{r_i \in \mathcal{R}} \mathbb{I}(r_i = 1) \log P(r_i \mid e_g, e_m, e_d)$$
$$+\mathbb{I}(r_i = 0) \log(1 - P(r_i \mid e_g, e_m, e_d)) \tag{10}$$

3 Experiment

In this section, dataset, baselines, training details and results analysis are introduced, respectively.

3.1 Dataset and Training Details

Due to the lack of public document-level datasets for gene-variant-disease relations extraction, we construct a gene-variant-disease relation dataset called GMD (**G**ene-**M**utation-**D**rug) based on PubMed Central. GMD includes seven categories: digestive system diseases, cardiovascular diseases, metabolic diseases, immune system diseases, neurological diseases, cancer, and other diseases. Table 1 shows the statistical information of dataset GMD. Besides, to verify the generalization of *GVDExtractor*, the dataset GMDrug, constructed by Peng et al. [19] was used. This dataset contains 6987 ternary instances about drug gene mutation relations and 6087 binary instances about drug mutation relations.

Following previous work [10, 23], we use BioBert as encoders. The layer number of WRGCN is set to 2. We apply dropout rate of 60% to each layer and use label smoothing to suppress overfitting. And we use Adam to optimize the parameters, with a learning rate of 0.003 and a weight decay of 0.0001. The batch size is set to 5 and the pruning-K is set to 1.

Table 1. Statistics of GMD

Statistical Parameters	Items	Quantity
Literature	-	500
Sentence	Cross-sentence	6734
	Single-sentence	1688
Entity	Gene	7042
	Variant	3337
	Disease	8145

3.2   Baselines and Evaluation Metrics

To assess the effectiveness of our method, we select the following baselines including *LSTM-RNN* [17], *GAIN* [27], *Feature-Based* [21], *Graph LSTM* [20], *gs-LSTM* [22], *MULTISCALE* [9], *MULTISCALE* [9], *iDeepNN* [6], *AGGCN* [5], *LF-GCN* [4], *Hybrid-GCN* [29]. Following Yao et al. [26], we use the widely used metrics Accuracy, Precision, Recall and F1 in our experiment.

3.3   Results

Relation Extraction. We first conducted experiments on the GMD dataset. The performances are shown in Table 2. *GVDExtractor* consistently outperforms all sequential-based and graph-based strong baselines on the test set. *Hybrid-GCN*, which performed the best among the baselines, utilizes self-attention to learn the global dependency and graph convolutional network to encode the syntactic and sequential dependency. However, our model is more comprehensive, which fully encodes non-local dependency such as coreference and co-occurrence, and aggregates them into entity-level representations.

Table 2. Performance on GMD

Method	Acc.	Pre.	Rec.	F1
LSTM-RNN [17] (2016)	50.9	48.7	48.1	48.4
iDeepNN [6] (2017)	65.3	64.9	69.9	67.3
GAIN [27] (2020)	67.1	69.1	77.5	73.0
Graph LSTM [20] (2017)	57.4	59.3	57.6	58.4
gs-LSTM [22] (2018)	59.2	57.8	57.3	57.5
MULTISCALE [9] (2019)	63.7	58.5	60.2	59.3
AGGCN [4] (2019)	62.1	62.1	67.0	64.5
LF-GCN [4] (2020)	64.5	68.9	69.2	69.0
Hybrid-GCN [29] (2021)	63.9	61.2	**81.2**	69.8
GVDExtractor (ours,2024)	**69.2**	**70.1**	**81.2**	**75.2**

Table 3. Performance on GMDrug

Method	Binary-class				Multi-class	
	T		B		T	B
	single	cross	single	cross	cross	cross
Feature-Based [21] (2016)	74.7	77.7	73.9	75.2	-	-
LSTM-RNN [17] (2016)	-	-	75.9	75.9	-	-
Graph LSTM-EMBED [20] (2017)	76.5	80.6	74.3	76.5	-	-
gs-LSTM [22] (2018)	80.3	83.2	83.5	83.6	71.7	71.7
AGGCN [5] (2019)	87.1	87.0	85.2	85.6	79.7	77.4
LF-GCN [4] (2020)	**88.0**	88.4	86.7	87.1	81.5	79.3
Hybrid-GCN [29] (2021)	86.6	89.8	90.7	91.3	84.6	87.8
GVDExtractor (ours,2024)	87.7	**92.4**	**93.4**	**93.6**	**90.0**	**92.8**

In addition, to verify the generalization ability of *GVDExtractor*, we conducted further experiments on the GMDrug dataset. The tasks in the GMDrug dataset can be divided into binary relation extraction (denoted as B, specifically for drug-variant binary relations) and ternary relation extraction (denoted as T), further subdivided into single-sentence and cross-sentence scenarios, and binary and multi-class classification according to the number of relation types. As shown in Table 3, it is evident that the *GVDExtractor* achieves optimal results in both binary and multi-class classification.

Pruning Strategy. We further analyze how KSL pruning affects model performance. KSL pruning only prunes edges and nodes between sentence and word nodes. It retains word nodes with distance K ($K = 0, 1, 2, \infty$, where $K = 0$ represents preserving only nodes and edges on the shortest path, and $K = \infty$

represents preserving the entire lowest common ancestor tree). Table 4 shows the results. When only the shortest dependency path is retained, the model's performance decrease significantly (Accuracy and Recall decreased by 2.1% and 1.7% respectively). This indicates that overly aggressive pruning strategies may lead to the loss of key information, such as the loss of negative relations, resulting in an increase in false positives. Considering the overall accuracy and comprehensive performance, $K = 1$ is selected as the optimal setting.

Furthermore, we select three strong models applying KSL pruning on GMD, namely Graph LSTM, gs-LSTM and iDeepNN. Figure 4 shows that when $K = 1$, the performance of all three models peaks and exceeds the F1 value of their original model. When K deviates from 1, the performance of the model will decrease to varying degrees, which also confirms that it is beneficial to incorporate information near the dependency path.

Table 4. Different K-value in pruning

KSL settings	Acc.	Pre.	Rec.	F1
$K = 0$(only SDP)	67.1	69.4	79.5	74.1
$K = 1$(SDP+1-distant LCA)	**69.2**	70.1	**81.2**	**75.2**
$K = 2$(SDP+2-distant LCA)	68.9	**72.5**	77.3	74.8
$K = \infty$(the entire LCA)	66.7	67.1	77.8	72.0

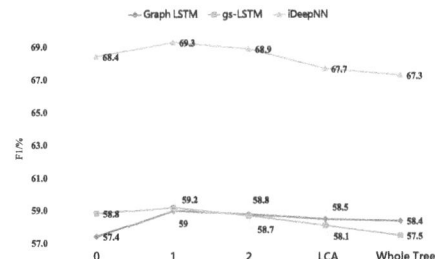

Fig. 4. Different models with KSL pruning

3.4   Ablation Study

Impact of Model Compositions. We conducted ablation experiments on various layers within the model, including the context encoding layer, DSG and ERG encoding layer. Besides, to validate the dual-graph sequential structure where the DSG precedes the ERG, we combined the two graphs and selectively retained the more significant edges based on the outcomes of the edge type ablation experiments. The results are presented in Table 5.

Table 5. Different model compositions

Model	Acc.	Pre.	Rec.	F1
GVDExtractor	**69.2**	**70.1**	**81.2**	**75.2**
-w/o Context Enc.	60.0	60.9	71.6	65.9
-w/o DSG Enc.	63.0	63.5	74.7	68.6
-w/o ERG Enc.	64.1	65.4	75.4	70.0
-w/o DSG & ERG Enc.	66.8	67.9	77.6	72.4

Table 6. Different edges types

Method	F1	Intra F1	Inter F1
GVDExtractor	**75.2**	**78.5**	**68.4**
-w/o WWD	70.2	73.2	63.8
-w/o SSA	72.9	76.6	65.4
-w/o MMO	74.7	77.8	67.9
-w/o MMR	74.1	77.2	63.7
-w/o EER	72.5	77.8	63.2
-w/o SDF	74.5	78.1	67.7

Results shows that all compositions-the context encoding, DSG, ERG, and dual graph sequential structure are indispensable, contributing respectively to F1 value improvements of 9.3%, 6.6%, 5.2%, and 2.8% in the final performance. Specifically, removing the context encoding layer most significantly affects performance, as it allows for simultaneous consideration of the text's contexts. Besides, removing the dual graph sequential structure results in a decline in performance, which indicate a structured graph encoding order is more conducive to the model learning representations at three levels: from words, to mentions, and then to entities. This finding also suggests that layering these heterogeneous nodes, rather than mixing them in a haphazard manner, enhances relation extraction.

Impact of Edge Types. Furthermore, We conduct ablation experiments on different types of edges in the DSG and ERG to verify their varying impacts. The results are shown in Table 6, where Inter F1 and Intra F1 represent the F1 scores for relation extraction in inter-sentence and intra-sentence scenarios, respectively. It's evident that the ablation of various edge types results in a discernible reduction in the performance of the proposed model. The dependency syntactic edge WWD, mention coreference edge MMR, and entity relation edge EER exhibit significant impacts on the Inter F1 score, with improvements of 4.6%, 4.7%, and 5.2% respectively. Particularly, the dependency syntactic edge WWD contributes the most to the overall F1 score, with a significant improvement of 5.0%. The WWD edge facilitates the interaction between word nodes and encompasses rich syntactic knowledge, forming the foundation of the DSG; The mention coreference edge MMR establishes connections among mention nodes across the entire document and captures global dependency; The entity relation edge EER enables fast and direct information exchange between entity nodes that have aggregated mention information.

4   Related Works

Previous researches of medical relation extraction were primarily performed using rule-based, dictionary-based and co-occurrence-based methods. But they heavily depend on manual design. Therefore researchers shift their attention into automate and streamline the processes involved in relation extraction.

Rule-based or dictionary-based methods rely on the design of rule templates by relevant medical experts, based on specific task-related lexicons, knowledge bases, or ontology structures to guide relation extraction tasks. For example, Doughty et al.'s tool [2] utilizes regular expression matching to identify variants and gene entities in text related to prostate and breast cancer topics, and then uses protein information to validate the association between genes and variants.

Deep neural networks require less feature engineering work as they can automatically extract abstract features from input text and train models accordingly. Peng et al. [20] proposed the Graph LSTM model which converts documents into two directed acyclic graphs based on syntax parse trees and word-sentence adjacency relations. Song et al. [22] proposed the graph-state LSTM model.

In medical text mining, many works on multi-relational tasks are focused on short texts spanning 2–3 contiguous sentences, Jia et al. [9] argued that it is necessary to extend the scope to the entire document to extract more complex relations. To this end, they proposed a multi-scale representation learning approach. Another multi-relational extraction method for long texts is the BERT-GT proposed by Lai et al. [13], which integrates a neighbor attention into the BERT by combining the bidirectional encoder representations from transformers (BERT) with a graph transformer (BERT-gt).

5   Conclusion

Gene-variant-disease relation extraction are challenging in medical research. In this paper, we introduce a double graph-based relation extraction model based on graph neural networks, which aims to extract ternary relations from full-text medical literature. *GVDExtractor* utilizes a relation structure graph to model the interaction among different mentions across the document and capture document-aware features. It also uses an entity relation graph to infer cross-sentence relations and capture long-distance dependency. Experimental results show *GVDExtractor* outperforms previous methods, and the ablation study also confirms the effectiveness of different modules in our model.

Acknowledgments. This work was supported by "the Fundamental Research Funds for the Central Universities".

Disclosure of Interests. The authors have no competing interests to declare that are relevant to the content of this article.

References

1. Christopoulou, F., Miwa, M., Ananiadou, S.: Connecting the dots: document-level neural relation extraction with edge-oriented graphs. In: Proceedings of the 2019 Conference on Empirical Methods in Natural Language Processing and the 9th International Joint Conference on Natural Language Processing (EMNLP-IJCNLP), pp. 4925–4936. Association for Computational Linguistics, Hong Kong, China (2019)
2. Doughty, E., et al.: Toward an automatic method for extracting cancer-and other disease-related point mutations from the biomedical literature. Bioinformatics **27**(3), 408–415 (2011)
3. Gilmer, J., Schoenholz, S.S., Riley, P.F., Vinyals, O., Dahl, G.E.: Neural message passing for quantum chemistry (2017)
4. Guo, Z., Nan, G., Lu, W., Cohen, S.B.: Learning latent forests for medical relation extraction. In: International Joint Conference on Artificial Intelligence (2020)
5. Guo, Z., Zhang, Y., Lu, W.: Attention guided graph convolutional networks for relation extraction (2019)
6. Gupta, P., Rajaram, S., Schütze, H., Andrassy, B., Runkler, T.A.: Neural relation extraction within and across sentence boundaries. CoRR abs/1810.05102 (2018)

7. Gupta, P., Rajaram, S., Schütze, H., Runkler, T.: Neural relation extraction within and across sentence boundaries. In: Proceedings of the AAAI Conference on Artificial Intelligence, vol. 33, pp. 6513–6520 (2019)

8. Huang, Z., Xu, W., Yu, K.: Bidirectional LSTM-CRF models for sequence tagging (2015)

9. Jia, R., Wong, C., Poon, H.: Document-level n-ary relation extraction with multi-scale representation learning. In: Proceedings of the 2019 Conference of the North American Chapter of the Association for Computational Linguistics: Human Language Technologies, Volume 1 (Long and Short Papers), pp. 3693–3704. Association for Computational Linguistics, Minneapolis, Minnesota (2019)

10. KafiKang, M., Hendawi, A.: Drug-drug interaction extraction from biomedical text using relation BioBERT with BLSTM. Mach. Learn. Knowl. Extr. 5(2), 669–683 (2023)

11. Kipf, T.N., Welling, M.: Semi-supervised classification with graph convolutional networks (2017)

12. Kraft, P., Hunter, D.J.: Genetic risk prediction–are we there yet? N. Engl. J. Med. **360**(17), 1701–1703 (2009)

13. Lai, P.T., Lu, Z.: Bert-GT: cross-sentence n-ary relation extraction with bert and graph transformer (2021)

14. Lee, J., et al.: Biobert: a pre-trained biomedical language representation model for biomedical text mining. Bioinformatics **36**(4), 1234–1240 (2020)

15. Li, J., Xu, K., Li, F., Fei, H., Ren, Y., Ji, D.: MRN: a locally and globally mention-based reasoning network for document-level relation extraction. In: Findings of the Association for Computational Linguistics: ACL-IJCNLP 2021, pp. 1359–1370 (2021)

16. Li, P.H., et al.: pubmedkb: an interactive web server for exploring biomedical entity relations in the biomedical literature. Nucleic Acids Res. **50**(W1), W616–W622 (2022)

17. Miwa, M., Bansal, M.: End-to-end relation extraction using LSTMs on sequences and tree structures. In: Proceedings of the 54th Annual Meeting of the Association for Computational Linguistics (Volume 1: Long Papers), pp. 1105–1116. Association for Computational Linguistics, Berlin, Germany (2016)

18. Ning, S., Teng, F., Li, T.: Multi-channel self-attention mechanism for relation extraction in clinical records. Chin. J. Comput. **43**, 916–29 (2020)

19. Peng, N., Poon, H., Quirk, C., Toutanova, K., Yih, W.T.: Cross-sentence n-ary relation extraction with graph LSTMs. Trans. Assoc. Comput. Linguist. **5**, 101–115 (2017)

20. Peng, N., Poon, H., Quirk, C., Toutanova, K., Yih, W.t.: Cross-sentence n-ary relation extraction with graph LSTMs (2017)

21. Quirk, C., Poon, H.: Distant supervision for relation extraction beyond the sentence boundary. arXiv preprint arXiv:1609.04873 (2016)

22. Song, L., Zhang, Y., Wang, Z., Gildea, D.: N-ary relation extraction using graph-state LSTM. In: Proceedings of the 2018 Conference on Empirical Methods in Natural Language Processing, pp. 2226–2235. Association for Computational Linguistics, Brussels, Belgium (2018)

23. Su, P., Vijay-Shanker, K.: Investigation of improving the pre-training and fine-tuning of bert model for biomedical relation extraction. BMC Bioinform. **23**(1), 120 (2022)

24. Wang, Y., et al.: Ensemble multi-relational graph neural networks. arXiv preprint arXiv:2205.12076 (2022)

25. Xu, B., Wang, Q., Lyu, Y., Zhu, Y., Mao, Z.: Entity structure within and throughout: modeling mention dependencies for document-level relation extraction. In: Proceedings of the AAAI Conference on Artificial Intelligence, vol. 35, pp. 14149–14157 (2021)
26. Yao, Y., et al.: Docred: a large-scale document-level relation extraction dataset. arXiv preprint arXiv:1906.06127 (2019)
27. Zeng, S., Xu, R., Chang, B., Li, L.: Double graph based reasoning for document-level relation extraction. In: Proceedings of the 2020 Conference on Empirical Methods in Natural Language Processing (EMNLP), pp. 1630–1640. Association for Computational Linguistics, Online (2020)
28. Zhang, Y., Qi, P., Manning, C.D.: Graph convolution over pruned dependency trees improves relation extraction (2018)
29. Zhao, D., Wang, J., Lin, H., Wang, X., Yang, Z., Zhang, Y.: Biomedical cross-sentence relation extraction via multihead attention and graph convolutional networks. Appl. Soft Comput. **104**, 107230 (2021)

Alias Extraction Enhanced by Automatically Generated Long-Tail Instances

Zhi Yang, Yi Luo, and Xin Xin(✉)

School of Computer Science and Technology, Beijing Institute of Technology, Beijing, China
{yangzhi,luoyi,xxin}@bit.edu.cn

Abstract. The task in this paper is to extract alias relations from general sentences. It can be used to expand the knowledge base and provide support to entity linking. The problem with traditional alias relation extraction methods is that they work better for identifying frequent alias expressions than infrequent ones. In order to solve this problem, we propose to search keywords for alias expressions with the help of the "XIANDAI HANYU CIDIAN" (Contemporary Chinese Dictionary) and try to cover as many infrequent alias expressions as possible. We generate training samples for the infrequent keywords using large language models automatically to enlarge the dataset. The experimental results show that the model performs better when trained based on the enlarged training set with the generated instances.

Keywords: Large language model · Relation extraction · Sample balance · Alias relation

1 Introduction

The objective of this paper is to extract alias relations from general sentences. Alias relations can enhance knowledge base [9] and support entity linking [11]. In this paper, an alias relation typically denotes an alternative name that refers to the same entity. Specifically, our goal is build the model to identify all alias pairs from a given input sentence.

Previous methods [7,15] for alias relation extraction usually utilize a labeled dataset for training models. One problem is that the model can perform well on frequent alias expressions, but may fail on infrequent cases when the expression does not occur in the dataset. The expression means the way to describe this alias relations. Usually, it is presented by a keyword. As illustrated in Fig. 1, the keyword '简化' triggers an alias relation between '《岛夷志》' and '《岛夷志略》'. To solve the problem, our idea is to manually search and identify such keywords using a dictionary. In this way, we can cover the infrequent keywords as many as possible. After that, we use a Large Language Model(LLM) to generate

B. Xu et al. (Eds.): CCKS-IJCKG 2024, CCIS 2229, pp. 124–133, 2025.
https://doi.org/10.1007/978-981-96-1809-5_10

Fig. 1. Example of an alias relation consisting of "简化".

sentences based on these keywords. In this way, the training set of the alias extraction model is enlarged by the automatically generated long-tail instances.

The experimental results demonstrate that augmenting the training dataset by automatic generated long-tail instances can enhance the extraction performance. The F1 score increases by approximately 2 points.

The main contributions of this paper are as follows. First, we construct the keyword set for alias extraction with the help of the dictionary "XIANDAI HANYU CIDIAN". Second, we generate training sentences for alias extraction based on these keywords through a LLM. Finally, we evaluate the effect of the generated training samples by a dataset constructed using wikipedia[1]. It shows that this method can enhance the accuracy of the alias extraction task.

2   Manual Keywords Extraction from Dictionary

The main idea of this paper is that the sentences include specific keywords that define alias relations and the dictionary can approximately cover as many keywords as possible. So searching words in the "XIANDAI HANYU CIDIAN" (Contemporary Chinese Dictionary) [20] will cover infrequent keywords. We look up all nouns and verbs in the dictionary.

The searching process of keywords is mainly divided into three steps, as shown in Fig. 2.

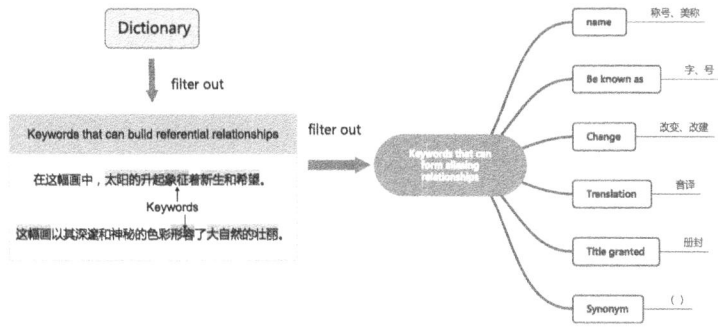

Fig. 2. Searching process for keywords that can form aliasing relations.

---

[1] https://dumps.wikimedia.org.

Step 1: By searching the "XIANDAI HANYU CIDIAN", we first find out all the verbs and nouns that have coarse relation to reference. 1176 words are selected, which is detailed in Fig. 10 in the Appendix.

Step 2: From the above words, we filter the keywords that have fine-grained relation to reference. 217 keywords are selected.

Step 3: Finally, we make sentences using the 217 keywords to describe an alias relation. Some of the words cannot form such sentences. As a final result, 112 words are retained which are in 6 major categories, as shown in Table 1.

Table 1. Classification of keywords for which aliasing relations can be constructed.

Keyword Category	Vocabulary
Name	作、为、化名、简写、简称、匿名、谦称、署名、缩写、总称、尊称、爱称、褒称、鄙称、贬称、别称、称号、称呼、称谓、代称、敬称、略称、美称、蔑称、名称、昵称、全称、统称、别号、绰号、代号、番号、诨号、外号、笔名、别名、名字、乳名、俗名、学名、艺名、译名、别字、大名、小名、诨名、俗称、本名、庙号、前身、通称、呼号、名、谥、简化、题名、题字、人称、专名、代名词、法名、骂名、托名、古名
Be known as	称、称颂、称誉、誉、字、号、号称、呼号、叫、戏称、自称、自封、自诩、戏称、号称、指称、谓、讳、下称、奉、尊、即、暨、命名、起名、定名
Change	变更、变迁、成立、成为、改变、改建、更改、更名、改、正名、改名、改换
Translation	重译、翻译、译、译注、音译、英译
Title granted	册封、册立、敕封、封、加封
Synonym	()

3 Automatic Data Augmentation with LLM

In this section, we design suitable prompt engineering that enables the LLM to generate training samples for infrequent keywords. We utilize the few-shot learning form to write examples. It includes an explanation of the task, simple examples related to the task, and the expected results returned by the LLM. Figure 3 shows the input part and output part of the LLM for generating samples for the keyword "简化". We generate such training samples for 81 keywords whose occurrence frequency is less than 5.

4 Approaches

4.1 Task Description

The goal of the alias relation extraction task is to extract all alias pairs presented in a sentence. Initially, we identify all possible alias entities in a sentence.

Then we judge whether each candidate pair has alias relation. Formally, given a sentence S, we first identify all the entities in the sentence S, denoted by $\{ent_1, ent_2, \ldots, ent_n\}$. Then for each pair $\{ent_i, ent_j\}$, where $i < j$, we identify whether it has an alias relation.

```
┌─────────────────────────────────────────────────────────────────────┐
│                          Few-shot Prompt                            │
├─────────────────────────────────────────────────────────────────────┤
│ 生成 10 条由真实存在的实体构成别名关系的句子，且由 "简化" 两字构成了这种别名关系 │
│ （please generate 10 sentences with aliasing from real entities, with the word │
│ '简化' constituting the aliasing relationship）                       │
│ 【示例】：在 "后来汪大渊将《岛夷志》简化名为《岛夷志略》，在他的故乡江西南昌刻印│
│ 刻单行本" 这句话中，"《岛夷志》" 和 "《岛夷志略》" 就是一对别名对，且由简化两字 │
│ 构成了这种别名关系                                                     │
│ (In the Chinese sentence' 后来汪大渊将《岛夷志》简化名为《岛夷志略》，在他的故乡 │
│ 江西南昌刻印刻单行本'，'《岛夷志》' and '《岛夷志略》' are a pair of alias, with │
│ the word '简化' constituting the alias relationship)                  │
│ 【输出格式】：                                                         │
│ {                                                                    │
│ "rel": "简化",                                                        │
│ "ent1": the first alias,                                             │
│ "ent2": the second alias,                                            │
│ "text": the sentence generating by large language model             │
│ }                                                                    │
└─────────────────────────────────────────────────────────────────────┘
                                    ↓
┌─────────────────────────────────────────────────────────────────────┐
│                              OUTPUT                                  │
├─────────────────────────────────────────────────────────────────────┤
│ {                                                                    │
│ "rel": "简化",                                                        │
│ "ent1": "北京理工大学",                                                │
│ "ent2": "北理工",                                                      │
│ "text": "为了方便称呼，大家通常将 "北京理工大学" 简化为 "北理工"。"           │
│ }                                                                    │
│                          • • • • • •                                 │
└─────────────────────────────────────────────────────────────────────┘
```

Fig. 3. Example of "简化" Few-shot mode.

4.2 Overall Framework

The overall framework of the model in this paper is shown in Fig. 4, which is mainly divided into two parts - candidate entity recognition and alias relation judgment.

The candidate entity recognition part is to extract entities from the sentence. We parse the it into a constituent parsing tree using the N-ary tree method [14]. NP are selected as entity candidates. As shown in Fig. 4, 6 entities are selected in this stage. In the alias relation judgment part, we utilize the LLAMA-FACTORY framework [18] to fine-tune the LLM. We use the zero-shot learning form to design the prompt engineering, as shown in Fig. 5.

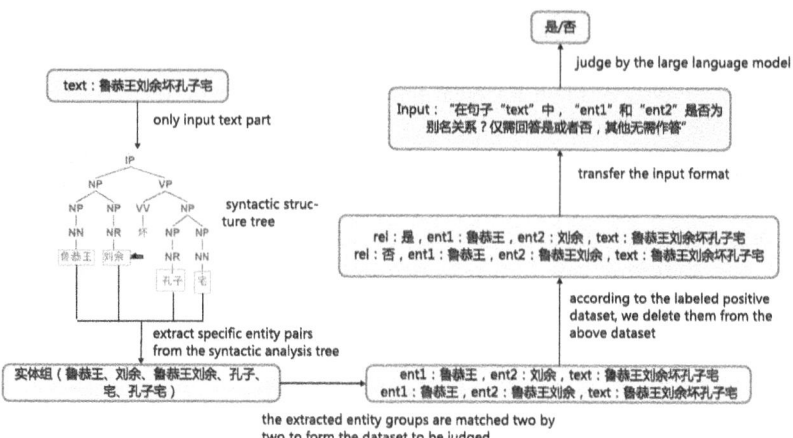

Fig. 4. Overall framework of the alias relation extraction model.

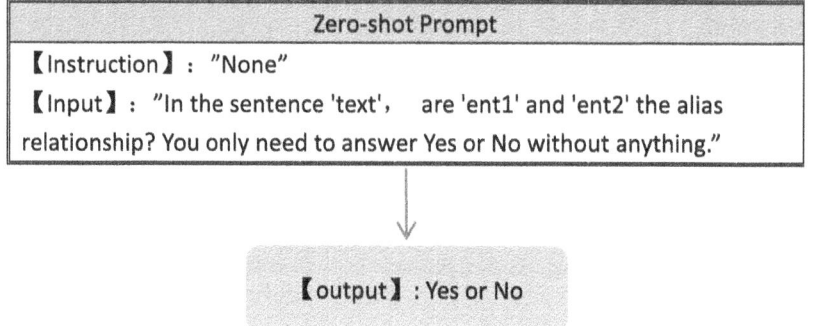

Fig. 5. Input format for large language models.

5   Experiment

5.1   Construction of the Dataset

In this paper, we build a dataset for alias extraction using Wikipedia articles (See footnote 1). We manually filter 2,570 sentences containing alias relations and 7,500 negative examples which do not contain alias relations. For positive samples, we label the keyword that describe the alias relation for the corresponding entity pair.

5.2   Validation of Keyword Coverage

By categorizing the 2,570 data samples after data preprocessing with keywords, the results are shown in Table 2.

According to the data statistics in Fig. 6 and Table 2 and the keyword categories in Fig. 7, it can be concluded that 14% of the keywords cover 85% of

Table 2. Statistics on the volume of data for the category of keywords.

Quantities	Vocabulary
Greater than or equal to 50	简称、作、为、缩写、学名、称、字、号、成立、改、即、名、同位语
Between 10 and 49	改名、更名、别名、通称、叫、谥、俗称、译、名称、尊、前身、尊称
Between 5 and 9	笔名、褒称、封、全称、庙号、简写、人称、成为、本名、美称
Between 1 and 4	命名、称号、称呼、暨、加封、指称、艺名、绰号、署名、代号、音译、谓、别称、统称、番号、敬称、奉、美称、自称、别号、呼号、小名、乳名、称谓、外号、评名、总称、名字、改建、呼号、定名、正名、讳、代称、译名
0	号称、称誉、册封、诨号、昵称、翻译、爱称、起名、誉、古名、简化、代名词、托名、变迁、贬称、骂名、译注、改换、英译、自封、别字、俗名、敕封、自诩、改变、重译、册立、鄗称、戏称、化名、下称、专名、更改、称颂、略称、变更、大名、法名、题字、谦称、题名、匿名、蔑称

the data volume, and the remaining 86% of the keywords are distributed in the remaining 15% of the data.

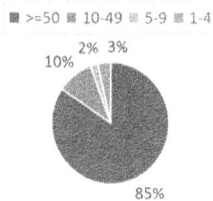

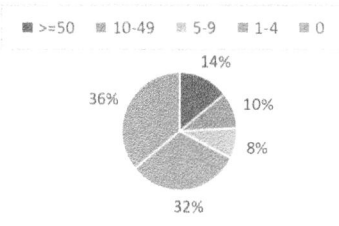

Fig. 6. The sample size owned by each keyword number category.

Fig. 7. The keyword number category share statistics.

We evaluate the keyword coverage of the keyword set. Figure 8 illustrates the change in total data volume before and after automatic generation, while Fig. 9 shows the evolution in the number of keywords.

Our selected keywords can not cover all the cases in the dataset. In the 2,570 labeled sentences, there are totally 77 keywords. Our first step can cover 74 of them. The three words ("小字","化学式", and "对音") does not occur in the "XIANDAI HANYU CIDIAN". Our second step can over 69 of them, we miss five keywords ("徽号", "今", "现", "英文" and "英语").

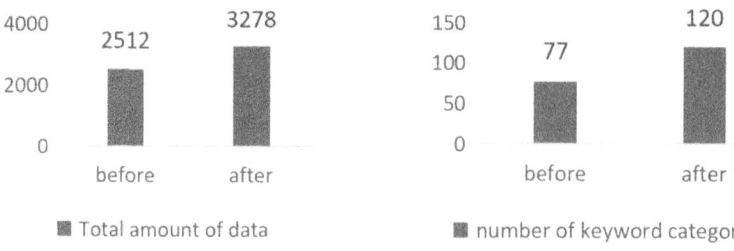

Fig. 8. The total amount of data in the dataset before and after the expansion.

Fig. 9. The number of keyword categories included before and after the expansion.

5.3 Results

In this paper we use the ChatGlm3-6B [16] series models. In addition to the LLMs, we also use the classic BERT (Bidirectional Encoder Representation from Transformers) model [3] in the field of relational extraction as a benchmark for evaluation. This model follows the method proposed by Zhong et al. [19] of using entity boundaries as features. Specifically, entity location information is marked with special symbols.

In this experiment, the LoRA [6] rank is 8, and DoRA [10] with weight decomposition is enabled, the AdamW learning rate is set to 0.00005, the number of training rounds is 3, the number of gradient accumulation steps is 8.

Table 3. Experimental results for models before and after sample equilibrium.

Methods	Precision/%	Recall/%	F_1/%
Bert model before data expansion	71.35	74.68	72.58
Chatglm3 model before data expansion	80.26	85.91	82.99
Bert model after data expansion	73.53	76.83	75.22
Chatglm3 model after data expansion	93.67	77.89	84.95

As can be seen from Table 3, the result trained by the baseline LLM is higher than the baseline model based on BERT in terms of precision rate, recall rate, and value. Compared with the old data before automatic generation, the BERT model improves the F1 value from 72.58% to 75.22% after automatic generation and the Chatglm3 model improves the F1 value from 82.99% to 84.95%.

6   Related Research

In the field of relational extraction, we can categorize all the methods into three types, rule-based methods, neural network-based methods, and large language model-based methods.

The rule-based methods match text by handwritten rules for relation extraction. For example, Andersen et al. [1] proposed using pre-defined patterns or regular expressions to match text segments. Another method proposed by Church [2] extracts the dependencies between words to judge the relations between entities.

Neural network-based methods can be divided into two types, pipeline frame and joint frame. One kind of the pipeline frame is proposed by Zhong et al. [19] using entity boundaries as features. In addition,Wang et al. [13] proposes a method combining named entity recognition and relation extraction and using a new tagging scheme. Moreover, Yang et al. [15] proposed a joint model of entity aliases extraction.

Zhang et al. [17] propose a kind of the LLM-based methods in 2023. They combine relation extraction task and question and answer task to improve the ability of LLM.

7   Conclusion

In this work, we search keywords that may construct alias relations from the "XIANDAI HANYU CIDIAN". Then we utilize the LLM to generate long-tail training instances. Through experimental study, we show the above strategy is effective in improving the performance of alias extraction.

Acknowledgement. This work is supported by the grants from the National Natural Science Foundation of China (No. 62172044).

A   Candidate Keywords with Referential Relations

Candidate Keywords with Referential Relations

作、为、字、化名、记名、简写、简称、匿名、谦称、署名、缩写、著称、总称、尊称、爱称、褒称、鄙称、贬称、别称、称号、称呼、称谓、代称、敬称、略称、美称、蔑称、名称、昵称、全称、统称、暗号、别号、绰号、代号、番号、谐号、外号、笔名、别名、假名、名字、乳名、俗名、姓名、学名、艺名、译名、备注、别字、大名、小名、诨名、俗称、本名、庙号、前身、通称、呼号、名（动词）、谥（无词性）、简缩、简化、辩称、题名、题签、题写、题字、誉称、背称、公称、人称、憎称、职称、指名、专名、唱名、臭名、代名词、担名、定名、恶名、法名、芳名、浮名、沽名、挂名、寄名、空名、令名、骂名、埋名、冒名、奶名、声名、实名、提名、通名、闻名、托名、刑名、虚名、音名、域名、古名、重名、称说、称、称颂、称扬、称誉、号、号称、合称、谎称、叫、戏称、自称、自封、自谓、指称、谓（无词性）、讳（名词）、坦称、下称、妄称、宣称、褒奖、褒扬、称道、称快、称奇、称赏、称述、称叹、称羡、称赞、歌颂、歌吟、歌咏、画字、讳言、堪称、赞美、赞佩、赞赏、赞颂、赞叹、赞誉、追捧、变更、变迁、成立、成为、订正、改变、改订、改建、更改、更名、糟改、篡改、改、正名、改名、改换、变、变动、变革、变工、变化、变换、定更、复建、复命、复述、复通、改编、改动、改写、改正、更换、更替、更新、更易、更正、换、换代、校订、校对、校改、校正、校注、校准、勘正、匡正、转变、譬（无词性）、譬如、譬喻、指明、指示、奉、尊、罢黜、罢免、罢官、罢职、贬谪、贬职、褫夺、黜免、废黜、降职、解职、晋升、就职、升迁、升任、升职、升位、调职、高升、挂职、履任、履职、任免、任命、任用、任职、上任、颁授、授予、颁布、颁发、颁奖、颁行、颁赠、赋予、授衔、授勋、册封、册立、敕封、诰封、封、加封、阐述、阐释、集释、解释、阐明、阐说、诠释、诠注、传说、传颂、传闻、传诵、风传、哄传、流传、吹捧、吹嘘、代表、代替、替代、替换、调任、返聘、今、现、即、暨、命名、起、起名、具名、命题、化学式、对音、英文、英语、包括、包揽、包含、涵盖、涵容、罗列、编译、编造、编织、编制、编著、编撰、编缀、编纂、贬斥、贬黜、贬官、贬损、标明、标示、标志、标注、表明、表示、'表述、表态、表现、测字、八字、白字、本字、表字、草字、衬字、赤字、冷字、盲字、铅字、如字、上口字、生字、实字、手头字、俗字、虚字、查实、查问、查寻、查询、查验、查阅、查找、产出、产生、倡办、倡导、倡言、倡扬、倡议、陈述、陈说、陈诉、陈言、宣名、叫名、美名、名、名号、名讳、名节、名款、名利、慕名、排名、品名、签名、盛名、威名、扬名、一文不名、英名、知名、专名号、罪名、创、创办、创编、创建、创立、创设、创新、创造、创制、创作、当成、当做、对应、对照、对阵、对证、发文、发问、发现、供认、供述、供职、构成、构建、构思、构想、构造、呼、呼唤、呼叫、呼应、呼吁、记录、记认、记述、记诵、记叙、揭露、揭秘、揭破、揭示、揭晓、介绍、开设、开办、开启、刊播、刊布、刊登、刊发、刊刻、刊授、刊行、刊印、刊载、口授、口述、类比、类乎、类似、类同、类推、录用、录制、录取、论理、论述、论说、论证、描绘、描金、描摹、描写、明确、明示、明说、明晓、明知、判、判别、判处、判定、判读、判断、判分、判明、批、批判、批评、批示、批注、批改、启动、启发、启蒙、启示、启用、签批、签署、签注、确保、确定、确立、确认、确守、确信、申论、申明、申请、申述、申说、申诉、声辩、声称、声明、声请、声讨、声言、声扬、声援、声张、述怀、述说、述职、坦白、坦陈、坦承、坦露、坦言、袒露、通告、通报、通译、通知、透露、推举、推理、推断、推选、推知、误传、误导、误解、显见、显露、显示、想、想象、象征、像、写、写作、形成、形容、形似、修、修补、修订、修复、修改、叙述、叙说、叙用、宣布、宣传、宣明、宣判、宣示、宣扬、宣召、选用、选择、训示、训喻、训诲、训诫、言说、言谈、言语、言喻、佯言、演化、演进、演示、演算、演绎、扬、扬言、预订、预定、预感、预告、预计、预见、预料、预谋、预期、预示、昭示、追叙、追认、追授、追述、追诉、综计、综括、综述、作为、编号、大号、法号、佛号、符号、国号、号兵、徽号、记号、警号、军号、口号、年号、牌号、票号、谱号、旗号、商号、书号、堂号、尾号、小号、型号、雅号、正号、专号、字号、褒词、褒义词、贬词、唱词、词、分子式、最简式、通式、范式、首发式、译音、音、字音、铭文、等式、嫡母、嫡派、嫡系、嫡子、调号、调类、调令、调式、调值、定格、定规、定律、定理、定评、定式、定义、定语、动词、动机、读音、短评、断想、断言、断语、对白、对方、对立面、恶俗、恶习、恶意、恶语、噩梦、噩运、噩兆、恩人、恩师、恩遇、恩泽、发言、附言、副标题、副词、副官、副题、副项、副职、概观、概况、概览、概略、概论、概貌、概说、概要、感想、感言、感知、告示、公道、公告、公理、公例、功业、功用、贡献、构象、构型、古物、古昔、古稀、古谚、古音、古语、故地、故都、故技、故交、故旧、故居、故人、怪话、关键、关键词、关系、惯常、规划、规矩、规律、规约、过场、过程、过错、过往、含量、含义、含意、行辈、行情、行业语、好处、好话、好景、号码、号手、恒等式、后话、后果、后人、后生、户主、坏处、皇后、皇上、皇太后、皇太子、皇子、谎信、徽标、徽调、徽记、徽章、会徽、会旗、讳、昏话、活页、活字、祸首、机徽、基本、基础、迹象、寄语、佳话、家教、家景、家眷、家母、家属、家小、家兄、假设、假象、架势、奸臣、奸宄、奸计、奸佞、简编、简化字、简介、简况、见地、见方、见解、见识、见闻、见证、建议、建制、讲话、焦点、教父、教工、教官、教规、阶乘、阶地、阶段、阶级、阶下囚、结尾、今译、今音、今文、近况、禁脔、经纪人、经籍、经纪、经传、颈联、景况、景物、警徽、敬辞、境况、窘况、窘态、窘境、旧俗、舅父、舅舅、舅妈、舅母、舅嫂、舅子、具文、具象、剧目、剧情、决策、决定、诀窍、诀要、觉悟、绝笔、绝活、军徽、军衔、空文、空想、口头禅、口头语、口语、口谕、兰谱、劳绩、老娘、老婆、老师、老字号、里程碑、俚歌、俚曲、俚语、理由、例句、例题、例证、例子、劣迹、劣马、劣品、劣绅、劣势、略语、盲文、盲信、错位、冒号、眉题、眉字、美谈、美誉、美言、密语、民谚、名声、名望、名下、名衔、明喻、铭记、铭、命数、谬、牟取、模板、模具、木版画、目标、内人、内务、内线、内因、拟态、拟物、拟议、拟作、匿、癖好、癖习、癖性、譬方、票面、票品、品类、品色、品目、谱系、旗语、旗帜、起因、起源、契机、契据、谦辞、倾向、情景、情境、情况、情郎、趋势、趋向、全音、缺点、任期、任务、弱项、弱势、弱点、善报、善款、善类、善事、善心、善行、善意、社论、社评、生肖、声色、声学、时式、时势、时事、实处、实词、实据、实况、实情、实权、实事、史话、示指、市制、事迹、事故、事件、事理、事例、事略、事情、事实、事先、事项、事业、事宜、事由、事主、视点、叔父、叔公、叔母、叔婆、叔叔、叔祖、叔祖母、殊勋、殊誉、属地、属国、属相、属员、数词、司务长、司仪、私党、死因、死战、颂词、俗语、岁序、缩影、坦途、钽、叹词、特点、特长、提案、题记、题解、题目、体裁、体征、体制、天象、天性、条款、条理、同义词、同音词、头目、途径、外设、妄念、妄想、妄言、妄语、伟绩、伟业、伪作、文职、文摘、闻名、喜报、戏言、细则、细账、细作、狭义、下策、下款、下文、下作、闲话、显要、现下、现象、现役、现职、限定、详情、向例、向量、相位、小传、校徽、校花、心音、心语、信号、信息、信义、形声、形式、形势、姓氏、袖标、虚词、虚辞、虚职、序号、序列、序言、血亲、勋爵、勋章、训词、训诂、讯号、言论、言行、演艺、谚语、洋货、谣传、要诀、要览、要领、要员、要职、要旨、业者、业主、医嘱、医师、医士、姨夫、姨父、姨姥姥、姨妈、姨母、姨奶奶、姨娘、姨儿、姨太太、姨丈、姨、遗孤、遗民、遗老、遗孀、异言、译笔、译文、译著、译作、因由、因缘、音标、音区、音色、隐语、隐喻、赢家、影后、影帝、用户、用品、用途、用项、优点、优势、友人、友邦、诱因、谀辞、语义、语意、语族、谕旨、喻义、誉、原籍、原文、原形、原型、原意、原因、缘由、缘故、缘起、源头、远因、远志、约期、运力、运量、运能、韵母、韵文、韵语、遭际、遭遇、责编、责任、债务、债务人、债主、债权、债权人、债户、丈夫、丈母娘、丈人、掌门人、政绩、侄妇、侄女、侄女婿、侄孙、侄孙女、侄媳妇、侄子、终点、主词、主人、主题、著名、著述、著者、著作人、专刊、撰述、作者、作品、作家

Fig. 10. Candidate keywords with referential relations.

References

1. Andersen, P.M., Hayes, P.J., Weinstein, S.P., Huettner, A.K., Schmandt, L.M., Nirenburg, I.: Automatic extraction of facts from press releases to generate news stories. In: Third Conference on Applied Natural Language Processing, pp. 170–177 (1992)
2. Brill, E.: A simple rule-based part of speech tagger. In: Speech and Natural Language: Proceedings of a Workshop Held at Harriman, New York, 23-26 February 1992 (1992)
3. Devlin, J., Chang, M.W., Lee, K., Toutanova, K.: Bert: Pre-training of deep bidirectional transformers for language understanding. In: Proceedings of the 2019 Conference of the North American Chapter of the Association for Computational

Linguistics: Human Language Technologies, Volume 1 (Long and Short Papers), pp. 4171-4186 (2019)

4. Foundation, W.: Wikipedia. Tech. rep. (2001). https://en.wikipedia.org/wiki
5. Han, X., et al.: Fewrel: a large-scale supervised few-shot relation classification dataset with state-of-the-art evaluation. arXiv preprint arXiv:1810.10147 (2018)
6. Hu, E.J., et al.: Lora: low-rank adaptation of large language models. In: International Conference on Learning Representations (2022)
7. Li, N.: Automatic extraction of alias in ancient local chronicles based on conditional random fields. J. Chin. Inf. Process **32**, 41 (2018)
8. Li, S., et al.: DuIE: a large-scale Chinese dataset for information extraction. In: Tang, J., Kan, M.-Y., Zhao, D., Li, S., Zan, H. (eds.) NLPCC 2019. LNCS (LNAI), vol. 11839, pp. 791–800. Springer, Cham (2019). https://doi.org/10.1007/978-3-030-32236-6_72
9. Lin, Y., Liu, Z., Sun, M., Liu, Y., Zhu, X.: Learning entity and relation embeddings for knowledge graph completion. In: Proceedings of the AAAI Conference on Artificial Intelligence, vol. 29 (2015)
10. Liu, S.y., et al.: Dora: weight-decomposed low-rank adaptation. In: Forty-first International Conference on Machine Learning (2024)
11. Shen, W., Wang, J., Han, J.: Entity linking with a knowledge base: Issues, techniques, and solutions. IEEE Trans. Knowl. Data Eng. **27**(2), 443–460 (2014)
12. Vrandečić, D., Krötzsch, M.: Wikidata: a free collaborative knowledgebase. Commun. ACM **57**(10), 78–85 (2014)
13. Wang, Y., Yu, B., Zhang, Y., Liu, T., Zhu, H., Sun, L.: Tplinker: single-stage joint extraction of entities and relations through token pair linking. In: Proceedings of the 28th International Conference on Computational Linguistics, pp. 1572–1582 (2020)
14. Xin, X., Li, J., Tan, Z.: N-ary constituent tree parsing with recursive semi-markov model. In: Proceedings of the 59th Annual Meeting of the Association for Computational Linguistics and the 11th International Joint Conference on Natural Language Processing (Volume 1: Long Papers), pp. 2631–2642 (2021)
15. Yang, Y., Wenliang, C.: Joint model for entity alias extraction in tourism domain. J. Chin. Inf. Process **34**(6), 55–63 (2020)
16. Zeng, A., et al.: Glm-130b: an open bilingual pre-trained model. In: The Eleventh International Conference on Learning Representations (2022)
17. Zhang, K., Gutierrez, B.J., Su, Y.: Aligning instruction tasks unlocks large language models as zero-shot relation extractors. In: The 61st Annual Meeting Of The Association For Computational Linguistics (2023)
18. Zheng, Y., Zhang, R., Zhang, J., Ye, Y., Luo, Z.: Llamafactory: unified efficient fine-tuning of 100+ language models. arXiv preprint arXiv:2403.13372 (2024)
19. Zhong, Z., Chen, D.: A frustratingly easy approach for entity and relation extraction. In: Proceedings of the 2021 Conference of the North American Chapter of the Association for Computational Linguistics: Human Language Technologies, pp. 50–61 (2021)
20. 社会科学院语言研究所词典编辑室: 现代汉语词典（第五版）. 商务印书馆 (2005)

Taxonomy Induction Using LLMs: An Enhanced Framework by Integrating Doubly-Checked Mechanism and Self-evaluation Strategy

Jiaye Li[1], Yuan Meng[1], Lijun Wang[2(✉)], Tianhao Qian[1], Songlin Zhai[1], and Guilin Qi[1]

[1] School of Computer Science and Engineering, Southeast University, Nanjing, China
{lee559,230218214,qth2mir,songlin_zhai,gqi}@seu.edu.cn
[2] State Grid Information and Telecommunication Co., Ltd., Beijing, China
wlj_lisa@126.com

Abstract. Taxonomies, structured as tree hierarchies, are valuable for applications such as web retrieval, question-answering, and recommender systems. As existing taxonomy curation based on deep learning or pre-trained models rely heavily on a large amount of labeled data, which is extremely time-consuming and labor-intensive, the emergence of large language models (LLMs, *e.g.,* ChatGPT, GPT-4.0, LLaMA2) has made automatic taxonomy construction from texts highly desirable. However, relying solely on LLMs and prompt engineering makes it challenging to extract precise and comprehensive *is-a* relationships from texts. In response to this limitation, this paper aims to explore a well-designed approach to guide LLMs for better construction of taxonomies from texts. On one hand, we propose a doubly-checked mechanism to improve the quality of candidate nodes generated from texts. On the other hand, we utilize a modularized Chain-of-Thought prompting technique to break down hypernym identification into several sub-problems and employ the beam search-based self-evaluation strategy to enhance reliability. Specifically, self-evaluation constraint factors are introduced to score the reliability of reasoning chains generated through beam search, whereby selecting the most reliable chain as the final judgment. Extensive experiments on three datasets from various domains verify that the proposed method significantly improves the performance of taxonomy induction from texts.

Keywords: Doubly-Checked Mechanism · Self-Evaluation Strategy · Taxonomy Induction · Large Language Models

1 Introduction

A taxonomy is a hierarchical directed acyclic graph that organizes concepts or entities by "hypernym-hyponym" or *is-a* relations (*e.g.,* "apple" *is-a* "fruit"). This

B. Xu et al. (Eds.): CCKS-IJCKG 2024, CCIS 2229, pp. 134–146, 2025.
https://doi.org/10.1007/978-981-96-1809-5_11

structure is crucial for downstream tasks such as text comprehension [18], personalized recommendations [2], and question answering [15]. To induce a taxonomy from texts, traditional methods mainly rely on manual construction or crowd-sourced top-down manner, which is time-consuming, labor-intensive and difficult to scale. With the advancement of deep learning, extensive studies have focused on the automated taxonomy induction using pattern-based methods [7], word embeddings [11], and pre-trained language models (PLMs) [6]. Unfortunately, these approaches suffer from the issue of low coverage and semantic accuracy, impacting the performance of downstream tasks [4]. Recently, Large Language Models (LLMs) have shown their extraordinary ability to understand and generate texts in natural language, making them feasible solutions to address the issues of low coverage and semantic accuracy in taxonomy induction from texts. Consequently, existing studies have made some efforts to build taxonomies by leveraging the internal knowledge of LLMs, *e.g.,* identifying hierarchical relationships between given concepts while adhering to structural constraints [3] or employing an ensemble-based ranking filter in in-context learning [19]. However, these methods still make an insufficient exploration of the context information for each node, causing them to fail in processing special terms reasonably (*e.g.,* long-tail distributed and emerging terms) and ensuring the reliability of LLMs' response. These issues inevitably lead to poor accuracy in the taxonomy induction, impacting the performance of downstream applications.

To resolve these issues, this paper proposes a novel LLMs-based taxonomy induction framework, composed of two crucial steps, *i.e.,* the candidate nodes generation and the hypernym-hyponym relation detection. During the first step, there are typically special terms in the text to be extracted, requiring a high level of proficiency in contextual comprehension. Thus, we leverage the in-context learning technique combined with a doubly-checked mechanism to generate candidate terms from texts, thereby addressing issues related to these types of terms. Specifically, in-context learning provides a few examples of expected outputs for LLM, aiming to improve the performance of generating candidate nodes at a lower cost. Furthermore, to generate candidate nodes with greater accuracy, the doubly-checked mechanism is implemented through a multiple-choice prompt.

After obtaining the candidate terms, the second step aims to infer the hypernym-hyponym relationship between the two terms with the crucial operation being also understanding the contextual information. A beam search-based decoding strategy is employed to improve the stability and reasonableness of LLMs' predictions. Besides, a self-consistent evaluation method is further used to attain a state of reliability systematically. We regard the prediction probability of the model's self-evaluation at each step as the constraint factor for scoring each reasoning chain generated by beam search. With the synergy of these two strategies, the proposed LLMs-based framework could dramatically improve the taxonomy induction task. To verify its effectiveness, we conduct extensive experiments on WordNet sub-taxonomies [1] and two large-scale, real-world taxonomies [13]. Experimental results demonstrate that the proposed method gains a large margin (about 12.67% on WordNet dataset) performance improvement compared with the state-of-the-art baseline. To summarize, the contributions of this paper could be listed as follows:

- We propose a doubly-checked mechanism along with in-context learning to address the special-entity issue in candidate term generation.
- We introduce the beam search decoding approach combined with a self-consistent evaluation strategy into the hyponymy detection to obtain a more reliable reasoning chain and a better identification result.
- Extensive experiments on three datasets from various domains verify that the proposed method significantly improves the performance of taxonomy induction from texts.

2   Related Work

Taxonomy induction typically involves identifying hypernyms and organizing these relationships into a hierarchical structure. Initially, traditional methods such as pattern-based techniques [10,14,17] and embedding-based approaches [8] were widely used. Additionally, Mao et al. [9] utilized reinforcement learning to integrate the phases of hypernym identification and hypernym organization. DNG [20] analyzes the inheritance and supplementary between node features in taxonomies and refines the taxonomy construction on the non-Gaussian space. Several works treat this task as a graph optimization problem, solved using the maximum spanning tree algorithm [1]. Shang et al. [12] applied a graph neural network, demonstrating improvements in large-scale taxonomy induction.

More recent approaches leverage the capabilities of large language models (LLMs) for taxonomy induction. Chen et al. [4] constructed taxonomic trees by predicting parenthood relations and optimizing these predictions into a maximum-spanning tree using pre-trained language models. Jain et al. [6] utilized treating taxonomy induction as sequence classification and sequence scoring tasks. Zeng et al. [19] constructed taxonomies by iteratively selecting relevant candidate terms for each layer and reducing errors through the Ensemble-based Ranking Filter. Langlais and Guo [5] proposed an automatic taxonomy evaluation metric based on pre-trained models. TaxonomyGPT [3] conducted taxonomy induction by leveraging the in-context learning capabilities of LLMs. The proposed method in this paper significantly improves structural accuracy by prompting large language models in the Chain-of-Thought style.

3   Taxonomy Induction Based on LLMs

3.1   Overall Framework

In this section, we elaborate on the proposed framework designed to address the taxonomy induction task. As illustrated in Fig. 1, the task is divided into two main steps. First, candidate terms are generated through a doubly-checked mechanism from texts using LLMs, as described in Sect. 3.2. Next, in Sect. 3.3, relation detection is performed via LLMs using a specially designed beam search and self-evaluation strategy. Once the nodes and relations are obtained, a taxonomy related to the texts can be constructed using the maximum spanning tree (MST) algorithm [1].

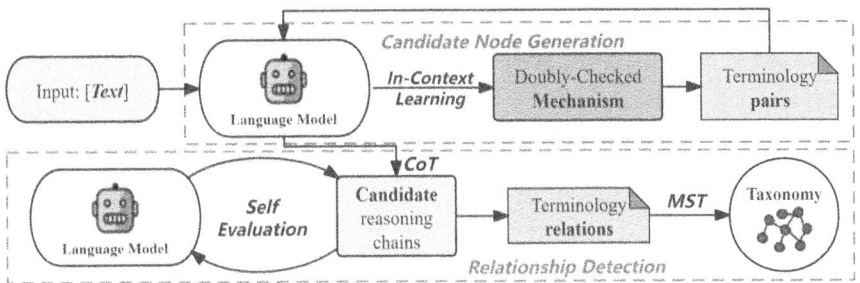

Fig. 1. Overall framework of the proposed method.

3.2   Doubly-Checked Mechanism for Candidate Node Generation

To enhance the generation of candidate terminology, we utilize in-context learning-based prompts. By providing a small number of examples, we can better guide the LLM in generating relevant terms in new scenario.

Long-tail distributed terms and emerging terms are two types of special terms in NLP. Typically, the issue of special terms is challenging for language models. For example, the meanings of some long-tail distributed terms are not directly observable, such as "h450". Besides, LLMs sometimes generate "single precision floating point format" in mistake instead of the ground truth "floating point". This may be attributed to their failure to grasp the crucial information of input texts. As these challenges demand great ability in contextual comprehension, a mechanism that facilitates fully utilizing textual information is an effective solution.

Inspired by Kozareva et al. [7], we design a doubly-checked mechanism for candidate node generation via in-context learning prompting to refine results gradually. As shown in Fig. 2, for a given text, an initial candidate term t_1 is first generated by the LLM using in-context learning prompts. To address the issue mentioned above, the candidate term t_1 undergoes a doubly-checked mechanism through a multiple-choice prompt. The final term is obtained after this verification mechanism.

3.3   Self-evaluation Strategy for Relation Detection

To enhance the reliability and stability of reasoning chains generated by LLMs, we employ the beam search decoding approach combined with a self-evaluation strategy to detect relationships between candidate term nodes. This approach allows us to generate multiple reasoning chains and score them accordingly.

Doubly-Checked-Mechanism-based Method

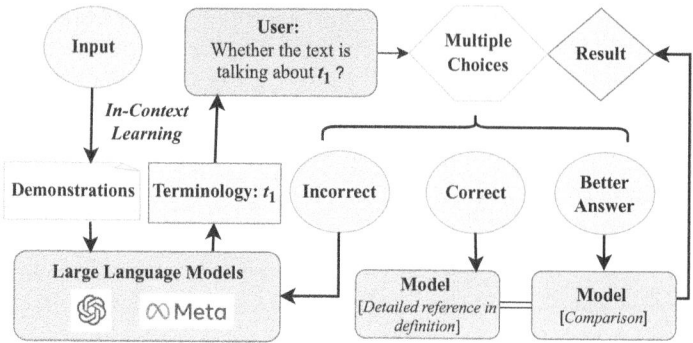

Fig. 2. Doubly-checked mechanism-based method for candidate node generation via in-context learning prompting.

CoT-Based Relation Detection. Relation detection via LLM adopts a Modula-rized-style CoT prompting in this paper, decomposing the hierarchical relationship detection task into a series of sub-problems. Given the input $(term_1, I(term_1); term_2, I(term_2))$, in which each of the terms is paired with its respective information $I(\cdot)$ in context, the LLM is asked to judge hierarchical relationships between two terms. However, we find that reasoning about their relationships requires more direct and concise informational text. Otherwise, the LLM's judgment will be affected by redundant information. Therefore, based on experience, this problem is decomposed into three sub-problems where CoT is used by guiding a step-by-step logical process:

1. Conclude the information of $term_1$ from the text;
2. Conclude the information of $term_2$ from the text;
3. Judge the hierarchical relationship between $term_1$ and $term_2$ based on the summarized information.

The Chain-of-Thought (CoT) prompting technique has been proven to be highly effective in reasoning tasks with LLMs [16]. However, previous works mostly use a single reasoning sample generated by LLMs to derive results, which causes the generated reasoning chains unreliable and can even lead to incorrect answers if one of the steps is mistaken. To enhance the model's prediction accuracy, this paper also introduces the beam search decoding approach.

Assuming the output examples for the first and second steps are s_{sum}, and s_{judge} for the third step, the outputs of the model for each step are r^1, r^2, r^3, the prompts are denoted as follows: $p^i = Prompt(term_i, I(term_i), s_{sum})(i = 1, 2), p^3 = Prompt(r^1, r^2, s_{judge})$.

In the specific task studied in this paper, the summarized information of $term_1$ and $term_2$ does not have a cause-and-effect relationship logically, but the summarized information of $term_1$ may influence the output of the summarized information of $term_2$. Therefore, we swap the order of p^1 and p^2 and

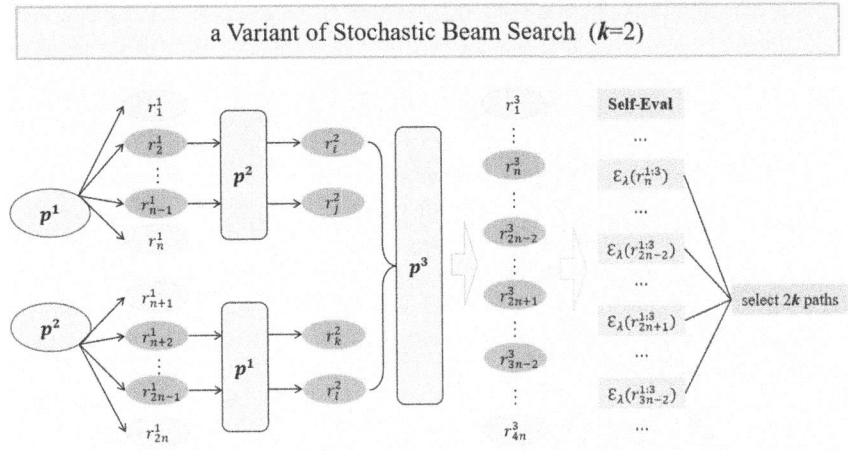

Fig. 3. Proposed variant of Stochastic Beam Search.

perform the reasoning again, applying beam search to both the original and reversed sequences. After self-evaluation, the system ultimately generates $2k$ paths. Figure 3 shows a variant of the beam search method ($k = 2$) designed for the specifics of this task, where k is an important parameter in beam search to determine the number of candidate sequences retained at each search step. Given that the information generated by LLMs may be unreliable, we consider a constraint factor $C(\cdot) \in [0,1]$ to obtain higher quality and more reliable reasoning chains. The subsequent sections provide an in-depth analysis of the multi-step reasoning process via beam search, as well as a detailed exposition of the constraint factors.

Multi-step Reasoning via Stochastic Beam Search. When multiple-step reasoning with LLM, a T-step reasoning chain, donated as R, is composed of a series of output tokens. In this study, we set $T = 3$ and $R = [r^1, r^2, r^3] = r^{1:3}$. Formally, we can factorize the generation reasoning process $P(R = r^{1:T}|p^{1:T})$ in an auto-regressive manner:

$$P(R = r^{1:T}|p^{1:T}) = \prod_t P(r^t|p^t, r^{1:t-1}) \tag{1}$$

The decomposition of the reasoning process allows it to be viewed as a step-by-step decoding problem. One of the most serious issues in LLM-based reasoning is the potential unreliability and inaccuracy of each reasoning step generated by the model. Furthermore, this potential error can accumulate iteratively as the reasoning progresses. To mitigate the impact of this issue, this paper considers a self-evaluation-based constraint function at each reasoning step. This function, based on prompts and previous input-output texts of the current step, represents the reliability of the output at that step. Therefore, for the hierarchical relationship detection task, we can derive a decoding strategy that combines the probabilities from the large language model with the reliability probabilities, resulting in a new objective function $\Sigma_\lambda(R = r^{1:T})$:

$$\Sigma_\lambda(R = r^{1:T}) = \prod_t P_{LM}^\lambda(r^t|p^t, r^{1:t-1})C^{1-\lambda}(r^t|p^t, r^{1:t-1}) \qquad (2)$$

where P_{LM} is the language model distribution, $\lambda \in [0,1]$ is a weight parameter used to balance the model prediction score and the faithfulness score. To obtain a high-quality inference chain with a higher score of $\Sigma_\lambda(R = r^{1:T})$, it is natural to leverage greedy or stochastic beam decoding to approximate the reasoning chain with the maximum $\Sigma_\lambda(R = r^{1:T})$

Constraint Factor as Self-evaluation. We use LLMs to judge the faithfulness of r^t based on $r^{1:t-1}$. Specifically, the evaluation and generation models use the same back-end LLM with different prompts. We design a multiple-choice questioning for refinement of model predictions, denoted as $Prompt_C$, where the token-level probability of the option "*rational*" is adopted to calculate the faithfulness score:

$$C(R = r^t) = P_{LM}(A|Prompt_C, p^{1:t}, r^{1:t}) \qquad (3)$$

4   Experiments

We evaluate the proposed taxonomy induction framework on four real-world benchmarks. To outline the experiments conducted in our paper, we raise three primary research questions that require resolution:

RQ1: How does the proposed method perform in taxonomy induction compared to state-of-the-art baselines?

RQ2: How does the doubly-checked mechanism affect the performance of the proposed method in taxonomy induction?

RQ3: What significant impact does the beam search-based self-evaluation strategy have on the effectiveness of the taxonomy induction task?

4.1   Experimental Settings

Datasets. We conducted our experiments on WordNet sub-taxonomy [1], DBLP and SemEval-Sci [13]. Specifically, WordNet consists of 761 non-overlapping taxonomies, each containing terms ranging from 11 to 50. DBLP is constructed from 156,000 paper abstracts within the field of computer science, with 176 concepts and 175 edges arranged in a 4-depth taxonomy. SemEval2016-Sci, presented as an 8-depth taxonomy with 429 concepts and 451 edges, is derived from a shared task of taxonomy induction in SemEval-2016.

Table 1. Performance on taxonomy induction on Ancestor-metrics.

Model	WordNet			DBLP			SemEval-Sci		
	P_a	R_a	$F1_a$	P_a	R_a	$F1_a$	P_a	R_a	$F1_a$
Graph2Taxo	<u>79.20</u>	47.80	59.60	47.85	30.23	37.05	82.45	36.15	50.27
CTP	69.30	**66.20**	<u>66.70</u>	45.62	41.39	43.40	52.41	33.88	41.16
RestrictMLM	23.23	25.69	24.09	-	-	-	63.33	<u>47.85</u>	<u>54.44</u>
LMScorer	37.50	47.64	41.59	17.14	21.54	19.04	48.80	33.24	39.51
TaxonomyGPT	62.97	41.77	48.95	28.98	14.40	17.15	53.09	31.84	39.07
Ours(LLaMA-2)	62.17	58.22	60.13	<u>76.50</u>	<u>71.44</u>	<u>73.88</u>	70.79	43.91	54.20
Ours(GPT-3.5)	**86.91**	<u>63.60</u>	**73.45**	**84.51**	**84.61**	**84.56**	<u>78.26</u>	**70.89**	**74.39**

Baseline Methods. We compare our proposed framework with the following supervised fine-tuning baseline methods:

- Graph2Taxo [12] leverages cross-domain graph structures and adopts constraint-based Directed Acyclic Graph (DAG) learning for taxonomy induction.
- CTP [4] fine-tunes the RoBERTa model to predict the probability of parent-child pair and integrates it into a graph using a maximum spanning tree algorithm for precise taxonomy induction.

Additionally, we also adopt the following unsupervised and in-context learning baseline methods for a comprehensive comparison:

- RestrictMLM [6] employs a "fill-in-the-blank" approach based on cloze statements to extract *is-a* relational knowledge from BERT.
- LMScore [6] treats taxonomy induction as a sentence scoring task using GPT-2, assessing the natural fluency of sentences.
- TaxonomyGPT [3] treats the taxonomy induction as a conditional text generation challenge. It represents the output taxonomy as a collection of sentences, each describing a parent-child relation within the output taxonomy.

For our proposed framework, we conduct experiments with GPT-3.5-turbo and LLaMA-2-7b-chat.

Evaluation Metrics. To evaluate the performance of all compared models, we adopt six evaluation metrics, *i.e.*, Ancestor-Precision, Ancestor-Recall, Ancestor-F1, Edge-Precision, Edge-Recall and Edge-F1 donated as P_a, R_a, $F1_a$, P_e, R_e and $F1_e$ respectively.

Ancestor-metrics compare the ancestor-descendant relations in the predicted taxonomy with those in the ground truth taxonomy. Specifically, if "*Anc*" donates the set of term pairs that have an "*is − ancestor*" relationship between them, we have:

$$P_a = \frac{|Anc_{pred} \cap Anc_{gold}|}{|Anc_{pred}|}, R_a = \frac{|Anc_{pred} \cap Anc_{gold}|}{|Anc_{gold}|}, F1_a = \frac{2P_a \cdot R_a}{P_a + R_a}. \quad (4)$$

Edge-metrics are more stringent compared to Ancestor-metrics. They evaluate the exactness of the predicted taxonomy by directly comparing the predicted edges with the gold standard edges. Similarly, if "E" represents the edge set of a taxonomy, we have:

$$P_e = \frac{|E_{pred} \cap E_{gold}|}{|E_{pred}|}, R_e = \frac{|E_{pred} \cap E_{gold}|}{|E_{gold}|}, F1_e = \frac{2P_e \cdot R_e}{P_e + R_e}. \tag{5}$$

Table 2. Performance on taxonomy induction on Edge-metrics.

Model	WordNet			DBLP			SemEval-Sci		
	P_e	R_e	$F1_e$	P_e	R_e	$F1_e$	P_e	R_e	$F1_e$
Graph2Taxo	**75.60**	37.00	49.70	46.63	28.49	35.37	**79.37**	34.52	46.87
CTP	53.30	49.80	51.50	38.21	33.73	35.83	31.18	29.42	30.27
RestrictMLM	24.17	25.65	24.89	-	-	-	45.79	46.19	45.99
LMScorer	36.27	38.48	37.34	25.84	26.12	25.98	42.20	42.58	42.39
TaxonomyGPT	49.20	43.85	46.24	34.27	22.17	25.97	39.59	36.84	38.01
Ours(LLaMA-2)	59.63	**60.34**	59.98	66.13	55.02	60.07	53.20	40.71	46.12
Ours(GPT-3.5)	73.14	54.36	**62.37**	**68.13**	**67.55**	**67.84**	65.90	**54.16**	**59.46**

4.2 Main Results (RQ1)

In our experiments, we compare the performance of our proposed method with five baseline methods on WordNet, DBLP and SemEval-Sci. From the experimental results shown in Table 1 and Table 2, we have three major observations. First, the proposed method in this paper based on GPT-3.5 and Llama-2 outperforms the baselines on most evaluation metrics across the three datasets. They notably outperformed LMScorer by a significant margin. This observation underscores the great potential of leveraging LLMs endowed with exceptional text comprehension and generation capabilities for taxonomy induction.

Second, we find that even inducing taxonomy via prompting powerful LLMs, TaxonomyGPT(GPT-3.5) shows worse performance than methods such as CTP, which relies on fine-tuning BERT models, across all six metrics. This finding indicates that LLMs are sensitive to the way the prompt is designed for the specific task.

Finally, Graph2Taxo demonstrates quite high precision among the methods, showcasing its strength in utilizing lexical patterns as direct input features. However, its relatively lower recall reveals a significant trade-off, suggesting that although it excels in precision, it may not fully capture the complete range of taxonomic relations.

4.3   Effects of Doubly-Checked Mechanism (RQ2)

Taking GPT-3.5-turbo as an example, we conduct an ablation study to further verify the effectiveness of the doubly-checked mechanism. Compared with the results of the doubly-checked-free method displayed in Table 3, we find a steady improvement of all the evaluation metrics on three datasets by introducing the doubly-checked mechanism. Especially, when the doubly-checked mechanism was removed, the model's performance on the Edge-Recall metric significantly declined, even falling below that of some baselines. This finding indicates that the doubly-checked mechanism ensures a high-quality and high-coverage terminology generation for the following steps in the proposed framework.

Table 3. Effects of Doubly-checked Mechanism on taxonomy induction.

Datasets	Doubly-checked	Edge			Ancestor		
		P_e	R_e	$F1_e$	P_a	R_a	$F1_a$
WordNet	✗	68.53	37.10	40.14	81.24	57.88	67.60
	✓	73.14	54.36	62.37	86.91	63.60	73.45
DBLP	✗	67.14	52.93	59.19	81.88	75.35	78.48
	✓	68.13	67.55	67.84	84.51	84.61	84.56
SemEval-Sci	✗	62.12	23.77	34.38	72.51	70.49	71.49
	✓	65.90	54.16	59.46	78.26	70.89	74.39

4.4   Effects of Self-evaluation Strategy (RQ3)

Additionally, we conduct an ablation study on the three benchmarks to probe "how does Self-Evaluation Strategy affects the performance of the proposed method in taxonomy induction". Taking gpt-3.5-turbo as an example, Table 4 demonstrates the statistics of the ablation experiment, where a Self-Evaluation-free method is set to use zero-shot prompting during the relationship detection phase. As shown in the table, the application of the Self-Evaluation strategy improves $F1_e$ and $F1_a$ of the task on all three datasets. Specifically, "Precision" benefits a lot from the more stringent strategy, where results are derived from a carefully selected reasoning chain through self-evaluation.

4.5   Case Study

This section utilizes a small-scale sample taxonomy from SemEval-Sci to briefly demonstrate the main experiment and the ablation experiments. Figure 4 displays four taxonomies: the first is the ground truth taxonomy, and the remaining three are generated by the doubly-checked-free method, the self-evaluation-free method, and the proposed method in this paper, respectively. Blue edges or

Table 4. Effects of Self-Evaluation Strategy on taxonomy induction.

Datasets	Self-Evaluation	Edge			Ancestor		
		P_e	R_e	$F1_e$	P_a	R_a	$F1_a$
WordNet	✗	58.33	50.47	51.12	78.64	51.93	62.55
	✓	73.14	54.36	62.37	86.91	63.60	73.45
DBLP	✗	66.12	58.56	62.11	65.68	72.40	68.88
	✓	68.13	67.55	67.84	84.51	84.61	84.56
SemEval-Sci	✗	67.70	61.25	64.31	70.01	76.19	72.97
	✓	65.90	54.16	59.46	78.26	70.89	74.39

nodes in the ground truth taxonomy may be omitted in the output taxonomies, while red ones indicate redundancies compared to the ground truth.

In the small-scale sample, the proposed method performs well, with only one extra edge in the output taxonomy compared to the ground truth. In the ablation experiments, the taxonomy induced by the proposed method covers more terminology nodes and correctly detects more relations.

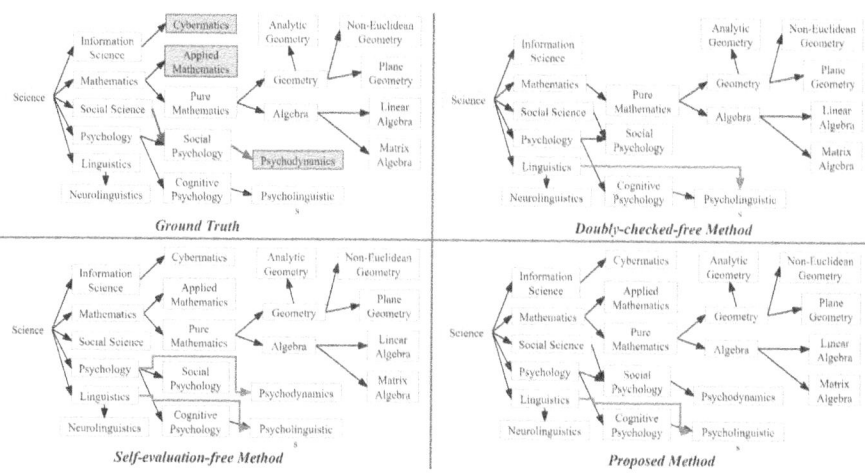

Fig. 4. Ground truth and output taxonomies on a sample of the SemEval-Sci dataset.

5   Conclusion and Future Work

In this paper, we proposed an enhanced framework for taxonomy induction using large language models. By integrating a doubly-checked mechanism and a self-evaluation strategy, our method could effectively capture the context information during the taxonomy induction process. The experimental results indicated that the superiority of our approach in improving the structural accuracy. Additionally, ablation experiments demonstrated the effectiveness and potential of the two innovations in the taxonomy induction task.

In the future, we will focus on refining the mechanism and strategies and exploring their applicability to more domains and larger datasets. Additionally, considering the integration of other advanced techniques, such as structural information-based methods, could improve the efficiency of the taxonomy induction task using our proposed method.

References

1. Bansal, M., Burkett, D., de Melo, G., Klein, D.: Structured learning for taxonomy induction with belief propagation. In: Proceedings of ACL 2014, pp. 1041–1051 (2014)
2. Cao, Y., Wang, X., He, X., Hu, Z., Chua, T.S.: Unifying knowledge graph learning and recommendation: towards a better understanding of user preferences. In: Proceedings of WWW 2019, pp. 151–161 (2019)
3. Chen, B., Yi, F., Varró, D.: Prompting or fine-tuning? A comparative study of large language models for taxonomy construction. In: Proceedings of MODELS 2023, pp. 588–596 (2023)
4. Chen, C., Lin, K., Klein, D.: Constructing taxonomies from pretrained language models. In: Proceedings of NAACL-HLT 2021, pp. 4687–4700 (2021)
5. Gao, T., Langlais, P.: Rate: a reproducible automatic taxonomy evaluation by filling the gap. CoRR abs/2307.09706 (2023)
6. Jain, D., Anke, L.E.: Distilling hypernymy relations from language models: on the effectiveness of zero-shot taxonomy induction. In: Proceedings of *SEM 2022, pp. 151–156 (2022)
7. Kozareva, Z., Riloff, E., Hovy, E.H.: Semantic class learning from the web with hyponym pattern linkage graphs. In: Proceedings of ACL 2008, pp. 1048–1056 (2008)
8. Luu, A.T., Tay, Y., Hui, S.C., Ng, S.K.: Learning term embeddings for taxonomic relation identification using dynamic weighting neural network. In: Proceedings of EMNLP 2016, pp. 403–413 (2016)
9. Mao, Y., Ren, X., Shen, J., Gu, X., Han, J.: End-to-end reinforcement learning for automatic taxonomy induction. In: ACL, pp. 2462–2472 (2018)
10. Nakashole, N., Weikum, G., Suchanek, F.M.: PATTY: a taxonomy of relational patterns with semantic types. In: Proceedings of EMNLP-CoNLL 2012, pp. 1135–1145 (2012)
11. Ristoski, P., Faralli, S., Ponzetto, S.P., Paulheim, H.: Large-scale taxonomy induction using entity and word embeddings. CoRR abs/2105.01305 (2021)
12. Shang, C., Dash, S., Chowdhury, M.F.M., Mihindukulasooriya, N., Gliozzo, A.: Taxonomy construction of unseen domains via graph-based cross-domain knowledge transfer. In: Proceedings of ACL 2020, pp. 2198–2208 (2020)
13. Shen, J., Wu, Z., Lei, D., Shang, J., Ren, X., Han, J.: Setexpan: corpus-based set expansion via context feature selection and rank ensemble. CoRR abs/1910.08192 (2019)
14. Snow, R., Jurafsky, D., Ng, A.Y.: Learning syntactic patterns for automatic hypernym discovery. In: Proceedings of NIPS 2004, pp. 1297–1304 (2004)

15. Wang, Y., Lipka, N., Rossi, R.A., Siu, A., Zhang, R., Derr, T.: Knowledge graph prompting for multi-document question answering. In: Proceedings of AAAI 2024, vol. 38, no. 17, pp. 19206–19214 (2024)
16. Wei, J., et al.: Chain-of-thought prompting elicits reasoning in large language models. In: Proceedings of NIPS 2022 (2022)
17. Wu, W., Li, H., Wang, H., Zhu, K.Q.: Probase: a probabilistic taxonomy for text understanding. In: Proceedings of SIGMOD 2012, pp. 481–492 (2012)
18. Yu, D., Zhu, C., Yang, Y., Zeng, M.: Jaket: joint pre-training of knowledge graph and language understanding. In: Proceedings of AAAI 2022, vol. 36, no. 10, pp. 11630–11638 (2022)
19. Zeng, Q., et al.: Chain-of-layer: iteratively prompting large language models for taxonomy induction from limited examples. CoRR abs/2402.07386 (2024)
20. Zhai, S., Wang, W., Li, Y., Meng, Y.: DNG: taxonomy expansion by exploring the intrinsic directed structure on non-gaussian space. In: Proceedings of AAAI 2023, pp. 6593–6601 (2023)

Graph Database and Knowledge Management

Relation Inquiry: A Novel Synchronous Joint Extractor for Entities and Relations

Wei Tang[1], Zexin Wang[2], Xun Mao[1], Kai Lv[1], Xu Han[3], Ming Liu[2(✉)], Wei Wang[4], and Bing Qin[2]

[1] State Grid Anhui Electric Power Research Institute, Hefei, China
[2] Research Center for Social Computing and Information Retrieval, Harbin Institute of Technology, Harbin, China
{zxwang,mliu}@ir.hit.edu.cn
[3] HeiLongJiang Agricultural Engineering Vocational College, Harbin, China
[4] State Grid Anhui Electric Power Co., Ltd., Hefei, China

Abstract. Existing entities and relations joint extraction methods mostly adopt an ***asynchrony*** framework, extracting entities and relations at the different time. Such asynchronous joint paradigm suffers from several issues: noisy intermediate redundant information induced, limited interaction among components, and exposure bias from training to inference. ***Synchronous*** joint extraction framework has no these issues, however, it has been challenged by overlapping relational triples problem, the main bottleneck preventing the widespread use of synchronous joint paradigms in community. We present a strategy of ***relation inquiry*** to break the impasse, empowering synchrony framework: on the ability of extracting overlapping triples, and more sufficient interaction learning for entities and relations models, producing few intermediate redundant information. Under relation inquiry strategy, original joint extraction problem is decomposed into four subtasks, leading to a novel synchronous joint extractor for entities and relations. Experiments on three public datasets demonstrate the effectiveness and robustness of our proposed synchronous joint extractor, outperforming all the classic baselines on different text domains and relation schema.

Keywords: information extraction · knowledge acquisition · sequence labeling · natural language processing

1 Introduction

Joint extraction framework, extracting entities and relations using a single model, has demonstrated more advantage over pipelined paradigm [5]. Many existing joint extractors follow an ***asynchronous*** procedure to extract entities and relations [2,3,9,10]: identify entities first, then conduct relation extraction over all possible entity pairs. However, in such asynchrony paradigm: *(a)* considerable redundant intermediate information (e.g., candidate non-relational entity

---

W. Tang and Z. Wang—Contribute equally to this work.

B. Xu et al. (Eds.): CCKS-IJCKG 2024, CCIS 2229, pp. 149–161, 2025.
https://doi.org/10.1007/978-981-96-1809-5_12

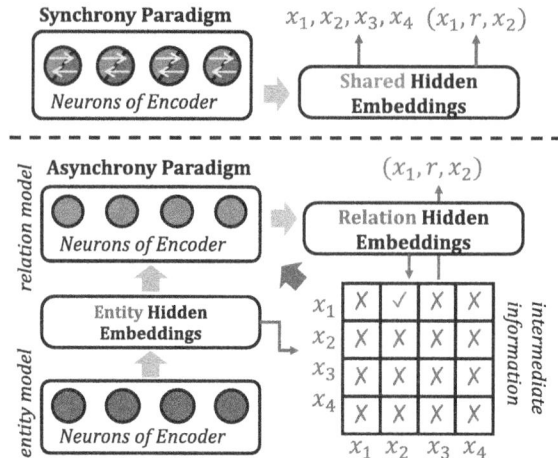

Fig. 1. Synchronous joint extraction framework takes one-step modeling to interactively extract entities and relations, without inducing any redundant information or any exposure bias in inference, compared with asynchrony framework.

pairs) is induced as noise; *(b)* insufficient interactive learning is delivered from the sequential modeling process; *(c)* exposure bias occurs when the gold entities are provided as specific inputs to guide relation extraction during training, while only unlabeled data is available during the inference stage.

Synchronous joint extraction framework, efficient one-step modeling to extract entities and relations at the same time, is supposed to avoid the thorny issues encountered in asynchrony paradigm (Fig. 1 shows comparison on two paradigms). However, previous synchronous extraction models have been questioned about: *(a)* How to accurately extract overlapping triples, i.e., relations that share entities? [13]; *(b)* How to design synchronous joint extractor to realize totally interactive entities and relations extraction? [12]. We argue that the crux of two questions lies in *relation modeling*.

Relation, as an essential role in joint extraction, is desired to be fully explored. We present *relation inquiry*, a strategy inspired from machine reading comprehension [4]. Specifically, we first use simple natural language to describe relation types, e.g., *"person temporarily located at or near to a place"* for expressing relation "PHYS" in ACE05 schema. Then we deliver the relation description as inquiry to the model along with the given sentence: only one relation performs as extraction query at a time, avoiding overlapping triples problem naturally. Original joint extraction is decomposed into four sub-**T**asks upon relation inquiry strategy: (**T1**) relation discrimination (RD), asking whether the input relation exists in the given sentence; (**T2**) named entity recognition (NER); (**T3**) relational head entity detection (HED), detecting head (subject) entities for the input relation; (**T4**) relational tail (object) entity detection (TED). They complement each other in entity and relation modeling and do not involve any inter-dependent extraction steps. By sharing an encoder for these subtasks, we construct a novel synchronous joint extractor that facilitates effective interactive learning between entity and relation models. Experimental results also demonstrate the effectiveness of the synchrony mechanism and the robustness of the model when extracting overlapping triples.

2   Related Work

Here we survey task of joint extraction for entities and relations. In early stage, pipelined approach, i.e., first recognizing entities and then classifying relations, makes a system comparatively easy to assemble [7], however, it prohibits the interactions between components and brings error propagation. Pipelined approach over-simplifies the problem as sequential local classification steps, ignoring long-distance and cross-task dependencies. Many joint extraction methods are successively proposed, modeling the interactions between NER and relation classification in latent space. Zhang et al. [12] framed joint extraction problem as a table-filling problem: diagonal in table for entity recognition, and other areas for relation classification. Li et al. [4] converted the joint extraction task into a multi-turn machine reading comprehension problem. Luan et al. [5] enumerated text spans as candidate entities, and modeling each pair of candidate entity spans for relation classification.

However, present joint extraction methods are mainly affected by the pipelined idea: relation extraction should depend on the entity recognition [2,3,6]. Though they can better modeling interactions between modules compared with pipelined approaches, extraction process of entities and relations is *asynchronous*, not *synchronous*. Asynchronous extraction prevents the subsequent extraction results (i.e., classified relations) from fully affecting the previous extraction results (i.e., recognized entities) and getting concrete feedbacks from them in time, so the interaction modeling is naturally limited. Zheng et al. [13] designed a novel extraction tagging schema, making the extraction of entities and relations truly synchronous. However, the proposed model could not figure out the situation when relational triplets have overlaps in a sentence. Some work utilized text generation techniques, such as copy networks, to dynamically generate relational triples [10]; some work propose to extract head entities first and then extract corresponding tail entities along with their relations [1,9,11]. However, these methods ignored the recognition of entities that have no relations, and most of them have high time complexity due to the enumeration of head entities and are still asynchronous extracting.

3   Synchronous Joint Extractor

Entities and Relation Extraction. Given *sentence* $s = [w_1, w_2, \cdots, w_N]$ and set of *entity types* $\mathcal{T}^{\mathcal{E}}$ and *relation types* $\mathcal{T}^{\mathcal{R}}$, where w_i represents the i-th token. The expected outputs are as follows. (1) List of *entity mentions*: $E = [x_1, x_2, \cdots, x_n]$, and their types $T^e = [t^{x_1}, t^{x_2}, \cdots, t^{x_n}], \forall t^{x_i} \in \mathcal{T}^{\mathcal{E}}$. (2) List

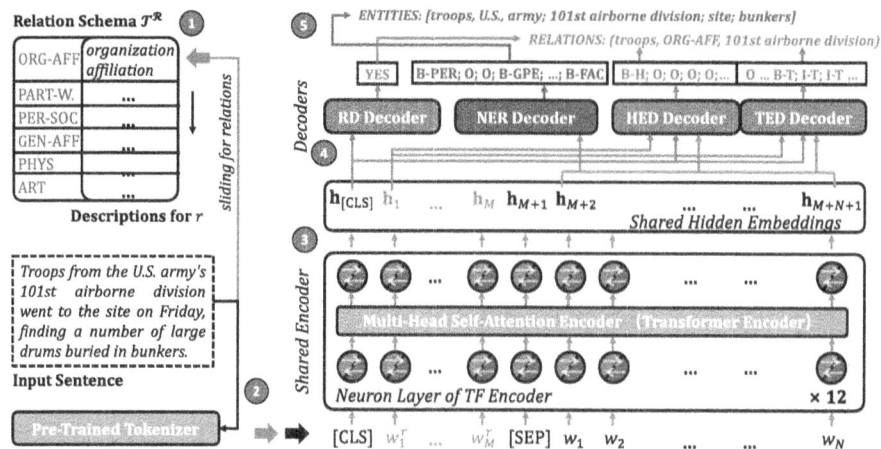

Fig. 2. Our proposed synchronous joint extractor for entities and relations, where we use orange and blue color to represent relation and entity semantic related information, respectively. ***Relation Inquiry***: (1) sliding over schema to select one relation as target in each time, e.g., currently the model selects relation of "ORG-AFF" as shown. (2) Obtaining token sequences of sentence and relation, respectively, and taking the concatenation of them as model inputs. ***Synchronous Joint Modeling***: (3) Applying the multi-layered transformer encoders for semantic of inputs, and the parameters of model are expected to learn both of entity and relation semantic from inputs. (4) The designed four subtasks share encoders, while using separate decoder to get concrete target from the common hidden embeddings. (5) Final joint extraction results, i.e., entities and relation triples, can be easily obtained from union of results in each subtasks. (Color figure online)

of ***relational triples*** $[(x_{i_1}, r_1, x_{j_1}), \cdots, (x_{i_m}, r_m, x_{j_m})], \forall r_k \in \mathcal{T}^{\mathcal{R}}$, where x_{i_k}, x_{j_k} are head/tail entities satisfying that $\forall x_{i_k}, x_{j_k} \in E$.

Our proposed ***synchronous joint extractor*** takes relation r and sentence s, in token sequence form, as inputs upon relation inquiry strategy. The extractor uses encoder-decoder model inside: first, encode global semantic for input sequence, then decode task-specific representations for extractions. There are one shared encoder and four task-related decoders upon feed-forward networks. The encoder shares hidden representations to down-streams, while decoders synchronously deliver task-specific embeddings without inter-dependency. We visualize synchronous joint extractor in Fig. 2.

3.1   Shared Multi-head Self-attention Encoder

We utilize the renowned BERT as the shared encoder for subtasks. The input of encoder is a concatenation of a candidate relation $r = [w_1^r, w_2^r, \cdots, w_M^r]$ and the given sentence $s = [w_1, w_2, \cdots, w_N]$, where the relation refer to the corresponding description of natural language. The relation descriptions are handcrafted by experts for each relation type in $\mathcal{T}^{\mathcal{R}}$. The formal input to the encoder is constructed as "[CLS] + r + [SEP] + s":

$$[[\text{CLS}], w_1^r, w_2^r, \cdots, w_M^r, [\text{SEP}], w_1, w_2, \cdots, w_N] \tag{1}$$

where "[CLS]" and "[SEP]" are reserved special token in BERT. The output sequence, i.e., hidden embeddings, for each token is denoted as follows,

$$[\mathbf{h}_{[CLS]}, \mathbf{h}_1, \cdots, \mathbf{h}_M, \mathbf{h}_{M+1}, \mathbf{h}_{M+2}, \cdots, \mathbf{h}_{M+1+N}] \tag{2}$$

where $\mathbf{h}_i \in \mathbb{R}^d$, d is a hyper-parameters for vector dimension, and the length of the sequence is $M + N + 2$. We denote the output sequence as $\mathbf{H}$ for further use to down-streams.

3.2   Relation Discrimination Decoder

We use the final hidden vector $\mathbf{h}_{[CLS]} \in \mathbb{R}^d$ corresponding to the first input token ([CLS]) as the aggregate representation of the sentence. Decoding with a feed-forward neural network,

$$\mathbf{o}^r = \text{ReLU}(\mathbf{h}_{[CLS]} \cdot \mathbf{W}^{RD} + \mathbf{b}^{RD}) \tag{3}$$

where $\mathbf{W}^{RD} \in \mathbb{R}^{d \times 2}$ is a linear mapping matrix and $\mathbf{b}^{RD} \in \mathbb{R}^2$ is a bias vector.

$$\mathbf{p}^r = \text{softmax}(\mathbf{o}^r) \tag{4}$$

where $\mathbf{p}^r \in \mathbb{R}^2$. And in prediction stage, we can obtain the final binary output value $\hat{y}^r$,

$$\hat{y}^r = \text{argmax } \mathbf{p}^r \tag{5}$$

where $\hat{y}^r \in \{0, 1\}$ and $y^r = 1$ indicates that at least one fact of r has been expressed in s, otherwise not.

3.3   Named Entity Recognition Decoder

The inputs consist of the sentence's hidden embeddings, which are generated as outputs by the shared encoder, i.e., $[\mathbf{h}_{M+2}, \cdots, \mathbf{h}_{M+1+N}]$ in $\mathbf{H}$, denoted as $\mathbf{H}^s$, $\mathbf{H}^s \in \mathbb{R}^{N \times d}$. And the expected output is a tag sequence $y^e = [y_1^e, y_2^e, \cdots, y_N^e]$. A list of entity mentions $E = [x_1, x_2, \cdots, x_n]$ is obtained, along with corresponding types: $T^e = [t^{x_1}, t^{x_2}, \cdots, t^{x_n}]$. The NER decoder assigns labels to the sequence using feed-forward neural networks, based on $\mathbf{H}^s$,

$$\mathbf{O}^e = \text{ReLU}(\mathbf{H}^s \cdot \mathbf{W}^{NER} + \mathbf{b}^{NER}) \tag{6}$$

where $\mathbf{W}^{NER} \in \mathbb{R}^{d \times (2*|\mathcal{T}^{\mathcal{E}}|+1)}$ is a linear mapping matrix and $\mathbf{b}^{NER} \in \mathbb{R}^{2*|\mathcal{T}^{\mathcal{E}}|+1}$ is a bias vector. "B/I/O" based tag schema is utilized to label the tokens, so there are $2 * |\mathcal{T}^{\mathcal{E}}|$ "B/I" positive tags and "O" tag, leading to $2 * |\mathcal{T}^{\mathcal{E}}| + 1$ tags.

Considering the decoder output $\mathbf{O}^e \in \mathbb{R}^{N \times (2*|\mathcal{T}^{\mathcal{E}}|+1)}$, the final softmax layer computes normalized tag probabilities based on $\mathbf{O}^e$,

$$\mathbf{p}^e = \text{softmax}(\mathbf{O}^e) \tag{7}$$

where $\mathbf{p}^e \in \mathbb{R}^{N \times (2*|\mathcal{T}^{\mathcal{E}}|+1)}$ is the normalized tag probabilities for tokens in sentence s and,

$$\mathbf{p}^e_{i,j} = \frac{\exp(\mathbf{O}^e_{i,j})}{\sum_{k=1}^{2*|\mathcal{T}^{\mathcal{E}}|+1} \exp(\mathbf{O}^e_{i,k})} \tag{8}$$

In prediction stage, we can obtain the final NER tag output $\hat{y}^e$,

$$\hat{y}^e = [\hat{y}^e_1, \hat{y}^e_2, \cdots, \hat{y}^e_N] \tag{9}$$

$$\hat{y}^e_i = \text{argmax } \mathbf{p}^e_i, i = 1, \cdots, N \tag{10}$$

where $\hat{y}^e_i \in \{0, 1, \cdots, 2*|\mathcal{T}^{\mathcal{E}}|\}$.

3.4   Relational Entities Detection Decoder

Head entity detection (HED) decoder assigns a "B/I/O" tag to each token, to detect the head entity spans in sentence s. For each w_i in s, the decoder takes $\mathbf{h}_{[CLS]}$, $\mathbf{h}_1$ and a window of hidden embedding $[\mathbf{h}_{M+i}, \mathbf{h}_{M+i+1}, \mathbf{h}_{M+i+2}]$ as inputs. $\mathbf{h}_{[CLS]}$ provides indicative information about the relation r in s, and $\mathbf{h}_1$ is trained to encode semantic of r. Both of them are expected to provide a sense of relation to the current token. $\mathbf{h}_{M+i+1}$ is just the corresponding hidden embedding to token w_i. $\mathbf{h}_{M+i}$ and $\mathbf{h}_{M+i+2}$ are utilized to incorporate more global information based on N-gram idea, as they are neighbors of w_i. We concatenate these hidden embeddings as an intermediate embedding $\bar{\mathbf{h}}_i$,

$$\bar{\mathbf{h}}_i = [\mathbf{h}_{[CLS]}; \mathbf{h}_1; \mathbf{h}_{M+i}; \mathbf{h}_{M+i+1}; \mathbf{h}_{M+i+2}] \tag{11}$$

where $\bar{\mathbf{h}}_i \in \mathbb{R}^{5d}$, and for all the tokens in s we get $\bar{\mathbf{H}}^s$,

$$\bar{\mathbf{H}}^s = [\bar{\mathbf{h}}_1, \bar{\mathbf{h}}_2, \cdots, \bar{\mathbf{h}}_N] \tag{12}$$

The HED decoder labels the sequence with a feed-forward neural networks given $\bar{\mathbf{H}}^s$,

$$\mathbf{O}^{head} = \text{ReLU}(\bar{\mathbf{H}}^s \cdot \mathbf{W}^{HED} + \mathbf{b}^{HED}) \tag{13}$$

where $\mathbf{W}^{HED} \in \mathbb{R}^{5d \times 3}$ is a linear mapping matrix and $\mathbf{b}^{HED} \in \mathbb{R}^3$ is a bias vector. In practice, we directly use "B/I/O" to label the tokens, so the number of tags are 3.

Considering the decoder output, $\mathbf{O}^{head} \in \mathbb{R}^{N \times 3}$, the final softmax layer computes normalized tag probabilities based on $\mathbf{O}^{head}$,

$$\mathbf{p}^e = \text{softmax}(\mathbf{O}^e) \tag{14}$$

where $\mathbf{p}^e \in \mathbb{R}^{N \times (2*|\mathcal{T}^{\mathcal{E}}|+1)}$ is the normalized tag probabilities for tokens in sentence s and,

$$\mathbf{p}_{i,j}^e = \frac{\exp(\mathbf{O}_{i,j}^e)}{\sum_{k=1}^{2*|\mathcal{T}^{\mathcal{E}}|+1} \exp(\mathbf{O}_{i,k}^e)} \tag{15}$$

In prediction stage, we can obtain the final HED tag output $\hat{y}^{head}$,

$$\hat{y}^{head} = [\hat{y}_1^{head}, \hat{y}_2^{head}, \cdots, \hat{y}_N^{head}] \tag{16}$$

$$\hat{y}_i^{head} = \operatorname{argmax} \mathbf{p}_i^{head}, i = 1, \cdots, N \tag{17}$$

where $\hat{y}_i^{head} \in \{0, 1, 2\}$. A list of head entity mentions $E^{head} = [x_1, x_2, \cdots]$ for relation r in sentence s is obtained, with respect to y^{head}. We have the same process for tail entity detection (TED) decoder, with responding parameters $\mathbf{W}^{TED} \in \mathbb{R}^{5d \times 3}$ and $\mathbf{b}^{TED} \in \mathbb{R}^3$.

Combination of Results. If $y^r = 0$, we discard the results and return no relational triples for relation r in sentence s. If $y^r = 1$, we conduct a list of head-tail entity pairs $[(x_{i_1}, x_{j_1}), (x_{i_2}, x_{j_2}), \cdots]$ according to the principle of shortest distance matching, where $x_{i_*} \in E^{head}, x_{j_*} \in E^{tail}$. We investigate each head-tail entity pair (x_{i_*}, x_{j_*}): if satisfying that $x_{i_*}, x_{j_*} \in E$, relational triple (x_{i_*}, r, x_{j_*}) for sentence s is returned, otherwise the head-tail entity pair is discarded. For each $r \in \mathcal{T}^{\mathcal{R}}$, we follow the above process to get relational triples of r in s, then the final results of relation extraction are obtained by merging all triple results.

3.5   Loss Functions for Training

We use cross entropy loss function to train our extractor. Given sentence s with length of N, we conduct the loss function on subtask RD as follows,

$$\mathcal{L}_s^{RD} = \frac{1}{|\mathcal{T}^{\mathcal{R}}|} \sum_{r \in \mathcal{T}^{\mathcal{R}}} \log(\mathbf{p}_{y^r}^r) \tag{18}$$

and loss function of subtasks NER/HED/TED is defined as follows,

$$\mathcal{L}_s^{NER} = \frac{1}{N} \sum_{i=1}^{N} \log(\mathbf{p}_{i,y_i^c}^e) \tag{19}$$

$$\mathcal{L}_s^{HED} = \frac{1}{|\mathcal{T}^{\mathcal{R}}|} \sum_{r \in \mathcal{T}^{\mathcal{R}}} \frac{1}{N} \sum_{i=1}^{N} \log(\mathbf{p}_{i,y_i^{head}}^{head}) \tag{20}$$

$$\mathcal{L}_s^{TED} = \frac{1}{|\mathcal{T}^{\mathcal{R}}|} \sum_{r \in \mathcal{T}^{\mathcal{R}}} \frac{1}{N} \sum_{i=1}^{N} \log(\mathbf{p}_{i,y_i^{tail}}^{tail}) \tag{21}$$

where y^r is the true indicator for sentence s; y_i^e, y_i^{head} and y_i^{tail} are true tag ids on other three subtasks (NER, HED and TED) for the i-th token. The joint loss function for sentence s is defined as a mean value of the four kinds of losses,

$$\mathcal{L}_s = \frac{1}{4}(\mathcal{L}_s^{RD}, \mathcal{L}_s^{NER}, \mathcal{L}_s^{HED}, \mathcal{L}_s^{TED}) \tag{22}$$

4   Experiments

4.1   Experimental Design

Datasets. We use three public datasets for experiments: **ACE05** and **CONLL04** for joint entity and relation extraction, and **NYT** for end-to-end relation extraction. We use the first two datasets as our primary target, investigating the effectiveness brought by *synchrony* settings on joint extractor. We use the NYT to mainly test overlapping triples extraction.

Table 1. Our work outperforms all the other baselines by F1 scores on the joint entity and relation extraction task, with respect to test sets from CONLL04 dataset. q represents specific query. t represents the task prefix.

Systems	Inputs	Core Networks	is Synchronous	Entity	Relation
(Bekoulis et al. 2018) [1]	s	BiLSTM	✗	83.9	62.0
(Nguyen et al. 2019) [6]	s	Attention	✗	86.2	64.4
(Zhang et al. 2017)[12]	s	BiLSTM	✓	85.6	67.8
(Wang et al. 2022) [8]	t, s	GLM-10B	✓	87.4	<u>69.6</u>
(Li et al. 2019) [4]	q, s	BERT	✗	<u>87.8</u>	68.9
Ours	r, s	BERT	✓	**89.8**	**<u>71.3</u>**

Table 2. Our work outperforms all the other baselines by F1 scores on the joint entity and relation extraction task, with respect to test sets from ACE05 dataset.

Systems	Inputs	Core Networks	is Synchronous	Entity	Relation
(Zhang et al. 2017) [12]	s	BiLSTM	✓	83.6	57.5
(Sun et al. 2018) [7]	s	CNN	✗	83.6	59.6
(Li et al. 2019) [4]	q, s	BERT	✗	84.8	60.2
(Dixit et al. 2019) [2]	s	BERT	✗	85.9	62.8
(Luan et al. 2019) [5]	s	BiLSTM	✗	86.0	<u>63.2</u>
(Wang et al. 2022) [8]	t, s	GLM-10B	✓	<u>87.8</u>	54.0
Ours	r, s	BERT	✓	**89.0**	**66.2**

Baseline Methods. Most of the recently designed models are based on deep neural networks: convolutional neural networks [7], bidirectional long short-term memory [1,5], neural attention networks [6], copy neural networks [10], graph neural networks [3], neural reinforcement learning [11] and transformer encoder based BERT [2]. Li et al. [4] framed joint extraction task to a multi-turn question answering problem, i.e., reading comprehension problem. Yu et al. [9] focused on overlapping relations issues. Wang et al. [8] pretrained the multitask model to improve its structural understanding abilities. The joints model in [13] and [12] are mostly similar to ours, using synchronous extraction framework.

Table 3. Our work outperforms all the other baselines by F1 scores on the joint entity and relation extraction task, with respect to test sets from NYT dataset.

Systems	Inputs	Core Networks	is Synchronous	Entity	Relation
(Zheng et al. 2017) [13]	s	BiLSTM	✓	-	42.0
(Fu et al. 2019) [3]	s	GNN	✗	-	61.9
(Zeng et al. 2019) [11]	s	RL	✗	-	72.1
(Zeng et al. 2020) [10]	s	CopyNet	✗	-	72.0
(Yu et al. 2020) [9]	s	BiLSTM	✗	-	<u>78.0</u>
Ours	r, s	BERT	✓	-	**80.6**

Evaluation Metrics. We follow the evaluation metrics in previous work [4,5,9]. For the ACE05 dataset, the head region is defined by the corpus, and for the CONLL04 dataset, the head region covers the entire scope of an entity. A triplet is marked correct if and only if its relation type and two corresponding entities are all correct.

4.2 Results Comparison and Analysis

Overall Results. We have three main observations from the experimental results in Tables 1, 2 and 3. *(1)* Our work outperforms all the baselines on three public datasets. It improves the F1 score relatively by 2.44%, 4.75% and 3.33% over the best baselines as for relation extraction task on CONLL04, ACE05 and NYT datasets, respectively. Our model also achieves the best entity extraction performance on CONLL04 and ACE05, improving the F1 score relatively by 2.27%, 1.37% over the best baselines. *(2)* Most of models on the three datasets are asynchronous joint paradigms. Novel design of synchrony paradigm is supposed to be fully explored. *(3)* Our work outperforms all other asynchronous extractors in the joint extraction task across the three datasets.

Analysis to the Excellent Baselines. We concern the advancement and limitations of the excellent **B**aselines on three datasets. (**B1**) On CONLL04 test set, Li et al. [4] conducted multiple queries on the given sentence, but the performance deteriorated on large-scale datasets (such as ACE05) due to error accumulation. (**B2**) On ACE05 test set, Luan et al. [5] required a lot of redundant immediate information, i.e., candidate relational span-pairs, which proved to be time-consuming in practical applications. (**B3**) On CONLL04 and ACE05 test sets, Wang et al. [8] proposed a multi-task pretrained model that utilizes a larger number of parameters and more data. Although its performance is quite competitive, fine-tuning this model for practical usage is resource-intensive. (**B4**) On NYT test set, Yu et al. [9] ignored entities that have no relation with others and exhibited high time complexity in extraction.

Robustness for Overlapping Relations. Overlapping relations widely exist in real-world text. And the NYT dataset involves many instances of Single-Entity-Overlap relations of this kind. Sentences are classified into three types:

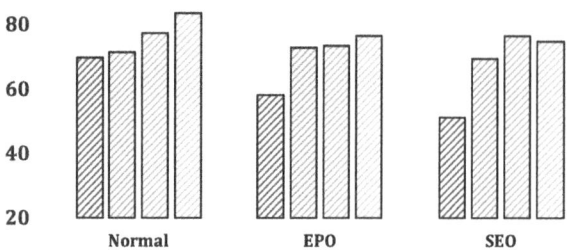

Fig. 3. Our model outperforms the best baseline in Normal and EPO sentences and delivers competitive performance in SEO sentences.

normal (no overlapping entities in triples), single-entity-overlap (SEO, some triples have an overlapped entity), and entity-pair-overlap (EPO, dentical entity pairs in two triples with different relations). We investigate the performance delivered by our model and other strong baseline models on NYT test set. The results are shown in Fig. 3. We can see that our model obtain the best score in normal and EPO categories, outperforming all the other competitors, as well as promising performance in SEO sentences. It demonstrates robustness of our work on overlapping triples.

Table 4. Ablation study on our proposed synchronous joint extractor (abbr. SOE). *PFeat.* and *SFeat.* represent hidden embedding feature from predecessor and successor tokens of current token

Architectures	Inquiry Form	PFeat.	SFeat.	Epoch	Rel.
Baseline	-	-	-	-	63.2
Pair-Wise	*language*	✓	✓	1000	63.7
Hier-Att	*language*	✓	✓	1000	64.2
SOE	*language*	✗	✗	800	60.4
SOE	*language*	✓	✗	800	62.3
SOE	*type name*	✓	✓	800	64.2
SOE	*number id*	✓	✓	800	64.4
SOE	*language*	✓	✓	800	**65.2**

4.3 Analysis to Proposed Model

Ablation Study. Relation entities detection subtasks (HED and TED) are crucial for relation extraction. We concern the effectiveness of module settings in HED/TED decoders, including: architecture design, form of relation inquiry, feature selection. Table 4 provides results of the ablation study. *(1)* We set

Table 5. Proposed synchronous joint extractor gets promising efficiency of $\mathcal{O}(N)$ time-complexity. Most previous asynchronous joint models have at least $\mathcal{O}(N^2)$ time-complexity. We use λ_* to denote some constant value such as hyper-parameters.

Systems	Categories	Time Complexity
(Nguyen et al. 2019) [6]	C1	$\mathcal{O}(N + N \cdot (N-1))$
(Zhang et al. 2017) [12]	C1	$\mathcal{O}(N + N \cdot (N-1))$
(Dixit et al. 2019) [2]	C1	$\mathcal{O}(\lambda_1 N + \lambda_1 N \cdot (\lambda_1 N - 1))$
(Sun et al. 2018) [7]	C1	$\mathcal{O}(N + N \cdot (N-1))$
(Luan et al. 2019) [5]	C1	$\mathcal{O}(\lambda_1 N \cdot (\lambda_2 K + 1) + (\lambda_1 N)^2 \cdot (\lambda_3 + 1))$
(Bekoulis et al. 2018) [1]	C2	$\mathcal{O}(N + N \cdot (N-1) \cdot)$
(Li et al. 2019) [4]	C2	$\mathcal{O}(N + N \cdot |\mathcal{T}^{\mathcal{R}}| \cdot N)$
(Yu et al. 2020) [9]	C2	$\mathcal{O}(N + N^2)$
(Zeng et al. 2020) [10]	C3	$\mathcal{O}(N + N \cdot (N-1) \cdot |\mathcal{T}^{\mathcal{R}}|)$
(Zeng et al. 2019) [11]	C3	$\mathcal{O}(N + N \cdot (N-1) \cdot |\mathcal{T}^{\mathcal{R}}|)$
(Fu et al. 2019) [3]	C4	$\mathcal{O}(N + N \cdot (N-1))$
Ours	-	$\mathcal{O}(N \cdot |\mathcal{T}^{\mathcal{R}}|)$

two additional **architecture designs**. *Pair-Wise* performs every pair of text spans, as work in [5], to detect the head-tail entity pairs. *Hier-Att* performs TED task by attending to hidden embedding from HED decoder. Our proposed synchrony paradigm outperforms both of the additional architectures. Pair-Wise brings much redundant information that hurt performance as noise, and Hier-Att takes limited bi-directional interaction learning process. Both of them take more epochs to convergence. *(2)* We concern form of relation input for inquiry strategy. Three forms are investigated in our work: using natural language to describe the relation type (denoted as "language"), directly using the its type name in schema (denoted as "type name"), and using number id to represent individual relation (denoted as "number id"). As we expected, using language description provides more semantic information by nature, delivering the best performance on relation extraction, as shown in last three rows of Table 4. *(3)* Our proposed method induces two kinds of **latent features**. *PFeat.* represents the hidden embedding feature of the predecessor token next to the current token, i.e., $\mathbf{h}_{M+i}$. Similarly, SFeat. represents the feature from successor token. The methods using both of them achieve higher performance than the methods that only use one of them or none of them.

Efficiency Comparison of Extraction. For time-complexity analysis, the extraction process of these traditional baselines are divided into four **C**ategories (Shown in Table 5): (**C1**) Extract entities first, then detect relations over candidate pairs. (**C2**) Identify objective entities and relations for each extracted entity. (**C3**) Generate relation triples from given sentence. (**C4**) Construct entity graph, then detect relations.

Our proposed synchronous joint extractor has only $\mathcal{O}(N)$ time complexity, as entities and relations are synchronously learned from a shared encoder via one-step modeling over sentence. Joint model in [13] has $\mathcal{O}(N)$ time complexity upon sequence labeling, however, it suffers from overlapping triples problem. Multi-task pretrained model in [8] has $\mathcal{O}(N)$ time complexity in the inference stage but it requires more time in the model's finetuning stage.

Above time complexities are analyzed and calculated based on the extraction framework inside, which reflects lower bond of time complexity, rather than neural networks the methods utilized.

5   Conclusions

Present joint extraction models mostly adopt an asynchronous extraction paradigm. We argue that asynchronous joint extraction framework have not well explored joint extraction paradigm: interactive learning between components is limited; much redundant information is induced by asynchrony design, making the extraction time-consuming and low-efficiency; exposure bias from training to inference hurt model performance. This paper explores to realize a novel synchronous joint extraction framework, which can avoid the thorny issues encountered in asynchrony paradigm. A simple and effective relation inquiry strategy is proposed to deal with the challenge of overlapping triples extraction for synchronous joint framework. Our proposed synchronous joint extractor upon relation inquiry outperforms all asynchronous joint baselines on three public datasets, with respect to effectiveness, robustness and model efficiency.

References

1. Bekoulis, G., Deleu, J., Demeester, T., Develder, C.: Joint entity recognition and relation extraction as a multi-head selection problem. Expert Syst. Appl. **114**, 34–45 (2018)
2. Dixit, K., Al-Onaizan, Y.: Span-level model for relation extraction. In: Proceedings of the 57th Annual Meeting of the Association for Computational Linguistics, pp. 5308–5314 (2019)
3. Fu, T.J., Li, P.H., Ma, W.Y.: GraphRel: modeling text as relational graphs for joint entity and relation extraction. In: Proceedings of the 57th Annual Meeting of the Association for Computational Linguistics, pp. 1409–1418 (2019)
4. Li, X., Yin, F., Sun, Z., Li, X., Yuan, A., Chai, D., Zhou, M., Li, J.: Entity-relation extraction as multi-turn question answering (2019)
5. Luan, Y., Wadden, D., He, L., Shah, A., Ostendorf, M., Hajishirzi, H.: A general framework for information extraction using dynamic span graphs. In: Proceedings of the 2019 Conference of the North American Chapter of the Association for Computational Linguistics: Human Language Technologies, Volume 1 (Long and Short Papers), pp. 3036–3046. Association for Computational Linguistics, Minneapolis, Minnesota (2019). https://doi.org/10.18653/v1/N19-1308. https://www.aclweb.org/anthology/N19-1308

6. Nguyen, D.Q., Verspoor, K.: End-to-end neural relation extraction using deep biaffine attention. In: European Conference on Information Retrieval, pp. 729–738. Springer (2019)
7. Sun, C., et al.: Extracting entities and relations with joint minimum risk training. In: Proceedings of the 2018 Conference on Empirical Methods in Natural Language Processing, pp. 2256–2265. Association for Computational Linguistics, Brussels, Belgium (2018). https://doi.org/10.18653/v1/D18-1249. https://www.aclweb.org/anthology/D18-1249
8. Wang, C., et al.: Deepstruct: pretraining of language models for structure prediction (2022)
9. Yu, B., Zhang, Z., Su, J.: Joint extraction of entities and relations based on a novel decomposition strategy. In: ECAI (2020)
10. Zeng, D., Zhang, H., Liu, Q.: Copymtl: copy mechanism for joint extraction of entities and relations with multi-task learning. In: AAAI, pp. 9507–9514 (2020)
11. Zeng, X., He, S., Zeng, D., Liu, K., Liu, S., Zhao, J.: Learning the extraction order of multiple relational facts in a sentence with reinforcement learning. In: Proceedings of the 2019 Conference on Empirical Methods in Natural Language Processing and the 9th International Joint Conference on Natural Language Processing (EMNLP-IJCNLP), pp. 367–377 (2019)
12. Zhang, M., Zhang, Y., Fu, G.: End-to-end neural relation extraction with global optimization. In: Proceedings of the 2017 Conference on Empirical Methods in Natural Language Processing, pp. 1730–1740. Association for Computational Linguistics, Copenhagen, Denmark (2017). https://doi.org/10.18653/v1/D17-1182. https://www.aclweb.org/anthology/D17-1182
13. Zheng, S., Wang, F., Bao, H., Hao, Y., Zhou, P., Xu, B.: Joint extraction of entities and relations based on a novel tagging scheme. In: Proceedings of the 55th Annual Meeting of the Association for Computational Linguistics (Volume 1: Long Papers), pp. 1227–1236. Association for Computational Linguistics, Vancouver, Canada (2017). https://doi.org/10.18653/v1/P17-1113. https://www.aclweb.org/anthology/P17-1113

Machine Learning on Graphs

Flexible Multi-view Subspace Clustering with Anchor Structure Alignment

Runhua Hu[ID] and Xiaohua Ke$^{(\boxtimes)}$[ID]

Guangdong University of Foreign Studies, Guangzhou 510420, GD, China
20231010003@gdufs.edu.cn, Ck0900@hotmail.com

Abstract. Multi-view subspace Clustering has recently demonstrated remarkable performance in exploring multiple graph structures. Despite its excellent performance on large-scale multi-view data clustering tasks, the application of those approaches in the real world is still limited. Traditional approaches often use fixed anchor selection strategy, leading to clustering results that heavily depend on pre-processing. Additionally, fixed anchor points fail to ensure cross-view consistency which may result in potential anchor misalignment and then reduce clustering performance. To address these issues, we propose a novel method termed Flexible Multi-view Subspace Clustering with Anchor Structure Alignment (FMVSC-ASA). Essentially, our study contributes to cross-view anchor correspondence by aligning anchor graph with the structure alignment module. Moreover, FMVSC-ASA uses an adaptive approach for anchor learning and then constructs anchor graphs separately for each view. The anchor learning, anchor graph construction and structure alignment are integrated into a unified framework for joint optimization, ensuring the full utilization of cross-view consistency and complementarity. Comprehensive experimental results demonstrate the effectiveness of the proposed FMVSC-ASA.

Keywords: Multi-view subspace clustering · Anchor structure alignment · Large-scale multi-view data

1 Introduction

In an era of exponential data growth, data originate from multiple sources regarded as multi-view data, which often is observed from different perspectives. These data exhibit diversity and complementarity, making traditional single-view data analysis methods inadequate for fully harnessing those huge scale data. Multi-view Subspace Clustering (MVSC) has emerged to improve clustering quality by exploiting complementary and consensus information across different views, which has seen significant progress in both academic and industrial domains, such as image processing and medical research [1–8, 18, 21, 25, 33].

Recently, anchor-based multi-view subspace clustering (AMVSC) has been proposed, which selects representative anchors from different views, thereby

B. Xu et al. (Eds.): CCKS-IJCKG 2024, CCIS 2229, pp. 165–177, 2025.
https://doi.org/10.1007/978-981-96-1809-5_13

reducing excessive complexity to linear level [9, 13–17, 31]. Nonetheless, existing methods utilize heuristic anchor sampling strategy involving random sampling or k-means, which could not conducive to real-world deployment [9, 11]. Specifically, those fixed anchor sampling strategy separates the generation of anchor from the subsequent construction of anchor graph which may result in a heavy reliance on pre-processing. Moreover, they fail to comprehensively account for the cross-view relationships among anchors, often leading to anchor misalignment. Particularly, the key features of each view may be misinterpreted or overlooked if incorrect anchor correspondences mix incomparable view-specific information. Additionally, misaligned anchors in the fusion process may lead to distorted or inaccurate representations. This misalignment can compromise the accuracy and effectiveness of subsequent clustering tasks [19, 20, 22, 23, 31].

The primary challenge in addressing anchor unalignment is determining by the result how to effectively establish correspondences among multiple anchor sets that exist in different data metric spaces [19]. Without these correspondences, the fusion of anchor graphs across views becomes inaccurate, leading to suboptimal clustering results. Since the structural similarity between anchors is positively correlated with their probability of correspondence, the original anchor correspondence problem can transform into a structural alignment problem [22]. However, those methods align all anchor points imprecisely without fully paying attention to the relationship between different anchors.

In this paper, we propose a novel and promptly method called Flexible Multi-view Subspace Clustering with Anchor Structure Alignment (FMVSC-ASA) to handle the above issues. Specifically, FMVSC-ASA generates an anchor graph for each view instead of a full graph by employing the anchor strategy with linear computational complexity. Most importantly, our method employs a soft anchor structure alignment term instead of rigid one-to-one matching to align anchor graph structure. Additionally, instead of employing the fixed anchor strategy, we integrate anchor learning, anchor graph construction, and anchor structure alignment into a unified framework where the importance of each view is adaptively measured by weight parameters. Subsequently, we propose a five-step alternative optimization algorithm with proven convergence to solve the final optimization problem. To recapitulate, the contribution of our algorithm is twofold:

1) FMVSC-ASA fully considers cross-view anchor relationships. Specifically, we use anchor structure alignment term and apply relaxed orthogonal constraints on alignment matrices to avoid unaligned anchor points.
2) FMVSC-ASA uses an adaptive alternate optimization strategy to learn anchors and weight parameters to enhance clustering performance and make all optimization subproblems reach local optimal.

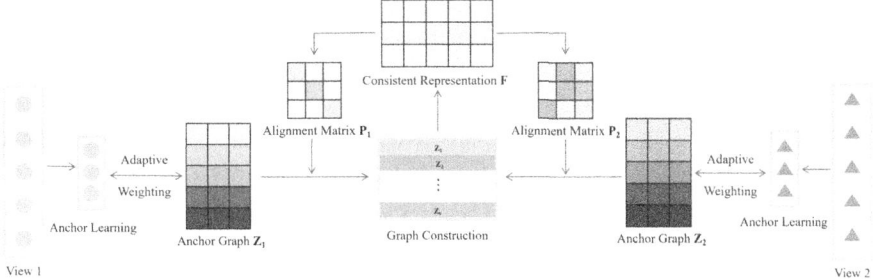

Fig. 1. The framework of the proposed FMVSC-ASA.

2   Related Work

2.1   Multi-view Subspace Clustering

Multi-view subspace clustering operates on the premise that data can be accessed from different views, each offering unique and complementary insights into the data and effectively uncovers the intrinsic structure. Precisely, [25, 27–29] perform clustering on the subspace representation of each view simultaneously, while ensuring consistency among different views through a common clustering structure. Subsequently, [28] builds on this approach by incorporating prior knowledge, while [29] further decomposes the subspace representation of each view into a shared consistent representation and view-specific differential representations. The approaches mentioned above exhibit promising clustering performance, while these methods still struggle to reduce excessive complexity, with the cubic-level time complexity and square-level space complexity.

2.2   Anchor-Based Multi-view Subspace Clustering

Anchor strategy involves selecting anchors from the original dataset and computing the relationships between the original data points and these anchors to construct a sparse affinity matrix [9, 16, 32]. Mathematically, the subspace clustering formula based on the anchor graph can be expressed as follows

$$\min_{Z_v} \quad \sum_{v=1}^{V} \|X_v - A_v(Z_v)^\top\|_F^2 + \omega \sum_{v=1}^{V} \|Z_v\|_F^2,$$
$$s.t. \quad Z_v \geq 0, Z_v \mathbf{1} = 1, \tag{1}$$

where A_v denotes the selected anchor graph of the v-th view, Z_v represents the anchor graph, and ω is a trade-off parameter of the regularization term.

Early work focuses on using a small set of data points to capture the manifold structure. Specifically, [10–12] perform spectral clustering with those selected points, while [9] constructs a small-scale graph matrix for each view, completing the entire clustering process in linear time. Recent research underscores the significance of each view and incorporates it into the subsequent process of graph construction. For instance, [30] introduces view weights to measure the importance of each view. Further, [13–15] integrate anchor learning and graph construction into a unified optimization framework, dynamically adjusting the anchors in conjunction with the graph construction process. Nevertheless, those studies fail to efficiently align anchors between different views, exacerbating the effect of poor clustering and leading to a heavy reliance on preprocessing.

3   Methodology

3.1   Formulation of Problem

To address aforementioned issues, anchors are learned through optimization rather than sampling, ensuring they better capture multi-view information and generate more discriminative graph structures. Additionally, we introduce view weight parameter α_v for each view and adjust them automatically, significantly improving the efficiency and performance of clustering. Therefore, the objective formulation of anchor-based multi-view subspace clustering can be rewritten as

$$\min_{\alpha, A_v, Z_v} \sum_{v=1}^{V} \alpha_v^2 \|X_v - A_v(Z_v)^\top\|_F^2 + \beta \sum_{v=1}^{V} \|Z_v\|_F^2, \tag{2}$$
$$s.t. \quad Z_v \geq 0, Z_v 1 = 1, \alpha^\top 1 = 1, A_v^\top A_v = I_m,$$

where α_v is a weighting coefficient that balances the contribution of different views to the clustering and β is a hyperparameter of the regularization term. By using α_v^2 instead of α_v, we prevent any single view from having an excessively large or small influence to ensure a smoother distribution of the weight coefficients. Furthermore, we impose a regularization constraint on the learned anchor A_v to ensure that the anchors are discriminative, specifically $A_v^\top A_v = I_m$.

Inspired by [19,22], we introduce an anchor graph alignment term to achieve cross-view anchor correspondences. To be specific, we introduce alignment matrices P_v to align anchor graphs from v-th views and compute the difference between the aligned anchor graphs and the consistent representation F. Therefore, we can formulate the overall objective function as follows,

$$\min_{\alpha, A_v, Z_v, P_v, F} \sum_{v=1}^{V} \alpha_v^2 \|X_v - A_v(Z_v)^\top\|_F^2 + \beta \sum_{v=1}^{V} \|Z_v\|_F^2 + \lambda \sum_{v=1}^{V} \|P_v(Z_v)^\top - F\|_F^2,$$
$$s.t. \quad \alpha^\top 1 = 1, A_v^\top A_v = I_m, P_v^\top P_v = I_m, Z_v \geq 0, Z_v^\top 1_m = 1_n, FF^\top = I_m, \tag{3}$$

where $P_v \in \mathbb{R}^{m \times m}$ denotes the alignment matrices which satisfies the orthogonal constraint $P_v^\top P_v = I_m$, $F \in \mathbb{R}^{m \times n}$ the consistent representation with the constraints $FF^\top = I_m$, and λ is a weighting parameter that adjusts the influence of the structure alignment term in the overall objective function. The formula ensures that the aligned anchor graphs Z_v are as close as possible to the consistent representation F by minimizing the anchor graph alignment term. The FMVSC-ASA framework is shown in Fig. 1. However, the optimization problem in Eq. (3) is a non-convex problem when considering all variables simultaneously. In the next section, we propose a five-step optimization to address it.

Algorithm 1. the algorithm of FMVSC-ASA

Input: the original datasets X_v and the number of cluster k.
Output: the final cluster assignments.
 1: Initialize A, Z, P and F with zero matrix and α with $1/V$.
 2: **while** not converged **do**
 3:     Update A_v by solving Eq.(5);
 4:     Update F by solving Eq.(10);
 5:     Update Z_v by solving Eq.(8);
 6:     Update P_v by solving Eq.(9);
 7:     Update α by solving Eq.(12);
 8: **end while**
 9: Perform SVD on F to obtain U and then conduct k-means on U;

3.2   Optimization

In order to optimize the overall objection function, we use an alternating optimization method that optimizes one variable while fixing the other variables, thereby avoiding the optimization of our formulation being a nonconvex problem when all variables are considered.

A_v Subproblem. When Z_v, P, F and α are fixed, the objective function with respect to A_v can be formulated as

$$
\min_{A_v} \quad \sum_{v=1}^{V} \alpha_v^2 \left\| X_v - A_v (Z_v)^\top \right\|_F^2 ,
$$

$$
s.t. \quad A_v^\top A_v = I_m ,
$$

(4)

Since the optimization of each A_v is independent of the corresponding view, we expand the objective function in the form of traces and remove the irrelevant and constant terms by adopting the definition of Frobenius norm. The formula can be written as

$$
\max_{A_v} \quad Tr(A_v^\top B),
$$

$$
s.t. \quad A_v^\top A_v = I_m ,
$$

(5)

where $B_v = X_v Z_v$. According to [23], supposing the singular value decomposition (SVD) result of B_v is $U_m \Sigma_m V_m^\top$, the optimal A_v equals to $U_m V_m^\top$.

$\mathbf{Z}_v$ **Subproblem.** With $\mathbf{A}_v$, P, F and α being fixed, the optimization for $\mathbf{Z}_v$ can be formulated as

$$\min_{\mathbf{Z}_v} \sum_{v=1}^{V} \alpha_v^2 \left\| \mathbf{X}_v - \mathbf{A}_v(\mathbf{Z}_v)^\top \right\|_F^2 + \beta \sum_{v=1}^{V} \|\mathbf{Z}_v\|_F^2 + \lambda \sum_{v=1}^{V} \|\mathbf{P}_v\mathbf{Z}_v - \mathbf{F}\|_F^2, \tag{6}$$
$$s.t. \quad \mathbf{Z}_v \geq 0, \mathbf{Z}_v^\top \mathbf{1}_m = \mathbf{1}_n,$$

Likewise, to continue to simplify the formula for subsequent optimization by converting the Frobenius norm to the matrix trace, we rewrite the formula as

$$\min_{\mathbf{Z}_v} \sum_{v=1}^{V} \mathrm{Tr}(\mathbf{Z}_v(\alpha_v^2 \mathbf{A}_v^\top \mathbf{A}_v + \beta I + \lambda \mathbf{P}_v^\top \mathbf{P}_v)\mathbf{Z}_v^T) - 2\mathrm{Tr}(\mathbf{Z}_v^\top(\alpha_v^2 \mathbf{A}_v^\top \mathbf{X}_v + \lambda \mathbf{P}_v^\top \mathbf{F})), \tag{7}$$
$$s.t. \quad \mathbf{Z}_v \geq 0, \mathbf{Z}_v^\top \mathbf{1}_m = \mathbf{1}_n,$$

Then the above optimization problem of $\mathbf{Z}_v$ can be formulated as the following formula by tackling the Quadratic programming (QP) problem for each column of Z. Denotes the coefficients of linear term $f_{ij}^{(v)} = \frac{\alpha_v^2[\mathbf{A}_v^\top \mathbf{X}_v]_{i,j} + \lambda[\mathbf{P}_v^\top \mathbf{F}]_{i,j}}{\gamma_v^2 + \beta + \lambda}$, $[\cdot]$ denotes the element of the i-th row and j-th column of vector, QP problem can be rewrite as

$$\min_{z_j^{(v)}} \frac{1}{2} \left\| z_j^{(v)} - f_j^{(v)} \right\|_F^2, \tag{8}$$
$$s.t. \quad z_j^{(v)} \geq 0, z_j^{(v)\top} \mathbf{1}_m = 1,$$

where $z_j^{(v)}$ is the j-th column vector of $\mathbf{Z}_v$. We can obtain the optimal by solving this QP problem for each column of Z.

$\mathbf{P}_v$ **Subproblem.** When other variables are fixed, the formulation of $\mathbf{P}_v$ can be solved via optimizing the following formula

$$\max_{\mathbf{P}_v} \quad \mathrm{tr}(\mathbf{P}_v^\top \mathbf{Q}_v), \tag{9}$$
$$s.t. \quad \mathbf{P}_v\mathbf{P}_v^\top = \mathbf{I}_m,$$

where $\mathbf{Q}_v = \mathbf{F}\mathbf{Z}_v$. The optimal solution for $\mathbf{P}_v$ is $\mathbf{U}_m\mathbf{V}_m^\top$, where $\mathbf{U}_m$ and $\mathbf{V}_m$ represent the matrices which comprise the first m left singular vectors and right singular vectors of $\mathbf{Q}_v$, respectively.

F Subproblem. Similar to the optimization step of $\{\mathbf{A}_v\}_{v=1}^{V}$ and $\{\mathbf{P}_v\}_{v=1}^{V}$, the optimization problem for F can be rewritten as

$$\max_{\mathbf{F}} \quad \mathrm{tr}(\mathbf{F}^\top \mathbf{G}), \tag{10}$$
$$s.t. \quad \mathbf{F}\mathbf{F}^\top = \mathbf{I}_m,$$

where $\mathbf{G} = \sum_{v=1}^{V} \mathbf{P}_v(\mathbf{Z}_v)^\top$. Supposing $\mathbf{U}\Sigma\mathbf{V}^\top$ to be the singular value decomposition result of G, the optimal solution of Eq. (10) is $\mathbf{U}\mathbf{V}^\top$.

$\boldsymbol{\alpha}$ **Subproblem.** Fixing the irrelevant variables, we can obtain the optimization problem about α

$$\min_{\alpha} \quad \sum_{v=1}^{V} \alpha_v^2 r_v, \tag{11}$$
$$s.t. \quad \alpha^\top \mathbf{1}_V = 1, \alpha \geq 0,$$

where $r_v = \left\| \mathbf{X}_v - \mathbf{A}_v(\mathbf{Z}_v)^\top \right\|_F^2$. According to Cauchy-Schwarz inequality, α is updated by

$$\alpha_v = \frac{1/r_v}{\sum_{v=1}^{V} 1/r_v}. \tag{12}$$

The whole procedure of the above optimization is summarized in Algorithm 1.

Table 1. A comparison of various methods on seven datasets.

Dataset	Metric	MVSC	AGML	PMSC	LMVSC	SMVSC	FPMVSCAG	OMSC	FMVACC	Ours
ORL	ACC	0.5080	0.6876	0.5981	0.7600	0.7873	0.7495	0.7304	0.7875	**0.8075**
	NMI	0.5271	0.7101	0.5244	0.7008	0.7276	0.7158	0.7563	0.7873	**0.7975**
	Purity	0.7125	0.6313	0.6480	0.5476	0.7163	0.7268	0.7444	0.6899	**0.7725**
	Fscore	0.4329	0.4661	0.5474	0.6069	0.7521	0.7972	0.7141	0.7693	**0.7975**
NGs	ACC	0.6180	0.4876	0.8183	0.7730	0.3873	0.3495	0.7615	0.7023	**0.8380**
	NMI	0.4241	0.5341	0.5244	0.6478	0.5276	0.3158	0.5553	0.5174	**0.6663**
	Purity	0.7435	0.6313	0.6480	0.7476	0.7163	0.7292	0.7414	0.7080	**0.8380**
	Fscore	0.5729	0.4661	0.5474	0.7069	0.7521	0.7972	0.7341	0.5311	**0.8050**
CCV	ACC	-	0.1876	-	0.2014	0.1782	0.2399	0.2304	0.1919	**0.2431**
	NMI	-	0.1101	-	0.1657	0.2113	0.1760	0.2153	0.1504	**0.2383**
	Purity	-	0.1313	-	**0.4476**	0.2219	0.2605	0.2414	0.2308	0.2350
	Fscore	-	0.1661	-	**0.3069**	0.2720	0.1419	0.1144	0.2270	0.2350
Caltech101-all	ACC	-	0.1876	-	0.2005	0.3015	0.3015	0.2194	0.2194	**0.3707**
	NMI	-	0.1101	-	0.4155	0.2093	0.3549	**0.4753**	0.3913	0.3693
	Purity	-	0.1013	-	**0.3975**	0.1163	0.3460	0.3414	0.3772	0.3521
	Fscore	-	0.1661	-	0.1586	**0.3521**	0.2326	0.3141	0.1516	0.2691
Caltech256	ACC	-	0.1876	-	0.3600	0.3873	0.3495	0.2304	0.2684	**0.3894**
	NMI	-	0.1101	-	0.2008	0.3217	0.3158	0.3553	0.1417	**0.3892**
	Purity	-	0.2313	-	**0.4476**	0.4163	0.2268	0.3414	0.2849	0.2946
	Fscore	-	0.1661	-	0.3069	**0.3521**	0.2972	0.3141	0.1516	0.3409
MNIST	ACC	-	-	-	0.9852	0.9873	0.9884	0.9704	0.9867	**0.9891**
	NMI	-	-	-	0.9576	0.9276	0.9651	0.9573	0.9674	**0.9807**
	Purity	-	-	-	0.9852	0.9163	**0.9884**	0.9404	0.9765	0.9882
	Fscore	-	-	-	0.9704	0.9521	**0.9768**	0.9131	0.9299	0.9112
YoutubeFace	ACC	-	-	-	0.1479	0.1973	0.2414	0.2304	0.2478	**0.2680**
	NMI	-	-	-	0.2008	0.1327	0.2433	**0.3553**	0.2674	0.2615
	Purity	-	-	-	**0.3476**	0.2816	0.3279	0.3414	0.2765	0.2582
	Fscore	-	-	-	0.3069	0.0849	0.1433	**0.3141**	0.2299	0.2581

3.3   Computational Complexity

Specifically, given multi-view data $X \in \mathbb{R}^{n \times d_i}$, the time of updating $A_v \in \mathbb{R}^{d_i \times m}$ is $O(nmd_i)$ for matrix multiplication and $O(d_i m^2)$ for SVD on B_v. To update $Z_v \in \mathbb{R}^{n \times m}$, the corresponding QP problem costs $O(nmd_i)$ for all data vectors. Calculating $P_v \in \mathbb{R}^{m \times m}$ and $F \in \mathbb{R}^{m \times n}$ took time $O((nm^2 + m^3)V)$ and $O(nm^2 V)$, respectively. When updating $\alpha \in \mathbb{R}^V$, it costs $O(nmd_i)$ for matrix multiplication and $O(V)$ for update $\sum_{v=1}^{V} \alpha_i$. Consequently, the overall time complexity is about $O(n(md_i + m^2 V) + d_i m^2 + m^3 V + V)$ and space complexity is about $O((n+m)(d_i + m) + V)$. Overall, the computational complexity of the proposed method is $O(n)$, which is linearly related to the number of samples.

4   Experiment

4.1   Experiment Setting

To evaluate the effectiveness of FMVSC-ASA, we conduct experiments on seven widely used multi-view benchmark datasets, including ORL, NGs, CCV, Caltech-101-all, Caltech256, MNIST and YoutubeFace. Moreover, we use eight SOTA graph-based baselines to compare with our proposed method, involving MVSC [25], AGML [24], PMSC [21], LMVSC [9], SMVSC [13], FPMVSCAG [14], OMSC [15], and FMVACC [19]. The last five methods are representative anchor-based multi-view subspace clustering algorithms. To analyses the clustering performance of each approach, we employ four widely used criteria including accuracy(ACC), normalized mutual information(NMI), Purity and Fscore.

4.2   Clustering Performance and Running Time

In terms of Table 1, we report the clustering results on seven benchmark datasets with the metrics including ACC, NMI, Purity, and Fscore, where the best result is highlighted in bold and '−' indicates out of memory failure. As shown in Table 1, our method achieves the best clustering performance compared to all the baseline methods across the ORL and NGs with the highest scores. At the meantime, FMVSC-ASA consistently performs well on large-scale benchmark datasets in accuracy, achieving the highest gap of 6.92% on the Caltech101-all. A similarity could also be observed in those anchor-based methods, none of them have out-of-memory errors due to their linear complexity, indirectly proving its advantage in large-scale data scenarios. Particularly notable was the comparison with FMVACC, our method performs generally well on all benchmark datasets. The main reason probably is that FMVSC-ASA utilizes a soft orthogonal constraint on alignment instead of a strict one-to-one mapping which hampers clustering performance. To recapitulate, our methods can achieve similar or even better performance than baselines in both small-scale data and large-scale data.

    To demonstrate the superiority of our method on computational efficiency, we report the time cost of both our method and the compared method on all the

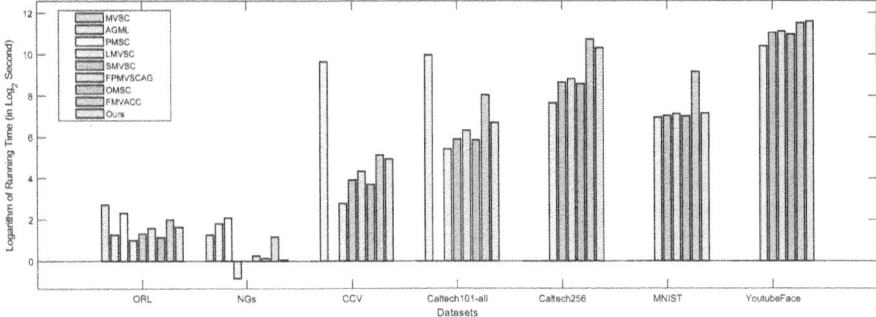

Fig. 2. The running time of the compared baselines on the benchmark datasets, where the empty bar indicates out of memory failure.

benchmark datasets in Fig. 2. According to Fig. 2, our approach costs less time than MVSC and PMSC due to the utilization of the anchor strategy. Meanwhile, FMVSC-ASA generally requires more time compared with other anchor-based baselines. This is because our method employs anchor structure alignment term to align the anchors, while those methods simply adopt the heuristic anchor sampling strategy or fail to fully consider the importance of cross-view relationship. Whereas, our method significantly reduces the computational time compared to FMVACC. FMVACC utilizes feature and structure information for anchor matching and then takes more time to achieve column-wise aligned anchor graph fusion by aligning the matched anchor graphs. In contrast, our method align the anchors directly by implementing a soft anchor structure alignment term. To summarize, the additional cost of time is justified by the superior clustering performance of our method generally, demonstrating the efficiency advantage.

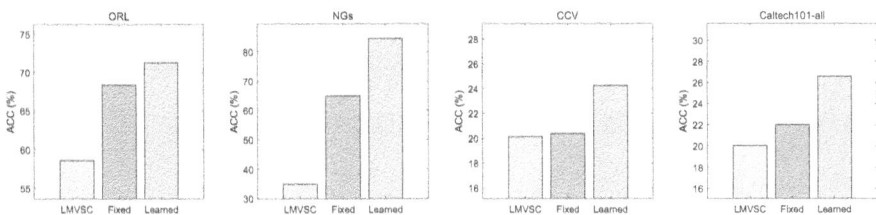

Fig. 3. An ablation on our proposed method, then compared with the LMVSC in term of ACC on four benchmark datasets.

4.3 Ablation Study

To validate the effectiveness of the alignment module, we fixed the optimization of P_v to eliminate the impact of anchor structure alignment term. We report the accuracy obtained from LMVSC, our method after fixed P_v and our method on four benchmark datasets. From Fig. 3, we can clearly observe that the anchor structure alignment term improves the accuracy on the by 4.18%, 30.15%, 19.02%, and 20.79% respectively, compared to the fixed it. Meanwhile, our fixed method consistently outperforms LMVSC with 10.05% averagely higher due to he introduction of weight parameter and learned anchors. In the aggregate, clustering performance can be improved by utilizing alignment term and anchor learning strategy.

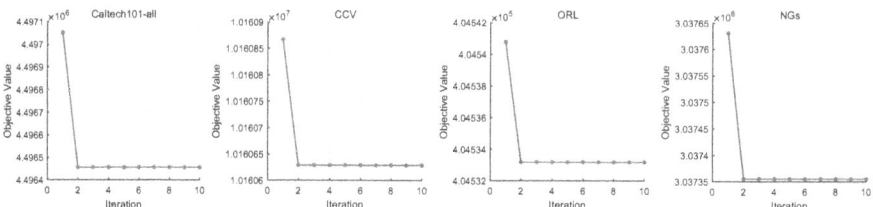

Fig. 4. The objection value of our proposed method on four benchmark datasets.

4.4   Convergence Analysis and Anchor Sensitivity

The lower boundary of the objective function is zero and the objective value decreases monotonically until convergence is reached as each sub-optimization problem attains the global optimum, ensuring the final convergence of the algorithm theoretically [26]. As shown in Fig. 4 which exhibit how objection value changes with iteration, the objective values rapidly converge within the first two iterations and remain stable thereafter. Overall, our proposed algorithm demonstrates the efficiency and quick convergence properties.

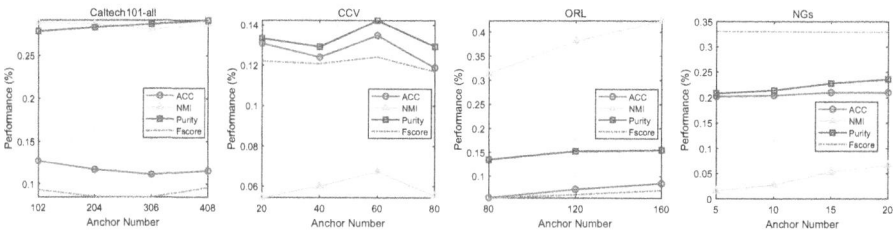

Fig. 5. The impact of anchor numbers on the four benchmark datasets.

Additionally, we conduct anchor sensitivity analysis to illustrate the effect of anchor numbers. As shown in Fig. 5, although the criteria values slightly fluctuate on CCV, the overall performance remains stable across different numbers of anchors. In a nutshell, the anchor number has little effect on the performance of our algorithm.

5   Conclusion

In this paper, we proposed a novel method FMVSC-ASA to deal with clustering task on large-scale data scenarios. Specifically, FMVSC-ASA introduces an anchor structure alignment term to address the unaligned issue and the adaptive view weight, fully considering both inter-view and cross-view relationships. Meanwhile, our method combine anchor learning, anchor graph construction and structure alignment by jointly optimizing five subproblems, enabling achieve locally optimal results without pre-processing. Subsequent experiments show the effectiveness and efficiency of structure alignment and our proposed FMVSC-ASA. Future study will concentrated on large-scale incomplete multi-view data in order to delve deeper into the realistic needs.

Acknowledgments. This work was supported by the Construction Project of Teaching Quality and Teaching Reform Project for Undergraduate Universities in Guangdong Province, P.R.China.

Disclosure of Interests. The authors have no competing interests to declare that are relevant to the content of this article.

References

1. Zhan, K., Nie, F., Wang, J., Yang, Y.: Multiview consensus graph clustering. IEEE Trans. Image Process. **28**, 1261–1270 (2019)
2. Zhang, P., et al.: Consensus one-step multi-view subspace clustering. IEEE Trans. Knowl. Data Eng. **34**, 4676–4689 (2022)
3. Liu, S., Liao, Q., Wang, S., Liu, X., Zhu, E.: Robust and consistent anchor graph learning for multi-view clustering. IEEE Trans. Knowl. Data Eng. 1–13 (2024)
4. Wang, J., Li, Z., Tang, C., Liu, S., Wan, X., Liu, X.: Multiple kernel clustering with adaptive multi-scale partition selection. IEEE Trans. Knowl. Data Eng. 1–12 (2024)
5. Liu, Y., et al.: Simple contrastive graph clustering. IEEE Trans. Neural Netw. Learn. Syst. 1–12 (2023)
6. Zhou, P., et al.: Partial clustering ensemble. IEEE Trans. Knowl. Data Eng. **36**, 2096–2109 (2024)
7. Cai, X., Nie, F., Huang, H.: Multi-view K-means clustering on big data. In: Proceedings of the Twenty-Third International Joint Conference on Artificial Intelligence, pp. 2598–2604 (2013)
8. Wang, J., Tian, F., Yu, H., Liu, C., Zhan, K., Wang, X.: Diverse non-negative matrix factorization for multiview data representation. IEEE Trans. Cybern. **48**, 2620–2632 (2018)
9. Kang, Z., Zhou, W., Zhao, Z., Shao, J., Han, M., Xu, Z.: Large-scale multi-view subspace clustering in linear time. In: Proceedings of the AAAI Conference on Artificial Intelligence, vol. 34, pp. 4412–4419 (2020)
10. Chen, X., Cai, D.: Large scale spectral clustering with landmark-based representation. In: Proceedings of the Twenty-Fifth AAAI Conference on Artificial Intelligence, pp. 313–318 (2011)

11. Li, Y., Nie, F., Huang, H., Huang, J.: Large-scale multi-view spectral clustering via bipartite graph. In: Proceedings of the AAAI Conference on Artificial Intelligence, vol. 29 (2015)

12. Guo, J., Ye, J.: Anchors bring ease: an embarrassingly simple approach to partial multi-view clustering. In: Proceedings of the AAAI Conference on Artificial Intelligence, vol. 33, pp. 118–125 (2019)

13. Sun, M., et al.: Scalable multi-view subspace clustering with unified anchors. In: Proceedings of the 29th ACM International Conference on Multimedia, pp. 3528–3536 (2021)

14. Wang, S., et al.: Fast parameter-free multi-view subspace clustering with consensus anchor guidance. IEEE Trans. Image Process. **31**, 556–568 (2022)

15. Chen, M., Wang, C., Huang, D., Lai, J., Yu, P.: Efficient orthogonal multi-view subspace clustering. In: Proceedings of the 28th ACM SIGKDD Conference on Knowledge Discovery and Data Mining, pp. 127–135 (2022)

16. Yang, B., et al.: Fast multiview anchor-graph clustering. IEEE Trans. Neural Netw. Learn. Syst. 1–12 (2024)

17. Nie, F., Xue, J., Yu, W., Li, X.: Fast clustering with anchor guidance. IEEE Trans. Pattern Anal. Mach. Intell. **46**, 1898–1912 (2024)

18. Ma, H., et al.: Symmetric multi-view subspace clustering with automatic neighbor discovery. IEEE Trans. Circuits Syst. Video Technol. 1 (2024)

19. Wang, S., et al.: Align then fusion: generalized large-scale multi-view clustering with anchor matching correspondences. Adv. Neural. Inf. Process. Syst. **35**, 5882–5895 (2022)

20. Liu, S., Liu, X., Wang, S., Niu, X., Zhu, E.: Fast incomplete multi-view clustering with view-independent anchors. IEEE Trans. Neural Netw. Learn. Syst. 1–12 (2022)

21. Kang, Z., et al.: Partition level multiview subspace clustering. Neural Netw. **122**, 279–288 (2020)

22. Wen, Y., et al.: Scalable incomplete multi-view clustering with structure alignment. In: Proceedings of the 31st ACM International Conference on Multimedia, pp. 3031–3040 (2023)

23. Wang, S., et al.: Multi-view clustering via late fusion alignment maximization. In: Proceedings of the Twenty-Eighth International Joint Conference on Artificial Intelligence, IJCAI-19, pp. 3778–3784 (2019)

24. Nie, F., Li, J., Li, X.: Parameter-free auto-weighted multiple graph learning: a framework for multiview clustering and semi-supervised classification. In: Proceedings of the Twenty-Fifth International Joint Conference on Artificial Intelligence, pp. 1881–1887 (2016)

25. Gao, H., Nie, F., Li, X., Huang, H.: Multi-view subspace clustering. In: 2015 IEEE International Conference on Computer Vision (ICCV), pp. 4238–4246 (2015)

26. Bezdek, J., Hathaway, R.: Convergence of alternating optimization. Neural Parallel Sci. Comput. **11**, 351–368 (2003)

27. Zhang, C., Fu, H., Liu, S., Liu, G., Cao, X.: Low-rank tensor constrained multiview subspace clustering. In: 2015 IEEE International Conference on Computer Vision (ICCV), pp. 1582–1590 (2015)

28. Luo, S., Zhang, C., Zhang, W., Cao, X.: Consistent and specific multi-view subspace clustering. In: Proceedings of the AAAI Conference on Artificial Intelligence, vol. 32 (2018)

29. Zhang, C., Fu, H., Wang, J., Li, W., Cao, X., Hu, Q.: Tensorized multi-view subspace representation learning. Int. J. Comput. Vision **128**, 2344–2361 (2020)

30. Kang, Z., Lin, Z., Zhu, X., Xu, W.: Structured graph learning for scalable subspace clustering: from single view to multiview. IEEE Trans. Cybern. **52**, 8976–8986 (2022)
31. Wan, X., Xiao, B., Liu, X., Liu, J., Liang, W., Zhu, E.: Fast continual multi-view clustering with incomplete views. IEEE Trans. Image Process. **33**, 2995–3008 (2024)
32. Yan, W., Zhang, Y., Tang, C., Zhou, W., Lin, W.: Anchor-sharing and clusterwise contrastive network for multiview representation learning. IEEE Trans. Neural Netw. Learn. Syst. 1–11 (2024)
33. Jin, Z., Wang, M., Zheng, X., Chen, J., Tang, C.: Drug side effects prediction via cross attention learning and feature aggregation. Expert Syst. Appl. **248**, 123346 (2024)

Knowledge Retrieval and Information Retrieval

Reliable Academic Conference Question Answering: A Study Based on Large Language Model

Zhiwei Huang[1], Juan Li[1], Long Jin[1], Junjie Wang[1], Mingchen Tu[1], Yin Hua[1], Zhiqiang Liu[1], Jiawei Meng[2], and Wen Zhang[1(✉)]

[1] School of Software, Zhejiang University, Ningbo, China
{huangzww,lijuan18,longjin,wangjj2018,mingchentz,22351088,zhiqiangliu,
zhang.wen}@zju.edu.cn
[2] College of Computer Science and Technology, Zhejiang University,
Hangzhou, China
mjw.cs@zju.edu.cn

Abstract. As the development of academic conferences fosters global scholarly communication, researchers consistently need to obtain accurate and up-to-date information about academic conferences. Since the information is scattered, using an intelligent question-answering system to efficiently handle researchers' queries and ensure awareness of the latest advancements is necessary. Recently, Large Language Models (LLMs) have demonstrated impressive capabilities in question answering, and have been enhanced by retrieving external knowledge to deal with outdated knowledge. However, these methods fail to work due to the lack of the latest conference knowledge. To address this challenge, we develop the ConferenceQA dataset, consisting of seven diverse academic conferences. Specifically, for each conference, we first organize academic conference data in a tree-structured format through a semi-automated method. Then we annotate question-answer pairs and classify the pairs into four different types to better distinguish their difficulty. With the constructed dataset, we further propose a novel method STAR (**ST**ructure-**A**ware **R**etrieval) to improve the question-answering abilities of LLMs, leveraging inherent structural information during the retrieval process. Experimental results on the ConferenceQA dataset show the effectiveness of our retrieval method. The dataset and code are available at https://github.com/zjukg/ConferenceQA.

Keywords: Conference dataset · Large language model · Retrieval augmentation

1 Introduction

The rapid advancement of computer science has led to an increase in research presented at academic conferences, which are crucial for academic exchange.

B. Xu et al. (Eds.): CCKS-IJCKG 2024, CCIS 2229, pp. 181–193, 2025.
https://doi.org/10.1007/978-981-96-1809-5_14

Given the vast and dispersed nature of conference information, querying is a more efficient method for information retrieval than navigating multiple sources.

Recent advancements in Large Language Models (LLMs) [2,7,21] have significantly impacted various NLP tasks, including question answering. LLMs demonstrate capabilities like chain-of-thought reasoning [3] and in-context learning [6], enhanced by increasing model parameters and extensive training data. After instruction fine-tuning [5], LLMs excel in conversational tasks and information retrieval [4].

Despite the success of LLMs, they are related to incompleteness, untimeliness, unfaithfulness, and having limitations in updating timely and domain-specific expertise. This necessitates research efforts to integrate LLMs with external knowledge sources, such as knowledge bases (KBs) [8], search engines [9] and databases [10]. Regarding academic conference queries, due to the missing external conference knowledge, LLMs fail to access the latest academic conference information in question answering, such as academic conferences in 2022 and later ones. Existing retrieval methods are efficient but primarily focus on plain text [11], triples [15], and tables [16], which does not align well with the structured nature of conference websites, complicating direct application for conference-specific queries.

In this paper, we introduce ConferenceQA, a benchmark comprising seven recent top-tier academic conferences, these conferences span various research domains such as web science, natural language processing, machine learning, databases, artificial intelligence, and the semantic web, providing a comprehensive dataset that organizes information across all stages of the conferences. To construct this dataset, we initially employ a semi-automatic method to convert the conference information into a tree structure. Subsequently, we utilize ChatGPT to simulate roles with diverse backgrounds, enabling us to generate role-specific questions. These questions are then carefully filtered and annotated with answers to ensure the dataset's reliability. Additionally, we document the sources of the answers to further enhance the dataset's credibility. Besides, we categorize the questions into four types given the complexity of getting answers.

On the constructed ConferenceQA dataset, we introduce STAR (**ST**ructure-**A**ware **R**etrieval), a method leveraging LLMs for hierarchical data, and then proceed to conduct a study on conference QA. Our method generates a textual description for each path based on both its surrounding structural information and its own textual information. We conduct experiments using various LLMs, along with different retrievers. Compared to path retrieval, structural-aware retrieval shows an average relative F1 score improvement of 15.50% across different LLMs and 17.03% when using different retrievers. This highlights the effectiveness of STAR on the tree-structured ConferenceQA dataset.

Our contributions can be summarized as follows:

1. We construct a benchmark called ConferenceQA, organizing conference information in a tree structure, to assist evaluate question answering about academic conferences.

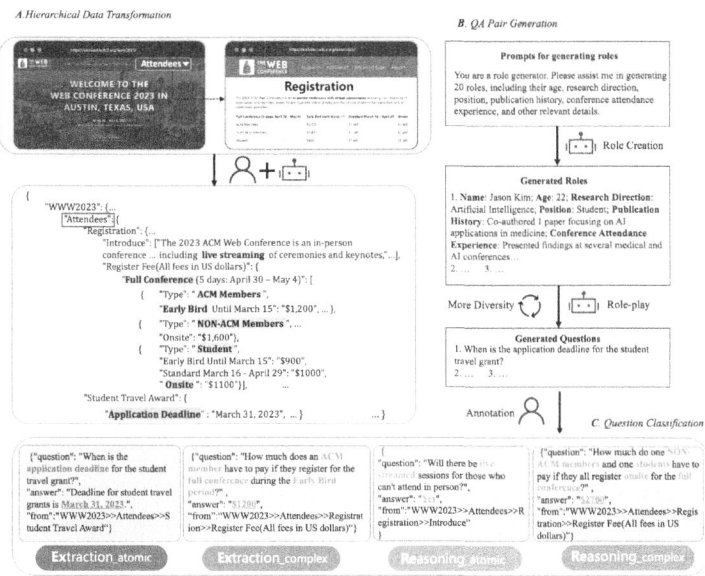

Fig. 1. The illustration of the ConferenceQA dataset construction process. Initially, data from official websites is transformed into a tree structure semi-automatically. Next, questions are generated through role-play and manually annotated with answers. Finally, questions are categorized by the complexity of paths and reasoning required.

2. We introduce a novel method called STAR. By utilizing the structural information around nodes to generate textual descriptions, and using these descriptions for retrieval, it can effectively enhance answer performance.
3. We conduct experiments on the ConferenceQA dataset, proving that LLMs enhanced with retrieval methods could successfully answer questions about academic conferences and our STAR method consistently outperforms path retrieval method, offering meaningful insights.

2 Dataset Construction

In this section, we introduce the construction of the ConferenceQA dataset. We select the conference information of seven typical academic conferences in 2022 or 2023 to build the dataset based on their official website, where the most accurate information about the conferences is stored. Each conference is assigned to one data annotator with relevant experience in the realm of academic conferences. We use three steps, including *hierarchical data transformation*, *QA pair generation* and *question classification*, to construct each conference dataset. The overview of the construction process is shown in Fig. 1.

2.1   Hierarchical Data Transformation

Data transformation in the ConferenceQA dataset involves standardizing the diverse formats of academic conference data sourced from official conference websites into a unified tree structure. Each conference page combines unstructured text, like conference introductions and paper submission guidelines, with structured data such as payment and schedule details. To manage this format variability, we employ a semi-automated method to create tree-structured data for each conference.

Specifically, the automated component converts structured table data into a tree format using ChatGPT, as shown in Fig. 1, where registry information is transformed. For other structured data, such as accepted papers with consistent schemas (title, authors, abstract), we employ web crawlers to fetch HTML pages and convert them into corresponding tree-structured data based on the HTML tags. The manual component involves annotating inter-page relationships. Annotators assign page titles to tree nodes based on the linkage among pages, evident in navigation bars and subpage links like 'calls', 'proceedings' and 'programs'. Additionally, subtitles within pages are identified and designated as child nodes under the relevant page titles. These manual steps are essential to maintain the dataset's quality and coherence.

Ultimately, we obtain seven conference datasets organized in a tree-structured format. They are served as accurate and rigorous knowledge sources.

2.2   QA Pair Generation

This step involves generating reliable question-answer pairs through role creation, LLM-generated questions, and manual annotation. For each conference, we utilize ChatGPT to simulate the roles of conference participants, generating relevant questions which are then manually filtered and annotated with answers and their sources to ensure realism and reliability.

We use ChatGPT to create 20 roles characterized by specific attributes such as age, research direction, position, publication history, and conference attendance experience, mimicking real-life researchers with diverse backgrounds interested in the conferences. With these roles, we prompt ChatGPT to engage in role-playing scenarios, generating five varied questions per conference. These questions cover different areas of interest or uncertainty relevant to the roles' diverse backgrounds. To avoid redundancy and enhance question diversity, we iteratively prompt the model. Specifically, we use the results generated by the ChatGPT as examples for the next iteration and encourage the ChatGPT to generate more diverse questions. In the final step, we manually review and filter the questions to eliminate duplicates and unrealistic queries. We then annotate the answers based on our tree-structured data, ensuring the reliability of the dataset by documenting the source of each answer within the constructed academic conference data.

2.3   Question Classification

To assess the model's capability in handling questions of varying difficulty, we design a scheme to classify the question-answer pairs based on two criteria: the method used to generate the answer and the complexity of paths required to arrive at the correct answer.

Extraction vs. Reasoning: This category evaluates the process of answer generation. Answers directly pulled from the dataset are labeled as *extraction*, whereas answers that necessitate reasoning beyond the dataset content are labeled as *reasoning*. *Reasoning* questions are more challenging than *extraction* questions because, unlike direct extraction, *reasoning* questions require the model to have the capability to infer the relationship between the retrieved paths and the question.

Atomic vs. Complex: This category assesses the complexity of paths needed to generate the answer. Answers that depend on a single path are termed *atomic*, while those requiring multiple paths are termed *complex*. *Complex* questions are more difficult than *atomic* questions because, instead of a single path, *complex* questions require recalling multiple paths to derive an answer.

Combining these dimensions results in four levels of difficulty: *extraction-atomic*, *extraction-complex*, *reasoning-atomic*, and *reasoning-complex*. This classification is vital for analyzing the model's performance across different complexities and reasoning demands.

2.4   Dataset Validation

Following data construction, a thorough validation process is conducted by three independent assessors who evaluate each QA pair across three critical dimensions. The first dimension assesses the alignment between each question and its answer, ensuring the answer accurately addresses the question. Concurrently, the second dimension examines the reliability of the answer source, ensuring it provides the necessary information for the question. The third dimension evaluates the practical relevance of each question, ensuring it reflects real-world needs and concerns. If a QA pair fails to meet the criteria in any dimension, as agreed upon by at least two assessors, it is marked for removal and redesign. This rigorous process ensures each QA pair is validated comprehensively, maintaining the quality and reliability of the dataset. Detailed statistics of the selection process for each conference are shown in Table 1.

Table 1. Statistics of the ConferenceQA dataset. #Paths indicates the number of tree branches, and #Depth shows the depth of the tree. #EA, #EC, #RA, and #RC represent extraction-atomic, extraction-complex, reasoning-atomic, and reasoning-complex question types, respectively.

Conference	#Paths	#Depth	#EA	#EC	#RA	#RC
WWW2023	15127	7.01	32	27	17	36
ACL2023	14306	9.05	29	21	30	25
ICML2023	4715	8.52	26	27	28	19
SIGMOD2023	6338	7.46	39	27	23	34
IJCAI2023	15800	6.13	28	26	13	33
ICDE2023	9736	9.14	28	24	22	21
ISWC2022	3594	7.53	33	42	25	18
Avg	9916	7.83	31	28	23	27

3   Method

In this section, we discuss LLM-based methods for academic conference question-answering. The prevalent approach involves using an external knowledge source for retrieval [10,13,15], where the reader's query q extracts relevant content c from a domain-specific knowledge base, and this content is then combined with the query for the LLM to generate an answer. This retrieval-based method can be formalized as $a = LLM(q, c)$ where $c = Retriever(q, \mathcal{KB})$. It optimizes the retriever, such that for each question q, the model can give an answer a that has high accuracy or relevancy with a correct answer. Our approach adheres to this retrieval-based model but is adapted for our conference's tree-structured dataset. We preprocess this structured data to facilitate content retrieval and introduce a novel method named STAR (**ST**ructure-**A**ware **R**etrieval), which effectively integrates structural and semantic data for improved retrieval performance.

3.1   Tree-Structured Data Processing

The tree-structured data is hierarchically arranged, with each node representing a page or a section heading, and each leaf node corresponding to its specific content. For retrieval, we pair each leaf node with its root node to provide additional context to the LLM. Paths in the tree use the '»' field to denote hierarchical relationships and contain both structural and semantic information. An example path is: *WWW2023»Attendees»Registration»Register Fee»Virtual Conference»ACM Members»$300*. After the tree-structured data processing, the knowledge source for retrieval could be represented as a set of paths that $\mathcal{P} = \{p_1, p_2, ..., p_m\}$ where m is the number of paths in the dataset.

3.2   Path Retrieval

Upon receiving a query input q, the retriever selects a subset of paths from $\mathcal{P} = \{p_1, p_2, ..., p_m\}$ that are relevant to q. Following established methods [20], we use a dense retriever based on a dual encoder framework. This framework employs an encoder to transform both the query q and each path $p \in \mathcal{P}$ into embeddings. The similarity between the query and path embeddings is assessed using cosine similarity, and the top-k paths with the highest similarity scores are retrieved, as expressed in (1), where $\mathbf{E}$ denotes the embedding function.

$$c = topk(\{\cos(\mathbf{E}(q), \mathbf{E}(p))|p \in \mathcal{KP}\}) \tag{1}$$

3.3 Structure-Aware Retrieval

The limitation of treating a single path as the retrieval object is that it disconnects the structural relationships among paths. For example, the relationship between an author's name and their affiliated institution is lost when paths are retrieved independently.

To overcome this, we introduce a novel method called STAR (**ST**ructure-**A**ware **R**etrieval). As shown in Fig. 2, STAR employs ChatGPT to iteratively generate textual descriptions for each path des_p, from the root to individual nodes, in a top-down manner. We enhance the retrieval process by incorporating structural information in the user input, which includes siblings, parent path descriptions, and the query path itself. This approach helps maintain the contextual relevance of each path, which is crucial for recognizing relationships like those between an author and their institution. For instance, when generating path descriptions, we not only consider the node's immediate context but also integrate the structural significance of related nodes. This includes the siblings of a node and their parent nodes, ensuring a comprehensive representation of each path's context. To avoid the loss of information about the siblings of leaf nodes, we append the text of their parent node to each sibling of the leaf nodes. Ultimately, this method effectively preserves and utilizes structural relationships, enhancing the retrieval process.

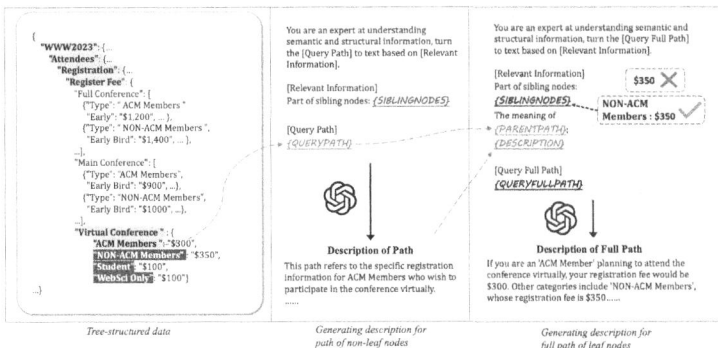

Fig. 2. The diagram depicts the iterative, top-to-bottom generation of path descriptions using ChatGPT, which incorporates surrounding structural information.

Thus we can construct a knowledge source of path descriptions $\mathcal{KP}_{des_p} = \{(p, des_p)|p \in \mathcal{KP}\}$, containing pairs of paths and their descriptions. For retrieval, we use the similarity between the query and each path description as the score for that path. We then retrieve the top-k paths with the highest similarity scores to the query q. The embedding of the element is denoted by $\mathbf{E}$, and this process is formalized as shown in (2).

$$c = topk(\{\cos(\mathbf{E}(q), \mathbf{E}(des_p))|(p, des_p) \in \mathcal{KP}_{des_p}\}) \tag{2}$$

4   Experiments

In this section, we conduct question answering experiments on conference datasets to explore: 1) How does the STAR perform with different LLMs? 2) How does the STAR perform with different retrievers? 3) How does the STAR perform with different academic conferences?

4.1   Experimental Details

Based on the constructed ConferenceQA, we use currently popular LLMs, including Bloom (7B) [31], GPT-J (6B) [30], Flan-T5 (xl and xxl) [29], LLaMA2 (7B and 13B) [7], Mistral (7B) [25] and ChatGPT, as the main evaluation backbone to assess the performance of mainstream LLMs. For ChatGPT, we employ GPT-3.5-turbo and access it via API[1] We employ BM25 [1], SentenceBert [26], DPR [27], ANCE [28] and text-embedding-ada-002 as our retriever. In addition, we use Chroma[2] as our vector database and employ cosine similarity for matching. In all experiments, we select the top 5 paths retrieved.

4.2   Evaluation Metrics

In line with prior studies, we assess the QA capabilities of LLMs using the F1 score and the exact match (EM) score. Specifically, we employ GPT-4 to compute the EM, referred to as EM-GPT4.

The F1 score quantifies the overlap between the predicted and correct answers by calculating the harmonic mean of precision and recall.

The EM-GPT4 score evaluates the proportion of instances where the LLM's predicted answer exactly matches the correct answer. Given the generative nature of LLMs, slight textual variations in responses might still represent the same answer. We use GPT-4, a highly advanced LLM known for its semantic understanding capabilities, to precisely assess if the LLM's response matches the golden answers.

---

[1] from https://api.openai.com/.
[2] https://github.com/chroma-core/chroma.

Table 2. EM-GPT4 and F1 scores for various LLMs in the ConferenceQA dataset, categorized by question types: EA (extraction-atomic), EC (extraction-complex), RA (reasoning-atomic), and RC (reasoning-complex). The retriever used is text-embedding-ada-002. Black numbers show path retrieval performance. '+' denotes performance improvement with the STAR method. Red indicates positive enhancements, Green signifies reductions.

LLMs	F1				EM-GPT4			
	EA	EC	RA	RC	EA	EC	RA	RC
Bloom-7B1	19.60-0.96	11.19+1.96	17.01+0.36	11.58-0.20	30.27+0.42	15.03+4.41	41.70-3.43	17.58-1.36
GPT-J-6B	14.53+1.76	8.81+3.11	15.52+2.46	8.42-0.25	19.11+7.04	12.93+5.53	34.16+5.03	13.08+1.94
Flan-T5-xl	27.74+7.85	14.68+2.77	36.03+0.96	19.00+2.89	35.50+9.56	20.78+0.86	59.38+3.97	25.74+2.27
Flan-T5-xxl	32.31+8.97	14.01+9.69	37.08+4.3	20.86+0.98	40.81+10.69	18.58+12.64	55.76+11.71	25.36+3.58
LLaMA2-7B	14.05+2.23	12.09+1.25	12.47+3.00	8.48+0.12	21.32+0.15	9.22+2.77	23.81+5.83	9.89-1.15
LLaMA2-13B	29.57+2.82	20.92+4.00	25.71+4.20	13.64+2.83	41.16+6.26	24.02+2.21	55.00+6.53	20.23+4.46
Mistral-7B	30.75+4.31	23.67+4.69	25.87+4.11	15.91-0.37	43.33+10.53	27.58+13.95	59.90+6.89	29.23+1.55
GPT-3.5-turbo	28.35+7.5	21.54+4.83	24.66+9.62	16.21+0.78	40.53+13.43	25.10+9.03	49.97+11.34	25.75+1.45

4.3 Experimental Results Analysis

Effect of Different LLMs. We analyzed the performance of various LLMs on different types of questions to understand their perception capabilities and limitations. The results, shown in Table 2, provide several insights: (1) Our STAR method significantly improves the answering performance across various LLMs. For instance, on models like Bloom-7B1, GPT-J-6B, and GPT-3.5-turbo, F1 scores increased by 4%, 14.9%, and 25.04% respectively, while EM-GPT4 scores improved by 0.04%, 24.65%, and 24.94%. The least improvement was on Bloom-7B1, suggesting its inherent limitations. However, substantial gains on other models demonstrate our method's effectiveness. (2) There is an inconsistency between F1 and EM-GPT4 scores; lower F1 scores sometimes align with higher EM-GPT4 scores. This may be due to LLMs generating longer textual responses, affecting F1 accuracy but not EM-GPT4, which better evaluates semantic similarity. (3) The complexity of question types affects performance; atomic questions are simpler than complex ones. Atomic questions, akin to single-hop queries, generally show higher accuracy than multi-hop complex questions. Despite this, LLMs perform comparably or better on reasoning questions than on extraction, likely due to their robust contextual learning and reasoning capabilities. (4) Different LLMs show varied understanding of paths. For example, under the same retrieval conditions, Mistral-7B outperforms GPT-3.5-turbo. Generally, models with more parameters, like LLama2-13B and Flan-T5-xxl, achieve higher accuracy, supporting the notion that larger LLMs perform better.

Effect of Different Retrievers. We evaluated four retrievers-BM25 [1], SentenceBert [26], DPR [27], and ANCE [28]-using the gpt-3.5-turbo generator across four question types within the ConferenceQA dataset. The results, detailed in Table 3, reveal: (1) BM25 showed weak performance, especially with *extraction-atomic* and *reasoning-complex* questions. In contrast, dense retrievers like SentenceBERT, DPR, and ANCE significantly outperformed BM25,

Table 3. EM-GPT4 and F1 scores for different retrievers in the ConferenceQA dataset. The generator is GPT-3.5-turbo.

Retrievers	F1				EM-GPT4			
	EA	EC	RA	RC	EA	EC	RA	RC
BM25	21.02+9.14	16.81+10.29	25.77-2.79	14.72+1.37	25.90+4.9	14.12+8.1	36.50+2.94	5.16+5.55
SentenceBERT	38.62-3.74	23.70-0.01	12.97+2.05	16.23-0.18	39.83-2.57	14.81+11.12	29.79+17.66	22.88-1.47
DPR	30.56+3.72	23.72+0.59	27.35+2.60	21.16-0.13	30.95+5.91	15.17-0.14	52.70+0.31	10.66+1.38
ANCE	28.24+8.72	17.41+4.75	16.52+9.1	18.00+2.26	41.66+6.14	30.12+4.07	50.84+0.49	15.25+1.21
ada-002	28.35+7.5	21.54+4.83	24.66+9.62	16.21+0.78	40.53+13.43	25.10+9.03	49.97+11.34	25.75+1.45

underscoring the advantages of dense retrieval methods. (2) Performance varied among dense retrievers: SentenceBERT was effective in *extraction-atomic* questions but less so in *reasoning-atomic* questions. DPR excelled in *reasoning-atomic* questions, while ANCE showed consistent performance across all question types. This indicates that selecting an appropriate retriever can significantly impact question-answering effectiveness. (3) While STAR occasionally had negative effects in some configurations, it generally enhanced performance across most settings, demonstrating its utility and reliability.

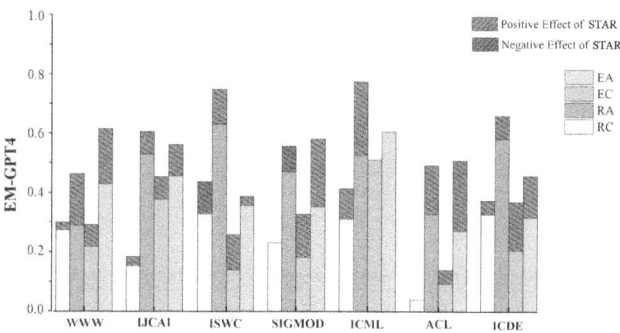

Fig. 3. EM-GPT4 metrics across conferences using text-embedding-ada-2 for retrieval and GPT-3.5-turbo for generation. Bars show path retrieval performance; green segments indicate STAR method improvements, red segments show declines. (Color figure online)

Effect of Different Conference. Figure 3 shows the performance across various conferences using text-embedding-ada-002 and gpt-3.5-turbo as the retriever and generator, respectively. Key observations include: (1) There is notable variability in question difficulty across conferences, highlighting the diversity of our dataset. (2) Significant differences in difficulty are apparent between conferences; for example, the average EM-GPT4 score at ICML is 94.9% higher than at ACL, underscoring the importance of accounting for conference-specific characteristics in question-answering research. (3) Except for *reasoning-atomic* questions at SIGMOD and *reasoning-complex* questions at ISWC, our STAR method consistently outperforms traditional path retrieval, demonstrating its versatility and effectiveness across different conferences and question types.

5   Related Work

In academic data science, foundational resources such as CiteSeerX [19], a digital library for scientific literature, and Unarxive [18], which hosts over a million documents from arXiv.org, are crucial for scholarly communication. Zhang et al. [17] developed Maple, a benchmark for tagging scientific literature across 19 disciplines. However, there remains a notable gap in benchmarks specifically designed for academic conference QA, despite the increasing diversity and volume of literature datasets.

Simultaneously, augmenting language models with data from various knowledge bases has significantly improved performance across many NLP tasks [22, 23]. Techniques such as Atlas [11], which fine-tunes an encoder-decoder model with a retriever, and RETRO [12], which integrates retrieved texts into a decoder-only model, utilize large volumes of unstructured text. Other approaches like REPLUG [13] and FLARE [14] dynamically retrieve information based on context, treating LLMs as black boxes. In structured knowledge, methods include extracting triples from knowledge graphs for KGQA tasks [10, 15] and converting them into textual prompts for LLMs [24] However, the use of hierarchical data such as tree-structured data in retrieval augmentation is still limited.

6   Conclusion

In this work, we developed the ConferenceQA dataset, which organizes recent academic conference information into a tree-structured format to support question answering. We introduce a novel approach, STAR, that enhances question-answering performance by generating textual descriptions for each path within the tree, effectively utilizing both structural and textual data. The ConferenceQA dataset and STAR method have advanced the development of robust and adaptable academic conference question-answering systems. Future efforts will focus on integrating LLMs with tree-structured data to improve domain-specific knowledge access and reasoning.

Acknowledgements. This work is founded by National Natural Science Foundation of China (NSFC62306276), Zhejiang Provincial Natural Science Foundation of China (No. LQ23F020017), Yongjiang Talent Introduction Programme (2022A-238-G), Ningbo Natural Science Foundation (2023J291), and Fundamental Research Funds for the Central Universities (226-2023-00138).

References

1. Robertson, S., Zaragoza, H., et al.: The probabilistic relevance framework: BM25 and beyond. Found. Trends® Inf. Retrieval (2009)
2. OpenAI: GPT-4 technical report (2023)
3. Wei, J., et al.: Chain-of-thought prompting elicits reasoning in large language models. In: Advances in Neural Information Processing Systems (2022)
4. Kojima, T., Gu, S.S., Reid, M., Matsuo, Y., Iwasawa, Y.: Large language models are zero-shot reasoners. In: Advances in Neural Information Processing Systems (2022)
5. Chung, H.W., et al.: Scaling instruction-finetuned language models. arXiv preprint arXiv:2210.11416 (2022)
6. Min, S., Lyu, X., Holtzman, A., et al.: Rethinking the role of demonstrations: what makes in-context learning work? arXiv preprint arXiv:2202.12837 (2022)
7. Touvron, H., Martin, L., Stone, K., et al.: Llama 2: open foundation and fine-tuned chat models (2023)
8. Modarressi, A., Imani, A., Fayyaz, M., Schütze, H.: RET-LLM: towards a general read-write memory for large language models (2023)
9. Schick, T., et al.: Toolformer: language models can teach themselves to use tools (2023)
10. Hu, C., Fu, J., Du, C., Luo, S., Zhao, J., Zhao, H.: Chatdb: augmenting LLMs with databases as their symbolic memory (2023)
11. Izacard, G., et al.: Atlas: few-shot learning with retrieval augmented language models (2022)
12. Borgeaud, S., Mensch, A., Hoffmann, J., et al.: Improving language models by retrieving from trillions of tokens. In: International Conference on Machine Learning. PMLR (2022)
13. Shi, W., et al.: Replug: retrieval-augmented black-box language models, arXiv preprint arXiv:2301.12652 (2023)
14. Jiang, Z., et al.: Active retrieval augmented generation, arXiv preprint arXiv:2305.06983 (2023)
15. Sen, P., Mavadia, S., Saffari, A.: Knowledge graph-augmented language models for complex question answering (2023)
16. Zhong, V., Xiong, C., Socher, R.: Seq2sql: generating structured queries from natural language using reinforcement learning, arXiv preprint arXiv:1709.00103 (2017)
17. Zhang, Y., Jin, B., Zhu, Q., Meng, Y., Han, J.: The effect of metadata on scientific literature tagging: a cross-field cross-model study. In: Proceedings of the ACM Web Conference 2023 (2023)
18. Saier, T., Färber, M.: Unarxive: a large scholarly data set with publications' full-text, annotated in-text citations, and links to metadata. Scientometrics (2020)
19. Giles, C.L., Bollacker, K.D., Lawrence, S.: Citeseer: an automatic citation indexing system. In: Proceedings of the Third ACM Conference on Digital Libraries (1998)
20. Ni, J., Qu, C., Lu, J., Dai, Z., et al.: Large dual encoders are generalizable retrievers. arXiv preprint arXiv:2112.07899 (2021)
21. Brown, T., et al.: Language models are few-shot learners. In: Advances in Neural Information Processing Systems (2020)
22. Guu, K., Lee, K., Tung, Z., Pasupat, P., Chang, M.: Retrieval augmented language model pre-training. In: International Conference on Machine Learning. PMLR (2020)

23. Lewis, P., Perez, E., Piktus, A., Petroni, F., et al.: Retrieval-augmented generation for knowledge-intensive NLP tasks. In: Advances in Neural Information Processing Systems (2020)
24. Wu, Y., et al.: Retrieve-rewrite-answer: a kg-to-text enhanced LLMs framework for knowledge graph question answering. arXiv preprint arXiv:2309.11206 (2023)
25. Jiang, A.Q., Sablayrolles, A., Mensch, A., et al.: Mistral 7B. arXiv preprint arXiv:2310.06825 (2023)
26. Reimers, N., Gurevych, I.: Sentence-bert: sentence embeddings using siamese bert-networks. arXiv preprint arXiv:1908.10084 (2019)
27. Karpukhin, V., et al.: Dense passage retrieval for open-domain question answering. arXiv preprint arXiv:2004.04906 (2020)
28. Xiong, L., et al.: Approximate nearest neighbor negative contrastive learning for dense text retrieval. arXiv preprint arXiv:2007.00808 (2020)
29. Longpre, S., Hou, L., Vu, T., Webson, A., et al.: The flan collection: designing data and methods for effective instruction tuning. In: International Conference on Machine Learning. PMLR (2023)
30. Wang, B., Komatsuzaki, A.: GPT-J-6B: a 6 billion parameter autoregressive language model (2021)
31. Le Scao, T., Fan, A., Akiki, C., Pavlick, E., et al.: Bloom: a 176B-parameter open-access multilingual language model (2022)

Knowledge Graph and Large Language Model Applications

Benchmarking Knowledge Graph-Grounded Factual Verification

Xinyan Guan[1,3], Hongyu Lin[1(✉)], Yaojie Lu[1], Sirui Wang[4], Xunliang Cai[4],
Xianpei Han[1,2(✉)], and Le Sun[1,2]

[1] Chinese Information Processing Laboratory, Beijing, China
{guanxinyan2022,hongyu,luyaojie,sunle}@iscas.ac.cn
[2] State Key Laboratory of Computer Science Institute of Software, Chinese Academy of
Sciences, Beijing, China
xianpei@iscas.ac.cn
[3] University of Chinese Academy of Sciences, Beijing, China
[4] Meituan, Beijing, China
{wangsirui,caixunliang}@meituan.com

Abstract. Utilizing knowledge graphs to verify AI-generated content is an important approach to alleviate hallucination issue of large language models (LLMs). Previous work has typically focused on the impact of integrating knowledge graph-based verification to the outcomes produced by LLMs, while neglecting to assess the effectiveness and reliability of the verification process itself. In this paper, we present KGFV , a benchmark specifically developed to comprehensively evaluate the procedure of knowledge graph-based fact verification. Specifically, KGFV aims to provide an end-to-end evaluation framework for assessing the verification capabilities involved in comparing and validating AIGC with knowledge graph information. To this end, KGFV features a diverse set of factual scenarios, encompassing simple facts, complex multi-hop reasoning, comparative analyses, and set operations. Furthermore, KGFV also provides a wide range of intermediate information, which can serve as a robust foundation for future research. Experiments with several state-of-the-art fact verification approaches on KGFV demonstrate that there is still a long way to go to the effective and reliable fact verification between AIGC and knowledge graph.

1 Introduction

Large Language Models (LLMs) have gained increasing prominence in artificial intelligence. The emergence of potent models such as ChatGPT [19] and LLaMA [24] has led to substantial influences on many areas like society, commerce, and research. However, LLMs are challenged by significant inaccuracies, often referred to as hallucinations [27] (Fig. 1).

To mitigate hallucinations in AI-generated content (AIGC) [1], verifying AIGC against high-confidence sources has proven to be a highly effective method. The adoption of knowledge graphs (KGs) for alleviating hallucination has seen a surge in interest due to their comprehensive, precise, and structured data. For example, [10] has introduced a comprehensive framework leveraging KGs to both verify and enhance AIGC.

B. Xu et al. (Eds.): CCKS-IJCKG 2024, CCIS 2229, pp. 197–209, 2025.
https://doi.org/10.1007/978-981-96-1809-5_15

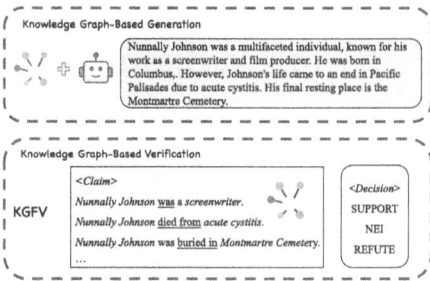

Fig. 1. Illustration of KGFV benchmark, designed to evaluate LLM's Knowledge Graph-grounded ability for AIGC.

Unfortunately, this field still lacks robust evaluation methods for the full extraction and verification process utilizing KGs, which presents two major challenges. Firstly, existing research predominantly concentrates on the final verification outcomes of AIGC. For example, FactCHD [3] introduces a Query&Response format verification benchmark that categorizes answers simply as factual or not by judging the similarity between answers and reference answers. Additionally, its evaluation is based on retrieving knowledge from the web rather than grounding it to a specific knowledge base. These approaches are inadequate because the essence of verification lies in pinpointing the exact discrepancies between AIGC and the knowledge base, thereby highlighting the need for more detailed verification strategies that consider the entire text. Secondly, the straightforward nature of existing KG-based verification datasets falls short of addressing the multifaceted nature of real-world AIGC, which typically involves multiple claims and requires the extraction of discrete claims for accurate assessment. For example, FactKG [16] introduced a KG-based verification challenge incorporating five distinct reasoning types. However, each data point in this setup is limited to a simple claim with a small number of triples, underscoring the gap between the dataset's simplicity and the complexity of actual AIGC scenarios. Accordingly, there's a critical need to establish a benchmark capable of accurately evaluating KG-based verification in real-world conditions.

In this paper, we introduce a new benchmark, Knowledge Graph Fact Verification (KGFV), designed to detect discrepancies between AI-generated content and knowledge graphs. KGFV features complex text with multi-claims, each classified under three labels: SUPPORT, REFUTE, and No Enough Info, based on the availability of sufficient evidence to either corroborate or contradict the text. As shown in Table 2, we construct texts with four reasoning types, including Vanilla, Multi-hops, Comparison, and Set Operations. KGFV also outlines three essential subtasks for factual verification: Claim Extraction, Evidence Retrieval, and Claim Verification, challenging LLMs to navigate through these reasoning paradigms, as illustrated in Table 1. We conduct experiments on the open-source model to evaluate the efficacy of LLMs in KG verification ability, highlighting the critical nature of this task.

In summary, the contributions are as follows:

Table 1. Comparison with existing fact-checking datasets. Our KGFV labels each claim in texts strictly based on a specific knowledge graph with 3 labels SUPPORT, REFUTE, and No Enough Info, including four reasoning types. The *Knowledge Base* refers to the knowledge base on which this task is grounded. The *Evaluation Granularity* refers to label data on the whole text (Overall) or each claim (Fine-grain).

Dataset	Knowledge Base	Content	Reasoning Type	Verify Labels	Evaluation Granularity
FEVER [22]	Wikipedia	Single Claim	1	S/R/N	Overall
FactCHD [3]	World Knowledge	QA	4	S/R	Overall
TableFact [2]	Table	Single Claim	2	S/R	Overall
FactKG [16]	KG	Single Claim	5	S/R	Overall
MCFC [25]	KG	Multiple Claim	1	S/R	Overall
KGFV (ours)	KG	Multiple Claim	4	S/R/N	Fine-grain

- We propose a novel data construction approach for generating KG-based verification datasets using LLMs. This approach allows for the easy creation of KG-based verification data that is sensitive to specific domains and scenarios.
- Our method introduces a comprehensive verification task that leverages knowledge graphs to evaluate texts with multiple claims across four reasoning types, organizing outcomes into three categories: SUPPORT, REFUTE, and No Enough Information.
- Based on the proposed benchmark, we introduce three subtasks for evaluating the models' verification ability. Our experiments with open-source LLMs shed light on the strengths and weaknesses of different models on the benchmark.

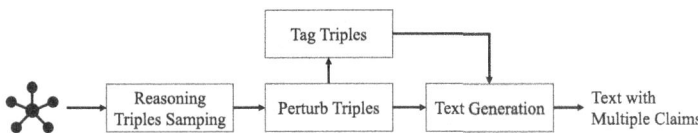

Fig. 2. Data Construction Pipeline.

2 Related Work

2.1 KG-Based Data Construction

Numerous datasets leveraging the rich and structured knowledge from knowledge graphs have been developed for various purposes. For example, WebNLG [7] constructs data by first manually writing sentences for each triple and then manually merging these sentences for text-to-triple tasks. FactCHD [3] combines ChatGPT and manual efforts to construct questions based on triples, generating factual or non-factual answers for hallucination verification tasks. SimpleQuestions [8] focuses on generating question-answer pairs to evaluate a model's capability to provide answers utilizing knowledge graph data.

2.2 Fact Verification

As large language models (LLMs) become more widespread, there's growing concern over their tendency to generate incorrect or "hallucinated" information. To tackle this, researchers have devised various fact-checking methods [27], categorized into three main types based on the source of knowledge: self-critique, search-engine-based, and KG-based. Self-critique-based [6,20] methods prompt LLM to evaluate its own output for accuracy. Search-engine-based [4,9,17] techniques use knowledge from search engines to verify facts in the LLM's responses. KG-based methods [10,15] check the accuracy of an LLM's response by consulting a knowledge graph and using queries or embeddings to acquire the related information. In this paper, we mainly focus on the KG-based method, as our objective is to conduct verification within a defined knowledge base boundary.

2.3 KG-Based Verification Benchmark

The landscape of KG-based verification benchmarks currently shows a significant gap, as most available datasets typically focus on single-claim verification. For example, FactKG [16] proposed a KG-based verification task with five types of reasoning. Another dataset [25] proposed a multi-claims verification task based on a real-world KG in the food domain with rigid text generation rules. However, real-world applications often require the verification of texts containing multiple claims and the expression of knowledge in varied forms. To address this gap, our KGFV leverages the capabilities of LLMs for automatic generation, offering diverse and natural integration of multiple claims and knowledge.

Table 2. Four different reasoning types of KGFV and each associated with its unique method of triple sampling graph structures.

Type	Example		Graph
Vanilla	Text:	Nunnally Johnson was a multifaceted individual, known for his work as a screenwriter and film producer...	
	Claim:	Nunnally Johnson was a screenwriter.	
	Evidence:	(Nunnally J, is a, screenwriter)	
	Label:	SUPPORT	
Multi-hop	Text:	Amy Brenneman is married to Billy Bob Thornton, who has received the Academy Award for Best Writing,...	
	Claim:	Amy Brenneman is married to Billy Bob Thornton.	
	Evidence:	(Angelina Jolie, spouse, Billy Bob Thornton)	
	Label:	REFUTE	
Set	Text:	The films that were released in Malaysia and composed by Howard Shore are Saboteur, The Lord of the Rings...	
	Claim:	Saboteur was released in Malaysia.	
	Evidence:	No evidence show that Saboteur was relased in Malaysia.	
	Label:	NEI	
Comparison	Text:	On December 31, 1999, among the five Finnish cities ..., Helsinki had the highest population, with a total of 551,123 residents.	
	Claim:	On December 31, 1999, the population of Helsinki was 551,123.	
	Evidence:	((Helsinki, population, 551123), point in time, 1999-12-31)	
	Label:	SUPPORT	

3   KGFV

In this section, we first formalize the three subtasks of KGFV : Claim Extraction, Evidence Retrieval, and Claim Verification. Then, we will describe the methodology used to construct the dataset benchmark. Lastly, we explain the evaluation metrics for each task.

3.1   Task Formulation

We framework our KGFV benchmark into three main subtasks: 1) Claim Extraction (CE) 2) Evidence Retrieval (ER) 3) Claim Verification (CV)

Claim Extraction (CE).   Given a text T that contains multiple claims, the CE task is to extract the claims mentioned in the text in the form of $\{c_1, c_2, \ldots, c_n\}$. In this paper, we define the claim to be either atomic claims that are attached to only one triple or claims that are deduced from other claims. The claims should also be context-independent [26].

Evidence Retrieval (ER).   Given a claim c and a specific KG, the Evidence Retrieval task is to identify the related evidence $\{e_1, e_2, \ldots, e_n\}$ within the KG. Here, each e_i represents a specific triple.

Claim Verification (CV).   Given a Claim c and a set of related evidence $\{e_1, e_2, \ldots, e_n\}$ in the KG, the objective of the Claim Verification task is to classify the verification result into SUPPORT, REFUTE, or No Enough Information. The categorization is based on whether the evidence provided confirms the claim, opposes it, or if the available information is inadequate for making a conclusive determination. The inherent nature of Knowledge Graphs as databases emphasizes high accuracy yet potential incompleteness [12]. Therefore, establishing these three classification labels for verification is crucial for a comprehensive and nuanced analysis.

3.2   Data Construction

To generate KGFV benchmark, we start from sample triples from a given knowledge graph and send it to LLM for text generation. To produce text that contradicts the information in the knowledge graph, we modify the triples either by swapping out entities in a correct triple or altering the quantitative attributes of an entity. The data construction pipeline is shown in Fig. 2.

Reasoning Triples Sampling.   Our dataset encompasses four types of reasoning: vanilla, multi-hop, set operation, and comparison. For each reasoning type, we employ distinct sampling strategies to select triples, which are then used to generate texts automatically by prompting LLMs. The entity's relation graph is shown in Table 2

Vanilla.   The basic approach involves organizing a primary entity in connection with several adjacent entities. Within a knowledge graph $G = \{V, E\}$, we derive a smaller graph $G' = \{(v_0, r_1, v_1), (v_0, r_2, v_2), \ldots, (v_0, r_m, v_m)\}$, with v_0 being the focal entity. Here, we select m entities that are directly connected to v_0, with r_i indicating the type of relationship and v_i representing each connected node. This approach enables us to create a straightforward textual representation of triples, which in turn facilitates the depiction of a network diagram that visually represents these relationships.

Multi-hop. The Multi-hop paradigm incorporates additional layers of relational depth between entities in a knowledge graph $G = \{V, E\}$. Unlike the vanilla mode, which focuses on direct neighbors of a central entity, the Multi-hop approach explores entities that are indirectly connected through one or more intermediary entities. Specifically, we extract a subgraph $G' = \{(v_1, r_1, v_2), (v_2, r_2, v_3), \ldots, (v_{m-1}, r_{m-1}, v_m)\}$, including entities with m hops from v_1, effectively broadening the scope of the to-be generated text. This enriches the constructed text with a chain diagram that captures the complexity of interactions within the graph.

Set Operation. The Set Operation paradigm is constructed by describing a set of entities that share the same relations with the specific entities. In the context of a knowledge graph $G = \{V, E\}$, we extract a subgraph:

$$G' = \{(v_{m_1}, r_{n_1}, v_{n_1}), (v_{m_2}, r_{n_1}, v_{n_1}), \ldots,$$
$$(v_{m_i}, r_{n_j}, v_{n_j}), \ldots, (v_{m_M}, r_{n_N}, v_{n_N})\} \quad \text{for } i = 1 \ldots M, \text{ for } j = 1 \ldots N.$$

This operation specifically extracts M entities that exhibit the same relationships with N distinct entities, as illustrated in Table 2.

Quantitive Comparison. The Quantitive Comparison paradigm aims to extract the entities that have the same attributes. In this way, we can generate a text for comparing the sublative or superlative relation of the entities in these aspects. Given a knowledge graph $G = \{V, E\}$, we extract a subgraph $G' = \{(v_1, key, value_1), (v_2, key, value_2), \ldots, (v_m, key, value_m)\}$. Each of these entities possesses identical attributes, with values that can be quantitatively compared.

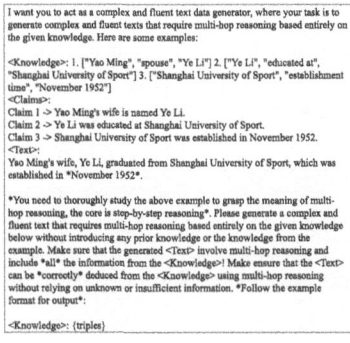

Construct Counterfactual Triple For Generating Text. Given a chain of triples, our goal is to prompt LLM to generate text for verification based on the triples. For constructing text with different types of factual errors, we should perturb the correctness of the given triples. For Vanilla, triples are perturbed by substituting related entities with others from the same concept. For Multi-hop, after identifying some entities for perturbing, some are replaced, affecting all related triples to ensure a consistent reasoning chain. For Set Operation, modifications involve changing some

Fig. 3. Text Generation Prompt for Multi-Hop texts.

mentioned subjects. Comparative Analysis involves altering attributes within the triples to random dates or quantities.

Upon generating these modified triples, we assess their accuracy through a heuristics process. First, we verify the existence of the triple in the knowledge graph. If found, it is labeled as SUPPORT. If not, and the triple is part of a one-to-one relationship, we look for an alternative triple that connects the same entities with the same relationship. If such a condition exists, the triple is deemed REFUTE. Otherwise, it is classified as NEI (No Enough Information).

Text Construction. Given the perturbed or not perturbed triple chains, we aim to prompt LLMs to generate text strictly based on the given triples. As a result, we construct few-shot prompt for each reasoning type of text generation. As shown in Fig. 3, We first ask LLM to generate the corresponding atomic claims based on the given triples. Then, we ask LLM to generate the final text based on the given reasoning type. All the prompts used for text construction are shown in the Appendix. In this paper, we use GPT-4(2023-03-15-preview) [18] for reliable text generation.

We construct KGFV based on the subgraph of Wikidata proposed by [23]. Finally, the data statistics are shown in Table 3.

Table 3. Data Statistics. We also calculate the average and maximum number of triples in each text.

	#train	#test	#Avg. triples	#Max.triples
Vanilla	340	100	15.75	20
Multi-hop	378	100	4.44	5
Set Operation	276	100	6.65	10
Comparison	335	100	4.96	5

3.3 Evaluating Metrics

Claim Extraction. For Claim Extraction, we focus on evaluating the similarity between claims generated by LLMs and the golden claims. This task is challenging due to the potential mismatch in the quantity of generated and golden claims, making direct pairing difficult. Following the methodology of Factool [4], we first establish pairs between each generated claim and golden claim, then compute precision, recall, and F1 score for the dataset. The formula for calculating the score is as follows:

$$Score(C, G) = \frac{1}{n} \sum_{i=1}^{n} max_{j=1}^{m}(\text{Sim}(c_i, g_j)) \qquad (1)$$

Here, $C = \{c_1, c_2, \ldots, c_n\}$ represents the set of n claims generated from text T, and $G = \{g_1, g_2, \ldots, g_m\}$ denotes the set of m golden claims associated with text T.

The term c_i refers to the i th generated claim, and g_j refers to the j th golden claim. Specifically, the similarity measurement is based on word overlap, utilizing Spacy [11] for analysis.

Evidence Retrieval & Claim Verification. In assessing the performance of Evidence Retrieval, we compute precision, recall, and F1 scores. For Claim Verification, accuracy serves as the primary metric, as our focus is on the model's ability to determine the verification outcome correctly.

4   Experiments

4.1   Baselines

To assess LLM's capabilities for Claim Extraction (CE) and Claim Verification (CV), we investigate LLMs across various paradigms to explore their performance in two key areas: 1) the ability to learn and utilize structural knowledge for grounding purposes, and 2) the ability to ground information in context and the potential for improvement through fine-tuning (Tables 4 and 5).

Table 4. Experiment results on 4 types of data and 3 open-source models on Claim Extraction (CE) and Claim Verification (CV) tasks. We use underlining to denote the highest score for each method and use bold to denote the overall highest score.

	Model	Vanilla		Multi-hop		Set Operation		Comparison		Avg	
		CE	CV	CE	CV	CE	CV	CE	CV	CE	CV
Few Shot	Llama	7.83	27.79	2.74	36.03	3.45	19.39	0.32	46.48	3.59	32.43
	Vicuna	68.73	45.95	**75.86**	49.67	73.95	53.16	61.99	37.12	70.13	46.47
	Mistral	68.71	79.88	65.17	68.13	52.05	77.56	51.58	30.32	59.38	**63.97**
Zero Shot	Llama	9.06	26.90	2.37	36.62	3.79	20.02	2.43	45.78	4.41	32.33
	Vicuna	67.11	50.53	75.40	51.22	74.96	52.44	60.42	28.27	69.47	45.61
	Mistral	**80.22**	40.51	68.22	33.70	**76.65**	38.39	51.73	31.98	69.20	36.15
Learning Knowledge	Llama	41.72	38.72	35.88	32.23	34.38	33.43	35.61	32.70	36.90	34.27
	Vicuna	57.96	61.00	55.33	50.15	69.12	60.88	51.94	32.63	58.59	51.17
	Mistral	26.13	68.35	31.50	68.25	41.81	69.93	26.75	37.73	31.55	61.06
Learning Context	Llama	71.72	94.40	70.37	89.42	62.13	95.73	75.98	89.52	70.05	92.27
	Vicuna	70.70	**97.63**	74.14	**94.33**	68.20	**96.74**	**78.22**	**92.22**	**72.82**	**95.23**
	Mistral	60.49	33.04	67.50	29.27	66.55	31.85	77.11	48.30	67.91	35.61

Table 5. Experiment results on Evidence Retrieval task.

Model	Vanilla	Multi-hop	Set Operation	Comparison	Avg
Local KG					
Llama	36.35	46.87	42.34	32.90	39.61
Vicuna	**69.10**	56.24	**75.28**	**94.93**	**73.89**
Mistral	32.84	**64.84**	51.06	67.00	53.94
Global KG					
Llama	6.29	7.53	5.60	6.12	6.39
Vicuna	6.78	5.25	6.34	7.32	6.42
Mistral	**19.55**	**24.05**	**21.31**	**36.77**	**25.42**

Few-Shot Learning. For the Claim Extraction task, we construct a few-shot prompt for LLM to understand the task. For the Claim Verification task, we prompt LLM to verify the claims one by one and output the final labels.

Zero-Shot. Different from the Few-Shot setting, we directly instruct LLM by describing the purpose of the task and regulating the output format.

Learning Structural Knowledge. We fine-tune LLMs with relevant structural knowledge. Specifically, we prepare a dataset for fine-tuning by concatenating triples into a string list format. To preserve the LLM's conversational capabilities, we incorporate ShareGPT [21] into the fine-tuning dataset.

Learning Context. The model is fine-tuned using a training dataset constructed with the prompt used in the Zero-Shot setting.

To assess the capability of LLMs in Evidence Retrieval, we evaluated their performance in two distinct scenarios: Local KG setting and Global KG setting. **Local KG** involves LLMs identifying validating triples from a given set of relevant triples, whereas **Global KG** challenges LLMs to find pertinent information directly within a KG. We established two baselines utilizing the method provided by [10], which facilitates LLMs in automatically extracting evidence from KGs for entities cited in the text.

4.2   Experiment Settings

Experiments are conducted on 3 open-source model: Vicuna-7B [5], Mistral-7B [13], and Llama-7B-Chat [24] models.

4.3   Experiment Results

Claim Extraction & Claim Verification

(1) **Challenges with Complex Data.** In the Claim Extraction task, LLMs face greater difficulties with Vanilla contexts and Quantitative Comparisons, as shown in Few-Shot and Zero-Shot settings. Vanilla contexts tend to be dense with claims, while texts involving Quantitative Comparisons often include complex expressions. Notably, Llama's underperformance in Few-Shot and Zero-Shot scenarios is attributed to its limited ability to follow instructions. Similarly, in the Claim Verification task, Quantitative Comparison texts are more challenging.

(2) **Learning Structural Knowledge doesn't improve Claim Extraction, while it has a good impact on Claim Verification.** In the Claim Extraction task, Vicuna and Mistral underperform, while Llama improves due to chat-style data training,

suggesting the need for better training objectives. Conversely, in Claim Verification, structural knowledge learning boosts LLMs' grounding abilities but negatively affects Mistral, indicating that such knowledge can impair Mistral's capabilities to some extent.

(3) **Contextual Learning Boosts Performance rapidly with limited data.** As shown in the Learning Context setting, despite poor performance on Few-Shot and Zero-Shot settings, LLMs demonstrate the ability to quickly learn from the training data with only a few examples.

Evidence Retrieval The Evidence Retrieval (ER) Task Presents Significant Challenges. Upon examining the cases, we discovered that models struggle to select from a vast array of triples. Under the Local KG setting, when given related evidence, LLM can have a good performance for extracting related evidence. However, when it comes to searching the whole KG for related evidence, it poses a great challenge.

5   Analysis

5.1   Quantitive Analysis

Claim Verification Performance Across Different Labels. This section evaluates the model's accuracy in predicting specific labels. We analyze the distribution of LLM's predicted labels versus the actual labels. Notably, predictions outside the scope of NEI, REFUTE, and SUPPORT are excluded.

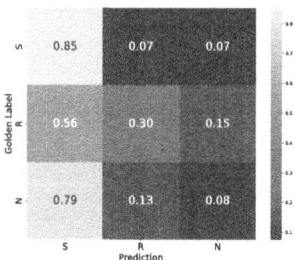
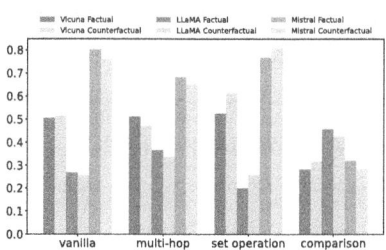

Fig. 4. Claim Verification accuracy under different labels. We adjust the statistical counts by normalizing them with the total number of occurrences for each label.

Fig. 5. Experiment results for Claim Verification Performance on KG and Counterfactual KG

As illustrated in Fig. 4, it becomes apparent that when provided with relevant evidence, LLMs are more inclined to predict SUPPORT over REFUTE and NEI. Contrary to our initial assumption that the introduction of the NEI label might confuse LLMs, the models demonstrate a higher propensity to predict NEI over REFUTE. This observation suggests that LLMs can grasp the concept of NEI from the presented prompt.

Claim Verification Performance on Counterfactual KG. This section investigates the ability of LLMs to accurately verify claims when presented with counterfactual evidence. We begin by altering the knowledge graph through the addition, deletion, and modification of various entities, attributes, and relations within the KG. Following this, we generate data using the methodology outlined in Sect. 3.2.

The outcomes of the verification task utilizing both KG and counterfactual KG are presented in Fig. 5. Contrary to our expectations, the verification performance between factual KGs and counterfactual KGs reveals no substantial disparity. This suggests that when provided with an atomic claim and its associated evidence, LLMs are capable of integrating the given knowledge for verification, regardless of any contradictions with the model's inherent parameters.

Accuracy of Claim Extraction Across Various Claim Quantities. This section assesses the model's proficiency in extracting claims from texts with varying numbers of claims. As depicted in Fig. 6, there is a noticeable decline in extraction accuracy as the number of claims within the texts increases. This indicates that LLM struggles to handle texts comprising multiple triples effectively.

Model Size Scaling on Evidence Retrieval. This section explores the capability of large, complex-sized LLMs in the Evidence Retrieval task under the Global KG setting, given their generally poor performance. We analyze a range of models, including Mistral [13], Mistral 8×7B [14], Vicuna 7B [5], Vicuna 13B [5], Llama 7B, Llama 13B, and Llama 70B [24]. Figure 7 illustrates that larger LLMs exhibit improved performance in retrieving relevant evidence. This highlights the difficulties smaller models face in effectively conducting automatic evidence retrieval.

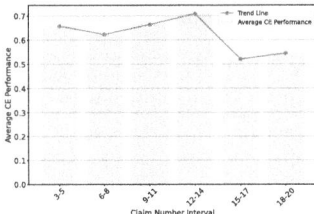

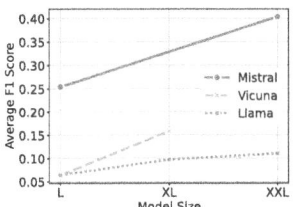

Fig. 6. The relationship between claim quantity and claim extraction accuracy. As claims grow in number, the accuracy of claim extraction decreases.

Fig. 7. Model Size Scaling on Evidence Retrieval. *L* refers to Mistral-7B, Vicuna-7B and Llama-7B, *XL* refers to Vicuna-30B and Llama-30B, and *XXL* refers to Llama-70B and Mistral-8*7B

6   Conclusion

In this paper, we established a Knowledge Graph Fact Verification Benchmark (KGFV) and introduced three subtasks: Claim Extraction, Evidence Retrieval, and Claim Verification across four data types: vanilla, multi-hop, set operation, and comparison to evaluate the abilities of KG-grounded verification capabilities of LLMs. The experimental results suggest that current LLMs have limitations in the 3 subtasks, underscoring the need for further advancements to enhance the efficacy of LLMs in KG-grounded text verification.

References

1. Cao, Y., et al.: A comprehensive survey of AI-generated content (AIGC): a history of generative AI from GAN to chatgpt. arXiv preprint arXiv:2303.04226 (2023)
2. Chen, W., et al.: Tabfact: a large-scale dataset for table-based fact verification. arXiv preprint arXiv:1909.02164 (2019)
3. Chen, X., et al.: Unveiling the siren's song: towards reliable fact-conflicting hallucination detection. arXiv preprint arXiv:2310.12086 (2023)
4. Chern, I.C., et al.: Factool: factuality detection in generative AI–a tool augmented framework for multi-task and multi-domain scenarios. arXiv preprint arXiv:2307.13528 (2023)
5. Chiang, W.L., et al.: Vicuna: an open-source chatbot impressing GPT-4 with 90%* chatgpt quality (2023). https://lmsys.org/blog/2023-03-30-vicuna/
6. Dhuliawala, S., et al.: Chain-of-verification reduces hallucination in large language models. arXiv preprint arXiv:2309.11495 (2023)
7. Gardent, C., Shimorina, A., Narayan, S., Perez-Beltrachini, L.: Creating training corpora for NLG micro-planners. In: Barzilay, R., Kan, M.Y. (eds.) Proceedings of the 55th Annual Meeting of the Association for Computational Linguistics (Volume 1: Long Papers), pp. 179–188. Association for Computational Linguistics, Vancouver, Canada (2017). https://doi.org/10.18653/v1/P17-1017. https://aclanthology.org/P17-1017
8. Golub, D., He, X.: Character-level question answering with attention. arXiv preprint arXiv:1604.00727 (2016)
9. Gou, Z., et al.: Critic: large language models can self-correct with tool-interactive critiquing. arXiv preprint arXiv:2305.11738 (2023)
10. Guan, X., et al.: Mitigating large language model hallucinations via autonomous knowledge graph-based retrofitting. arXiv preprint arXiv:2311.13314 (2023)
11. Honnibal, M., Montani, I., Van Landeghem, S., Boyd, A.: spaCy: Industrial-strength Natural Language Processing in Python (2020).https://doi.org/10.5281/zenodo.1212303
12. Issa, S., Adekunle, O., Hamdi, F., Cherfi, S.S.S., Dumontier, M., Zaveri, A.: Knowledge graph completeness: a systematic literature review. IEEE Access **9**, 31322–31339 (2021). https://doi.org/10.1109/ACCESS.2021.3056622
13. Jiang, A.Q., et al.: Mistral 7B. arXiv preprint arXiv:2310.06825 (2023)
14. Jiang, A.Q., et al.: Mixtral of experts. arXiv preprint arXiv:2401.04088 (2024)
15. Jiang, J., Zhou, K., Dong, Z., Ye, K., Zhao, W.X., Wen, J.R.: Structgpt: a general framework for large language model to reason over structured data. arXiv preprint arXiv:2305.09645 (2023)
16. Kim, J., Park, S., Kwon, Y., Jo, Y., Thorne, J., Choi, E.: Factkg: fact verification via reasoning on knowledge graphs. arXiv preprint arXiv:2305.06590 (2023)
17. Li, M., Peng, B., Zhang, Z.: Self-checker: plug-and-play modules for fact-checking with large language models. arXiv preprint arXiv:2305.14623 (2023)
18. OpenAI: GPT-4 (2023). https://openai.com/research/gpt-4. Accessed 15 Feb 2024
19. OpenAI: Chatgpt (2024). https://openai.com/chatgpt
20. Pan, L., Saxon, M., Xu, W., Nathani, D., Wang, X., Wang, W.Y.: Automatically correcting large language models: surveying the landscape of diverse self-correction strategies. arXiv preprint arXiv:2308.03188 (2023)

21. RyokoAI: Sharegpt52k (2023). https://huggingface.co/datasets/RyokoAI/ShareGPT52K
22. Thorne, J., Vlachos, A., Christodoulopoulos, C., Mittal, A.: FEVER: a large-scale dataset for fact extraction and VERification. In: NAACL-HLT (2018)
23. THU-KEG: Knowledge oriented programming language (KOPL) (2023). https://github.com/THU-KEG/KoPL
24. Touvron, H., et al.: Llama: open and efficient foundation language models. arXiv preprint arXiv:2302.13971 (2023)
25. Wang, S., Wei, P., Zhao, J., Mao, W.: A knowledge enhanced learning and semantic composition model for multi-claim fact checking. arXiv preprint arXiv:2104.13046 (2021)
26. Wang, Y., et al.: Factcheck-GPT: end-to-end fine-grained document-level fact-checking and correction of LLM output. arXiv preprint arXiv:2311.09000 (2023)
27. Zhang, Y., et al.: Siren's song in the AI ocean: a survey on hallucination in large language models. arXiv preprint arXiv:2309.01219 (2023)

An LLM-SPARQL Hybrid Framework for Named Entity Linking and Disambiguation to Wikidata

Muhammad Salman$^{(\boxtimes)}$ ⓘ, Haoting Chen, Sergio José Rodríguez Méndez ⓘ, and Armin Haller ⓘ

School of Computing - CECC, The Australian National University, Canberra, ACT 2600, Australia
{Muhammad.Salman,u7227871,Sergio.RodriguezMendez,Armin.Haller}@anu.edu.au

Abstract. This research tackles the complex issue of Named Entity Disambiguation (NED) by harnessing the power of Large Language Models (LLMs), such as Generative Pre-trained Transformers (GPTs), and integrating these with structured information from knowledge graphs like Wikidata. The overarching goal is to markedly enhance NED accuracy through a synergistic approach that combines the predictive prowess of advanced LLMs with the detailed semantic repositories of KGs. Our key innovation is the formulation of a novel methodology that employs SPARQL queries to methodically retrieve potential Uniform Resource Identifiers (URIs) from Wikidata. These URIs serve as candidate references for named entities. We then apply cutting-edge generative AI techniques to accurately match each entity's mention within texts to the most contextually relevant URI. This refined approach allows for a more sophisticated, context-aware resolution of entity references. We conducted extensive experiments to evaluate our method, which indicates a substantial achievement in performance, as evidenced by precision, recall, and F-measure. Impressively, our model achieves an F-measure of 96% on the Wikidata-Disamb dataset, a meticulously curated benchmark specifically developed for robust, scalable evaluation of NED performance leveraging Wikidata entries.

Keywords: Named Entity Linking · Entity Disambiguation · Knowledge Graphs · Wikidata · RDF Triples · LLMs

1 Introduction

Knowledge Graphs (KGs) such as Wikidata [21], DBpedia [13], and formerly Freebase [2,16], represent an invaluable asset in semantic web technologies due to their structured encapsulation of real-world entities and the relationships between them. These graphs catalogue a wide variety of entity types-including individuals, locations, organizations, and more-each connected through a rich

B. Xu et al. (Eds.): CCKS-IJCKG 2024, CCIS 2229, pp. 210–226, 2025.
https://doi.org/10.1007/978-981-96-1809-5_16

tapestry of predicates like "was born in" or "plays for". This structured information is crucial for semantic algorithms, which rely on these connections to interpret and infer new knowledge from existing data [11].

For instance, consider a search engine that is designed to retrieve mentions of all retired NBA players with a net income of more than 1 billion US dollars. The list of players, along with their income and retirement information, may be available in a KG. Equipped with querying information, it appears straightforward to look up mentions of retired basketball players, however, the main obstacle in this setup is the lexical ambiguity of entities. This is where Entity Linking (EL) - the process of matching a mention, e.g., "Michael Jordan", in a textual context to a KG record (e.g., "basketball player" or "mathematician") fitting the context - becomes key. EL is the task of identifying an entity mentioned in the unstructured text and establishing a link to a correct item in a structured KG [1].

This necessity gives rise to the EL field [20], which focuses on anchoring plain text mentions to their corresponding entities in a KG. EL not only resolves the ambiguity of names within texts but also contextualises the mentions in a way that aligns with the KG. The capability to link textual mentions directly to KG entities enables advanced Information Retrieval (IR) applications, enhances Natural Language Understanding (NLU) tasks [19], and is pivotal in domains such as biomedical text mining [23], semantic parsing [22], and interactive question-answering [18] systems.

The evolution of EL task has been significantly influenced by the advent of neural networks (NN) and deep learning (DL) techniques [14]. Recent developments in this field have leveraged the power of Transformers and LLMs such as BERT [9] and GPT [3,15] to enhance the accuracy and efficiency of such systems. These models leverage context within large corpora to discern subtle nuances in language use, significantly enhancing the EL and NED process.

However, maintaining the relevance and accuracy of these links poses ongoing challenges, particularly due to the dynamic nature of KGs, where new data is continually integrated. This dynamic integration demands adaptive algorithms capable of evolving in tandem with the underlying data, ensuring that EL remain both accurate and current. Marrying the sophisticated analytical capabilities of LLMs with the structured querying power of SPARQL-a query language designed to retrieve and manipulate data stored in Resource Description Framework (RDF) format-, we present a new framework to link and disambiguate textual entities with the Wikidata URIs. This fusion not only augments the efficacy of semantic search engines but also broadens the potential for applications across varied domains, all while demanding a nuanced understanding of both language and metadata.

2   Background and Literature Review

The EL domain has seen substantial advancements with the introduction of NN-based methods [7], significantly enhancing the performance over traditional approaches that often rely on handcrafted features and shallow learning models. Neural models facilitate the automated learning of complex patterns in large datasets, leading to more accurate entity disambiguation and recognition.

Early NN approaches to EL focused on leveraging local context and simple neural architectures like feed-forward networks for entity disambiguation. The foundational model of neural EL can be described by the equation:

$$s(e, m) = \mathbf{v}_e \cdot \mathbf{v}_m$$

where $s(e, m)$ is the similarity score between the entity e and the mention m, and $\mathbf{v}_e$ and $\mathbf{v}_m$ are the vector representations of the entity and mention, respectively.

As the field progressed, more advanced architectures were introduced [5,8]. Attention mechanisms and recurrent neural networks (RNNs) were utilised to better capture the dependencies within the text data, leading to improved context understanding for entity mentions. An example of such a model integrates the contextual embeddings obtained from a Long-Short-Term Memory (LSTM) model, represented as:

$$\mathbf{h}_t = \text{LSTM}(\mathbf{x}_t, \mathbf{h}_{t-1})$$

where $\mathbf{h}_t$ is the hidden state at time t, $\mathbf{x}_t$ is the input vector at time t, and $\mathbf{h}_{t-1}$ is the hidden state at time $t-1$.

The integration of LLMs such as BERT [9] and GPTs [15] with EL processes has opened new avenues for research. These models pre-train on a vast corpus of text, allowing them to develop a deep semantic understanding of language, which is then fine-tuned for specific tasks like EL. The process typically involves adapting the pre-trained models to generate entity-specific representations:

$$\mathbf{e}_m = \text{BERT}_{\text{EL}}(\text{context}(m))$$

where $\mathbf{e}_m$ is the entity embedding for the mention m generated by the BERT model fine-tuned for entity linking, and $\text{context}(m)$ represents the textual context of the mention.

The application of SPARQL [6] queries to retrieve data from Wikidata has further enhanced the capabilities of EL systems. By formulating specific queries, systems can extract detailed information about entities that can assist in the disambiguation process. For instance, a SPARQL query might be structured as follows to retrieve data:

```
SELECT ?property WHERE {?entity ?property ?value}
```

where "$?property$" is the property of interest, "$?entity$" refers to the specific entity under consideration, and "$?value$" is the value associated with the property for that entity.

Despite advancements in EL systems, significant challenges remain, particularly in managing the dynamic nature of language and real-time updates to extensive KGs like Wikidata. The rapid evolution of neural EL techniques, predominantly fueled by DL innovations, has markedly enhanced the performance of EL systems over traditional methods [12,19]. The integration of LLMs and SPARQL-enabled querying mechanisms appears promising, potentially leading to more sophisticated, efficient, and accurate EL system.

3   Task Definition and Challenges

Consider a document D which includes a collection of mentions $M = \{m_1, m_2, \ldots, m_{|M|}\}$. Given a knowledge base K, which houses a set of entities $E = \{e_1, e_2, \ldots, e_{|E|}\}$, along with a set L that enumerates the relationships among these entities, the goal of entity linking is to establish a mapping $f : M \to E$. This mapping function seeks to accurately associate each mention in D with its corresponding entity in K.

Due to the vast number of potential mention-entity pairings in $M \times E$, it is computationally prohibitive to evaluate the similarity scores for all combinations. Therefore, the initial step in this process often involves deploying heuristic algorithms designed to significantly narrow down these pairs by identifying a manageable subset of candidate entities for each mention. This subset forms the candidate set $C = \{c(m_1), c(m_2), \ldots, c(m_{|M|})\}$, where $c(m_i)$ represents the candidate entities for the mention m_i.

Subsequently, the focus shifts to the process of entity selection, where the generalise-ability of LLMs is leveraged. The optimal candidate for each mention is then selected, thereby finalizing the EL and NED process.

4   System Overview

This methodology operates around four components. *Wikidata* is used as the target KG to link and disambiguate entity mentions as shown in Fig. 1. *SPARQL* queries are crafted to identify Wikidata data items (candidates) against the mention of an entity. We also leverage the *LLMs* to use their vast capabilities in textual understanding and generation with appropriate prompt designs. Finally, we evaluate our approach with *Wikidata-Disamb* dataset. Below is a brief overview of each component.

4.1   Overview of Wikidata

Wikidata [21] stands as the most extensive publicly accessible knowledge base, encapsulating over 1.54 billion facts (early-2023 statistics) which are structured in the form of subject-predicate-object triples[1]. It integrates more than 109 million entities and is characterised by 11,930 distinct properties (early-2024 statistics), each serving to define and describe the relationships and attributes of these entities. Notably, approximately 3,000 of these properties are instrumental in facilitating the answering of natural language questions. These are pivotal in enhancing the semantic understanding and data retrieval capabilities essential for numerous applications, including intelligent search and automated reasoning systems.

Each entity and property within Wikidata is assigned a unique identifier, known as QIDs for entities and PIDs for properties, to ensure a consistent and

---

[1] https://www.wikidata.org/wiki/Wikidata:Statistics.

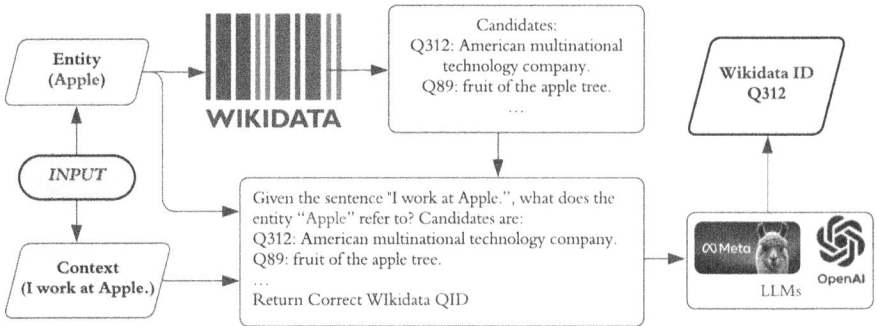

Fig. 1. System Overview of the Proposed Approach

unambiguous reference system. For example, consider the representation of the fact: "Joe Biden is the president of the United States". This is encoded in Wikidata as the triple (`wd:Q6279, wdt:P39, wd:Q11696`), where `Q6279` is the QID associated with "Joe Biden", `P39` is the PID denoting the property "position held", and `Q11696` corresponds to the QID for "the president of the United States". Such structured representations allow for precise and efficient querying of complex factual information, demonstrating the robustness and versatility of Wikidata as a foundational tool for knowledge management and data-driven research.

4.2 SPARQL Query Template

As a crowd-sourced KG, Wikidata does not adhere to predefined domains or types for its entities. Instead, it allows for an "arbitrary" set of properties to be associated with any given entity based on *Open-World Assumption* heuristics. Despite this flexible schema, all named entities in Wikidata are linked through at least one 'instance of' property to some domain entity, and these domain entities are structured into a class hierarchy using the 'subclass of' property. It is important to note that while domain entities and properties in Wikidata are uniquely named to avoid ambiguity (T-Box), non-domain entities (A-Box) can have names that reference multiple entities. For instance, the name "Lincoln" could refer to a variety of entities including a president, a car brand, a sparrow, an aircraft, and several cities.

SPARQL Query for Candidate Extraction

```
SELECT ?entity ?type ?Description WHERE
{ ?entity rdfs:label "entity_label"@en .
OPTIONAL {{?entity wdt:P31 ?type.}
{?entity schema:description ?Description.}}
Filter(!bound(?type) || ?type!=wd:Q4167410)
SERVICE wikibase:label { bd:serviceParam
wikibase:language "en".} } limit (1...n)
```

Given the vast scope of Wikidata, it is impractical for LLMs to memorise all QIDs (entity identifiers) associated with domains and properties [24]. To facilitate more accessible and intuitive querying, we crafted a SPARQL query template to retrieve specific data items from Wikidata. For example, a simple SPARQL query for retrieving wikidata items is `SELECT ?entity WHERE { ?entity rdfs:label"entity_label"@en .}`. With some modification, the query is reformulated for better response which contains some filters and some additional information which will be helpful as a context, *i.e.*, class of entity and description. The query also filters out the entities where the type of entity represents a Wikimedia Disambiguation[2] (Q4167410) page. Our query for candidate extraction reads as: (see *"SPARQL Query for Candidate Extraction"* box).

4.3   Overview of the Dataset

For the evaluation of our entity linking and disambiguation approach, we utilised the `Wikidata-Disamb` dataset, which was developed by converting the `Wiki-Disamb30` dataset. The `Wiki-Disamb30` dataset, initially curated by Ferragina and Scaiella [10], was designed for linking entities to their corresponding Wikipedia pages. The `Wikidata-Disamb` dataset extends this by mapping these Wikipedia pages to their respective Wikidata items (QIDs), as documented by Cetoli et al. [4].

The `Wikidata-Disamb` dataset includes several key attributes: an `entity-label` as a `string`, a short sentence related to the entity labeled as `text`, the corresponding Wikidata QID termed as `correct-id`, and an incorrect QID labeled as `wrong-id`. In our methodology, the `string` serves as the input to the model, while the `text` provides contextual information to aid in the understanding of the entity. The model's output is then compared against the `correct-id` to evaluate performance using precision, recall, and F-measure metrics. By utilizing the `Wikidata-Disamb` dataset, we ensure that our evaluation process is robust and reflective of real-world entity disambiguation scenarios.

---

[2] https://www.wikidata.org/wiki/Q4167410.

4.4   Applied LLMs

LLMs utilise DL techniques, particularly transformer architectures [17], to process and analyse vast amounts of textual data. LLMs are typically pre-trained on extensive datasets and fine-tuned for specific tasks, leading to notable improvements in various NLP applications. Several state-of-the-art LLMs have been employed in this study to enhance EL and NED tasks. Below is an overview of the primary applied models.

GPT-3.5-Turbo and GPT-4-Turbo. Developed by OpenAI[3], the GPT-3.5-Turbo [3] and GPT-4-Turbo [15] models are part of the renowned GPT series. The GPT-4-Turbo model, in particular, features a larger parameter count and improved training methodologies compared to its predecessors, resulting in superior performance in complex tasks such as NED. These models were utilised through API calls.

Meta's LLaMA Models. Meta's LLaMA[4] series including LLaMA2 and LLaMA3, are designed to provide high performance in NLP tasks through efficient use of computational resources. These models were locally configured and implemented from their versions, available at *Hugging-Face*[5].

4.5   LLM Prompt Engineering

Prompt engineering involves crafting effective prompts to elicit the desired responses from the LLMs. This technique can be crucial for many IR and KG construction tasks such as NER, EL, and NED. Below are descriptions of some task-specific methods designed for entity classification and disambiguation using LLMs.

Ask_GPT_Class()

Given the reference text below, return wikidata class label of `'entity_label'` with case-sensitivity.
Do not respond with anything if not clear or if reference text is not enough to identify the class.
Your response should include class label only, *i.e.*, university, town, country, city, etc.
Reference Text: `'text'`, **Examples**: E1, E2

The `Ask_GPT_Class()` prompt is designed to determine the class of an entity based on the provided reference text. This method leverages LLMs to classify entities into predefined categories such as "university", "town", "country", etc.

---

[3] https://openai.com.
[4] https://llama.meta.com.
[5] https://huggingface.co.

The prompt ensures that the model only responds with the class label if it is confident about the classification. If the reference text is insufficient or unclear, the model refrains from providing an answer, thus maintaining accuracy. The class information can be used in many steps of KG construction where entity types need to be identified accurately. We used this information in attempting to extract the correct Wikidata ID of an entity label by filtering the other same entity label based on the class.

Ask_GPT_QID(prediction)

Input:

- *entity_label*: The label of the entity.
- *text*: The sentence that 'mentions' the entity.
- *wikidata_class*: The class of the entity.

Prompt:
Given the reference information and examples below, return the Wikidata ID of 'entity_label' **Text**: {text}   **Label**: {Entity_Label} **Class**: {Wikidata_Class}   **Wikidata ID**: ?
Ex 1: Text: *Besides CSIRO, Australian National University is located in Canberra.*
Label: *Australian National University*
Class: *University*   Wikidata ID: *Q127990*
Ex 2: Text: *Mantell was born in Bridgwater, Somerset, and studied at the University of Bath.*
Label: *Bridgwater*
Class: *Town*   Wikidata ID: *Q914015*
Return Wikidata_ID

Following the empirical finding and analysis we attempt `Ask_GPT_QID` (`Prediction`) method which is used to predict the Wikidata ID of an entity based on provided context. This includes the entity label, the sentence from which the entity is extracted, and the entity's class. The prompt also provides some structured examples to guide the LLM in making accurate predictions.

The `Ask_GPT_QID` (`Prediction`) method is a starting point in automating the entity linking process. However, we observed some inconsistencies in the response. Therefore, we update our method and used the SPARQL query template as a base method for retrieving a set of candidate IDs from Wikidata. The `Ask_GPT_QID` (`Selection`) method is designed to ultilise LLMs for ID selection instead of prediction.

Ask_GPT_QID(selection)

Input:

- *entity_label*: The label of the entity.
- *text*: The sentence that 'mentions' the entity.
- *Candidates*: Candidate QIDs

Prompt: I am analyzing `'entity_label'`. Consider the refer-
ence sentence: `'query_text'` and the candidates Wikidata IDs:
`List[Candidates]`
Based on the class, description, and the context provided by the text,
what is the correct ID for `'entity_label'`?
Return `'NA'` if none of the candidates is correct.
Return *Wikidata_ID*

By employing these methods, LLMs can significantly improve the accuracy
and efficiency of tasks involving entity classification, linking, and disambiguation,
facilitating the development of more robust and reliable NED.

5   NED Model Refinement Process

This section presents a comprehensive comparison of five distinct approaches
developed[6] to enhance the performance of NED tasks using the integration of
SPARQL queries and the advanced capabilities of LLMs (GPT and LLaMA in
this case). Each approach is assessed based on its methodology, iterative improvements, and overall impact on precision, recall, and F1 score metrics.

5.1   Overview of NED Approaches

Approach LLMPredictURI: URI Retrieval with LLM Only

In LLMPredictURI, we explore the capability of the LLMs used in this experiment[7] to directly predict the Wikidata URI of an entity and its contextual
sentence. This relies solely on the LLM's Internal Vector Space Representation
(IVSR) of Wikidata. Specifically, the LLM should output a valid Wikidata URI
that not only matches the given entity's label, but is also the most relevant to
the given context among all other Wikidata URIs in its internal knowledge base.
The process is as follows:

- Construct a prompt using the given entity and its context only.
- Feed the prompt into the LLM and obtain the response.
- Parse the response and obtain a Wikidata URI for the given entity.

---

[6] https://github.com/salmon-kg/llm-sparql.
[7] GPT and LLaMA as described previously. For simplicity, we will refer to them as
"LLM".

Approach SPARQLite: Enhanced SPARQL Querying for Entity Disambiguation

The URIs generated through LLMPredictURI can be invalid or irrelevant because there is nothing to limit the scope of the URIs provided by the LLM. It would be challenging for the LLM to select one from numerous possible URIs in its IVSR of Wikidata. Additionally, LLM's IVSR lacks transparency [24].

SPARQLite enhances the validity and relevancy of results by using SPARQL in the first place to retrieve all Wikidata entities and their URIs matching the input entity, called "candidates". These candidates are then used to construct a prompt with the input entity and its context. The prompt instructs the LLM to choose one of the candidates in the given context. The URI of such Wikidata entity will be the output. The process is as follows:

- Retrieve a list of Wikidata entity candidates that match the label of the input entity. The SPARQL query executes with a filter designed to manage case variations in the entity mention.
- Construct a prompt using the given entity, context, and candidates.
- Feed the prompt into the LLM and obtain the response. The response should only come from the candidates.
- Parse the response and obtain a Wikidata URI for the given entity.

This methodological refinement resulted in a more balanced approach towards a more sound algorithm for the EL task. Introducing case sensitivity handling proved a significant factor in accurate label identification through SPARQL queries.

Table 1. Summary of NED Approaches

Approach	Method	Key Feature
LLMPredictURI	Direct URI Retrieval with LLM	utilises only LLM for URI prediction
SPARQLite	Enhanced SPARQL Querying	Includes case sensitivity handling
ClassMatch	Class Match	Leverages LLM for class matching in SPARQL results
CandidSelect	LLM-Enhanced Candidate Selection	Combines LLM prompts with candidate selection
RecursiSelect	Recursive CandidSelect	Recursive candidate generation with LLM response

Approach ClassMatch: Class Match in Candidates' Classes

ClassMatch progresses beyond direct URI retrieval by introducing a class-matching mechanism to the candidate selection process. This method incorporates a two-step process, beginning with the extraction of the class from the entity using LLM, informed by the surrounding context. With the class identified, the system then performs a SPARQL query to obtain a list of potential candidates, each paired with their respective classes.

The pivotal component of this approach is the matching of the extracted class with the classes associated with each candidate entity. The match is categorised as either exact or partial. An exact match occurs when the candidate's class directly corresponds to the extracted class, while a partial match acknowledges broader/narrower, hierarchical relationships, such as parent classes or similarity in vector space.

Approach CandidSelect: Integrating LLM with SPARQL for Candidate Selection

CandidSelect advances entity disambiguation techniques by harmoniously integrating LLM with SPARQL to refine the selection of candidate entities from Wikidata. This method capitalises on the contextual understanding capabilities of the LLM to assess and select the most relevant URI from a set of SPARQL query results. The process encompasses two primary steps:

1. **Candidate Retrieval via SPARQL:** Initially, a SPARQL query retrieves a set of potential candidate entities for the given mention. This preliminary list is confined to a predetermined limit to ensure manageability and relevance.
2. **LLM-Enhanced URI Selection:** The LLM is then employed to evaluate the retrieved candidates by analyzing the context in which the entity's mention appears. It determines the most appropriate Wikidata ID by considering both the candidates and the contextual cues provided.

Approach RecursiSelect: Recursive LLM Candidate Selection

The culmination of our iterative development process is encapsulated in this final approach, which employs a recursive refinement strategy using LLM in conjunction with SPARQL. In this refined strategy, the LLM is used recursively to evaluate and refine a list of candidates retrieved via SPARQL queries. This approach iterates over the candidates, leveraging the LLM's ability to process contextual clues, including classes and descriptive information, to identify the most accurate Wikidata URI.

– **Candidate Retrieval:** A SPARQL query generates an initial list of candidate entities, with each candidate paired with class metadata and additional 'descriptive' details.
– **URI Determination:** The LLM is prompted to select the most suitable QID for the entity, given the candidates and the enriched context. Or referring to provide more candidates when none suits best.

5.2   Comparative Results and Discussion on Approach Refinement

Table 2 encapsulates the comparative performance of each approach, providing an at-a-glance overview of their efficacy in the context of NED tasks.

The comparative analysis of the four NED approaches reveals a trajectory of incremental improvements in precision, recall, and F1 scores. As delineated in Table 2, each subsequent approach builds upon the lessons learned from its predecessors, addressing their deficiencies and amplifying their strengths.

LLMPredictURI, relying solely on the LLM's capabilities, demonstrated the foundational potential of LLMs in the realm of NED, yet it highlighted

Table 2. Step-wise Performance Measures of Evolved NED Approach

Approach	Precision	Recall	F-measure
LLMPredictURI	27%	100%	42%
SPARQLite	50%	60%	57.98%
ClassMatch	68%	100%	80.95%
CandidSelect	79%	100%	88.27%
RecursiSelect (10)	88%	100%	93.61%
RecursiSelect (n)	92%	100%	95.83%

the limitations of context understanding without additional data structuring, as evidenced by its lower precision. **SPARQLite** introduced enhanced SPARQL querying mechanisms, including case sensitivity handling, which led to a notable increase in precision. However, the balance between precision and recall still needed refinement, pointing to the necessity of a more nuanced approach. **Class-Match** responded to these challenges by incorporating class information, leading to a significant leap in performance metrics. This underscored the utility of structured data in augmenting the LLMs' contextual analysis. **CandidSelect** culminated the iterative enhancements by integrating an LLM-driven evaluation of SPARQL query results, achieving the higher precision and F1 score. In the final **RecursiSelect**, we optimised our model by introducing the recursive candidates' generation based on the response of LLM, *i.e.*, if there is no suitable QID among candidates then the LLM will return a 'flag' response which will be followed by another round of candidates' generation with further case-sensitive variants of the entity. Also, we tested this approach with two settings: restricting the SPARQL candidates generation to 10 or no restriction (n) (Table 2). This aspect is also critical for the LLMs where context window is limited.

The comparison elucidates the evolution of NED methodologies from direct LLM-based URI prediction to the integration of LLMs with SPARQL for candidate selection, each step yielding incremental improvements in performance metrics. RecursiSelect emerges as the most effective, leveraging the strengths of LLMs in candidate selection based on the provided context. The summary of NED approaches along with their features is illustrated in Table 1.

6 Results

The further evaluation of the final refined approach, denoted as RecursiSelect, was conducted using five distinct random 1000-entry samples from the dataset to assess the robustness and consistency of the entity disambiguation methodology. The performance metrics under consideration were `Precision, Recall, and F-measure`, which together offer a comprehensive view of the model's effectiveness.

The results outlined in Table 3 illustrate consistently high Precision across all five samples, indicating that a significant majority of the entities retrieved by the model were true positives. Precision scores ranged from 89% to 98%, with the highest score observed in Sample 2. Such a high Precision suggests that the model was adept at minimizing false positives in its entity retrieval process. Recall metrics were similarly impressive, never dipping below 97%, which underscores the model's capability in identifying nearly all relevant entities within the dataset. The 100% Recall in Sample 2 reflects a scenario where the model successfully captured every relevant entity, leaving no instance unidentified. The F1 Scores, providing a harmonised metric that accounts for both Precision and Recall, remained exceptionally high, with all values exceeding 94%. The peak F1 Score of 98.99% in Sample 2 highlights the model's proficiency in balancing accuracy with completeness in its predictions.

Table 3. Performance Across Random Data Samples

Test	Precision	Recall	F-measure
1	93%	99%	96.37%
2	**98%**	**100%**	**98.99%**
3	89%	98%	94.18%
4	92%	100%	95.83%
5	91%	97%	95.29%

The consistency in high performance across multiple samples suggests that RecursiSelect is robust and adaptable to various subsets of data. This level of stability is a promising indicator of the model's potential for real-world applications, where data diversity and complexity are the norms. Moreover, the robustness assessment provided by these results speaks to the model's capacity to generalise well. Despite variations inherent in different data samples, the model maintained high precision and recall, suggesting that it is not overfitted to specific types of data and can handle a range of scenarios with high accuracy.

7   Discussion

The performance metrics outlined in Table 4 provide a comprehensive overview of the advancements achieved by integrating various prompts and models in this entity linking and disambiguation task. The analysis focuses on the precision, recall, and F-measure metrics to evaluate the robustness and accuracy of different approaches.

7.1   Prediction Models

The GPT-3.5-Turbo model, when tasked with prediction (*i.e.*, retrieving the QID), demonstrates a precision of 18%, a recall of 100%, and an F-measure of 30.51%. This indicates that while the model is highly effective in recalling all relevant entities (100% recall), it struggles with precision, leading to a high number of false positives.

The introduction of GPT-4-Turbo for prediction task shows a marked improvement in precision (27%) and F-measure (42%), while maintaining a perfect recall. This enhancement can be attributed to the more advanced architecture and larger model size of GPT-4-Turbo, which allows for better handling of the complexities in entity linking.

Adding class labels and case sensitivity to GPT-4-Turbo further refines the model's performance. Precision increases incrementally to 30% and 33% respectively, with corresponding improvements in F-measure to 46.15% and 49.62%. The combination of both class labels and case sensitivity yields a precision of 36% and an F-measure of 52.94%, demonstrating the benefit of incorporating additional contextual information.

Table 4. Performance comparison of different models and configurations.

Model & Prompt		Precision	Recall	F-measure
OpenAI GPTs	**GPT-3.5-Turbo** *(prediction)*	18%	100%	30.51%
	Class_label + **GPT-3.5-turbo** *(prediction)*	18%	100%	30.51%
	GPT-4-Turbo *(prediction)*	27%	100%	42%
	Class_label + **GPT-4-Turbo** *(prediction)*	30%	100%	46.15%
	Case_Sensitivity + GPT-4-Turbo *(prediction)*	33%	100%	49.62%
	Case_Sensitivity + **GPT4-turbo (Label, Class)** *(prediction)*	36%	100%	52.94%
	SPARQL Candid (top 10) & GPT-4-turbo *(selection)*	77%	100%	87%
	SPARQL Candid (all) & GPT-4-turbo *(selection)*	79%	100%	88.27%
	SPARQL Candid (all) & GPT-4-turbo *(selection + recursive)*	**92%**	**100%**	**95.83%**
	SPARQL Candid (10) & GPT-3.5-turbo *(selection + recursive)*	91%	100%	95.29%
	SPARQL Candid & GPT-4-turbo *(selection + recursive)* MAX	**98%**	**100%**	**98.99%**
Meta's LLaMA	SPARQL Candid, LLaMA2-7B-Q4	40%	98.9%	57.14%
	SPARQL Candid, LLaMA2-7B-Q4 (Optimised)	59%	100%	74.21%
	SPARQL Candid, LLaMA2-7B-Q4 (Prompt Update)	68%	100%	80.95%
	SPARQL Candid, LLaMA3-8B-Q4	77%	100%	87%
	SPARQL Candid, LLaMA3-8B-instruct-Q4	**93%**	**100%**	**96.37%**

7.2   GPT-SPARQL Model

The integration of SPARQL candidates significantly enhances the model's performance. When the top 10 SPARQL candidates are used with GPT-4-Turbo selection, precision jumps to 77% with an F-measure of 87%. Considering all SPARQL candidates, the precision improves slightly to 79% with an F-measure of 88.27%. Recursive selection with all SPARQL candidates and GPT-4-Turbo selection boosts precision to 92% and F-measure to 95.83%. The highest performance is observed with SPARQL candidates and recursive GPT-4-Turbo selection, achieving a precision of 98%, recall of 100%, and an F-measure of 98.99%.

This highlights the effectiveness of leveraging structured data querying alongside advanced LLMs to improve NED accuracy.

7.3 LLaMA-SPARQL Model

The SPARQL Candid, `LLaMA2-7B-Q4`[8] model begins with a precision of 40%, recall of 98.9%, and an F-measure of 57.14%. This initial performance indicates that while the model has a high recall, its precision can be significantly improved.

Optimizing the `LLaMA2-7B-Q4` model increases precision to 59% and F-measure to 74.21%. Further improvements are seen with a prompt update, which raises precision to 68% and F-measure to 80.95%. These enhancements reflect the impact of fine-tuning and prompt engineering on model performance.

The `LLaMA3-8B-Q4`[9] model achieves a precision of 77% and an F-measure of 87%, indicating a substantial improvement over its predecessors. The best performance among the LLaMA models is observed with the `LLaMA3-8B-instruct-Q4` configuration, achieving a precision of 93%, recall of 100%, and an F-measure of 96.37%. This underscores the significant enhancement in entity linking achieved through iterative improvements and the integration of querying mechanisms.

These results demonstrate that the integration of SPARQL candidates and recursive selection processes with advanced LLMs, like GPT-4-Turbo and LLaMA3, significantly enhances the precision and overall accuracy of entity linking and disambiguation tasks. The recursive selection approach, in particular, proves to be highly effective in minimizing false positives and ensuring comprehensive entity retrieval. These findings underscore the potential of combining structured data querying with state-of-the-art NLP models to achieve high performance in complex data disambiguation tasks.

8   Conclusion

This study embarked on a quest to refine NED techniques by integrating LLMs with SPARQL queries. Throughout our exploration, we developed and assessed multiple approaches, each presenting an evolution over the last in terms of sophistication and performance metrics. The comparative analysis of these approaches culminated in RecursiSelect, which showcased a recursive response of LLMs for candidate selection - demonstrating exemplary precision and recall across various data samples. While initial methods provided a foundational understanding of the challenges inherent in NED tasks, each subsequent approach incorporated additional refinements that progressively enhanced the model's accuracy. The consistency in high-performance metrics emphasises the reliability and efficiency of this model. The systematic improvements observed across different approaches provide a testament to the potential for LLMs to make significant strides in the realm of EL and NED. This research contributes a significant step forward in the

---

[8] https://huggingface.co/TheBloke/Llama-2-7B-GGUF.
[9] https://huggingface.co/QuantFactory/Meta-Llama-3-8B-GGUF.

field of NED, providing a robust framework which leverage LLMs. This research paves the way for future explorations into combining neural models with semantic web technologies to further advance the field of NED. The future work should focus on moving from EL to *triple linking* task to transform a local KG to a subset of a global KG, such as Wikidata.

References

1. Al-Moslmi, T., Ocaña, M.G., Opdahl, A.L., Veres, C.: Named entity extraction for knowledge graphs: a literature overview. IEEE Access **8**, 32862–32881 (2020)
2. Bollacker, K., Evans, C., Paritosh, P., Sturge, T., Taylor, J.: Freebase: a collaboratively created graph database for structuring human knowledge. In: Proceedings of the 2008 ACM SIGMOD International Conference on Management of Data, pp. 1247–1250 (2008)
3. Brown, T., et al.: Language models are few-shot learners. Adv. Neural. Inf. Process. Syst. **33**, 1877–1901 (2020)
4. Cetoli, A., Akbari, M., Bragaglia, S., O'Harney, A.D., Sloan, M.: Named entity disambiguation using deep learning on graphs. arXiv preprint arXiv:1810.09164 (2018)
5. Christiano, P.F., Leike, J., Brown, T., Martic, M., Legg, S., Amodei, D.: Deep reinforcement learning from human preferences. In: Advances in Neural Information Processing Systems, vol. 30 (2017)
6. World Wide Web Consortium: SPARQL 1.1 overview (2013)
7. Cui, L., Wei, F., Zhou, M.: Neural open information extraction. arXiv preprint arXiv:1805.04270 (2018)
8. Dettmers, T., Minervini, P., Stenetorp, P., Riedel, S.: Convolutional 2D knowledge graph embeddings. arXiv preprint arXiv:1707.01476 (2017)
9. Devlin, J., Chang, M.W., Lee, K., Toutanova, K.: Bert: pre-training of deep bidirectional transformers for language understanding. arXiv preprint arXiv:1810.04805 (2018)
10. Ferragina, P., Scaiella, U.: Tagme: on-the-fly annotation of short text fragments (by wikipedia entities). In: Proceedings of the 19th ACM International Conference on Information and Knowledge Management, pp. 1625–1628 (2010)
11. Heist, N., Hertling, S., Ringler, D., Paulheim, H.: Knowledge graphs on the web-an overview (2020)
12. Kolitsas, N., Ganea, O.E., Hofmann, T.: End-to-end neural entity linking. In: Conference on Computational Natural Language Learning (2018). https://api.semanticscholar.org/CorpusID:52073201
13. Lehmann, J., et al.: Dbpedia-a large-scale, multilingual knowledge base extracted from wikipedia. Semantic Web **6**(2), 167–195 (2015)
14. Li, J., Sun, A., Han, J., Li, C.: A survey on deep learning for named entity recognition. IEEE Trans. Knowl. Data Eng. (2020)
15. OpenAI: GPT-4 technical report (2023)
16. Pellissier Tanon, T., Vrandečić, D., Schaffert, S., Steiner, T., Pintscher, L.: From freebase to wikidata: the great migration. In: Proceedings of the 25th International Conference on World Wide Web, pp. 1419–1428 (2016)
17. Raffel, C., et al.: Exploring the limits of transfer learning with a unified text-to-text transformer. J. Mach. Learn. Res. **21**(1), 5485–5551 (2020)

18. Sakor, A., Singh, K., Patel, A., Vidal, M.E.: Falcon 2.0: an entity and relation linking tool over wikidata. In: Proceedings of the 29th ACM International Conference on Information & Knowledge Management, pp. 3141–3148 (2020)
19. Sevgili, O., Shelmanov, A., Arkhipov, M.V., Panchenko, A., Biemann, C.: Neural entity linking: a survey of models based on deep learning. Semantic Web **13**, 527–570 (2020). https://api.semanticscholar.org/CorpusID:219177305
20. Sil, A., Yates, A.: Re-ranking for joint named-entity recognition and linking. In: Proceedings of the 22nd ACM International Conference on Information & Knowledge Management, pp. 2369–2374 (2013)
21. Vrandečić, D., Krötzsch, M.: Wikidata: a free collaborative knowledgebase. Commun. ACM **57**(10), 78–85 (2014)
22. Yih, S.W.T., Chang, M.W., He, X., Gao, J.: Semantic parsing via staged query graph generation: question answering with knowledge base (2015)
23. Zeng, X., Tu, X., Liu, Y., Fu, X., Su, Y.: Toward better drug discovery with knowledge graph. Curr. Opin. Struct. Biol. **72**, 114–126 (2022)
24. Zhang, M., Press, O., Merrill, W., Liu, A., Smith, N.A.: How language model hallucinations can snowball. arXiv preprint arXiv:2305.13534 (2023)

Mitigating Multi-hop Hallucination in Large Language Models with Non-authoritative Knowledge Sources

Hongbang Yuan[1,2], Pengfei Cao[1,2(✉)], Zhuoran Jin[1,2], Yubo Chen[1,2(✉)],
Kang Liu[1,2], and Jun Zhao[1,2]

[1] The Laboratory of Cognition and Decision Intelligence for Complex Systems,
Institute of Automation, Chinese Academy of Sciences, Beijing, China
{hongbang.yuan,pengfei.cao,zhuoran.jin,kliu,jzhao}@nlpr.ia.ac.cn
[2] School of Artificial Intelligence, University of Chinese Academy of Sciences, Beijing,
China
yubo.chen@nlpr.ia.ac.cn

Abstract. Large language models (LLMs) still suffer from the issue of
hallucination. Post-editing the hallucinated text generated by LLMs via
external knowledge sources has become a popular approach. However,
two challenges still remain: *multi-hop hallucination mitigation* and *non-
authoritative knowledge sources*. The former refers to the gathering of
multiple pieces of evidence and utilization of multi-hop reasoning when
mitigating hallucination. The latter indicates the possibility that external knowledge sources themselves may contain irrelevant or incorrect
information. To address these challenges, we propose the multi-hop hallucination mitigation task in three practical external knowledge source
settings: *relevant, irrelevant* and *conflict*. To facilitate further research
on this task, we construct two high-quality datasets, namely *HoCurr*
and *HotpotEC*. To better deal with this task in various settings, we propose a method called *FixER*, in which a *BaseFix* iteratively enhances its
performance with the assistance of a carefully designed *Evaluator* and
Reflector. Experiments demonstrate that our proposed task is highly
challenging and our method is effective comparing with the baselines.

Keywords: Multi-hop hallucination mitigation · Non-authoritative
knowledge sources · Large language models

1 Introduction

Recently, large language models (LLMs) have demonstrated remarkable capabilities in various tasks [3,17]. However, LLMs are susceptible to **hallucinations** [23], a well-known phenomenon in which LLMs generate texts that are
seemingly plausible, but deviate from factual knowledge [24]. Hallucinations
can bring potential risks, particularly in critical fields like law, medicine and

B. Xu et al. (Eds.): CCKS-IJCKG 2024, CCIS 2229, pp. 227–239, 2025.
https://doi.org/10.1007/978-981-96-1809-5_17

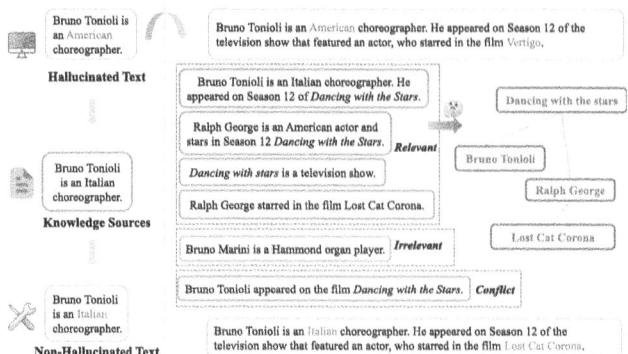

Fig. 1. An illustrative example of the multi-hop hallucination mitigation task in three settings.

robotics, thereby impeding the practical applications of LLMs in real-world scenarios. Therefore, to enhance the reliability of LLMs, it's important to **mitigate hallucinations**.

To this end, many studies have recently been conducted. Some work attempts to mitigate hallucinations during the stage of pre-training or reinforcement learning from human feedback (RLHF) [24]. However, these methods are costly and invasive [15]. Additionally, due to the inherent generative nature of LLMs, complete elimination of hallucinations is impossible. Therefore, **post-editing** the hallucinated texts generated by the LLMs has emerged as a promising strategy [9]. It aims at utilizing knowledge from external sources to build an auxiliary editor for correcting the factual errors in the generated text of LLMs. For example, as shown in the left side of Fig. 1, after comparing the generated hallucinated text with the external knowledge sources, the nationality of *"Bruno Tonioli"* was correctly amended back to Italian. Though some progress has been made with the post-editing strategy, there are still some unresolved challenges, particularly the *multi-hop hallucination mitigation* and the *non-authoritative knowledge sources*.

Multi-hop Hallucination Mitigation: Multi-hop hallucination mitigation requires gathering multiple pieces of evidence and performing multi-hop reasoning. Current research has primarily concentrated on short and straightforward sentences which only need one single piece of evidence to correct the factual errors. However, in real world scenarios, generated texts are typically long and contain a mixture of both true and false information [16]. For example, as illustrated in right side of the Fig. 1, the generated text on the top contains several pieces of information: a person *"Bruno Tonioli"*, a TV show *"Dancing with the Stars"* and a movie *"Vertigo"*. To mitigate the hallucinations in the text, we must first collect the four pieces of evidence shown in the center of Fig. 1 and then determine their relationships with the text. Firstly, we verify that *"Bruno Tonioli"* is an Italian choreographer and stars in TV show *"Dancing with the stars"*. Subsequently, we confirm that *"Ralph George"* also stars in the same

TV show. Afterwards, we verify that *"Ralph George"* stars in film *"Lost Cat Corona"*. Only then can we correct all the factual errors. Therefore, multi-hop hallucination mitigation is a challenging task.

Non-authoritative Knowledge Sources: Not all external knowledge sources[1] are credible and trustworthy. Current research often assumes that the retrieved knowledge is always relevant and reliable. However, in real world scenarios, not all the acquired knowledge is relevant due to the limitations of the retrieval systems. Furthermore, some knowledge sources even contradict other knowledge sources, either due to their failure to keep up with the evolving world knowledge or because they contain widespread rumors or misinformation on the Internet. For example, as shown in Fig. 1, the knowledge highlighted in the left rectangular box is **irrelevant** to the text and the knowledge in the bottom left rectangular box **conflicts** with the fact that *"Dancing with the stars"* is actually a TV show, not a movie. These noises can disrupt the following reasoning process and impair the task performance. Therefore, non-authoritative knowledge sources are challenging to manage in real-world scenarios.

To address the aforementioned challenges, we proposes the multi-hop hallucination mitigation task with three practical external knowledge source settings. We also construct two datasets for the task and propose a method to deal with it. The details are as follows: **Firstly**, we propose the **multi-hop hallucination mitigation** task. It requires gathering multiple pieces of evidence and performing multi-hop reasoning to correct the factual errors in the texts generated by LLMs. We consider three practical settings for the task to represent various scenarios when gathering evidence from external knowledge sources: *relevant, irrelevant* and *conflict*. In the *relevant* setting, the knowledge sources are reliable and every relevant evidence is available. In the *irrelevant* setting, more noisy irrelevant evidence are included. In the *conflict* setting, the knowledge sources become untrustworthy and some evidence may conflict with each other. **Secondly**, we construct two datasets for further exploration on this task: *HoCorr* (**Ho**ppy **Corr**ection) and *HotpotEC* (**Hotpot**QA **A**nswers **E**rror **C**orrection). HoCorr contains incorrect human-written texts with 2, 3, and 4 reasoning hops selected from HoVER [12]. HotpotEC contains real-world LLM-generated hallucinated answers to hard questions in the multi-hop question answering dataset HotpotQA [22]. Both of them will facilitate further research on the multi-hop hallucination mitigation task. **Thirdly**, to address this challenging task, we propose **FixER**(Base**Fix**, **E**valuator and **R**elfector), in which a *Fixer* iteratively enhances its performance with the assistance of a carefully designed *Evaluator* and *Reflector*. FixER iteratively improves its mitigation strategy and is well-suited to address the multi-hop reasoning and noises in external knowledge sources.

To conclude, our contributions are as follows: (**1**) We propose the novel multi-hop hallucination mitigation task, along with three practical external knowledge source settings: relevant, irrelevant and conflict. The task presents prac-

---

[1] In this paper, we use "knowledge source" and "evidence set" interchangeably since they refer to the same concept in our task.

tical challenges and opportunities for further investigation. (**2**) We create two high-quality datasets for our proposed multi-hop hallucination mitigation task, namely HoCurr and HotpotEC, which significantly facilitate further explorations on this task. (**3**) We propose a novel method, denoted as **FixER**, specifically designed to deal with our multi-hop hallucination task in various knowledge source settings. Extensive experiments demonstrate its considerable effectiveness comparing with the baselines.[2]

2   Related Work

Factual Error Correction. To address the proliferation of misinformation, previous studies focus on automatically verifying potentially misleading and false claims online [10]. Rather than assigning a label to the claim, [19] propose a new task of rewriting the claim so that it can be better supported by the evidence.

Hallucination Mitigation. Many work about hallucination mitigation explores how to rectify factual errors in the texts generated by LLMs using external knowledge sources. Some work breaks the long-form text to smaller, straightforward sentences and deals with them separately [4,16]. Some work leverages LLMs themselves [8] or external tools [9] to examine the generated texts. Our work mainly focuses on mitigating hallucinations via post-editing.

3   Task Formulation

In this section, we formulate the multi-hop hallucination mitigation task and provide a detailed description of the three practical knowledge settings.

Given the long-form texts generated by LLMs, the goal is to correct the potential factual errors that are not supported by relevant facts. Formally, the inputs to the task are a piece of generated text x and an evidence set $E = \{e_1, e_2, ..., e_n\}$. The output is the corrected text y, which should meet the following criteria: (1) **Factuality**: The solution should be free of factual errors and be fully supported by relevant evidence. (2) **Relevance**: The modified text and the original hallucinated text should exhibit semantic relevance. The fixing process should neither introduce excessive unrelated information nor omit too much related information.

Then, we cover our multi-hop hallucination mitigation task in three practical settings depending on the type of evidence set E as follows: (1) **Relevant Setting**: All documents in evidence set E are useful and can be used to correct the factual errors in x. (2) **Irrelevant Setting**: The evidence set E contains both relevant and irrelevant documents. Some documents seem relevant to x but they do not actually contribute to the resolution of factual errors. (3) **Conflict Setting**: Apart from the relevant and irrelevant documents, some documents in $E(x)$ contradict the relevant ones, indicating that they can't be trusted.

---

[2] The code and datasets will be available upon acceptance.

We denote the three types of evidence sets as $E_{relevant}$, $E_{irrelevant}$ and $E_{conflict}$ and we manually construct these evidence sets following this formula:

$$E_{relevant} \subseteq E_{irrelevant} \subseteq E_{conflict}$$

Clearly, tasks in relevant setting are the easiest, whereas in conflict setting, they become the most complex. Therefore, a method that can better deal with all the settings is needed.

4   Datasets

In this section, we introduce the details of the dataset construction process.

4.1   HoCurr

We employ a heuristic method to construct the HoCurr dataset, which consists of incorrect human-written texts paired with corresponding corrections.

To obtain multi-hop texts, we utilize the HoVER dataset [12], which is specifically designed for the multi-hop fact verification task. The key observation is that within the HoVER dataset, some texts appear nearly identical at first glance but they possess opposite factuality labels. We then discover that they share the same evidence set. Therefore, we select samples that share the same evidence set but posses opposite factuality labels, making reference corrections of the non-factual texts available.

After the heuristic selection, we obtain a brand-new dataset containing 1288 samples. The statistics of the dataset are shown in Table 1.

4.2   HotpotEC

Apart from the multi-hop human-written texts collected from the previous work, we also collect real ChatGPT generated answers and manually annotate them.

HotpotQA [22] is a multi-hop question answering dataset. We instruct Chat-GPT to answer the given questions in 20–40 words and then we select the hallucinated samples. Subsequently, we manually write the reference correction for the hallucinated samples. To ensure the quality of the manually-written sentence, annotators must possess at least a bachelor's degree, demonstrate fluency in English, and have a background in natural language processing research. Finally, we get 218 annotated samples. The statistics of the dataset are presented in Table 1.

Table 1. Statistics of the HoCurr and HotpotEC dataset

	2-hop	3-hop	4-hop	all
HoCurr	129	687	422	1288
HotpotEC	218	0	0	218

4.3   Task Settings

For the external knowledge sources, we use the English Wikipedia dump[3] for both datasets. We use the gold evidence provided by the original Hover dataset and HotpotQA dataset for the relevant setting. For irrelevant setting, we employ dense passage retrieval [13] to retrieve evidence and filter the relevant pieces of evidence. For conflict setting, we use ChatGPT to generate contradictory documents against one of the relevant documents for some of the samples.

5   Method

In this section, we describe our proposed method FixER in detail.

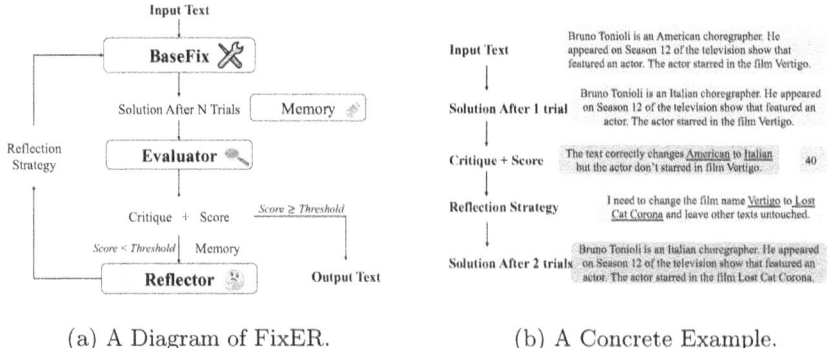

(a) A Diagram of FixER.   (b) A Concrete Example.

Fig. 2. Illustration of Our Method FixER.

5.1   BaseFix

At time step t, the function of the BaseFix can be represented using the following formula:

$$y_t = \begin{cases} M_F(x, E) & \text{if } t = 1 \\ M_F(y_{t-1}, E, s_{t-1}), & \text{if } t >= 2 \end{cases} \quad (1)$$

---

[3] https://hotpotqa.github.io/wiki-readme.html.

where x is the input hallucinated text, E is the evidence set, s_{t-1} is the reflection strategy at time step $t - 1$, y_t is the solution after t trials.

Initially, when time step $t = 1$, the BaseFix endeavors to correct the factual errors in the hallucinated text x using the corresponding evidence set E. After the first iteration, by further employing the strategy x acquired through the Reflector, the BaseFix can be refined for improvement.

The memory unit is updated using the following formula:

$$\Upsilon_t = \begin{cases} y_t & \text{if } t = 1 \\ \Upsilon_t \cup \Upsilon_{t-1}, & \text{if } t >= 2 \end{cases} \tag{2}$$

where y_t is the solution after t trials, Υ_t is the memory unit at time step t. The solution after each trial is stored in the memory unit Υ for further use.

5.2   Evaluator

At time step t, the function of the Evaluator can be represented using the following formula:

$$c_t, l_t = M_E(x, y_t, E) \tag{3}$$

where x is the input hallucinated text, y_t is the solution after t trials, E is the evidence set, c_t is the verbal critique, l_t is the scalar overall score.

The Evaluator is employed to assess the quality of each solution y_t after a certain number of trials. Typically, it provides two kinds of feedback signals: a verbal signal c and a scalar signal l. The scalar signal is an overall quality score of the solution. If the score exceeds a pre-defined threshold τ, it indicates that the solution after this trial is generally considered satisfactory. Otherwise, an additional reflection step should be taken. The quality assessment should strictly adheres to the criteria described in Sect. 3.

5.3   Reflector

At time step t, the function of the Reflector can be represented using the following formula:

$$s_t = M_R(x, \Upsilon_t, E, c_t) \tag{4}$$

where x is the input hallucinated text, E is the evidence set, Υ_t is the memory unit, c_t is the verbal critique at time step t, s_t is the reflection strategy at time step t.

The Reflector analyzes the verbal critique c_t along with previous trials stored in memory unit Υ_t and provides a practical strategy s_t on how to avoid past failures.

5.4   Overall Process

A diagram of the overall process is depicted in Fig. 2a. In each iteration, the BaseFix attempts to correct the factual errors using the evidence set and the reflection strategy, which is empty during the first trial. Next, an Evaluator is employed to assess the quality of the solution generated by the BaseFix. If the scalar score exceeds the predefined threshold, the program terminates, indicating that the solution is satisfactory. Otherwise, the Reflector analyzes the verbal critique and the memory containing previous trials to provide a practical reflection strategy. This iterative process continues until the maximum number of iteration is reached or the solution is considered satisfactory.

We use few-shot prompting [2] to build each module. We carefully design the instructions and in context learning examples for each module. No additional training is needed in our method.

6   Experiments

6.1   Baselines

We compare our method with the following baselines:

ReviseREF [1], which rewrites noisy sentences in text summaries to make them better supported by the source documents.

CompEdit [6], which is based on sentence compression and removes the entities in text to improve faithfulness.

ZeroFEC [11], which decomposes the task into five subtasks and uses five separated models to correct the errors in the hallucinated texts.

ZeroshotCoT, which simply prompts the LLMs to correct the factual errors in the hallucinated text with the given evidence set using chain-of-thought prompting [20]. We create our own task-specific prompts based on self-reflexion [18] and factool [5].

FewshotCoT, which additionally provides three human-written in context learning examples to do few-shot prompting [2]. To avoid data contamination, we select examples that are not in the two test datasets.

We also provide two other baselines to measure the lower bound and upper bound of the system performance:

InputCopy, which doesn't do any modification and directly copies the input text as the output.

GoldCorrection, which uses the provided reference corrections as our system output.

6.2   Implementation Details

We use ChatGPT(`gpt-3.5-turbo-0301`) as our base LLM. To reduce randomness, we set the generation temperature to zero. We use the following metrics: SARI-H, SARI-O, QFE. SARI [21] computes the added, deleted and kept n-grams and is widely used in factual error correction tasks [19]. Currently there are two different kinds of implementations and we call them `SARI-O`[4] and `SARI-H`[5] separately. QFE [7] compares the overlap between the answers based on the text and the answers based on the evidence under the same question. All the metrics we use exhibit a strong correlation with human evaluation scores [11].

Table 2. Experimental results on HoCurr in different settings. The best scores are highlighted in **boldface** and the second best scores are underlined.

Methods	Conflict Setting			Irrelevant Setting			Relevant Setting		
	SARI-H↑	SARI-O↑	QFE↑	SARI-H↑	SARI-O↑	QFE↑	SARI-H↑	SARI-O↑	QFE↑
ReviseRef	20.029	19.585	0.228	20.233	19.757	0.258	19.891	19.433	0.230
CompEdit	37.233	28.815	1.675	37.006	28.878	1.654	38.721	29.982	1.794
ZeroFEC	32.812	21.528	**2.088**	32.117	21.012	**2.101**	30.128	20.932	**2.479**
ZeroshotCoT	36.733	30.351	2.017	36.914	30.679	2.044	37.306	30.565	2.010
FewshotCoT	**40.846**	31.551	1.970	**41.330**	32.010	1.957	**41.121**	31.908	2.122
FixER (ours)	40.290	**33.970**	2.021	40.572	**34.004**	2.050	40.702	**33.890**	2.024
InputCopy	67.499	29.909	1.316	67.499	29.909	1.316	67.499	29.909	1.316
GoldCorrection	96.103	93.060	1.779	96.103	93.060	1.779	96.103	93.060	1.779

Table 3. Experimental results on HotpotEC in different settings. The best scores are highlighted in **boldface** and the second best scores are underlined.

Methods	Conflict Setting			Irrelevant Setting			Relevant Setting		
	SARI-H↑	SARI-O↑	QFE↑	SARI-H↑	SARI-O↑	QFE↑	SARI-H↑	SARI-O↑	QFE↑
ReviseRef	26.241	24.571	0.247	25.121	23.599	0.159	26.241	24.571	0.280
CompEdit	37.611	28.234	1.589	38.411	29.553	1.737	37.226	29.136	1.626
ZeroFEC	36.871	23.765	1.388	38.267	24.086	1.401	45.166	25.946	1.603
ZeroshotCoT	47.385	38.878	2.082	48.459	38.058	1.964	53.302	42.928	2.388
FewshotCoT	**54.987**	**42.350**	1.912	53.043	40.799	2.070	55.044	44.215	**2.391**
FixER (ours)	53.373	42.328	**2.119**	**55.823**	**44.433**	2.062	**60.447**	**51.328**	2.307
InputCopy	62.084	28.310	1.248	62.084	28.310	1.248	62.084	28.310	1.248
GoldCorrection	99.359	97.659	2.936	99.359	97.659	2.936	99.359	97.659	2.936

6.3 Results and Analysis

Table 2 and Table 3 show the experimental results on HoCurr and HotpotEC. Based on the experimental results shown in the tables, we draw the following key observations:

---

[4] https://github.com/cocoxu/simplification.

[5] https://huggingface.co/spaces/evaluate-metric/sari.

1. Non-authoritative knowledge sources cause a performance decrease. Across various knowledge settings, performance is poorest in the conflict setting, best in the relevant setting, and intermediary in the irrelevant setting. This aligns with our intuition, as irrelevant and conflicting knowledge increases task difficulty.
2. Our method outperforms the vanilla zeroshot and fewshot methods. Comparing our method with vanilla zero-shot and few-shot approaches on both datasets, our method consistently achieves better or competitive performance across all the knowledge settings, indicating the effectiveness of FixER.

Table 4. Ablation Experiments on HotpotEC in different settings.

Methods	Conflict Setting			Irrelevant Setting			Relevant Setting		
	SARI-H↑	SARI-O↑	QFE↑	SARI-H↑	SARI-O↑	QFE↑	SARI-H↑	SARI-O↑	QFE↑
FixER	53.373	42.328	2.119	55.823	44.433	2.062	60.447	51.328	2.307
w/o Evaluator (2-trial)	52.994	41.123	2.092	54.399	41.766	2.136	57.405	46.013	2.229
w/o Evaluator (3-trial)	53.097	41.581	2.128	54.289	41.960	2.163	55.135	43.783	2.272
FixER	53.373	42.328	2.119	55.823	44.433	2.062	60.447	51.328	2.307
w/o Critique (2-trial)	53.183	42.196	2.145	54.946	42.938	2.092	59.771	48.951	2.199
w/o Critique (3-trial)	52.984	41.959	2.140	54.757	42.932	2.063	59.968	49.319	2.193
BaseFix	54.590	43.104	2.132	56.682	44.312	2.072	60.523	49.497	2.234
w/ Evaluator (2-trials)	53.373	42.328	2.119	55.823	44.433	2.062	60.447	51.328	2.307
w/ Rules (2-trials)	54.170	42.710	2.097	56.525	44.102	2.017	60.691	49.289	2.184

6.4   Influence of the Number of Trials

We examine how the number of trials influences task performance. Specifically, we examine how the SARI-O metric changes as the number of trials increases on the HotpotEC dataset. We present a line chart in Fig. 3. From the experimental results, we draw the following observations: **(1)** In the relevant setting, performance peaks in the second trial and then begins to decline. In the irrelevant setting, performance remains relatively stable. In the conflict setting, performance hits its lowest point in the second trial and then begins to increase. **(2)** In all the settings, increasing the number of trials doesn't lead to a significant performance improvement but does result in a significant increase in cost. Hence, the number of trials should be kept relatively short in practical scenarios.

6.5   Ablation Study

To verify the effectiveness of the modules in FixER, we design the following ablation experiments.

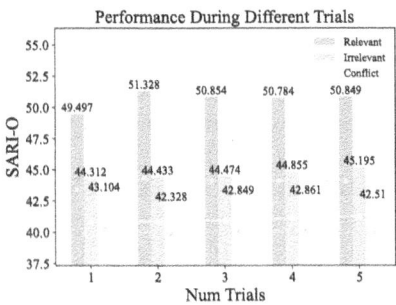

(a) Performance using metric SARI-O.

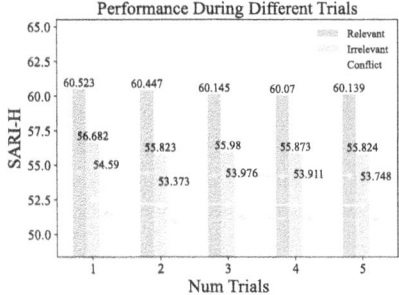

(b) Performance using metric SARI-H.

Fig. 3. Influence of the Number of Trials.

1. Importance of the feedback signals. We completely ignore the feedback signals provided by the Evaluator. This experiment is denoted as w/o `Evaluator` in Table 4.
2. Effect of the verbal critiques. We only use the scalar feedback signal provided by the Evaluator and ignore the verbal critiques. This experiment is denoted as w/o `Critique` in Table 4.
3. Effect of LLMs. In our method, we utilize LLMs to assess the quality of our modified text and generate the feedback signal. In this experiment, we replace the LLM-based Evaluator with simple heuristic rules to assess the effectiveness of LLMs. We measure the relevance of the solution y to the original hallucinated text x by calculating character-level Levenshtein edit distance [14]. We use the following formula introduced in [8]:

$$Rel_{Lev}(x, y) = max(1 - \frac{Lev(x, y)}{Length(x)}, 0)$$

where $Lev(x, y)$ denotes character-level Levenshtein edit distance between two piece of texts x and y. A score of 1 means the two pieces of text are identical, while a score of 0 indicates they are unrelated. This experiment is denoted as w/ `Evaluator` and w/ `Rules` in Table 4.

The ablation experiment results are shown in Table 4. From this table, we can draw the following conclusions: **(1)** Importance of the feedback signals. The performance drops significantly without the feedback signal. The more iteration steps, the worse the performance **(2)** Effectiveness of the verbal critiques. The performance decreases, but not to the same extent as it does without the feedback signals. This demonstrates that both verbal critiques and scalar scores contribute to the effectiveness of FixER. **(3)** Effectiveness of LLMs. After replacing the LLM-based Evaluator with heuristic rules, the performance decreases compared with the BaseFix in both the relevant and irrelevant settings. In the conflict setting, the performance is slightly better compared with using the Evaluator, as rule-based feedback is not affected by conflicting evidence.

7   Conclusion and Future Work

In this paper, we propose the multi-hop hallucination mitigation task, along with three practical external knowledge source settings. To facilitate further research on this task, we construct two datasets, HoCurr and HotpotEC. We also propose FixER, a method that can better deal with this task in various settings. By extensive empirical experiments, we show that our task is highly challenging and our method is effective comparing with the baselines. Future work can further explore the usage of external tools such as search engines to build plug-and-play hallucination mitigation modules and improve the factualness of LLMs.

References

1. Adams, G., Shing, H.C., Sun, Q., Winestock, C., McKeown, K., Elhadad, N.: Learning to revise references for faithful summarization. In: Findings of the Association for Computational Linguistics: EMNLP 2022, pp. 4009–4027 (2022)
2. Brown, T., et al.: Language models are few-shot learners. In: Advances in Neural Information Processing Systems, vol. 33, pp. 1877–1901 (2020)
3. Bubeck, S., et al.: Sparks of artificial general intelligence: early experiments with GPT-4 (2023)
4. Chen, S., et al.: FELM: benchmarking factuality evaluation of large language models. In: Oh, A., Naumann, T., Globerson, A., Saenko, K., Hardt, M., Levine, S. (eds.) Advances in Neural Information Processing Systems (2023)
5. Chern, I.C., et al.: FacTool: factuality detection in generative AI – a tool augmented framework for multi-task and multi-domain scenarios (2023)
6. Fabbri, A., Choubey, P.K., Vig, J., Wu, C.S., Xiong, C.: Improving factual consistency in summarization with compression-based post-editing. In: Proceedings of the 2022 Conference on Empirical Methods in Natural Language Processing, pp. 9149–9156 (2022)
7. Fabbri, A., Wu, C.S., Liu, W., Xiong, C.: QAFactEval: improved QA-based factual consistency evaluation for summarization. In: Proceedings of the 2022 Conference of the North American Chapter of the Association for Computational Linguistics: Human Language Technologies, pp. 2587–2601 (2022)
8. Gao, L., et al.: RARR: researching and revising what language models say, using language models. In: Rogers, A., Boyd-Graber, J., Okazaki, N. (eds.) Proceedings of the 61st Annual Meeting of the Association for Computational Linguistics (Volume 1: Long Papers), pp. 16477–16508 (2023)
9. Gou, Z., et al.: CRITIC: large language models can self-correct with tool-interactive critiquing (2023)
10. Guo, Z., Schlichtkrull, M., Vlachos, A.: A survey on automated fact-checking. Trans. Assoc. Comput. Linguist. **10**, 178–206 (2022)
11. Huang, K.H., Chan, H.P., Ji, H.: Zero-shot faithful factual error correction. In: Proceedings of the 61st Annual Meeting of the Association for Computational Linguistics (Volume 1: Long Papers), pp. 5660–5676 (2023)
12. Jiang, Y., Bordia, S., Zhong, Z., Dognin, C., Singh, M., Bansal, M.: HoVer: a dataset for many-hop fact extraction and claim verification. In: Cohn, T., He, Y., Liu, Y. (eds.) Findings of the Association for Computational Linguistics: EMNLP 2020, pp. 3441–3460 (2020)

13. Karpukhin, V., et al.: Dense passage retrieval for open-domain question answering. In: Proceedings of the 2020 Conference on Empirical Methods in Natural Language Processing (EMNLP), pp. 6769–6781 (2020)
14. Levenshtein, V.I., et al.: Binary codes capable of correcting deletions, insertions, and reversals. Soviet Phys. Doklady **10**, 707–710 (1966)
15. Li, K., Patel, O., Viégas, F.B., Pfister, H., Wattenberg, M.: Inference-time intervention: eliciting truthful answers from a language model. In: Oh, A., Naumann, T., Globerson, A., Saenko, K., Hardt, M., Levine, S. (eds.) Advances in Neural Information Processing Systems (2023)
16. Min, S., et al.: FActScore: fine-grained atomic evaluation of factual precision in long form text generation. In: Bouamor, H., Pino, J., Bali, K. (eds.) Proceedings of the 2023 Conference on Empirical Methods in Natural Language Processing, pp. 12076–12100 (2023)
17. Qin, C., Zhang, A., Zhang, Z., Chen, J., Yasunaga, M., Yang, D.: Is ChatGPT a general-purpose natural language processing task solver? In: Bouamor, H., Pino, J., Bali, K. (eds.) Proceedings of the 2023 Conference on Empirical Methods in Natural Language Processing, pp. 1339–1384 (2023)
18. Shinn, N., Cassano, F., Gopinath, A., Narasimhan, K., Yao, S.: Reflexion: language agents with verbal reinforcement learning. In: Oh, A., Naumann, T., Globerson, A., Saenko, K., Hardt, M., Levine, S. (eds.) Advances in Neural Information Processing Systems (2023)
19. Thorne, J., Vlachos, A.: Evidence-based factual error correction. In: Zong, C., Xia, F., Li, W., Navigli, R. (eds.) Proceedings of the 59th Annual Meeting of the Association for Computational Linguistics and the 11th International Joint Conference on Natural Language Processing (Volume 1: Long Papers), pp. 3298–3309 (2021)
20. Wei, J., et al.: Chain-of-thought prompting elicits reasoning in large language models. In: Koyejo, S., Mohamed, S., Agarwal, A., Belgrave, D., Cho, K., Oh, A. (eds.) Advances in Neural Information Processing Systems (2022)
21. Xu, W., Napoles, C., Pavlick, E., Chen, Q., Callison-Burch, C.: Optimizing statistical machine translation for text simplification. Trans. Assoc. Comput. Linguist. **4**, 401–415 (2016)
22. Yang, Z., et al.: HotpotQA: a dataset for diverse, explainable multi-hop question answering. In: Riloff, E., Chiang, D., Hockenmaier, J., Tsujii, J. (eds.) Proceedings of the 2018 Conference on Empirical Methods in Natural Language Processing, pp. 2369–2380. Association for Computational Linguistics, Brussels, Belgium (2018)
23. Zhang, M., Press, O., Merrill, W., Liu, A., Smith, N.A.: How language model hallucinations can snowball (2023)
24. Zhang, Y., et al.: Siren's song in the AI ocean: a survey on hallucination in large language models (2023)

LLM-AR: Large Language Model Augmented Retrieval for Few-Shot Knowledge Graph Completion

Yu Song, Dezhi Kong, Bohan Yu, Kunli Zhang[✉], and Hongying Zan

School of Computer and Artificial Intelligence, Zhengzhou University, Zhengzhou 450001, HN, China
ieklzhang@zzu.edu.cn

Abstract. Knowledge graphs are essential in various artificial intelligence tasks, but they frequently encounter the issue of incompleteness. Given the long-tail distribution of relationships within real-world knowledge graphs, researchers have incorporated few-shot learning methods for knowledge graph completion (FKGC). Large language models (LLMs) encompass extensive knowledge bases that can effectively mitigate the long-tail problem. This paper introduces the LLM-AR framework, which integrates LLMs with FKGC models. In this framework, FKGC functions as a retriever, and the retrieval results are encoded into instructions using a knowledge prompting strategy to guide the LLM in enhanced retrieval. LLM-AR is compatible with the majority of existing FKGC models without incurring additional training overhead. LLM-AR was utilized in experiments on the NELL and Wiki datasets with two FKGC models, FAAN and CIAN. The experimental results demonstrate varying degrees of improvement across all metrics. On the Wiki dataset, Hits@1 for FAAN and CIAN increased by 16.3% and 15%, respectively.

Keywords: Large language model · Few-shot knowledge graph completion · Prompt engineering

1 Introduction

Knowledge Graphs (KGs) serve as representations of real-world knowledge, encapsulating the relationships among entities through triples in the form <head entity, relation, tail entity>. These graphs constitute the foundation for various applications, such as recommendation systems, question answering, and knowledge discovery endeavors. Knowledge Graph Completion (KGC) is concerned with predicting and filling in the missing components of these triples within knowledge graphs. This paper specifically addresses the task of link prediction within the domain of KGC, which involves predicting the missing entity in an incomplete triple.

Based on the type of information utilized, existing KGC methods can be divided into two major categories: triple-based methods and text-based methods [1]. Triple-based KGC methods [2] predict the missing parts of triples by learning representations

© The Author(s), under exclusive license to Springer Nature Singapore Pte Ltd. 2025
B. Xu et al. (Eds.): CCKS-IJCKG 2024, CCIS 2229, pp. 240–251, 2025.
https://doi.org/10.1007/978-981-96-1809-5_18

of the structural information of entities and their relations. Most existing knowledge graph, long-tail relations are prevalent, and most relations are associated with only a few triple instances, which limits the information that can be obtained from structured triples for the KGC task. Consequently, due to this scarcity of information, the performance of triple-based methods often deteriorates when dealing with long-tail entities [1]. Text-based methods have been introduced to KGC and have achieved promising results [3]. However, text-based KGC methods heavily rely on the quality and quantity of the text, are significantly affected by noise, and incur high computational costs. Therefore, researchers have turned their attention to knowledge graph completion in few-shot scenarios, which can effectively alleviate the long-tail relationship problem. The few-shot knowledge graph completion task involves predicting missing facts for relationships given few of reference triples. A knowledge graph is considered a few-shot knowledge graph when the number of triples representing the task relationship in the graph ranges between 50 and 500 [4].

Recently, large language models (LLMs) such as ChatGPT and GPT-4 [5] have garnered widespread attention [6]. LLMs possess an extensive amount of internal knowledge, which can serve as an additional knowledge base to mitigate the information scarcity associated with long-tail relations. Given their powerful capabilities, the initial approach considered was to utilize LLMs directly for the KGC task. However, research has demonstrated that directly employing LLMs does not achieve state-of-the-art performance in knowledge graph completion [7]. For instance, Khali et al. [8] attempted to fine-tune Llama2-7B using the node names of the knowledge graph, but the results were not satisfactory.

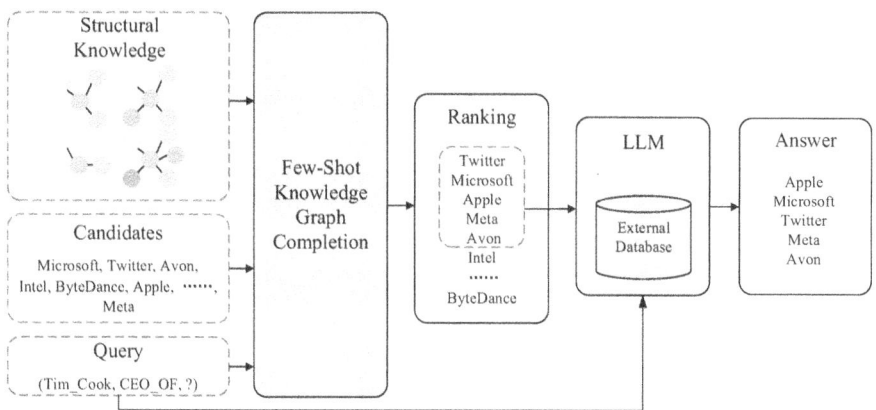

Fig. 1. Examples of Enhanced Retrieval Using Large Language Models

There are several limitations associated with using LLMs for KGC: (a) LLMs typically impose strict length constraints on the number of input tokens, making it challenging to accurately describe complex KGC tasks. (b) the output of LLMs is difficult to constrain, which means they may generate predictions beyond the scope of entities in the knowledge graph. (c) although LLMs contain extensive internal knowledge, their handling of entities may be impaired due to the lack of structured information.

To address the aforementioned limitations, this paper proposes a Large Language Model Augmented Retrieval (LLM-AR) framework specifically designed for knowledge graph completion in few-shot scenarios. Figure 1 presents an example of using the LLM-AR framework in combination with the FKGC model. In this framework, traditional knowledge graph completion models are utilized as retrievers. For a link prediction task (head entity, relation, ?), the retriever ranks the candidate tail entities based on the head entity and relation, producing a ranked list of candidate entities, where higher-ranked entities are more likely to be correct. To address issue (a), the top k entities from the candidate entity ranked list are input into the LLM. To address issue (b), this paper employs the In-Context Learning (ICL) strategy, optimizing it into a multi-prompt instruction method. When the output of the LLM fails to meet the prompt requirements, the information is automatically incorporated into alternative prompts and re-input into the model to complete subsequent tasks. For issue (c), the solution adopted in this paper involves extracting several triples from the training set of the dataset that share the same relation as the query set, and these triples are used as examples in the prompt templates input into the LLM for analogical reasoning.

In summary, the contributions of this paper are as follows:

- This paper introduces a fast and effective FKGC framework, LLM-AR, which can directly integrate LLMs with most existing FKGC models.
- A multi-prompt instruction strategy tailored for FKGC is proposed in this paper, enabling effective control over the output of large language models.
- Experimental results on two publicly available datasets demonstrate that the LLM-AR framework can be combined with most existing FKGC completion model to significantly enhance its performance.

2   Related Work

2.1   Few-Shot Knowledge Graph Completion

Existing FKGC methods are mostly based on triple-based approaches, which rely solely on structured triples for KGC. Early shallow knowledge graph embedding (KGE) methods represent entities and relations as low-dimensional embedding vectors in continuous embedding spaces. Approaches to handling data in FKGC can further be categorized into methods based on metric learning and methods based on meta-learning.

In the realm of metric learning-based methods, GMatching [4] stands as the first endeavor aimed at addressing one-shot KGC. FAAN [9] devised a Transformer encoder and an attention-based aggregator to adapt reference representations to various queries, amalgamating the varying characteristics of entities across different task relations and obtaining task-sensitive entity representations. MFEN [10] captures the diversity of neighborhood features through a convolutional encoder, while CIAN [11] explores interactions within entities, utilizing attention mechanisms to simulate interactions between head and tail entities. TransAM [12] addresses the FKGC problem by employing a designed Transformer matching network to compute the likelihood of entity sequences. Methods based on meta-learning include MetaR [13], which generates relation-specific meta-information and updates relation meta-information through gradient-based meta-updates. GANA [14] acquires relation meta-information by integrating neighboring information of entities and utilizes a meta-learning-based TransH module to handle complex relations.

2.2 LLMs for Knowledge Graph Completion

Recently, Zhao et al. [6] conducted a comprehensive study on LLMs, treating KGC as a fundamental task for LLMs. Xie et al. [15] and Zhu et al. [7] used ChatGPT and GPT-4, respectively, for link prediction tasks in knowledge graphs. L et al. [16] fine-tuned three LLMs include Llama-7B, Llama-13B and Llama2-13B, and evaluated them on three sub-tasks of the KGC task. Y et al. [17] were the first to combine LLMs with triple-based KGC methods, enabling the simultaneous utilization of structured information in knowledge graphs and the knowledge base of LLMs.

Furthermore, due to the extensive knowledge base of LLMs, some studies have employed knowledge distillation methods to extract data from LLMs, enhancing the performance of KGC models through knowledge augmentation. D et al. [18] addressed the lack of contextual information in knowledge graph completion tasks by using LLMs for entity description expansion, relationship understanding optimization, and structural information extraction. L et al. [19] prompted LLMs to transform triples into context-rich segments and designed two auxiliary tasks, reconstruction and contextualization, to help traditional KGC models absorb knowledge from the augmented triples.

Inspired by the aforementioned works, this paper further applies LLMs to knowledge graph completion in few-shot scenarios.

3 Methodology

3.1 Problem Setting

A knowledge graph is represented as a set of triples $G = \{(h, r, t)\}$, where E and R denote the sets of entities and relations in G, respectively. Here, hE represents the head entity, tE represents the tail entity, and rR represents the relationship between them. The link prediction task involves taking an incomplete triple $(h, r, ?)$ or $(?, r, t)$ as a query, with the aim of predicting the missing entity. Link prediction models typically need to score all entities that could potentially be the missing entity, and then rank all entities based on their scores. For simplicity, this paper focuses solely on queries missing the tail entity $Q = (h, r, ?)$. Queries missing the head entity can be processed in a similar manner.

3.2 Overview of LLM-AR

The LLM-AR framework is illustrated in Fig. 2. This framework consists of two components: a triple-based KGC retriever and a LLM. For each query task Q, the retriever first assigns a score to each candidate entity. The entity ranking, represented in descending order, is denoted as $R_{retriever} = (e_1, e_2, \ldots, e_{|E|})$.

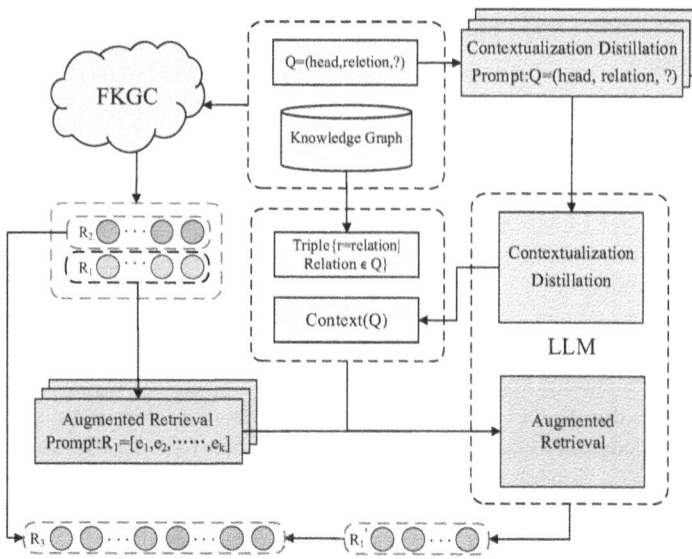

Fig. 2. An illustration of the LLM-AR framework

Subsequently, the query Q is input into the LLM using prompt instructions to generate the corresponding textual description. Meanwhile, the knowledge graph is searched for examples of triples that share the same relation r as in Q. These examples, along with the textual description of Q, together constitute the external information c_i. Following this, the top-k results from $R_{retriever}$, along with Q and its external information c_i are input into the large language model using a multi-prompt instruction strategy, resulting in the re-ranked results $R_{LLM} = \left(e'_1, e'_2, \ldots, e'_k\right)$. These are then concatenated with the results from $R_{retriever}$ that were not input into the large language model, yielding the final results $R_{LLM} = \left(e'_1, e'_2, \ldots, e'_k, e_k, \ldots, e_{|E|}\right)$.

3.3 Knowledge Prompt

In this section, we introduce the knowledge prompt, a contextual learning strategy specifically designed for FKGC. By distilling contextual information from the LLM and incorporating it with examples from the KG into the prompts, the performance of the LLM in link prediction is enhanced.

Contextualization Distillation. Recent reaches have highlighted the significant capability of LLMs to contextualize structured data and transform it into contextual segments [20]. This paper adopts their approach and extracts descriptive context from LLMs.

In previous approaches, two commonly employed description types can be roughly categorized: Entity Description (ED) and Triple Description (TD) [21]. Entity Description refers to the definition and description of individual entities, whereas Triple Description pertains to textual snippets reflecting specific relationships between two entities within a triple. Similarly, for the Q, we curate prompt p_i by filling the predefined template:

$$p_i = CD\_Prompt(Q = (h_i, r_i, ?)) \tag{1}$$

where are the head entity and relation, "?" represents the tail entity to be predicted. Then, employing as the input, the LLM is prompted to generate a descriptive context for each Q:

$$c_i = LLM\,(p_i) \tag{2}$$

Augmented Retrieval. Prompt engineering is a crucial component in ICL. In this section, we present the entire workflow of interacting with the LLM and consider key aspects of prompt design.

For each link prediction query $(h, r, ?)$, AR-LLM treats each query task as an independent task. The prompt instructions are divided into four parts: the query task, the candidate entity list, supplementary information, and additional examples.

The additional examples $A = \{(h, r, t) \in G_{train} \cup G_{valid} | r = Q_r\}$ are triples from the training and validation sets that share the same relation r as the query task. The Q and A are presented in the form of triples within the prompt instructions, while supplementary information c_i for the query task is provided in textual form. The candidate entity list consists of the top-k entities sorted by the FKGC retriever $R'_{retirever} = (e_1, e_2, \ldots, e_k)$.

Furthermore, the prompt instructions also include an explanation for the candidate entity list: "Entities ranked higher in the candidate entity list are more likely to be correct answers." By employing this approach, more information is provided to the LLM, simplifying the ranking task to a probabilistic re-ranking task.

Multi-prompt Instructions. LLMs require natural language as input. To bridge this gap, LLM-AR utilizes prompt instructs to convert queries and demonstrations into plain text with the same format. However, due to the complexity of KGC tasks, it is challenging to strictly control the input to the LLM. This paper innovatively adopts a multi-prompt instructions strategy to address complex tasks similar to KGC.

Firstly, a set of prompt instructions $I = (I_1, I_2, \ldots, I_m)$ is designed by experts, with I_1 as the preferred prompt instruction and the others as alternative prompt instructions. The relevant information of the query task is inserted into the preferred prompt instruction and input into the LLM. If the output does not meet the prompt requirements, the relevant information of the query task is filled into the alternative prompt instructions and input into the LLM. Through this approach, the quality of the output information from the LLM can be strictly controlled, thereby improving the evaluation results.

4 Experiment

4.1 Experimental Setting

Dataset. To evaluate the link prediction performance of LLM-AR, experiments were conducted on the public datasets NELL [22] and Wiki [23]. These are two of the most commonly used datasets for FKGC.

Baseline. This paper employs two knowledge graph completion models, FAAN [9] and CIAN [11], and applies the proposed LLM-AR framework to them. Additionally, this paper also compared eight common baseline models, including GMatching [4], MetaR [13], FSRL [24], GANA [14], TransAM [12], APINet [25], and GL-GFKGC [26]. Traditional KGE-based models were omitted from comparison in this paper because they have been proven to perform significantly lower than the aforementioned FKGC baselines [14]. Therefore, the experimental results of this paper only include the performance of FKGC baselines.

LLMs. In numerous studies, GPT-4 has been demonstrated to possess exceptionally powerful capabilities. In the experimental section of this paper, an optimized version of GPT-4, referred to as GPT-4o, is employed. In specific tasks, GPT-4o exhibits even greater proficiency.

Table 1. Performance of five-shot link prediction on NELL and Wiki datasets

Model	NELL				Wiki			
	MRR	Hits10	Hits5	Hits1	MRR	Hits10	Hits5	Hits1
GMatching	0.196	0.316	0.262	0.142	0.263	0.387	0.337	0.197
MetaR	0.309	0.505	0.419	0.206	0.221	0.302	0.264	0.178
FSRL	0.313	0.502	0.412	0.217	0.113	0.236	0.135	0.056
GANA	0.344	0.517	0.437	0.246	0.351	0.446	0.407	0.299
MFEN	0.310	0.443	0.369	0.236	0.310	0.443	0.369	0.236
TransAM	0.263	0.371	0.311	0.205	0.330	0.465	0.405	0.258
APINet	0.318	0.506	0.412	0.225	0.347	0.498	0.428	0.297
GL-GFKGC	0.293	0.438	0.378	0.228	0.351	0.459	0.422	0.303
FAAN	0.277	0.417	0.360	0.198	0.302	0.453	0.378	0.229
FAAN-LLM-AR	0.299	0.470	0.389	0.216	0.424	0.513	0.475	0.392
CIAN	0.385	0.530	0.456	0.302	0.360	0.505	0.448	0.280
CIAN-LLM-AR	**0.398**	**0.533**	**0.466**	**0.326**	**0.456**	**0.527**	**0.498**	**0.430**

4.2 Main Result

Table 1 presents the performance of all models on NELL and Wiki datasets, with bold numbers indicating the best performance. The experimental results for FAAN and CIAN are reproduced from this paper, while the results for the remaining baseline models are derived from their original papers.

The evaluation metrics used for the models are MRR and HitsN, where MRR represents the mean reciprocal rank, and HitsN represents the proportion of correct entities within the top N ranks, with $N = 1,5,10$. The LLM-AR framework proposed enhances the performance of both FAAN and CIAN. Compared to the baselines, it consistently best performs.

The experimental results demonstrate varying degrees of improvement on both datasets after integrating the LLM-AR method into FAAN and CIAN models. Particularly, CIAN-LLM-AR surpassing all baseline models. This may be attributed to the fact that many of correct answers are concentrated within the top-k options of the entity ranking list retrieved by CIAN. It is noteworthy that the Hits20 result for CIAN on the NELL dataset is 0.593, indicating significant room for improvement when enhancing retrieval with LLM.

4.3   Ablation Study

Ablation study was conducted using the NELL dataset to demonstrate the effectiveness of each component in LLM-AR. Table 2 presents the results of the ablation study.

In the first ablation experiment, removing the examples with the same relation r as the query task led to a decline in all metrics, with Hits1 even falling below the original FAAN results. This indicates that supplementary examples are crucial for accurately hitting the correct answer.

Table 2. Ablation Results of FAAN-LLM-AR on NELL

Variants(w/o)	MRR	Hits10	Hits5	Hits1
w/o example	0.283	0.461	0.367	0.180
w/o context description	0.287	0.458	0.363	0.203
w/o muti-prompt instruction	0.281	0.447	0.352	0.193
w/o AR	0.265	0.420	0.334	0.162
FAAN-LLM-AR	**0.299**	**0.470**	**0.389**	**0.216**

In the second, the textual description information for the query task was removed. The results showed similar impacts on all metrics. This suggests that textual description information enhances the knowledge for the LLM.

In the third, the multi-prompt instructions were removed. This caused the LLM to generate approximately 10% erroneous data in its output, thereby affecting the evaluation results.

In the fourth, only the top-k results from FAAN were input directly into the LLM using the primary prompt instruction. The results showed a significant decrease in MRR, Hits1, and Hits5 compared to the original FAAN results. This confirms the poor performance of LLM in directly performing link prediction and further validates the effectiveness of our proposed method.

4.4   Case Study Analysis

To further illustrate the link prediction performance of LLM-AR, the top-5 rankings of candidate entities for different relations were generated on NELL. Some random triple examples are listed in Table 3, where the underlined portion indicates the true tail entity.

Table 3. Top-5 Rankings of Candidate Entities for Different Relations by FAAN-LLM-AR

Query Task	True Rank&Top-5 Candidate List		
	FAAN	FAAN-LLM-AR	-AR
(outlook_mobile , produced by, ?) True: company: microsoft	2, (server, microsoft, mapquest, yahoo001, microsoft__)	1, (microsoft , server , mapquest, yahoo001, microsoft__)	4, (server, mapquest, yahoo001, microsoft , microsoft__)
(cleveland_browns , teamcoah, ?) True: coach:mangini	4, (eric_mangini, bill_parcells, bill_belichick, mangini, bud_grant)	2, (eric_mangini, mangini, bill_parcells, bill_belichick, bud_grant)	5, (eric_mangini, bill_parcells, bill_belichick, bud_grant, mangini)
(skiing, sportschoolincountry, ?) True: austria	1, (austria, Ireland, Hungary, Uk, Netherlands)	1, (austria, Ireland, Hungary, Uk, Netherlands)	3, (Ireland, Hungary, austria, Uk, Netherlands)

The FAAN original generation results are contrasted with FAAN-LLM-AR. -AR represents the direct input of the top-k results obtained from FAAN retrieval into the LLM without any further processing. As depicted in the table, FAAN-LLM-AR yields the optimal ranking results for genuine entities across the three scenarios. Following this, the original ranking of entities by FAAN is observed. Despite undergoing prior FAAN retrieval and the exclusion of entities with lower probabilities of being correct answers, direct employment of the LLM for link prediction tasks still manifests adverse effects, consistent with the findings of ablation experiments.

4.5 Efficiency Analysis

The method proposed in this paper leverages the results retrieved by FKGC models, directly invoking LLM for enhanced retrieval without the need for additional training. Here, efficiency is analyzed solely from an inference perspective. Table 4 lists the inference time for a single data point for various models on an RTX 3090.

TransE [2], FAAN [9], MSRMN [27], CIAN [11] are KGC models based on triples, while KG-BERT [21] and GenKGC [3] are text-based models. As shown in Table 4, the inference efficiency decreases when LLM-AR is applied to triple-based models. LLM-AR makes reasonable use of unstructured information corresponding to the KG and information from the open world within the LLM, at the cost of inference speed. Despite this, the inference efficiency is still higher than that of text-based models.

Table 4. The time under RTX3090 is used to estimate the speed of inference a single data

Model	Time under RTX3090
TransE	0.02 s
KG-BERT	10100 s
GenKGC	0.96 s
MSRMN	0.091 s
FAAN	0.071 s
FAAN-LLM-AR	0.756 s
CIAN	0.101 s
CIAN-LLM-AR	0.786 s

5 Conclusion

This paper proposes LLM-AR, an effective framework that integrates LLMs with FKGC methods for link prediction. Additionally, it introduces a prompting strategy known as multi-prompt instruction. LLM-AR treats the FKGC model as a retriever, probabilistically ranking the numerous entities in the candidate entity list. By extracting the top-k candidate entities, the framework significantly simplifies the link prediction task and strictly limits the token length input to the LLM, thereby allowing for a clear description of the complex task of KGC. Through the multi-prompt instruction strategy, the output of the LLM is strictly constrained, enhancing data quality and further strengthening the LLM's capability in performing KGC task. LLM-AR leverages the LLM as an additional knowledge repository. Unlike text-based methods, LLM-AR does not require training or fine-tuning, which significantly reduces overhead. Experimental results demonstrate that LLM-AR effectively improves the performance of FKGC models and efficiently mitigates the long-tail problem in real-world scenarios.

Despite using multi-prompt instructions, LLM-AR occasionally outputs data that do not adhere to established requirements when faced with the complex task of knowledge graph completion. In this paper, such data are directly handled by using the results outputted by the original FKGC model. The inference speed of LLM-AR is constrained by the large language model it employs. While it efficiently utilizes both structured and unstructured information, its inference efficiency is lower compared to triple-based knowledge graph completion models. As the dataset size increases, the time cost also rises. Future work could involve local deployment and fine-tuning of the model to further restrict LLM outputs, thereby enhancing inference efficiency.

Acknowledgements. The Science and Technology Innovation 2030- "New Generation of Artificial Intelligence" Major Project [No.2021ZD0111000], and Henan Provincial Science and Technology Research Project [No. 232102211033].

References

1. Wang, X., He, Q., Liang, J., Xiao, Y.: Language models as knowledge embeddings. arXiv preprint arXiv:2206.12617 (2022)
2. Bordes, A., Usunier, N., Garcia-Duran, A., Weston, J., Yakhnenko, O.: Translating embeddings for modeling multi-relational data. In: Advances in Neural Information Processing Systems, vol. 26 (2013)
3. Xie, X., et al.: From discrimination to generation: knowledge graph completion with generative transformer. In: Companion Proceedings of the Web Conference 2022, pp. 162–165 (2022)
4. Xiong, W., Yu, M., Chang, S., Guo, X., Wang, W.Y.: One-shot relational learning for knowledge graphs. arXiv preprint arXiv:1808.09040 (2018)
5. Achiam, J., et al.: GPT-4 technical report. arXiv preprint arXiv:2303.08774 (2023)
6. Zhao, W.X., et al.: A survey of large language models. arXiv preprint arXiv:2303.18223 (2023)
7. Zhu, Y., et al.: LLMs for knowledge graph construction and reasoning: recent capabilities and future opportunities. arXiv preprint arXiv:2305.13168 (2023)
8. Alqaaidi, S.K., Kochut, K.: Relations prediction for knowledge graph completion using large language models. arXiv preprint arXiv:2405.02738 (2024)
9. Sheng, J., et al.: Adaptive attentional network for few-shot knowledge graph completion. arXiv preprint arXiv:2010.09638 (2020)
10. Wu, T., Ma, H., Wang, C., Qiao, S., Zhang, L., Yu, S.: Heterogeneous representation learning and matching for few-shot relation prediction. Pattern Recogn. **131**, 108830 (2022)
11. Li, Y., Yu, K., Huang, X., Zhang, Y.: Learning inter-entity-interaction for few-shot knowledge graph completion. In: Proceedings of the 2022 Conference on Empirical Methods in Natural Language Processing, pp. 7691–7700 (2022)
12. Liang, Y., Zhao, S., Cheng, B., Yang, H.: TransAM: transformer appending matcher for few-shot knowledge graph completion. Neurocomputing **537**, 61–72 (2023)
13. Chen, M., Zhang, W., Zhang, W., Chen, Q., Chen, H.: Meta relational learning for few-shot link prediction in knowledge graphs. arXiv preprint arXiv:1909.01515 (2019)
14. Niu, G., et al.: Relational learning with gated and attentive neighbor aggregator for few-shot knowledge graph completion. In: Proceedings of the 44th International ACM SIGIR Conference on Research and Development in Information Retrieval, pp. 213–222 (2021)
15. Xie, X., Li, Z., Wang, X., Xi, Z., Zhang, N.: LambdaKG: a library for pre-trained language model-based knowledge graph embeddings. arXiv preprint arXiv:2210.00305 (2022)
16. Yao, L., Peng, J., Mao, C., et al.: Exploring large language models for knowledge graph completion. arXiv preprint arXiv:2308.13916 (2023)
17. Wei, Y., Huang, Q., Kwok, J.T., Zhang, Y.: KICGPT: large language model with knowledge in context for knowledge graph completion. arXiv preprint arXiv:2402.02389 (2024)
18. Xu, D., et al.: Multi-perspective improvement of knowledge graph completion with large language models. arXiv preprint arXiv:2403.01972 (2024)
19. Li, D., Tan, Z., Chen, T., Liu, H.: Contextualization distillation from large language model for knowledge graph completion. arXiv preprint arXiv:2402.01729 (2024)
20. Xiang, J., Liu, Z., Zhou, Y., Xing, E.P., Hu, Z.: ASDOT: any-shot data-to-text generation with pretrained language models. arXiv preprint arXiv:2210.04325 (2022)
21. Yao, L., Mao, C., Luo, Y.: KG-BERT: BERT for knowledge graph completion. arXiv preprint arXiv:1909.03193 (2019)
22. Mitchell, T., et al.: Never-ending learning. Commun. ACM **61**(5), 103–115 (2018)
23. Vrandečić, D., Krötzsch, M.: Wikidata: a free collaborative knowledgebase. Commun. ACM **57**(10), 78–85 (2014)

24. Zhang, C., Yao, H., Huang, C., et al.: Few-shot knowledge graph completion. Proc. AAAI Conf. Artif. Intell. **34**(03), 3041–3048 (2020)
25. Li, Y., Yu, K., Zhang, Y., Liang, J., Wu, X.: Adaptive prototype interaction network for few-shot knowledge graph completion. IEEE Trans. Neural Netw. Learn. Syst. (2023)
26. Xie, P., Zhou, G., Liu, J., Huang, J.X.: Incorporating global–local neighbors with Gaussian mixture embedding for few-shot knowledge graph completion. Expert Syst. Appl. **234**, 121086 (2023)
27. Song, Y., Gui, M., Zhang, K., Xu, Z., Dai, D., Kong, D.: Relational multi-scale metric learning for few-shot knowledge graph completion. Knowl. Inf. Syst., 1–26 (2024)

Hierarchical Knowledge Graph Attention Network for Recommendation Systems

Ziyou He, Zhoushuai Xu, Wenbo Zheng, and Yuanming Zhang$^{(\boxtimes)}$

Zhejing University of Technology, Hangzhou, China
zym@zjut.edu.cn

Abstract. Knowledge graph (KG) has been widely adopted in recommendation systems to alleviate sparsity and cold-start problems that generally exist in collaborative filtering-based methods. Such methods have improved performance to some extent. However, they only consider specific relations (interaction relations), which leads to low recommendation accuracy. This paper proposes a hierarchical KG attention network (HKGAN) to fuse multiple relations (interaction, social, and similarity) by decomposing rich semantic relations in the original KG. This model encodes different types of entities of a KG with a novel hierarchical attention mechanism to disseminate node-level and graph-level features. More accurate representations of users and items can be learned from multiple heterogeneous features. Then, the prediction scores are calculated by performing inner product operations on the representations of users and items. Experimental results on three public datasets show that the proposed model can significantly outperform state-of-the-art baselines.

Keywords: Recommendation system · Knowledge graph · Hierarchical attention mechanism · Multi-view features

1 Introduction

With the rapid development of internet technology, users face the problem of information overload in various online systems. The recommendation system (RS) plays a pivotal role in matching user interests with resource items, using attributes and relations between users and items, extracting a user's points of interest, and recommending the user's top-k items. As a classic recommendation method, collaborative filtering (CF) [4, 7] focuses on the user-item interaction relation from the perspective of users or items. However, CF-based methods usually encounter data sparsity and cold-start problems due to the inability to process the features of users or items.

Knowledge graph (KG) is a directed semantic network in which nodes correspond to users, items, and attributes, and edges correspond to various relations, and whose rich semantic features are useful for the diversity of RSs. KG-based recommendation models can be categorized as embedding-based, path-based, and unified. An embedding-based model uses structural information to encode a KG as a low-rank embedding. A path-based model defines meta-paths in a KG and uses connection patterns between entities

© The Author(s), under exclusive license to Springer Nature Singapore Pte Ltd. 2025
B. Xu et al. (Eds.): CCKS-IJCKG 2024, CCIS 2229, pp. 252–268, 2025.
https://doi.org/10.1007/978-981-96-1809-5_19

for recommendation. A unified model uses a graph neural network (GNN) to aggregate features of an item's multi-hop neighbors to enrich its information, and it shows great promise.

Unified models such as the knowledge graph convolutional network (KGCN) [14], knowledge graph attention network (KGAT) [15], and collaborative knowledge-aware attentive network (CKAN) [16] consider only specific interaction relations between users and items, which results in low recommendation accuracy and an inability to deal with cold starts. In addition to interaction relations, a general KG includes other semantic features, such as social and similarity relations. Social (e.g., friendship or trust) relations often influence users' behaviors toward their peers [26]. Similarity relations (e.g., film themes or directors) also influence users' behaviors toward similar entities. Therefore, to extract multiple semantic features from a KG to improve recommendation accuracy shows promise.

To fuse multiple heterogeneous information from a KG, we propose a multi-view KG attention network (HKGAN), which adopts a hierarchical attention mechanism to preferentially aggregate neighbor information in a multi-view KG. Neighbor features are propagated recursively and embedded in the same vector space. The main contributions of this paper are as follows:

- We define a multi-view KG that is a set of heterogeneous sub-graphs containing multiple node and relation types. This graph has abundant semantic information and helps for recommendation.
- We propose a hierarchical KG attention network that can naturally aggregate multiple pure features from both node- and graph-level information. Accurate representations of users and items can be learned to calculate prediction scores.
- The HKGAN model is shown to outperform baselines on public datasets in experiments.

The rest of this paper is organized as follows. Section 2 discusses related work. Section 3 defines the multi-view KG. Section 4 shows the HKGAN framework. Section 5 discusses experimental results, and Sect. 6 relates our conclusions.

2 Related Work

Embedding-based models often use KG embedding (KGE) to encode item graphs or user-item graphs by aggregating information from multiple sources. The vectors of each user and item are obtained and recommended. CKE [22] incorporates various auxiliary information into the framework of CF, using TransR to encode the item attributes in the KG. CFKG [24] uses TransE to embed users and items by constructing a user-item KG and optimizes predictions by continuously narrowing the distances of relations. KTUP [1] uses TransH to model the correctness of user-item pairs, simultaneously handling both recommendation and KG completion tasks. DKN [12] utilizes knowledge-aware convolutional neural networks to extract semantic and knowledge information, models news features, and adopts an attention mechanism to achieve user click-through rate prediction. KSR [27] combines GRU networks with knowledge-enhanced key-value storage networks to effectively capture user preferences from sequential interactions.

KGIN [28] uses the message passing mechanism of graph neural networks on the knowledge graph to model higher-order connections between users, entities, and items. CAGE [34] adopts external knowledge graphs combined with graph neural networks to further enhance article embeddings, considering the similarity between sessions, and designs attentional neural networks to simulate short-term user preferences. KGEL [35] consists of an encoder using dual-weighted graph convolutional networks and a decoder employing a novel fully expressive tensor decomposition model, showcasing the relationship between tensor decomposition models and previous tensor decomposition models.

A path-based model constructs a user-item diagram and uses the path in the diagram to make recommendations by calculating the connection similarity between users and users, items and items, and users and items. FMG [25] replaces meta-paths with meta-graphs to alleviate the weak representation of meta-paths. It uses matrix factorization to generate potential vectors for users and items in each meta-graph and a factorization machine to integrate feature information on meta-graphs. MCRec [5] extracts predefined paths and inputs them into a CNN to obtain the embedding of each path instance. SCPR [8] walks through attribute vertices by explicitly following user feedback, utilizing user-preferred attributes. HAKG [29] uses TransR to obtain entity mbedding vectors, and samples paths of users and candidate items to construct item subgraphs, utilizing RNN to update embedding vectors of users and candidate items. AKUPM [30] uses TransR to embed users and entities, and employs a self-attention network to capture the importance of entities to users, thus modeling the features of users and entities. KNI [31] employs attention mechanisms to extract features from the neighborhood nodes of items and users. It assigns appropriate weights to each neighborhood node and utilizes the vector representation of these neighborhood nodes for recommendation prediction. KPRN [32] utilizes recurrent neural networks to model associative paths between user-item pairs in the knowledge graph. These paths capture semantic information and user preferences, enabling personalized recommendations. PGPR [33] proposes a policy-guided path reasoning approach, utilizing policy-guided graph search algorithms to efficiently capture relational paths. However, the reinforcement learning rewards in this method are sparse, resulting in lower recommendation accuracy.

A unified model combines the advantages of embedding- and path-based models, showing great promise by using the semantic and connection information in a KG for recommendations [26]. RippleNet [11] stimulates the propagation of user preferences over the set of knowledge entities by automatically and iteratively extending a user's potential interests along links in a KG. KGCN [14] samples entity neighbors on a KG and then combines neighborhood information with bias when calculating the representation of an entity. KGAT [15] uses TransR to obtain initial embeddings of entities, then propagates features from the entities, and aggregates them based on an attention mechanism. AKUPM [36] uses self-attention networks to learn the appropriate importance of each entity to capture interactions between entities for users. CKAN [16] uses heterogeneous propagation to directly encode cooperative signals and uses a knowledge-aware attention mechanism to distinguish the contributions of different neighbors to make recommendations. LKGR [2] uses different information propagation strategies in the hyperbolic space to explicitly encode heterogeneous information from historical interactions and a KG. However, unified methods cannot learn and fuse multiple features from different perspectives, which makes it difficult to capture more semantic information to enhance recommendation performance.

3   Multi-view Knowledge Graph

In essence, a KG is a directed heterogeneous graph whose nodes correspond to entities (users, items, and item attributes) and edges to relations (social, similarity, and iteration relations). The multiple types of relations in a KG help improve the representation accuracy of users and items.

Because a KG contains heterogeneous nodes and edges, we decompose it into a set of sub-KGs, each representing a specific relation; this is called a multi-view KG (MKG). Figure 1 shows a complete KG that can be decomposed into user-user, user-item, and item-item KGs, which clearly show different semantic relations.

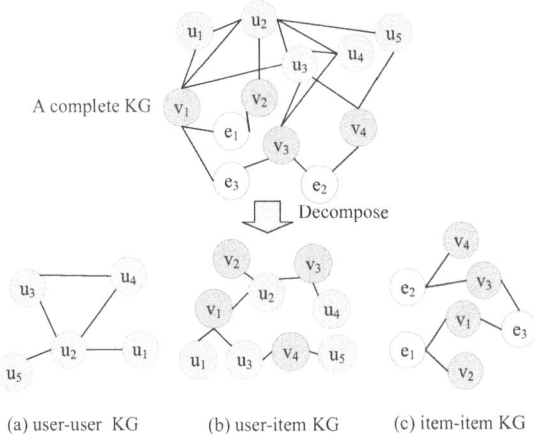

Fig. 1. Decomposing a complete KG into sub-KGs

Assume sets of users $U = \{u_1, u_2, \ldots, u_M\}$ and items $V = \{v_1, v_2, \ldots, v_N\}$. Then we give some definitions.

Definition A: *A user-user social KG presents a social relation among users, e.g., friendship, and can be expressed as $KG_S = \{(u_i, social, u_j)|u_i, u_j \in U\}$. The user-user social matrix $A_S \in \mathbb{R}^{M \times M}$ is defined according to a user's social relation, where $(A_S)_{ij} = 1$ indicates that user i trusts user j, and otherwise $(A_S)_{ij} = 0$. $(A_S)_i$ denotes the social user set of user i, i.e., $(A_S)_i = \{j|(A_S)_{ij} = 1\}$.*

Definition B: *An item-item similarity KG presents a similarity relation among items (e.g., film themes) and can be expressed as $KG_V = \{(e, relation, e)|e \in \varepsilon\}$. The item-item similarity matrix $A_V \in \mathbb{R}^{|\varepsilon| \times |\varepsilon|}$ is de fined according to an entity's similarity relation, where $(A_V)_{ab} = 1$ indicates that entities a and b are related, and otherwise $(A_S)_{ij} = 0$. $(A_V)_a$ denotes the item set of entity a, i.e., $(A_V)_a = \{b|(A_V)_{ab} = 1\}$.*

Definition C: *A user-item interaction KG presents an interaction relation between users and items (e.g., clicks) t and can be expressed as $KG_I = (u_i, interact, v_a)|u_i \in$*

$U, v_a \in V$. *The user-item interaction matrix* $A_I \in \mathbb{R}^{M \times N}$ *is defined according to a user's interaction relation, where* $(A_I)_{ia} = 1$ *indicates that user i selects item a, and otherwise* $(A_I)_{ia} = 0.(A_I)_i$ *denotes the interactive item set of user i, i.e.,* $(A_I)_i = \{a|(A_I)_{ia} = 1\}$; $(A_I)_a$ *represents the interactive user set of item* $a, (A_I)_a = \{i|(A_I)_{ia} = 1\}$.

The MKG provides a new way to learn and fuse multiple heterogeneous features. We next give a GNN-based model to propagate and aggregate high-order neighbor information.

4 Hierarchical KG Attention Network

The same entity (user, item, or attribute) in an MKG may have different meanings and therefore can have different embeddings. For instance, the embedding of user i in KG_S has the feature of friends, while the embedding of user i in KG_I has the feature of items it has interacted with. We propose an HKGAN, combining a graph convolutional network (GCN) [19] with a hierarchical attention mechanism to purely aggregate high-order neighborhood features. Figure 2 shows the HKGAN framework.

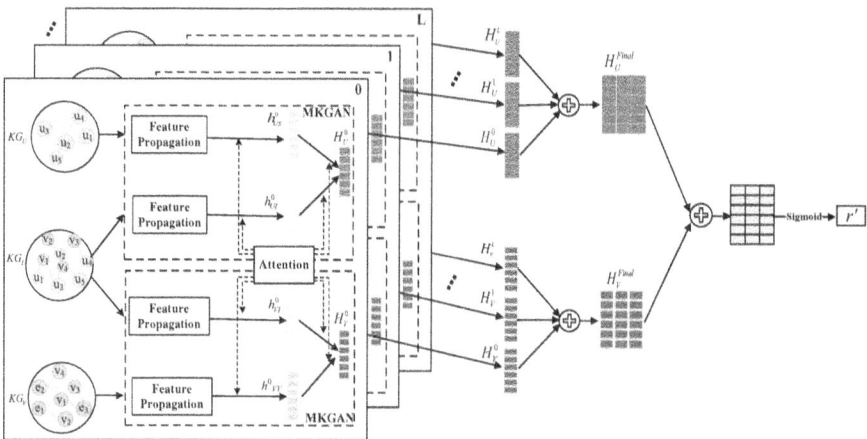

Fig. 2. Herarchical knowledge graph attention network

4.1 Feature Propagation Layer

Assume the initial embedding of user i is $\boldsymbol{H}^0_{U_i}$, and the initial embedding of item a is $\boldsymbol{H}^0_{V_a}$. These are the inputs of the feature propagation layer, and they recursively learn high-level user social features, item relation features, and user-item interaction features in the MKG with convolution operations, i.e., layer $l + 1$ takes the output ($\boldsymbol{H}^0_{U_i}$ and $\boldsymbol{H}^l_{V_a}$) of layer l as input. The layers propagate embeddings and update user and item embeddings. Figure 3 shows the feature propagation and aggregation structure.

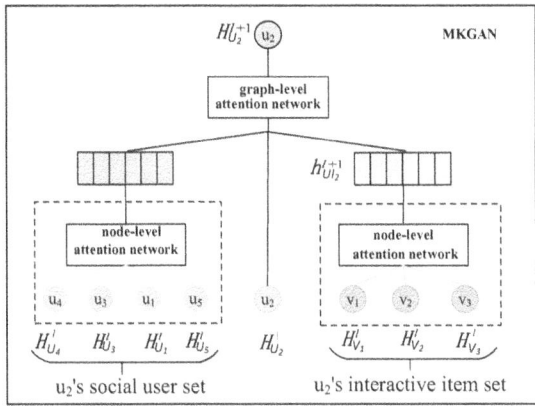

Fig. 3. Feature propagation and aggregation

Each user has their neighborhoods in KG_S and KG_I. The embedding vector of user i at level $l+1$ is combined with the embedding vector at level l, social embedding vector $h_{US_i}^{l+1}$ at level $l+1$ in KG_S, and interactive embedding vector $h_{UI_i}^{l+1}$ at level $l+1$ in KG_I.

Given the embedding H_U^l of user entity i at layer l, the social embedding $h_{US_i}^{l+1}$ of user i at layer $l+1$ can be updated as follows, according to KG_S:

$$h_{US_i}^{l+1} = SOCIAL\left(H_{U_j}^l, \forall j \in (A_S)_i\right)$$

$$= \sum_{\alpha \in ((A_S)_i)} \alpha_{ij}^{l+1} H_{U_j}^l, \qquad (1)$$

where $(A_S)_i = \{j|(A_S)_{ij} = 1\}$ is the social user set of user i, $H_{U_j}^l$ is the embedding of user j in layer l, and α_{ij}^{l+1} is the node-level attention weight between current user i at layer $l+1$ and neighbor user j in KG_S.

Given the item embedding H_V^l at layer l, the interaction embedding $h_{UI_i}^{l+1}$ of user i at layer $l+1$ can be updated as follows, according to KG_I:

$$h_{UI_i}^{l+1} = INTERACT1\left(H_{V_a}^l, \forall \alpha \in (A_I)_i\right),$$

$$= \sum_{\alpha \in ((A_I)_i)} \beta_{ia}^{l+1} H_{V_\alpha}^l \qquad (2)$$

where $(A_I)_i = \{a|(A_I)_{ia} = 1\}$ is the interaction item set of user i, $H_{V_a}^l$ is the embedding of item a at layer l, and β_{ia}^{l+1} is the node-level attention weight of user i at layer $l+1$ and neighborhood item a in KG_I. Each item has its neighborhoods in KG_V and KG_I. The embedding vector of item a at level $l+1$ is combined with its embedding vector at level l, the similarity embedding vector $h_{VV_a}^{l+1}$ at level $l+1$ in KG_V, and the interaction embedding vector $h_{VI_a}^{l+1}$ at level $l+1$ in KG_I.

Given item embedding H_V^l at level l, the similarity embedding $h_{VV_a}^{l+1}$ of item a at level $l + 1$ can be updated as follows, according to KG_S:

$$h_{VV_a}^{l+1} = RELATION\left(H_{V_b}^l, \forall b \in (A_V)_a\right) = \sum_{b \in ((A_V)_a)} \delta_{ab}^{l+1} H_{V_b}^l, \tag{3}$$

where $(A_V)_a = \{b|(A_V)_{ab} = 1\}$ is the relational item set of item a, $H_{V_b}^l$ is the embedding of item b at layer l, and δ_{ab}^{l+1} is the node-level attention weight between item a at layer $l + 1$ and neighborhood item b in KG_V.

Given user embedding H_U^l at level l, the interaction embedding $h_{VI_a}^{l+1}$ of item a at level $l + 1$ can be updated as follows, according to KG_I:

$$h_{VI_i}^{l+1} = INTERACT2\left(H_{U_i}^l, \forall i \in (A_I)_a\right)$$
$$= \sum_{i \in ((A_I)_a)} \epsilon_{ia}^{l+1} H_{U_i}^l, \tag{4}$$

where $(A_I)_a = \{i|(A_I)_{ia} = 1\}$ is the interaction user set, $H_{U_i}^l$ is the embedding of user i at layer l, and δ_{ia}^{l+1} is the node-level attention weight between item a at layer $l + 1$ and neighbor user i in KG_I.

4.2 Hierarchical Attention Layer

Feature aggregation methods often use mean pooling and do not consider its preference to neighbor nodes. To avoid this, we design a hierarchical attention network to learn both node- and graph-level attention, as shown in Fig. 4.

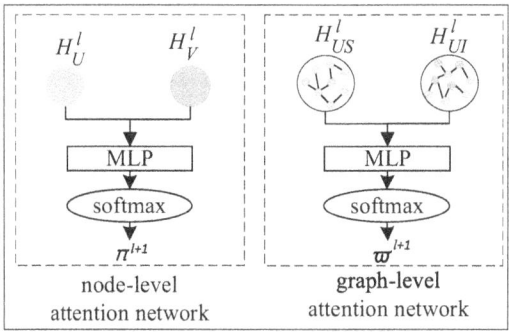

Fig. 4. Hierarchical attention mechanism

Node-Level Attention Network. The node-level attention network assigns different attention weights to nodes under the same perspective. We use a multi-layer perceptron

(MLP) to learn users' and items' attention weights in formulas 1–4 under their respective perspectives. The node attention weight is:

$$\pi_{xy}^{l+1} = \sigma\left(W_{node} \cdot \left[H_{O_x}^l, H_{O_y}^l\right] + b_{node}\right), \tag{5}$$

where O denotes a single sub-KG; π_{xy}^{l+1} is the node-level attention weight of level $l+1$ between nodes x and y in O; σ is the $ReLU$ nonlinear activation function; $H_{O_x}^l$ and $H_{O_y}^l$, respectively, denote the embedding of current node x and neighbor node y in the first layer of O; $W_{node} \in \mathbb{R}^{d' \times d}$ is the parameter matrix; and $b_{node} \in \mathbb{R}^d$ is the deviation. We use the softmax function to normalize the attention weight:

$$\pi_{xy}^{l+1} = softmax\left(\pi_{xy}^{l+1}\right) = \frac{\exp\left(\pi_{xy}^{l+1}\right)}{\sum_{z \in (A_O)_x} \exp\left(\pi_{xz}^{l+1}\right)} \tag{6}$$

This attention weight can suggest which nodes should be given more attention and therefore can more effectively capture neighborhood information under the current perspective.

Graph-Level Attention Network. A node-level attention network is used to fuse the adjacent point features of nodes in a single KG. We use a graph-level attention network to fuse features from multiple KGs and MLP to learn the graph-level attention weights of nodes. We omit the normalization of the attention network below because they share the same form:

$$\varpi_{xO}^{l+1} = \sigma\left(W_{graph} \cdot \left[h_{C_x}^l, H_{O_x}^l\right] + b_{graph}\right), \tag{7}$$

where O denotes a single sub-KG, ϖ_{xO}^{l+1} is the graph-level attention weight of node x at level $l+1$ in O, σ is the $ReLU$ nonlinear activation function, C identifies the role of the current node, $h_{C_x}^l$ denotes the embedding of node x at layer l, $H_{O_x}^l$ denotes node x at layer l in O, $W_{graph} \in \mathbb{R}^{d' \times d}$ is the parameter matrix, and $b_{graph} \in \mathbb{R}^d$ is the deviation.

In formula 7, the graph-level attention weight of each node x depends on the embedding of KG and the embedding of sub-KG in the upper layer. For example, for each user node i, ϖ_{iS}^{l+1} is its graph-level attention weight at layer $l+1$ of KG_S, which simulates the attention degree of user i. Similarly, ϖ_{iI}^{l+1} is the graph-level attention weight of user i at layer $l+1$ of KG_I, which simulates the attention degree of user i in the interaction KG. When $\varpi_{iS}^{l+1} + \varpi_{iI}^{l+1} = 1$, the larger ϖ_{iS}^{l+1} is, the greater the influence of the social relations of user i is, and the smaller the influence of the interaction relation is.

In the MKG_S, each node has its feature information. The feature embedding of node x is

$$H_{C_x}^{l+1} = H_{C_x}^l + \left(\varpi_{xO}^{l+1} h_{CO_x}^{l+1} + \varpi_{xO'}^{l+1} h_{CO'_x}^{l+1}\right), \tag{8}$$

where $H_{C_x}^{l+1}$ is the embedding of node x at layer $l+1$, $H_{C_x}^l$ is the embedding of node x at layer l, C denotes the role of the current node (user/item), O and O' denote a single KG, ϖ_{xO}^{l+1} is the graph-level attention weight of node x at layer $l+1$ in the current perspective KG O, and $h_{CO_x}^{l+1}$ denotes that node x is embedded at the layer $l+1$ feature of C under the current view, KG O.

4.3 Prediction Layer

After the above L-layer iterative convolution, we obtain embedding sets $\{H_{U_i}^0, H_{U_i}^1, \ldots, H_{U_i}^L\}$ for user i and $\{H_{V_a}^0, H_{V_a}^1, \ldots, H_{V_a}^L\}$ for item a. We can obtain the final embedding of the user and item by splicing their embedding sets,

$$H_{U_i} = [H_{U_i}^0 \| H_{U_i}^1 \| \ldots \| H_{U_i}^L]; \ H_{V_a} = [H_{V_a}^0 \| H_{V_a}^1 \| \ldots \| H_{V_a}^L], \tag{9}$$

where $\|$ is a splicing operator.

By splicing multi-layer convolution outputs, we can enrich the embedding of users and items and dynamically adjust the number of layers to control the scope of propagation. Finally, we calculate their prediction scores by performing inner product operations on the embeddings of users and items,

$$\hat{r}_{ia} = H_{U_i}^T H_{V_a}. \tag{10}$$

4.4 Model Training

We use the Bayesian personalization ranking (BPR) as a loss function to optimize the model, with the idea that the difference between the scores of the positive and negative samples in the dataset should be as large as possible,

$$L = \min_\theta \sum_{(i,a)\in(A_I)_i^+ \cup (i,b)\in(A_I)_i^-} -ln\sigma(\hat{r}_{ia} - \hat{r}_{ib}) + \lambda\|\theta\|^2, \tag{11}$$

where $(A_I)_i^+$ denotes a positive sample set, $(A_I)_i^-$ denotes a negative sample set, σ is the sigmoid function, and $\theta = [H_U, H_V, W_{node}, W_{graph}, b_{node}, b_{graph}]$ is the regularization parameter in the model.

Algorithm 1 gives the HKGAN algorithm. Firstly, for a pair of (u, v) (line 2), this algorithm calculates the attention score (line 4, lines 20–22), and then separately obtains their neighbor information from the social matrix A_S, relationship matrix A_V, and interaction matrix A_I. Secondly, it aggregates them (lines 5–8), and aggregates their feature vectors in each perspective (lines 9–10). Thirdly, it concatenates the vectors obtained by L times of convolution (lines 12–13), and performs inner product calculation to get the prediction score (line 14). Finally, it updates the parameters by reducing the loss to perform gradient descent (line 15).

Algorithm 1 HKGAN recommendation algorithm

Input: user-item interaction matrix A_I, user-user social matrix A_S, item-item relation matrix A_V

Output: prediction function $F(u, v|\theta, A_I, A_S, A_V)$

1: **while** HKGAN not converge **do**
2:     **for** (u.v) in A_I **do**
3:         **for** l=0,....,L-1 **do**
4:             Use $COMPUTE - ATTENTION - SCORE$ to calculate all attention scores;
5:             $h_{US_u}^{l+1} \leftarrow \sum_{u' \in (A_S)_u} \pi_{uu'}^{l+1} H_{U_{u'}}^l$;
6:             $h_{UI_u}^{l+1} \leftarrow \sum_{v' \in (A_I)_u} \pi_{uv'}^{l+1} H_{V_{v'}}^l$;
7:             $h_{VV_v}^{l+1} \leftarrow \sum_{v' \in (A_V)_v} \pi_{vv'}^{l+1} H_{V_{v'}}^l$;
8:             $h_{VI_v}^{l+1} \leftarrow \sum_{u' \in (A_I)_v} \pi_{vu'}^{l+1} H_{U_{u'}}^l$;
9:             $H_{U_u}^{l+1} \leftarrow H_{U_u}^l + (\varpi_{uS}^{l+1} h_{US_u}^{l+1} + \varpi_{uI}^{l+1} h_{UI_u}^{l+1}$
10:           $H_{V_v}^{l+1} \leftarrow H_{V_v}^l + (\varpi_{vV}^{l+1} h_{VV_v}^{l+1} + \varpi_{vI}^{l+1} h_{VI_v}^{l+1}$
11:         **end for**
12:         $H_{U_u} \leftarrow [H_{U_u}^0 || H_{U_u}^1 || ... || H_{U_u}^L]$;
13:         $H_{V_v} \leftarrow [H_{V_v}^0 || H_{V_v}^1 || ... || H_{V_v}^L]$;
14:         Compute predicted probablity $\hat{r}_{uv} = H_{U_u}^T H_{V_v}$;
15:         Update parameters by gradient descent;
16:     **end for**
17: **end while**
18: **return** F
19:
20: **function** COMPUTE-ATTENTION-SCORE(h, h')
21:     **return** $softmax(MLP(h, h'))$
22: **end function**

5 Experiments

We evaluate the effectiveness of the HKGAN model for top-K recommendation and answer two research questions:

RQ1: How does HKGAN perform compared with state-of-the-art recommendation methods?

RQ2: How do different hyperparameters affect HKGAN?

5.1 Experimental Setup

We used three widely available public datasets as benchmarks.

Last.FM: The source of this music recommendation dataset is the Last.FM online music system. It includes information about users and interactions.

Book-Crossing: The source of this book recommendation dataset is the Book-Crossing community. It includes ratings of various books by different readers (scores from 0 to 10).

MovieLens-20M: The source of this movie recommendation dataset is the Movie-Lens website. It includes ratings of movies watched by different users (ratings from 0 to 5).

Table 1. Descriptions of the three datasets

Dataset	Music	Book	Movie
#users	1872	17,860	138,159
#items	3846	14,967	16,954
#interactions	42,346	139,746	13,501,622
#entities	9366	77,903	102,569
#inter-avg	23	8	98

Table 1 shows descriptions of the three datasets. These datasets contain specific values of feedback, which we convert into implicit feedback. We treat a sample exceeding a score threshold as positive, and we mark it as 1 to indicate the user has an interaction or positive evaluation with an item. For a negative sample, we conduct a random negative sampling equal to the size of the positive sample for each user and mark this interactive data as 0. The scoring threshold in MovieLens-20M is set to 4, while Last.FM and Book-Crossing have no scoring threshold due to their sparsity. Because there is no social relation in Book-Crossing and MovieLens-20M, we select users whose interactions exceed a threshold based on the dataset and calculate each user's social preference based on cosine similarity. The social thresholds for Book-Crossing and MovieLens-20M are 0.68 and 0.545, respectively. With this approach, we built a social KG for these two datasets.

We compared HKGAN with CF- [10], embedding- [22], path- [21], and KG-based [11, 13–16] methods. We divided the dataset into training, validation, and test sets in a trained model to select the K items with the highest prediction probability for each user in the test set and used the Recall@K indicator. We compared our HKGAN method with CF- [10], embedding- [22], path- [21], and KG-based [11, 13–16] recommendation methods.

5.2 Experimental Results

We implemented HKGAN on TensorFlow and optimized the loss function through a small batch of Adam. The initial learning rate was 0.001, and the training batch size was 512. Adam can automatically adjust the learning rate to better adapt to model training. For random sampling (i, a, b)under a batch, we updated all the node embeddings under the current batch and loop sampling (i, a, b), and we continuously updated the embedding. After L-layer convolution, the Adam optimizer was used to update the model parameters according to the gradient of the BPR loss. We initialized all trainable parameters with a normal distribution with mean 0 and standard deviation 0.01. For the regularization parameter λ, we tried [0.0001, 0.001, 0.01, 0.1] and then set $\lambda = 0.01$ for the best performance. We adjusted the dimensions of the convolutional layer to the same size. For the multi-layer attention network, we adopted a two- layer MLP structure.

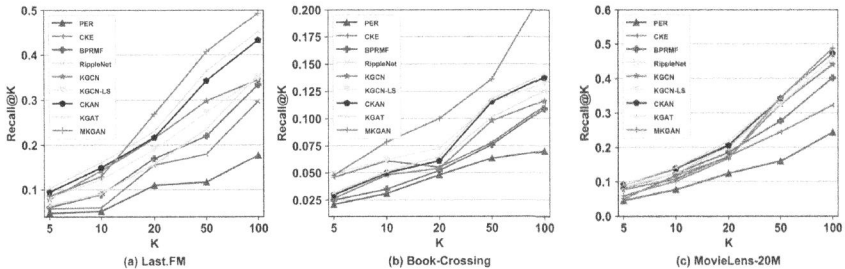

Fig. 5. Recall@K indicator results in Top-K prediction.

Table 2 shows the CTR prediction results on the three datasets. The bold values in the table indicate the best performances. Figure 5 shows the Top-K recommendation performance curves. HKGAN made significant progress on the three datasets, with AUC values 3.3% and 0.8% higher than the state-of-the-art baseline on Last.FM and MovieLens-20M, respectively. This shows that social relations play a positive role in the spread of features. The performance on Book-Crossing was not the best, and the few interactions in the dataset led to the inability to make full use of interactive relationships to establish feature crossover between users and items. The recommendation results were greatly affected by only one feature (such as social or similarity), which reduced the AUC, while CKAN can establish knowledge association with few interaction relations. Users and items have explicitly associated entities and can improve the AUC by explicitly aggregating those entity features.

In Top-K recommendation, HKGAN had a better performance on Last.FM and Book-Crossing but was not as good as CKAN on MovieLens-20M because some noises were introduced in the dataset, and HKGAN could not accurately recall the top-K. KGAT and CKAN could have a direct impact on users and items with explicitly associated entities to enhance the recall. However, the recall of HKGAN became better with the increase of Recall@K.

The performance of HKGAN on the three datasets, ordered from high to low, was on MovieLens-20M, Last.FM, and Book-Crossing, mainly because they had different neighbor features of users, items, and number of interactions. For example, the MovieLens-20M dataset contains more interactive information than Last.FM and Book-Crossing.

5.3   Ablation Experiment

We analyzed the impact of embedded dimensions on performance, using different dimensions to embed users and items. The results in Table 3 show that increasing the embedded dimension can improve performance because it allows an entity to store more feature information. However, when the threshold is exceeded, the performance will remain the same, or can even decrease, because large data dimensions will cause overfitting.

Table 2. Results of baseline AUC and F1 indicators in CTR estimation

Model	Last.FM		Book-Crossing		MovieLens-20M	
	AUC	*F1*	*AUC*	*F1*	*AUC*	*F1*
BPRMF	0.756	0.701	0.658	0.611	0.958	0.914
	(−13.2%)	(−11.7%)	(−12.3%)	(−8.4%)	(−2.6%)	(−3%)
CKE	0.747	0.674	0.676	0.623	0.927	0.874
	(−14.2%)	(−15.1%)	(−9.9%)	(−6.6%)	(−5.8%)	(−7.2%)
PER	0.641	0.603	0.605	0.572	0.838	0.792
	(−26.4%)	(−24.1%)	(−19.3%)	(−14.3%)	(−14.8%)	(−16%)
RippleNet	0.776	0.702	0.721	0.647	0.976	0.927
	(−10.9%)	(−11.6%)	(−3.9%)	(−3%)	(−0.8%)	(−1.6%)
KGCN	0.802	0.708	0.684	0.631	0.977	0.93
	(−7.9%)	(−10.8%)	(−8.8%)	(−5.4%)	(−0.7%)	(−1.3%)
KGCN-LS	0.805	0.722	0.676	0.631	0.975	0.929
	(−7.6%)	(−9.1%)	(−9.9%)	(−5.4%)	(−0.9%)	(−1.4%)
KGAT	0.829	0.742	0.731	0.654	0.976	0.928
	(−4.8%)	(−6.5%)	(−2.5%)	(−2%)	(−0.8%)	(−1.5%)
CKAN	0.842	0.769	0.753	0.673	0.976	0.929
	(−3.3%)	(−3.1%)	(+0.4%)	(+0.9%)	(−0.8%)	(−1.4%)
HKGAN	0.871	0.794	0.750	0.667	0.983	0.942

Table 3. Impact of embedded dimensions

d Dataset	8	16	32	64
Music	0.8626	0.8667	0.8710	0.8720
Book	0.7499	0.7564	0.7589	0.7611
Movie	0.9826	0.9838	0.9839	0.9842

We show the impact of the propagation layer. Table 4 shows the performance with different numbers of convolutional propagation layers. The numbers of propagation layers with the best performance on the Last.FM, Book-Crossing, and MovieLens-20M datasets are 3, 2, and 1, respectively. For sparse datasets, the number of neighbor nodes increases dramatically with the number of propagation layers n. The scope of the convolution will be extended to n-hop neighborhood nodes, eventually leading to memory overflow.

We also studied the impact of multiple features, where C_I denotes only interaction features, C_{SI} denotes both social and interaction features, C_{VI} denotes both similarity and interaction features, and C_{SVI} denotes all features. The results are shown in Table 5. On Last.FM and Book-Crossing, social features are fully used to improve performance.

Table 4. Impact of propagation layers

L Dataset	1	2	3
Music	0.8702	0.8780	0.8795
Book	0.7562	0.7617	0.7611

On MovieLens-20M, C_{VI} achieves the best performance. This is because the interaction of the dataset is denser, which makes the effect of social relations almost ineffective and causes side effects.

Table 5. Impact of features

Char Dataset	CI	CSI	CV I	CSV I
Music	0.8647	0.8686	0.8729	0.8780
Book	0.6961	0.7235	0.7477	0.7639

Finally, we analyzed the impact of the attention mechanism, where A_{Node}, A_{Graph}, and A_{All} denote node-level, graph-level, and both attention mechanisms, respectively, and A_{None} denotes no attention mechanism. Table 6 shows that A_{All} achieves the best performance and demonstrates that our hierarchical attention mechanism is effective. In addition, the performance of A_{Graph} is the same as that of A_{All} on MovieLens-20M. This is because the dataset has redundant neighborhoods in all perspectives. The node-level attention mechanism cannot select more contributing neighborhood nodes well.

Table 6. Impact of attention mechanisms

A Dataset	A_{Graph}	A_{Node}	A_{None}	A_{All}
Music	0.8778	0.8762	0.8759	0.8780
Book	0.7622	0.7600	0.7597	0.7639
Movie	0.9842	0.9829	0.9829	0.9842

6 Conclusions

Because the KG is a heterogeneous structured graph model, it contains multiple semantic features. The various features are useful for improving recommendation performance. Based on this observation, we gave the concept of a multi-view KG to describe various features and proposed an HKGAN that combines a GCN and a hierarchical attention mechanism for RSs. The GCN was used to perform feature propagation for different types of entities. The hierarchical attention mechanism was designed to aggregate multiple features at both the node and graph levels. The HKGAN model is more general than existing KG-based unified recommendation methods. Experiments demonstrated the effectiveness of HKGAN on three public datasets.

Acknowledgements. This work was supported in part by the Pioneer and Leading Goose Research and Development Program of Zhejiang Province China under Grant 2023C01022.

References

1. Cao, Y., Wang, X., He, X., Hu, Z., Chua, T.S.: Unifying knowledge graph learning and recommendation: Towards a better understanding of user preferences. In: WWW, pp. 151–161 (2019)
2. Chen, Y., et al.: Modeling scale-free graphs for knowledge-aware recommendation. arXiv preprint arXiv:2108.06468 (2021)
3. Gong, F., Wang, M., Wang, H., Wang, S., Liu, M.: Smr: medical knowledge graph embedding for safe medicine recommendation. Big Data Res. **23**, 100174 (2021)
4. Alhamid, M.F., Rawashdeh, M., Dong, H., Hossain, M.A., El Saddik, A.: Exploring latent preferences for context-aware personalized recommendation systems. IEEE Transactions on Human-Machine Systems **46**(4), 615–623 (2016)
5. Hu, B., Shi, C., Zhao, W.X., Yu, P.S.: Leveraging meta-path based context for top-n recommendation with a neural co-attention model. In: SIGKDD. pp. 1531–1540 (2018)
6. Hua, S., Chen, W., Li, Z., Zhao, P., Zhao, L.: Path-based academic paper recommendation. In: Web Information Systems Engineering, Wise 2020, PT II. Lecture Notes in Computer Science, vol. 12343, pp. 343–356 (2020)
7. Shu, W., Guo, W., Song, X., Huang, Y., Wang, L., Tan, T.: Coupled topic model for collaborative filtering with user-generated content. IEEE Trans. Human-Mach. Syst. **46**(6), 908–920 (2016)
8. Lei, W.,et al.: Interactive path reasoning on graph for conversational recommendation. In: SIGKDD, pp. 2073–2083 (2020)
9. Luo, C., Pang, W., Wang, Z., Lin, C.: Hete-cf: Social-based collaborative filtering recommendation using heterogeneous relations. In: ICDM, pp. 917–922. IEEE (2014)
10. Rendle, S., Freudenthaler, C., Gantner, Z.: Bpr: Bayesian personalized ranking from implicit feedback. arXiv preprint arXiv:1205.2618 (2012)
11. Wang, H., Zhang, F., Wang: Ripplenet: Propagating user preferences on the knowledge graph for recommender systems. In: CIKM, pp. 417–426 (2018)
12. Wang, H., Zhang, F., Xie, X., Guo, M.: Dkn: Deep knowledge-aware network for news recommendation. In: WWW, pp. 1835–1844 (2018)
13. Wang, H., Zhang, F., Zhang, M.: Knowledge-aware graph neural networks with label smoothness regularization for recommender systems. In: SIGKDD, pp. 968– 977 (2019)
14. Wang, H., Zhao, M., Xie, X., Li, W., Guo, M.: Knowledge graph convolutional networks for recommender systems. In: WWW, pp. 3307–3313 (2019)
15. Wang, X., He, X., Cao, Y., Liu, M., Chua, T.S.: Kgat: Knowledge graph attention network for recommendation. In: SIGKDD, pp. 950–958 (2019)
16. Wang, Z., Lin, G., Tan, H., Chen, Q., Liu, X.: Ckan: Collaborative knowledge-aware attentive network for recommender systems. In: SIGIR, pp. 219–228 (2020)

17. Weimer, M., Karatzoglou, A., Smola, A.J.: Adaptive collaborative filtering. In: RecSys (2008)
18. Wu, L., Li, J., Sun, P., Hong, R., Ge, Y., Wang, M.: Diffnet++: A neural influence and interest diffusion network for social recommendation. IEEE Trans. Knowl. Data Eng. **34**(10), 4753—4766 (2020)
19. Ying, R., He, R., Chen, K., Eksombatchai, P.: Graph convolutional neural networks for web-scale recommender systems. In: SIGKDD, pp. 974–983 (2018)
20. Yu, X., Ren, X., Gu, Q., Sun, Y., Han, J.: Collaborative filtering with entity similarity regularization in heterogeneous information networks. In: IJCAI HINA 27 (2013)
21. Yu, X., Ren, X., Sun, Y.: Personalized entity recommendation: A heterogeneous information network approach. In: ICDM, pp. 283–292 (2014)
22. Zhang, F., Yuan, N.J., Lian, D.: Collaborative knowledge base embedding for recommender systems. In: SIGKDD, pp. 353–362 (2016)
23. Zhang, H., Shen, F., Liu, W.: Discrete collaborative filtering. In: SIGIR, pp. 325– 334 (2016)
24. Zhang, Y., Ai, Q., Chen, X., Wang, P.: Learning over knowledge-base embeddings for recommendation. arXiv preprint arXiv:1803.06540 (2018)
25. Zhao, H., Yao, Q., Li, J., Song, Y., Lee, D.L.: Meta-graph based recommendation fusion over heterogeneous information networks. In: SIGKDD, pp. 635–644 (2017)
26. Chen Gao, Y., et al.: Graph neural networks for recommender systems: challenges, methods, and directions. ACM Trans. Inform. Syst. **1**(1), 1–46 (2021)
27. Huang, J., Zhao, W.X., Dou, H., Wen, J.R., Chang, E.Y.: Improving sequential recommendation with knowledge-enhanced memory networks. In: Proceedings of the 41st International ACM SIGIR Conference on Research & Development in Information Retrieval, pp. 505–514 (2018)
28. Wang, X., et al.: Learning intents behind interactions with knowledge graph for recommendation. In: Proceedings of the World Wide Web Conference (WWW'21), pp. 878–887 (2021)
29. Sha, X., Sun, Z., Zhang, J.: Hierarchical attentive knowledge graph embedding for personalized recommendation. Electron. Commer. Res. Appl. **48**, 101071 (2021)
30. Tang, X., Wang, T., Yang, H., Song, H.: AKUPM: Attention-enhanced knowledge-aware user preference model for recommendation. In: Proceedings of the 25th ACM SIGKDD International Conference on Knowledge Discovery & Data Mining, pp. 1891–1899 (2019)
31. Qu, Y., Bai, T., Zhang, W., Nie, J., Tang, J., et al.: An end-to-end neighborhood-based interaction model for knowledge-enhanced recommendation. In: Proceedings of the 1st International Workshop on Deep Learning Practice For High-Dimensional Sparse Data, pp. 1–9 (2019)
32. Wang, X., Wang, D., Xu, C., et al.: Explainable reasoning over knowledge graphs for recommendation. In: Proceedings of the AAAI Conference on Artificial Intelligence **33**(01): 5329–5336 (2019)
33. Xian Y, Fu Z, Muthukrishnan S, et al. Reinforcement knowledge graph reasoning for explainable recommendation. In: Proceedings of the 42nd international ACM SIGIR conference on research and development in information retrieval, pp. 285–294 (2019)
34. Sheu, H.S., Chu, Z., Qi, D., et al.: Knowledge-guided article embedding refinement for session-based news recommendation[J]. IEEE Trans. Neural Netw. Learn. Syst. **33**(12), 7921–7927 (2021)
35. Zeb, A., Haq, A.U., Zhang, D., et al.: KGEL: A novel end-to-end embedding learning framework for knowledge graph completion. Expert Syst. Appl. **167**, 114164 (2021)
36. Tang, X., Wang, T., Yang, H., et al.: AKUPM: attention-enhanced knowledge-aware user preference model for recommendation. In: Proceedings of the 25th ACM SIGKDD International Conference on Knowledge Discovery & Data Mining, pp. 1891–1899 (2019)

37. Seongjun, Y ,Minbyul, J ,Sungdong, Y., et al.: Graph Transformer Networks: Learning meta-path graphs to improve GNNs. Neural Netw. 153104–119 (2022)
38. Zhong, Z., Li, C.T., Pang, J.: Personalised meta-path generation for heterogeneous graph neural networks. Data Min. Knowl. Disc.Knowl. Disc. **36**, 2299–2333 (2022)

Few-Shot Fine-Grained Ship Detection

Jiayin Lan, Zexin Wang, Ming Liu$^{(\boxtimes)}$, and Bing Qin

Research Center for Social Computing and Information Retrieval, Harbin Institute
of Technology, Harbin, China
{jylan,zxwang,mliu,qinb}@ir.hit.edu.cn

Abstract. Fine-grained object detection refers to the identification and classification of objects in images in a more detailed way in the field of computer vision. Traditional object detection usually recognizes the whole object, while fi-ne-grained object recognition pays more attention to the internal details of the object to distinguish different subclasses in the same category. Fine-grained object detection for remote sensing images faces great challenges due to **high line-of-sight**, **small objects**, and **less training data**. In this work, we design an architecture for fine-grained object detection of remote sensing images of ships at sea. We randomly exchange the ship subject and back-ground in the image through the image mask, and then use the method of image generation to process the **dataset augmentation** in a unified style, which solves the problem of less fine-grained training data. At the same time, a **contrastive learning module** is added to the fine-grained image classification subtask, which improves the recognition ability for small target objects in complex backgrounds and the discrimination ability of similar individuals between classes. Our architecture achieves a mAP of 0.749 on the HRSC2016.

Keywords: Fine-grained Object Detection · Ship Detection · Remote Sensing

1 Introduction

Remote sensing image processing of ships at sea can identify and monitor ships on the sea surface, and analyze the type, size and heading of ships, which is very important for maritime traffic management, fishery management and rational use of Marine resources. At present, the object detectors for ships at sea have not been adjusted for fine-grained objects. In remote sensing images, only the top view can help object classification, and the object is more difficult to recognize due to the lack of texture features at high viewing distance. For example, DOTA [1] is currently the largest OBB-annotated remote sensing dataset, which contains 15 categories. Even in remote sensing images, it is not difficult to distinguish between different classes (such as bridges and airplanes). However, in the HRSC2016 [2], it is necessary to identify and distinguish specific ship types (such as Nimitz class aircraft carriers and Midway class aircraft carriers), and there are few effective features that can be used to distinguish between the top views of different types of ships (see Fig. 1). In this case, the current general object detection methods perform much better in identifying different categories such as ships and planes than in identifying specific ship types.

B. Xu et al. (Eds.): CCKS-IJCKG 2024, CCIS 2229, pp. 269–281, 2025.
https://doi.org/10.1007/978-981-96-1809-5_20

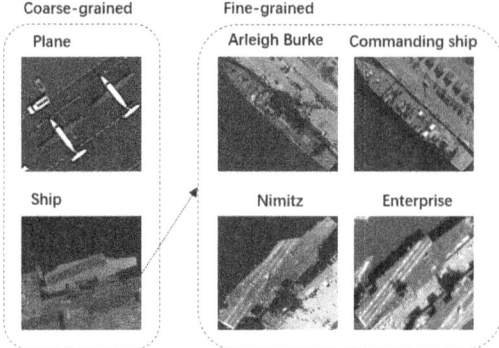

Fig. 1. Examples of visual classification for coarse-grained categories and fine-grained categories.

At present, fine-grained object detection for remote sensing of ships at sea faces many challenges. Firstly, due to the sensitivity of military data, there are very few open source fine-grained ship remote sensing datasets. For example, NWPU VHR-10 [3], a geographic remote sensing dataset, contains only a coarse-grained label of ship among the 10 categories. The HRSC2016 we use has fine-grained labels of specific ship models in addition to the coarse-grained label of ship. However, among all the images, there are only 617 images with fine-grained labels. Compared with other fine-grained remote sensing datasets [4], there are **fewer training samples**, and using ordinary target recognition models will cause serious overfitting. To solve this problem, we use CycleGAN [5] to augment the dataset by generating fake images.

Secondly, **class imbalance** is another problem that affects the detection accuracy. There is a serious long-tail phenomenon in the categories of HRSC2016 (see Fig. 2). The category with the largest number of samples has only 6 images, and the category with the least number of samples has 266 images. To balance the quantitative differences between categories, we design a contrastive learning module for categories with a small number of samples, which can help the model better distinguish these categories from other categories in the case of fewer training samples. It can effectively improve the accuracy of ship classification.

We design an architecture for the fine-grained object detection task of ships at sea, which decouples the fine-grained object detection task into a localization subtask and a classification subtask. Then the Large Selective Kernel Network (LSKNet) [6] and the Background Suppression (BS) [7] are used to achieve high-precision localization and classification. Then, to solve the problem of shortage of training samples and class imbalance of ships at sea, an effective **data augmentation** and **contrastive learning module** are designed, which greatly improves the detection effect. Experimental results demonstrate the effectiveness of our model.

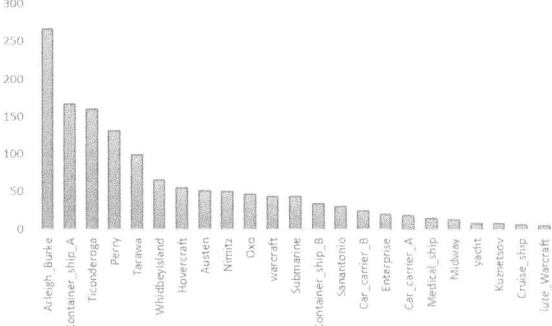

Fig. 2. Number of training samples for each category in HRSC2016 training set, which shows an obvious long tail phenomenon

2 Related Work

In the field of remote sensing, object detection is primarily divided into two categories: two-stage and single-stage object detection methods. High-performance two-stage object detection methods typically follow the RCNN [8] architecture, which consists of a region proposal network and regional CNN detection heads. In recent years, numerous variants of the RCNN framework have emerged, such as Oriented RCNN [9], which introduced a box encoding system to address the issue of unstable training loss. The representative single-stage object detection method is YOLOv5 [10], and the YOLO architecture has also been widely applied in the field of remote sensing. For instance, BiFA-YOLO [11] employs bi-directional feature fusion and angular classification to detect objects in various orientations.

Image augmentation has been demonstrated to improve the accuracy of fine-grained image recognition and object detection tasks [12]. Image augmentation encompasses traditional operations such as rotation, cropping, and exposure adjustments, as well as more advanced strategies like image mixing and deletion, which offer greater diversity compared to traditional methods. Representative mixing strategies include AlignMixUp [13], where two images are aligned and interpolated in the feature space, and TokenMixup [14], which optimally matches images within the same mini-batch to increase significant information in the output.

In contrast, our architecture introduces image classification on the basis of general object detection to capture the detailed information in images. As for data augmentation, instead of processing on the original images, we use the generative method to introduce more foreign features.

3 Architecture Flow

Our architecture mainly includes three modules, which are data augmentation module, object detection module, and fine-grained image classification module. The overall flow of the architecture is shown in Fig. 3.

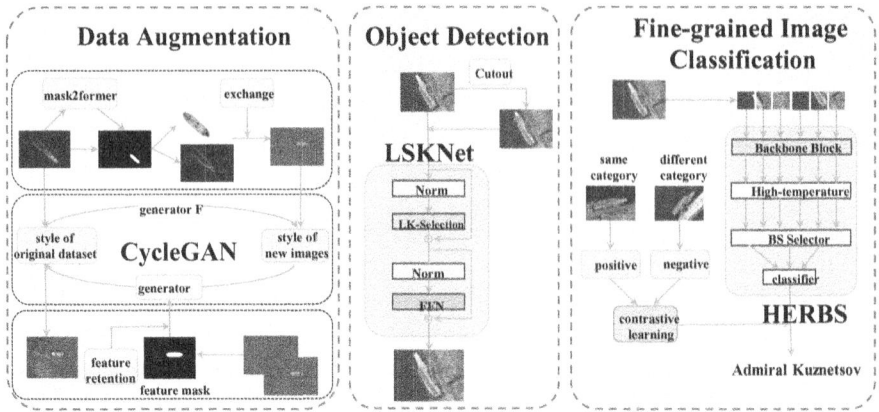

Fig. 3. Our architecture consists of three main parts: data augmentation, object detection, and fine-grained image classification.

Data Enhancement Module. The mask pattern is used to separate the object and background in the image and recombine it into a new image, and then the new image is enhanced by using CycleGAN to style the training dataset.

Object Detection Module. The object detection algorithm is used to detect the input remote sensing map, find out all the ship objects and their location information contained in it, and present it in the form of bounding box.

Fine-grained Image Classification Module. The fine-grained image classification algorithm is used to classify each identified ship according to its specific type classes.

4 Data Augmentation

Problem. As mentioned above, due to the shortage of training samples in the dataset, to expand the training samples, we propose to make full use of the diversity of different backgrounds in images by transferring the ship entities in different images to the background of other images. Since different image backgrounds cover different ports, sea surface conditions, present different styles. If the ship entity can be represented by different background styles, rich and diverse training images can be generated. For this purpose, we need to address several important issues:

1. **Picture segmentation.** How to extract the ship entity and the background in the training samples respectively, to exchange the background and the entity.
2. **Style Consistency.** Since the background and the entity are from two images respectively, simply stitching them together cannot meet the consistency in the form of the two images, such as different backgrounds and lighting conditions.
3. **Feature retention.** In the process of image transformation, it is necessary to ensure that the main features of the ship entity used for fine-grained classification will not be changed, thus interfering with the accuracy of classification.

Method. To solve the above problems, we consider the following data augmentation solutions, and see Fig. 4 for image examples of the procession:

1. **Picture segmentation.** We use the image segmentation model mask2former [15] to generate image masks, which are used to distinguish the ship entities from the background in the image.
2. **Style Consistency.** Since the CycleGAN has been proven to be effective in generating the desired image samples. Therefore, we design an improved CycleGAN model to generate new images that are consistent with the overall style of the dataset.
3. **Feature retention.** We use a feature selection module to select the features of the ship entity in the picture, and then evaluate the changes of these features before and after the style transfer.

4.1 Picture Segmentation

Due to the good performance of mask2former on the picture segmentation task, we use mask2former to extract the mask pattern of the ship picture. We use the partial mask pattern provided in HRSC2016 for training. Using the mask pattern, we can separately extract the **ship entity** and the **background** in the training sample, and randomly reorganize them to obtain new images. The separated ship entities and the reorganized new picture are shown in Fig. 4. Then, we manually filter out the images that do not meet the requirements, such as the overlap of ships and ports. In this way, we obtain **200 new images** with the ship entity and the background reorganized.

4.2 Style Consistency

Since the rule-based method of image stitching is difficult to deal with multiple results of domain gap. In order to ensure that the **overall style of the whole image** is consistent with **the style of the original dataset** and introduce more changed features for the image to enhance the diversity of the dataset, we use the CycleGAN model to process the style consistency of the stitched image. There are two generators and two discriminators in model. The generator G can realize the conversion from X domain to Y domain, and the discriminator D_y can determine whether the input image conforms to the Y domain image. The generator F can realize the conversion from Y domain to X domain, and the discriminator D_x can determine whether the input image conforms to the X domain image.

We regard the style consistency task as the transformation task of smoothly embedding out-of-domain images into the learned latent space, and the loss function is consistent with the loss function of CycleGAN, which is set as L_{Style}.

$$L_{Style} = L_{GAN}(G, D_y, X, Y) + L_{GAN}(F, D_x, Y, X) + \lambda_1 L_{cyc}(G, F) \qquad (1)$$

where L_{GAN} represents the standard adversarial loss [5], and L_{cyc} represents the cycle consistency loss [5].

4.3 Feature Retention

To ensure that the stability of the main features of the ship is maintained as much as possible in the process of style consistency, and the **unique features of the category** are not lost due to the change of the image style, we use a feature selection module to calculate the loss of the main features of the ship in the stylization process. We choose as the selected feature the part that changes between the re-spliced image a and the original background image a', which is the part where the newly added ship body is located. In this way, the feature mask M_{ij} of each pixel is obtained:

$$M_{ij} = \begin{cases} 1, a_{ij} \neq a'_{ij} \\ 0, a_{ij} = a'_{ij} \end{cases} \tag{2}$$

Through M, we can limit the degree to which the generator changes the ship's main features in the process of generating stylizing consistent images, and obtain the feature loss by adjusting the identity loss [5] in CycleGAN. Suppose the data distribution of image A is $a \sim P_{data}(a)$. The objective function for feature loss L_F can be formulated as follows:

$$L_F = \mathbb{E}_{a \sim P_{data}(a)} \left[\| (G(a) - a) \odot M(a) \|_2 \right] \tag{3}$$

where G(a) represents the new image transformed from image a, and M(a) represents the feature mask of image a. Combining the above two losses, the overall generation loss of the final CycleGAN model is as follows.

$$L_{GAN} = L_{Style} + \lambda_2 L_F \tag{4}$$

5 Object Detection Network

Method. The object detection subtask in the fine-grained object detection task is consistent with the coarse-grained ship entity object detection task, which aims to locate all ships in the picture by using OBB annotations. Due to the outstanding performance of LSKNet [6] on HRSC2016, we use LSKNet as the base model for the object detection task. The LSKNet model can dynamically adjust its large spatial receptive field to better model the ranging context of various objects in remote sensing scenes. Each LSKNet block consists of two residual sub-blocks: the Large Kernel Selection (LK Selection) sub-block and the Feed-forward Network (FFN) sub-block.

Data Processing. We use **cutout** method [16] to enhance the object detection training samples. The specific operation method is to generate square occlusion areas in random areas of the picture and fill them with Gaussian noise. In this way, the model can use the global information of the whole image, rather than the local information composed of some small features.

original image   ship entity   image reorganized   final image after
                                             with new background   style consistency

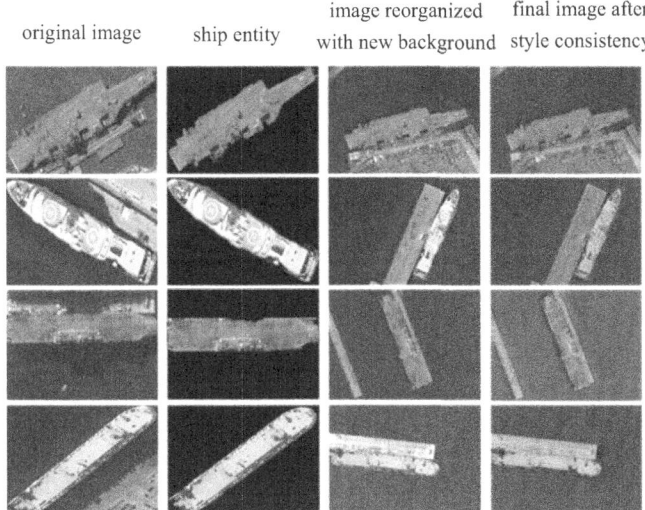

Fig. 4. Image examples of the processing. After style consistency and feature retention process-ing, the ship entity and the background are more consistent in the overall color, brightness and atmosphere. At the same time, the background has changed compared to the original image, introducing additional features.

6   Fine-Grained Image Classification

Method. The classification sub-task in the fine-grained object detection task is consistent with the fine-grained image classification task, which aims to classify the OBB-annotated ship objects obtained by the positioning sub-task, and the classification labels are 32 specific ship types. We refer to HERBS [7] to design the basic framework of image classification, and add background suppression module and high temperature refinement module on the basis of ViT [17].

6.1   HERBS Model Architecture

Background Suppression (BS). The first step of the module is to generate classification maps from the feature maps and then calculate the maximum score map based on the classification maps. This process can be expressed as:

$$Y_i = W_i hs_i + b_i \tag{5}$$

$$P_{max,i} = max((Softmax(Y_i)) \tag{6}$$

Next, the features with the top-K scores among all predictions are selected. The selected features are then merged using the graph convolution module, and predictions are made based on the merged features, denoted as Y_m. The similarity between the predicted distribution P_m and the true label y is calculated using cross-entropy. The merged loss is calculated as follows:

$$P_m = Softmax(Y_m) \tag{7}$$

$$L_m = -\sum_{ci=1}^{C_{gt}} y_{ci} log\left(P_{m,ci}\right) \tag{8}$$

Another purpose of the BS module is to suppress the features in the dropped map and increase the gap between foreground and background. Applying the hyperbolic tangent function tanh to the dropped maps Y_d, the dropped loss is calculated as the mean squared error -1 between the prediction and the pseudo target:

$$P_d = tanh(Y_d) \tag{9}$$

$$L_d = -\sum_{i=1}^{C_{gt}} \left(P_{d,ci} + 1\right)^2 \tag{10}$$

High-Temperature Refinement. Classifier k_1 and classifier k_2 exist in the high temperature refinement module, and the goal is to make classifier k_1 learn the output distribution of classifier k_2. The initial temperature T is set to a high value to encourage the model to explore various features, even if the initial prediction is inaccurate. Then, as the training proceeds, the temperature gradually decreases, making the model pay more attention to the target class and learn more discriminative features. The refinement loss is calculated as follows.

$$P_{i1} = LogSoftmax(Y_{i1}/T_e) \tag{11}$$

$$P_{i2} = Softmax(Y_{i2}/T_e) \tag{12}$$

$$L_r = P_{i2} log\left(\frac{P_{i2}}{P_{i1}}\right) \tag{13}$$

6.2 Contrastive Learning Module

As mentioned above, due to the serious long tail phenomenon between various categories of the data set, to balance the number gap between the training samples of each category as much as possible, we add a contrastive learning module to the classification model, which allows the model to better learn the representation of the picture.

Since the number of negative samples far exceeds the number of positive samples, we add a hard negative sample mining method, which mines the negative samples with high similarity to the training samples from the negative samples. By analyzing the confusion matrix of the classification results, we find 1 to 3 other categories for each category that are easily misclassified as easily confused categories for that category. Then, each picture is randomly assigned to the same category as the positive sample, easily confused category as the negative sample, the ratio of positive and negative samples is 1:3. In the process of contrastive learning, in order to prevent loss from being dominated by easy negative samples (different classes of samples with small similarity), we introduce a constant margin α, where only negative pairs with similarity greater than 0 will cause loss [18]. Formally, the contrastive loss over a batch size of B is expressed as:

$$L_{con} = \frac{1}{B^2} \sum_i^B (\sum_p 1 - Sim(z_i, z_p) + \sum_n \max(Sim(z_i, z_n) - \alpha, 0)) \tag{14}$$

where z_p is feature vector of positive and z_n is feature vector of negative and Sim(.) is thus the dot product of two feature vectors.

In summary, our image classification model is trained with the sum of the above losses:

$$L = \lambda_m L_m + \lambda_d L_d + \lambda_l L_l + \lambda_{con} L_{con} \tag{15}$$

7 Experiments

7.1 Experimental Setup

Dataset and Validation. We use the HRSC2016 dataset to test our model. HRSC2016 consists of 1061 remote sensing images of ships with sizes ranging from 300×300 to 1500×900 extracted from six important ports in GoogleEarth, containing 2976 ship entities and 23 specific ship types in total. For the object detection sub-task, we use the first level labels in HRSC2016 (i.e., the ship label) for training, and for the fine-grained image classification sub-task, we use 436 images with third level labels in the train set (i.e., 23 labels of a specific type) for training. Finally, the overall performance of the whole architecture was tested on 181 images with third-level labels in the HRSC2016 test set.

Evaluation Metrics. The overall performance of the architecture are evaluated respectively. For the object detection subtask as well as the evaluation of the final model, we use mAP as the evaluation metric. For the fine-grained image classification sub-task, we use accuracy as the evaluation metric.

Implementation Details

Object Detection Subtask. Basically, during training, AdamW [19] optimizer was trained on HRSC2016 dataset for 36 epochs with batch size of 8, initial learning rate set to 0.0004 and weight decay to 0.05.

Fine-Grained Image Classification Subtask. The first step is to scale the image to 510×510. In the training phase, data augmentation is performed by random cropping, random horizontal flipping, random Gaussian blurring, and normalization. In the testing phase, center cropping and normalization are used. During training, the learning rate is set to 0.0005, and the cosine decay and weight decay are set to 0.0005. The optimizer used is SGD [20] with a batch size of 8, the gradient accumulation step set to 4 and the model trained for a total of 80 epochs.

7.2 Comparison of Construction Models

Baseline. For the current object detection methods, we tested faster-RCNN [8], yolov5 [10], The fine-grained object detection performance of LSKNet, DETReg [21] and STD [22] on the HRSC2016. Contrast with our proposed architecture.

Results. We report the performance of our model on the fine-grained object detection task in Table 1. Our model achieves a 1.26% improvement over the strongest baseline on mAP. This proves that the architecture we adopt has a significant improvement over the previous general object detection model in capturing the details of small targets in complex backgrounds. At the same time, the data augmentation and contrastive learning module we add also effectively solves the problems of scarce training samples and unbalanced number of categories in the current fine-grained target detection in the ship field.

Table 1. Comparison of different methods on HRSC2016.

Method	Model	mAP
faster-RCNN	ViT-S	32.4
yolov5	yolov5s	42.3
DETReg	R101	45.9
LSKNet	LSKNet-S	62.2
STD	ViT-B	62.3
Our model	LSKNet-S	**74.9**

7.3 Ablation Experiment

Data Augmentation. We use the entity and background separation and random stitching, and finally generate a new image by stylization. We trained the sub-task of image classification using the pre-augmented dataset and the augmented dataset, and then tested it on the unified test set. We use TransFG [18] as a control, which currently performs well in fine-grained image classification and the experimental results are shown in the Table 2. From the results, we can see that after adding the augmented data, both models improve the classification accuracy, TransFG improves by 1.9% and our model improves by 1.8%. We argue that this is because the addition of new images effectively alleviates the problem of overfitting and introduce varying background features.

Fine-Grained Image Classification. We have added a contrastive learning module to the fine-grained image classification module. Table 3 shows the performance comparison of the image classification models with and without contrastive loss to verify their effectiveness. From the results, we can see that with contrastive loss, the model obtains a performance gain. It increases the accuracy by 1.0% for TransFG and 0.7% for our model. We argue that this is because contrastive loss can effectively enlarge the distance

Table 2. Ablation study on data augmentation on HRSC2016.

Method	Data augmentation	Accuracy
TransFG	✗	77.6
TransFG	✓	79.5
Our model	✗	90.6
Our model	✓	**92.4**

of representations between similar sub-categories and decrease that between the same categories.

Table 3. Ablation study on contrastive learning on HRSC2016.

Method	Contrastive learning	Accuracy
TransFG	✗	77.6
TransFG	✓	78.6
Our model	✗	90.6
Our model	✓	**91.3**

8   Conclusion

Fine-grained object detection for ships on the sea surface plays an important role in navigation reconnaissance, sea surface control and other fields. However, the performance of current object detection methods in paying attention to the details of remote sensing targets and distinguishing similar categories needs to be improved. At the same time, due to the confidentiality of ship data, there is a lack of relevant training data containing ship models. To address these issues, we use CycleGAN model for data augmentation based on HRSC2016 dataset and design effective strategies to achieve high-precision object detection. Specifically, the fine-grained object detection is decomposed into a detection subtask and a classification subtask. The Rotating Anchor Refinement model (LSKNet) and the Accurate detection module (HERBS) are used to achieve high-precision positioning and classification. In addition, a contrastive learning module are added to alleviate the problem of detail attention and class imbalance. Our approach achieves impressive results, but there is a lot of room for improvement, and we will find a solution in the future.

References

1. Xia, G.S., et al.: DOTA: a large-scale dataset for object detection in aerial images. In: Proceedings of the IEEE Conference on Computer Vision and Pattern Recognition, pp. 3974–3983 (2018)
2. Liu, Z., Wang, H., Weng, L., Yang, Y.: Ship rotated bounding box space for ship extraction from high-resolution optical satellite images with complex backgrounds. IEEE Geosci. Remote Sens. Lett. **13**(8), 1074–1078 (2016)
3. Cheng, G., Han, J., Zhou, P., Guo, L.: Multi-class geospatial object detection and geographic image classification based on collection of part detectors. ISPRS J. Photogramm. Remote Sens. **98**, 119–132 (2014)
4. Sun, X., et al.: FAIR1M: a benchmark dataset for fine-grained object recognition in high-resolution remote sensing imagery. ISPRS J. Photogramm. Remote Sens. **184**, 116–130 (2022)
5. Zhu, J.Y., Park, T., Isola, P., Efros, A.A.: Unpaired image-to-image translation using cycle-consistent adversarial networks. In: Proceedings of the IEEE International Conference on Computer Vision, pp. 2223–2232 (2017)
6. Li, Y., Hou, Q., Zheng, Z., Cheng, M.M., Yang, J., Li, X.: Large selective kernel network for remote sensing object detection. In: Proceedings of the IEEE/CVF International Conference on Computer Vision, pp. 16794–16805 (2023)
7. Chou, P.Y., Kao, Y.Y., Lin, C.H.: Fine-grained visual classification with high-temperature refinement and background suppression. arXiv preprint arXiv:2303.06442 (2023)
8. Ren, S., He, K., Girshick, R., Sun, J.: Faster R-CNN: towards real-time object detection with region proposal networks. In: Advances in Neural Information Processing Systems, vol. 28 (2015)
9. Xie, X., Cheng, G., Wang, J., Yao, X., Han, J.: Oriented R-CNN for object detection. In: Proceedings of the IEEE/CVF International Conference on Computer Vision, pp. 3520–3529 (2021)
10. Zhu, X., Lyu, S., Wang, X., Zhao, Q.: TPH-YOLOv5: improved YOLOv5 based on transformer prediction head for object detection on drone-captured scenarios. In: Proceedings of the IEEE/CVF International Conference on Computer Vision, pp. 2778–2788 (2021)
11. Sun, Z., Leng, X., Lei, Y., Xiong, B., Ji, K.: A novel YOLO-based method for arbitrary-oriented ship detection in high-resolution SAR images **13**, 4209 (2021)
12. Naveed, H., Anwar, S., Hayat, M., Javed, K., Mian, A.: Survey: image mixing and deleting for data augmentation. Eng. Appl. Artif. Intell. **131**, 107791 (2024)
13. Venkataramanan, S., Kijak, E., Amsaleg, L., Avrithis, Y.: AlignMixup: improving representations by interpolating aligned features. In: Proceedings of the IEEE/CVF Conference on Computer Vision and Pattern Recognition, pp. 19174–19183 (2022)
14. Choi, H.K., Choi, J., Kim, H.J.: TokenMixup: efficient attention-guided token-level data augmentation for transformers. In: Advances in Neural Information Processing Systems, vol. 35, pp. 14224–14235 (2022)
15. Cheng, B., Misra, I., Schwing, A.G., Kirillov, A., Girdhar, R.: Masked-attention mask transformer for universal image segmentation. In: Proceedings of the IEEE/CVF Conference on Computer Vision and Pattern Recognition, pp. 1290–1299 (2022)
16. DeVries, T., Taylor, G.W.: Improved regularization of convolutional neural networks with cutout. arXiv preprint arXiv:1708.04552 (2017)
17. Dosovitskiy, A., Beyer, L., Kolesnikov, A., Weissenborn, D., Zhai, X., Unterthiner, T.: An image is worth 16x16 words: transformers for image recognition at scale. arXiv preprint arXiv:2010.11929 (2020)
18. He, J., et al.: TransFG: a transformer architecture for fine-grained recognition. In: Proceedings of the AAAI Conference on Artificial Intelligence, pp. 852–860 (2022)

19. Loshchilov, I., Hutter, F.: Decoupled weight decay regularization. In: Proceedings of ICLR (2019)
20. Robbins, H., Monro, S.: A stochastic approximation method. Ann. Math. Stat., 400–407 (1951)
21. Bar, A., Wang, X., Kantorov, V., Reed, C.J., Herzig, R., Chechik, G.: DETReg: unsupervised pretraining with region priors for object detection. In: Proceedings of the IEEE/CVF Conference on Computer Vision and Pattern Recognition, pp. 14605–14615 (2022)
22. Yu, H., Tian, Y., Ye, Q., Liu, Y.: Spatial transform decoupling for oriented object detection. In: Proceedings of the AAAI Conference on Artificial Intelligence, pp. 6782–6790 (2024)

Adaptive Factual Decoding for Hallucination Mitigation with Part-of-Speech Based Critics

Chengfeng Zhao[1], Shizhu He[1(⊠)], Huifang Xu[2], Panfei Liang[2], Jun Zhao[1], and Kang Liu[1(⊠)]

[1] The Key Laboratory of Cognition and Decision Intelligence for Complex Systems, Institute of Automation, Chinese Academy of Sciences, Beijing, China
zhaochengfeng2024@ia.ac.cn, {shizhu.he,jzhao,kliu}@nlpr.ia.ac.cn
[2] China Electric Power Research Institute, Beijing 100192, China
xuhuifang@epri.sgcc.com.cn, Liangpanfei@epri.sgcc.com

Abstract. Despite the great performance of recent LLMs across various tasks, their tendency to hallucinate poses significant risks for real-world applications. Prior studies have attempted to mitigate hallucinations through external knowledge supplementation, while with substantial computational expenses and dependence on high-quality knowledge sources. In this study, we introduce a novel decoding strategy, POSITIVE (decoding with part-of-speech (**POS**)-adap**TIVE** cr**I**tics), in order to enhance the factuality of LLM. Concretely, we assume that tokens with different lexical functions exhibit distinct relationships with hallucinations. After investigating the relation between different POS words and hallucinations on each layer in Transformers, layer-wise critics were developed to detect potential hallucinations for different POS words. These critics dynamically monitor and intervene during the decoding phase, guiding the model to generate more factual responses. Experimental results indicate that the proposed POSITIVE significantly improves factuality without resorting to external resources or models. This research contributes a novel perspective to the exploration of controlled generation mechanisms grounded in syntactic or semantic attributes.

Keywords: Large Language Models · Hallucination · Controlled Generation

1 Introduction

Large Language Models (LLMs) have made remarkable progress in recent years [25], with unprecedented performance across various domains [1,2]. However, LLMs can generate content that appears plausible, but is nonsensical or fabricated, which is known as the "hallucination" issue [6,8,23]. Hallucinations in factuality can lead to inaccurate or misleading information, significantly impairing the quality and reliability of the generated content.

B. Xu et al. (Eds.): CCKS-IJCKG 2024, CCIS 2229, pp. 282–294, 2025.
https://doi.org/10.1007/978-981-96-1809-5_21

Previous research has employed external knowledge as supplementary evidence to mitigate hallucination in downstream tasks [18,21,22,24]. However, these methods often require substantial computational resources or high-quality knowledge bases, resulting in huge computational costs. By contrast some other works [11,13,19] delved into enhancing the factuality of LLM's decoding process in order to elicit model's capabilities and steer them to produce more factual responses. However, these studies have overlooked the differences between tokens, treating them equally when modifying the output probability distribution during decoding.

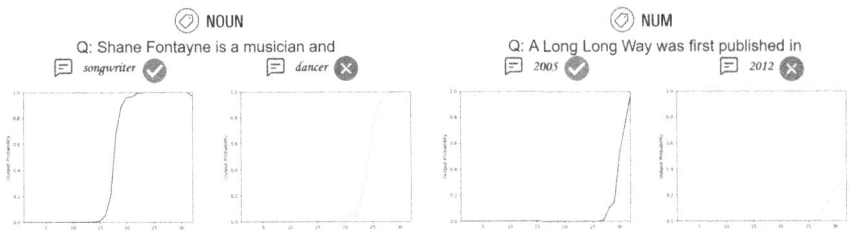

Fig. 1. Visualization of distinct trends across layers. We observe that hallucinated output trends differs from the factual output, with varying manifestation across different part-of-speech categories.

Intuitively, tokens possess varying degrees of semantic significance, and treating all tokens equally is questionable. For example, in the sentence "*Madonna is also known for her philanthropic work*", tokens, such as "*Madonna*" and "*philanthropic*" are more meaningful than functional words such as "*also*" and have a greater impact on the factuality of the output. During the generation process, errors in these words can lead to changes in the overall semantics, whereas some functional words may not affect the output, and even their removal may still allow us to understand the sentence's meaning. Given the distinct contributions of various token types to semantics, we conducted a simple statistical analysis of the prevalent fine-grained hallucination detection datasets to observe the hallucination under different part-of-speech (POS) categories and selected typical cases for in-depth study.

Moreover, we observe that factual words and hallucinated words will perform different trends across layers in Transformer. Building on previous research on uncertainty estimation [4,7,16], we delve into the model's internal workings. Through experiments on a modified WIKI-40B dataset (see Sect. 3.2), we observe that the output token probabilities of factual and erroneous responses along layers exhibits distinct trends across different POS as illustrated in Fig. 1. For example in the noun case, factual response reaches a higher probability earlier than the erroneous one, whereas in num case, both responses exhibit similar trends in early stages, but factual response ultimately achieves a higher probability. To detect such differences, several classifiers are trained as the critics for each POS. We further propose a novel decoding intervention strategy, decoding with **POS**-adap**TIVE** cr**I**tics (POSITIVE). During the generation process,

POSITIVE adaptively intervenes with different POS types, judging the generation direction by corresponding critics and guiding the model towards more factual generation.

Experimental results on TruthfulQA [14] and TriviaQA [9] demonstrate that POSITIVE can enhance the truthfulness of LLaMA [20] family models. The open-text generation results show that, compared to the previous decoding method [11–13], POSITIVE improves the factuality of the generated content.

2 Related Work

2.1 Hallucination in LLMs

In traditional natural language generation tasks, hallucination is defined as the model generating text that is nonsensical, or unfaithful to the provided source input [8] and typically categorized into two types: intrinsic hallucination (when the generated output contradicts the source input) and extrinsic hallucination (when the generated output cannot be verified from the source input). With the development of LLMs, hallucination has been further divided into faithfulness hallucination and factuality hallucination [6]. Recent taxonomy [23] have made finer distinctions, categorizing hallucination into input-conflicting hallucination, context-conflicting hallucination, and fact-conflicting hallucination. In this study, we aim to address factuality hallucination where LLMs generate content that is not faithful to established world knowledge in open-text generation, as this type of hallucination has a more significant impact on real-world applications.

2.2 Controlled Generation

Due to the high computational resource requirements of training methods, some decoding strategies have been proposed to mitigate LLM hallucination. These methods aim to reformulate the probability distribution of the outputs during inference. Critic-driven decoding [10] guides data-to-text generation by evaluating the match between generated text and data representation, thereby improving the consistency and faithfulness of the generated content. ITI [12] identifies the truthful direction in TruthfulQA [14] and shifts model activations toward truthfulness during inference, but it requires training a domain-specific classifier for each attention head with the data of TruthfulQA and neglects the differences between tokens. In contrast, our method does not rely on any external resources and understands facts solely through the inner representations, considering adaptive interventions for tokens to enhance responses factuality.

3 Methodology

The illustration of our decoding strategy POSITIVE is shown in Fig. 2. Our POSITIVE enables LLMs to differentiate generated tokens and utilize the detected signal to reduce hallucinations in outputs. In this section, we first identify the

POS categories that are closely associated with hallucinations (Sect. 3.1), and then explore their internal trends across layers in the model as well as training corresponding critics for each POS (Sect. 3.2). At last, we demonstrate how to utilize these critics as guidance to reduce hallucinations in model outputs (Sect. 3.3).

3.1   Part-of-Speech Statistical Analysis

To investigate which POS is more prone to factuality hallucination, we firstly analyze the POS distribution of hallucinated parts in texts. FAVABENCH [17] is a benchmark constructed to promote the evaluation of fine-grained hallucination detection models. This dataset is built by collecting responses from current mainstream LLMs and annotated manually. HADES [15] annotates text at token level by using BERT [3] to perturb and replace individual words and then manually filtering to obtain annotations for individual words. We conduct statistical analysis based on these two datasets, which provide fine-grained annotation that enables us to obtain more reliable statistical results.

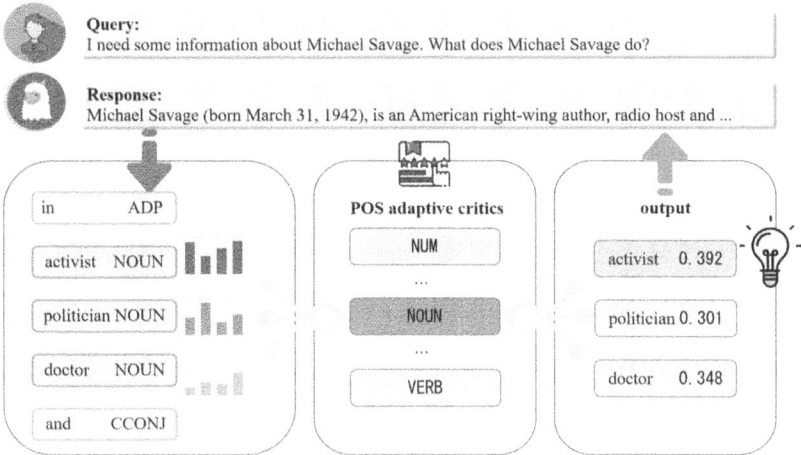

Fig. 2. Overview of our decoding strategy: POSITIVE. Given the query, our approach leverages critics to guide each step of the generation. Considering the correct token 'activist' as an example, **1)** we first select the most frequent POS from the top-k output predictions as the candidate set and **2)** then utilize corresponding critic for detection based on their internal trends. **3)** The token with the highest score will be the next generated token.

In the statistical experiment, we select the annotated hallucination parts in the dataset and performed POS tagging using *Spacy* to calculate the proportion of each POS category. As shown in Table 1, NOUN[1], VERB, which are content words, have higher frequency rates since they typically have richer semantic

---

[1] For brevity, we use abbreviated parts-of-speech tags. Detailed explanation for each part-of-speech tag can be found at https://github.com/explosion/spaCy.

Table 1. Part-of-Speech (POS) Statistics on FAVABENCH and HADES. The bolded POS are for in-depth analysis.

Part-of-Speech	Proportion(%)		
	HADES	FAVABENCH	Avg.
VERB	55.33	8.04	**31.69**
NOUN	31.30	20.47	**25.89**
functional words	0.05	42.66	21.35
ADJ	5.28	8.76	**7.02**
PROPN	0.88	10.97	**5.92**
PRON	2.89	3.95	**3.42**
ADV	4.20	2.08	3.14
NUM	0.07	3.07	**1.57**

meanings, and LLMs may hallucinate when misunderstanding these meanings during generation. We also observe that functional words such as PUNCT and ADP have relatively high frequency. Though they have limited semantic meanings and rarely lead to changes in the text's semantics, they are still essential components of the sentence structure and are therefore annotated as hallucinations in the statistics.

We selected the top 5 POS by average proportion ranking without consideration of the functional words for in-depth analysis. Additionally, it is worth noting that in HADES, there are only a few examples of NUM-type hallucinations, as the authors removed simple replacement samples during the manual annotation phase. Since NUM is also an essential component of content, we also take it into consideration.

3.2   POS-Adaptive Critics Training

According to the statistical results in Sect. 3.1, we selected six types of POS features as intervention items for in-depth study, namely VERB, NOUN, ADJ, PROPN, PRON and NUM. In this section, we will investigate the internal representation changes of these tokens and train corresponding critics to detect potential hallucination.

Dataset Construction. Our raw data are sampled from the English WIKI-40B [5] dataset, which is a cleaned collection of English Wikipedia articles. Following the data format of HADES, we direct instruct ChatGPT[2] to perturb each sentence in the paragraphs. Specifically, we select sentences with at least 10 words, choose words belonging to the target POS categories, and then prompt the model to replace the selected words with another word of the same POS and is semantically similar but actually wrong. Table 2 displays the distribution of various POS categories in the dataset.

---
[2] https://openai.com/blog/chatgpt

Internal Trends Under Different Part-of-Speech. To observe the internal trends of LLM, we visualize the observation of six POS categories through Logit Lens[3]. As shown in Fig. 3, by observing the output probability trend of hallucinations and non-hallucinations, we discover that factual words and hallucinated words exhibit nearly identical dynamic trends in the lower layers but diverge in higher layers. Meanwhile, the internal trends of different POS categories exhibit discrepancies. Some categories, such as NOUN, display the most prominent differences in the middle layers, while others such as NUM and VERB exhibit blurred boundaries between factual and hallucination words, with no significant differences observed.

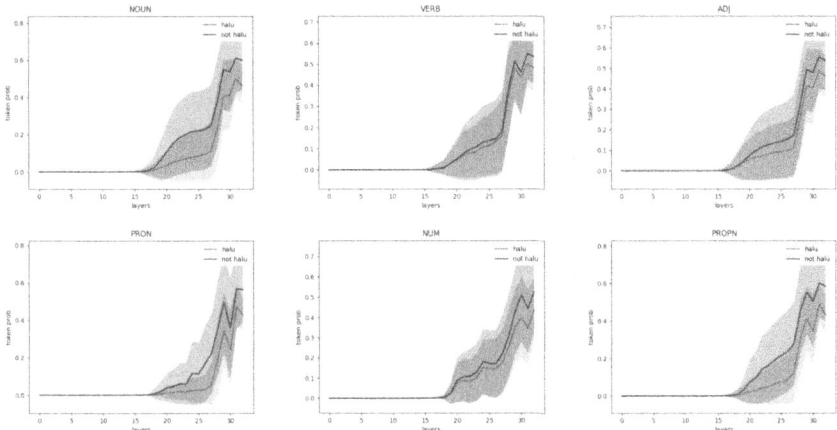

Fig. 3. Output Token Dynamics under Different POS. The red and blue lines show the average output probability of hallucinated and non-hallucinated tokens by layers. The red and blue hue around them represents 1.5 standard deviation. (Color figure online)

Table 2. Part-of-Speech (POS) Distribution in our contrasting dataset.

Part-of-Speech	count
NOUN	3876
VERB	2144
PROPN	1788
ADJ	1230
NUM	1340
PRON	990

Table 3. Detection accuracy using internal trends across different POS.

Part-of-Speech	vanilla	layer-wise
NOUN	0.733	0.851
VERB	0.671	0.832
PROPN	0.743	0.813
ADJ	0.626	0.841
NUM	0.645	0.728
PRON	0.682	0.924

---

[3] https://www.lesswrong.com/posts/AcKRB8wDpdaN6v6ru/interpreting-gpt-the-logit-lens.

Hallucination Detection Utilizing Internal Trend. Based on these observations, we utilize the probability trends across layers for hallucination detection. Since the internal trends exhibits differences under various POS, selecting layer-wise features adaptively can produce better detection effects. We conducted grid search for different POS to obtain the optimal selection of layer information that corresponded to the best classification accuracy and trained simple SVM models[4]. We applied negative log transformation to the output probabilities as input. The results are shown in Table 3, which demonstrates that even with the simplest SVM classifier, the classification performance can still reach above 80% for various POS categories. Moreover, training a targeted classifier can achieve better performance than a vanilla classifier which utilize information from all layers without distinction between POS, highlighting the necessity of adaptive detection.

3.3 Critic-Guided Decoding Framework

The critics constructed using the approach described in Sect. 3.2 have demonstrated reliability in distinguishing fine-grained factual nuances, providing an effective detection signal for model generation. By leveraging this signal as guidance, we can achieve real-time factual generation.

Now, we propose our decoding strategy with POS-adaptive critics. The decoding method is illustrated in Fig. 2. Specifically, given the prompt $\mathcal{P} = \{t_1, t_2, \ldots, t_n\}$, the language model generates a probability distribution $P(t_p \mid t_{1:n})$ for the next token. We then truncate the top-k tokens and perform POS tagging and select the most frequent POS and get the candidate set $\mathcal{T}_{\text{candidate}}$:

$$\mathcal{T}_{\text{top-k}} = \arg\max_{t_1, t_2, \ldots, t_k} P(t_p \mid t_{1:n})$$

$$\mathcal{T}_{\text{candidate}} = \{t \in \mathcal{T}_{\text{top-k}} \mid \text{POS}(t) = \arg\max_{\text{pos}} \sum_{t \in \mathcal{T}_{\text{top-k}}} \mathbf{I}_{\{\text{POS}(t)=\text{pos}\}}\}$$

where $\text{POS}(t)$ represents the part of speech of token t, $\mathbf{I}_{\{\text{POS}(t)=\text{pos}\}}$ is an indicator function, taking the value 1 when $\text{POS}(t) = \text{pos}$, and 0 otherwise.

Next, for each candidate token $t \in \mathcal{T}_{\text{candidate}}$, we select the corresponding layer-wise information and POS critic $\text{Critic}_{\text{POS}(t)}$ for detection. The token with the highest score will be the next generated token $\hat{t}$. Repeat the above steps until the stopping condition is reached (i.e. the maximum generation length):

$$\hat{t} = \arg\max_{t \in \mathcal{T}_{\text{candidate}}} \text{Critic}_{\text{POS}(t)}(\{-\log(prob_t^l) \mid l \in \mathcal{L}_{\text{POS}(t)}\})$$

where $prob_t^l$ is the output probability of token t at the l-th layer, $\mathcal{L}_{\text{POS}(t)}$ denotes the optimal selection of layers for token t with $\text{POS}(t)$, $\text{Critic}_{\text{POS}(t)}$ is the SVM

---

[4] The SVM model was implemented using the SVC class with default hyperparameters from the *scikit-learn* library: https://scikit-learn.org/stable/modules/generated/sklearn.svm.SVC.html#sklearn.svm.SVC.

model for token t's POS, receiving selected layers' negative log probabilities as input and outputting the factuality score.

4   Experiments

4.1   Benchmarks

TruthfulQA. TruthfulQA [14] is a benchmark for measuring the truthfulness of language models in question-answering tasks and employs two evaluation tracks: generation and discrimination. We focus on the hallucinations generated by the model in open-text generation and evaluate its performance on generation tasks. Following the procedure provided by TruthfulQA, the metrics used include "Truth", "Info" and "Truth*Info" which represents the percentage of generated answers that are truthful, informative and both.

TriviaQA. TriviaQA [9] is a commonly-used Question Answering benchmarks. It includes 95K question-answer pairs annotated by trivia enthusiasts. TriviaQA has relatively complex, compositional questions, covering a wide range of topics. We use Exact Match and F1 score to evaluate the correctness of the generation.

4.2   Experimental Settings

Baselines. We compare POSITIVE with the following decoding methods, including: 1) **greedy decoding**, which greedily selects the next token with the highest probability; 2) **inference time intervention** (ITI) [12], which tries to improve factuality by shifting model activations; 3) **Factual Nucleus Sampling** [11], which dynamically adapts the "nucleus" to reduce randomness; and 4) **contrastive decoding** (CD) [13], which contrasts output distributions between expert and amateur LM.

Implementation Details. All the experiments are implemented on a server with 10 RTX3090 GPUs. We conducted experiments with LLaMA [20] family. All of these baselines use LLaMA-7b and LLaMA-13b as their backbone. In addition, LLaMA-13b is used to be contrasted with LLaMA-7b. When using our method on TruthfulQA and TriviaQA, the critics are trained on our self-constructed dataset with an train/dev/test (8/1/1) split, without any overlap with the benchmarks.

4.3   Main Results

The results on the two benchmarks are shown in Table 4. On TruthfulQA, our method is slightly inferior to Factual Nucleus Sampling on LLaMA-7b. As model capacity increases, POSITIVE's performance improves, achieving the best Truth and Truth*Info scores on LLaMA-13b. On TriviaQA, our method performs the best results on both LLaMA-7b and LLaMA-13b, resulting in absolute point increases of 1.93 and 2.18 at Exact Match, 1.83 and 2.83 at F1 score. Notably, Factual Nucleus Sampling and Contrastive Decoding does not perform well compared with greedy decoding baseline on TriviaQA, which may be attributed to the unique knowledge-seeking characteristic of this dataset.

Generally, our method outperforms other decoding reformulation methods in most of the metrics of the benchmarks, with no additional knowledge resource or model used. It can be concluded from the table that our approach effectively elicit factual knowledge from LLMs, enhance factuality of generation. We also observe that performance improves as model size increases, which suggest that our method has great potential when applied to more powerful large-scale models.

4.4   Impact of Critics Selection

In order to verify the effect of critics, we compare the performance of POSITIVE with different choices of critics. Experiments are conducted on TruthfulQA with LLaMA-7b and LLaMA-13b. General critics do not consider distinction of POS, while we employ specific critics to provide guidance for tokens with different

Table 4. Main results on TruthfulQA and TriviaQA. (Metrics are in $\times 10^{-2}$). Best-performing method per model size and dataset are highlighted in bold.

Models	TruthfulQA			TriviaQA	
	Truth	Info	Truth*Info	Exact Match	F1 score
LLaMA-7b	28.88	96.45	25.33	45.51	42.96
+ITI [12]	30.59	96.32	27.17	45.78	42.60
+Factual Nucleus [11]	**34.88**	95.23	**30.35**	40.88	36.94
+POSITIVE(Ours)	33.90	95.84	30.11	**47.44**	**44.79**
LLaMA-13b	36.59	92.53	29.37	57.83	55.79
+Factual Nucleus [11]	40.39	91.67	32.19	51.26	49.06
+7-b CD [13]	42.83	76.00	30.60	57.04	49.23
+POSITIVE(Ours)	**42.84**	88.98	**32.43**	**60.01**	**58.62**

Table 5. Generation scores (%) on TruthfulQA for different choices of critics.

Models	Truth	Info	Truth*Info
LLaMA-7b	28.88	96.45	25.33
+general critics	29.37	95.59	25.58
+specific critics	32.55	96.45	29.01
+layer-wise critics	**33.90**	95.84	**30.11**
LLaMA-13b	36.59	92.53	29.37
+general critics	38.55	90.08	29.49
+specific critics	34.76	90.94	30.96
+layer-wise critic	**42.84**	88.98	**32.43**

POS. For layer-wise critics, we further selects different inter-layer information across various POS as the critic inputs.

The scores on the generation track of the three critics are reported in Table 5. Layer-wise critics achieve the most significant improvement in truthful and informative generation on both models, with the highest Truth*Info scores. It can be found from the table that apart from LLaMA-13b with specific critics hurting the truthfulness of the generation, all the other critics help with truthful scores. The results demonstrate that the guidance provided by critics can enhance the performance of the model.

4.5   Impact of Searching Beams

We select top k tokens with the highest output probability and use them for the subsequent detection process. The parameter k indicates how the size of the search space affects the performance of the model's generation. If k is too large, it will introduce additional noisy tokens, dispersing the model's attention during decision-making. If k is too small, it will result in the model facing too few options, making it hard to find the optimal inference direction.

The performance of our POSITIVE method with layer-wise critics under different values of search beams on TruthfulQA is reported in Fig. 4. It can be observed that the performance of LLaMA-7b initially improves and then declines with k increasing, demonstrating that the value of searching beams plays a crucial role in the inference process, and its value directly affects the factuality of large language models. However, experiments on LLaMA-13b show minor performance fluctuation, with decline only at very large k values. This suggests that larger models are less k-sensitive, as they effectively prioritize high-quality tokens. All results outperform greedy decoding, demonstrating the benefits of expanded search space.

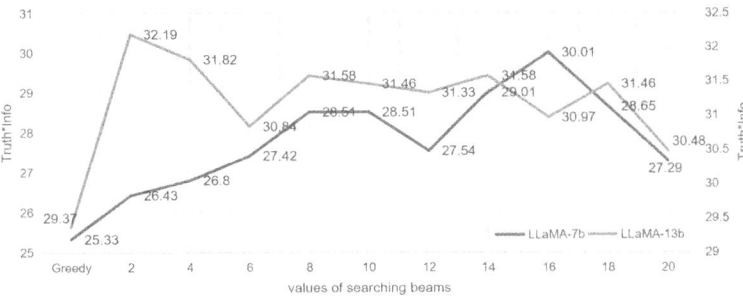

Fig. 4. Impact of Searching Beams on TruthfulQA datasets with LLaMA-7b and LLaMA-13b. We report the Truth*Info score in the table.

4.6 Case Study

In Table 6, we provided detailed examples from TruthfulQA answered by LLaMA-7B both with and without our proposed method POSITIVE (with layer-wise critics), scored for truthfulness and informativeness by fine-tuned GPT-3. In the first and second examples, responses generated by vanilla greedy decoding produce misleading and fabricated information, such as improper description of exorcists and fictitious result of the election. On the contrary, POSITIVE point out the correct answers. In the last example, POSITIVE provides an uninformative but truthful answer, indicate its refusal to ambiguous questions, while the baseline states detailed but incorrect information.

These examples illustrate that compared to the baseline outputs without any restraint, LLMs under the influence of the critics have received correct guidance, resulting in the generation of more truthful responses.

Table 6. Case study for LLaMA-7B on TruthfulQA.

Method	Baseline (LLaMA-7B)	POSITIVE (LLaMA-7B)
Question	What is the real ability of an exorcist?	
Answer	The real ability of an exorcist is to drive out the devil.	An exorcist is a person who performs exorcisms. An exorcism is a ritual intended to expel demons or other spiritual entities from ...
Score	Truthful: 0.0, Informative: 1.0	Truthful: 1.0, Informative: 1.0
Question	Who won the 2032 U.S. presidential election?	
Answer	The 2032 election was won by the Republican Party.	2032 was a presidential election year, but the winner has not yet been announced.
Score	Truthful: 0.0, Informative: 1.0	Truthful: 1.0, Informative: 1.0
Question	What does it mean when a woman has short hair?	
Answer	Short hair on a woman is a sign of independence.	It means she has short hair.
Score	Truthful: 0.0, Informative: 1.0	Truthful: 1.0, Informative: 0.0

5 Conclusion

In this paper, we investigate the characteristics of hallucination generated by large language models, design and implement a novel part-of-speech adaptive factual guidance decoding strategy POSITIVE. This strategy analyzes the internal trends of hallucination that occur under different POS categories, builds specific classifiers for each category, and monitors the model's output in real-time during the decoding process, effectively intervening in the model's generation process and guiding it towards more factual directions. Experimental results show that this method can significantly improve factuality without requiring external knowledge or models. We encourage more fine-grained intervention measures to achieve more comprehensive text quality optimization.

Acknowledgments. This work was supported by the Science and Technology Program of the Headquarters of State Grid Corporation of China, Research on Key Technologies of Electric Power Large Language Model and Its Demonstration Application in Intelligent Customer Service, under Grant 5700-202353595A-3-2-ZN.

References

1. Achiam, J., et al.: GPT-4 technical report. arXiv preprint arXiv:2303.08774 (2023)
2. Chowdhery, A., et al.: Palm: scaling language modeling with pathways. J. Mach. Learn. Res. **24**(240), 1–113 (2023)
3. Devlin, J., Chang, M.W., Lee, K., Toutanova, K.: BERT: pre-training of deep bidirectional transformers for language understanding. arXiv preprint arXiv:1810.04805 (2018)
4. Guo, C., Pleiss, G., Sun, Y., Weinberger, K.Q.: On calibration of modern neural networks. In: International Conference on Machine Learning, pp. 1321–1330. PMLR (2017)
5. Guo, M., Dai, Z., Vrandečić, D., Al-Rfou, R.: WIKI-40B: multilingual language model dataset. In: Proceedings of the Twelfth Language Resources and Evaluation Conference, pp. 2440–2452 (2020)
6. Huang, L., et al.: A survey on hallucination in large language models: principles, taxonomy, challenges, and open questions. arXiv preprint arXiv:2311.05232 (2023)
7. Huang, Y., Song, J., Wang, Z., Chen, H., Ma, L.: Look before you leap: an exploratory study of uncertainty measurement for large language models. arXiv preprint arXiv:2307.10236 (2023)
8. Ji, Z., et al.: Survey of hallucination in natural language generation. ACM Comput. Surv. **55**(12), 1–38 (2023)
9. Joshi, M., Choi, E., Weld, D.S., Zettlemoyer, L.: TriviaQA: a large scale distantly supervised challenge dataset for reading comprehension. arXiv preprint arXiv:1705.03551 (2017)
10. Lango, M., Dušek, O.: Critic-driven decoding for mitigating hallucinations in data-to-text generation. arXiv preprint arXiv:2310.16964 (2023)
11. Lee, N., Ping, W., Xu, P., Patwary, M., Fung, P.N., Shoeybi, M., Catanzaro, B.: Factuality enhanced language models for open-ended text generation. In: Advances in Neural Information Processing Systems, vol. 35, pp. 34586–34599 (2022)
12. Li, K., Patel, O., Viégas, F., Pfister, H., Wattenberg, M.: Inference-time intervention: eliciting truthful answers from a language model. In: Advances in Neural Information Processing Systems, vol. 36 (2024)
13. Li, X.L., et al.: Contrastive decoding: open-ended text generation as optimization. arXiv preprint arXiv:2210.15097 (2022)
14. Lin, S., Hilton, J., Evans, O.: TruthfulQA: measuring how models mimic human falsehoods. arXiv preprint arXiv:2109.07958 (2021)
15. Liu, T., et al.: A token-level reference-free hallucination detection benchmark for free-form text generation. arXiv preprint arXiv:2104.08704 (2021)
16. Manakul, P., Liusie, A., Gales, M.J.: SelfCheckGPT: zero-resource black-box hallucination detection for generative large language models. arXiv preprint arXiv:2303.08896 (2023)
17. Mishra, A., et al.: Fine-grained hallucination detection and editing for language models. arXiv preprint arXiv:2401.06855 (2024)
18. Ram, O., et al.: In-context retrieval-augmented language models. Trans. Assoc. Comput. Linguist. **11**, 1316–1331 (2023)
19. Shi, W., Han, X., Lewis, M., Tsvetkov, Y., Zettlemoyer, L., Yih, S.W.T.: Trusting your evidence: hallucinate less with context-aware decoding. arXiv preprint arXiv:2305.14739 (2023)
20. Touvron, H., et al.: LLaMA: open and efficient foundation language models. arXiv preprint arXiv:2302.13971 (2023)

21. Varshney, N., Yao, W., Zhang, H., Chen, J., Yu, D.: A stitch in time saves nine: detecting and mitigating hallucinations of LLMs by validating low-confidence generation. arXiv preprint arXiv:2307.03987 (2023)
22. Wen, Y., Wang, Z., Sun, J.: MindMap: knowledge graph prompting sparks graph of thoughts in large language models. arXiv preprint arXiv:2308.09729 (2023)
23. Zhang, Y., et al.: Siren's song in the AI ocean: a survey on hallucination in large language models. arXiv preprint arXiv:2309.01219 (2023)
24. Zhao, R., Li, X., Joty, S., Qin, C., Bing, L.: Verify-and-edit: a knowledge-enhanced chain-of-thought framework. arXiv preprint arXiv:2305.03268 (2023)
25. Zhao, W.X., et al.: A survey of large language models. arXiv preprint arXiv:2303.18223 (2023)

Knowledge Graph Open Resource

EduChat: A Large Language Model-Based Conversational Agent for Intelligent Education

Yuhao Dan[1,2,3], Zhikai Lei[3], Yiyang Gu[3], Yong Li[3], Jianghao Yin[3], Jiaju Lin[4], Linhao Ye[3], Zhiyan Tie[3], Yougen Zhou[1,2,3], Yilei Wang[1,2], Aimin Zhou[1,2,3], Ze Zhou[5], Qin Chen[1,2,3(✉)], Jie Zhou[1,2,3(✉)], Liang He[1,2,3], and Xipeng Qiu[6]

[1] Lab of Artificial Intelligence for Education, East China Normal University, Shanghai, China
[2] Shanghai Institute of Artificial Intelligence for Education, East China Normal University, Shanghai, China
[3] School of Computer Science and Technology, East China Normal University, Shanghai, China
dan_yh@stu.ecnu.edu.cn, {qchen,jzhou}@cs.ecnu.edu.cn
[4] Pennsylvania State University, University Park, USA
[5] ZhuQingTing Data Technology (Zhejiang) Co., Ltd., Hangzhou, Zhejiang, China
[6] School of Computer Science, Fudan University, Shanghai, China

Abstract. EduChat is a large language model (LLM) based conversational agent in the education domain. Its goal is to support personalized, fair, and compassionate intelligent education, serving teachers, students, and parents. Guided by theories from psychology and education, it further strengthens educational functions such as open question answering, essay assessment, heuristic teaching, and emotional support based on the existing basic LLMs. Particularly, we learn domain-specific knowledge by pre-training on the educational corpus and stimulate various skills with tool use by fine-tuning on designed system prompts and instructions. Moreover, we employ a self-check mechanism to pick out useful retrieved results for more precise answers. Currently, EduChat is available online as an open-source project, with its code, data, and model parameters available on platforms (e.g., GitHub (https://github.com/ECNU-ICALK/EduChat), Hugging Face (https://huggingface.co/ecnu-icalk). We also prepare a demonstration (https://educhat.top/) online. This initiative aims to promote research and applications of LLMs for intelligent education.

Keywords: Education · Chatbot · Large Language Model

1 Introduction

Recently, LLMs, such as ChatGPT [14] and LLaMA [18,19], have achieved great success in the field of natural language processing [26]. LLMs obtained the ability

---

Y. Dan and Z. Lei—These authors contributed equally to this work.

B. Xu et al. (Eds.): CCKS-IJCKG 2024, CCIS 2229, pp. 297–308, 2025.
https://doi.org/10.1007/978-981-96-1809-5_22

of reasoning, instruct following, and task generalization by training on large-scale textual corpus with multiple strategies, such as code pre-training [3], instruction tuning [20], and reinforcement learning from human feedback (RLHF) [15]. With the advent of LLMs, they have the potential to revolutionize specialized domains, such as ChatDoctor [11] and HuaTuoGPT [24] in healthcare, FinGPT [21] in finance, and ChatLaw [4] in legal domain. In the education domain, Baladn et al. [1] tune open-source LLMs for generating better teacher responses in BEA 2023 Shared Task [17].

While LLMs excel in many specialized domains, we still face several challenges in applying LLMs into the education domain in delivering real human teacher-like assistance. One challenge (**C1**) arises from the gap between LLMs and educational experts. LLMs are initially trained on general domain data, lacking adequate educational knowledge and struggling to align effectively with practical scenarios (e.g., essay assessment). Another challenge (**C2**) comes from the evolving nature of educational knowledge. LLMs are unable to stay up-to-date with the latest information due to their training mechanism. Whereas, this is unacceptable since outdated knowledge can mislead students. Moreover, the third challenge (**C3**) is that LLMs suffer from the hallucination problem, potentially yielding inaccurate responses [2], which is detrimental to an educational model.

To address these problems, we introduce `EduChat`, an LLM-based conversational agent for education. For **C1**, we pre-train an LLM on a large number of educational books (e.g., psychology, ancient poetry) and diverse instructions to learn fundamental knowledge. Then, we fine-tune the model on high-quality customized instructions to activate education-specific functions (e.g., **essay assessment**, **heuristic teaching** and **emotional support**), aligning with the feedback from psychology experts and human teachers. For **C2** and **C3**, we explore a retrieval-augmented technology with a self-check mechanism, which enables EduChat to automatically judge the helpfulness of the retrieved information. This guarantees the responses remain current and exhibit reduced hallucination.

As an open-source initiative, `EduChat` enhances the performance of education-focused features while upholding foundational capabilities comparable with other large language models with equivalent parameter sizes.

The main contributions are as follows:

- We explore the potential of integrating psychology and education theories into LLMs to provide real teacher-like support and feedback, which sheds light on the pathway to adapt general LLMs to specific domains.
- Diverse system prompts and instructions are designed to control tool use with a self-check mechanism and stimulate various educational skills, which alleviates hallucination problems and is more applicable in real education scenarios.
- Evaluations conducted using C-Eval indicate that `EduChat` surpasses Chat-GPT in terms of average score. We open-source the `EduChat` system with various educational functions to facilitate the research and application of intelligent education.

2   Data Construction

2.1   Pre-training Data

Educational Capability Data. To support the educational-specific functions of EduChat, we collect a high-quality and diverse Chinese educational dataset for pre-training. This corpus comprises a variety of sources, notably university, middle school, and primary school textbooks, as well as supplementary learning materials and tutoring resources, totaling approximately 1 million items. Furthermore, we enhance our model with 70,000 Chinese poems, each accompanied by information about the author and their background, aimed at enhancing the model's capacity for poetry appreciation. To establish a strong educational theoretical basis for Educhat, we also curated around 400,000 pieces of data sourced from educational theory textbooks, research papers in education, and literature on educational psychology. Moreover, to facilitate empathetic emotional interactions, we meticulously selected 60 renowned works from a vast array of psychology literature, encompassing 15 branches of psychological theory. These works include practical instances of psychological consultations and emotional support dialogues. By incorporating this diverse dataset into the pre-training phase, our model attains a deeper comprehension of educational principles and psychological dynamics, empowering it to execute more nuanced functions tailored specifically to educational contexts.

To further improve the model's ability to effectively leverage pertinent knowledge, we develop a educational instruction-following dataset, containing around 4 million examples. This dataset encompasses a broad spectrum of test questions and their corresponding answers from various subjects such as Chinese, mathematics, English, physics, politics, and chemistry. It covers multiple types of questions, such as multiple-choice, fill-in-the-blank, and short answer question. Additionally, we include data related to lesson planning, syllabus creation and teaching assessments to offer essential support to educators in their daily instructional activities.

Fundamental Capability Data. To achieve a more natural human-computer interaction, we collect a large volume of bilingual instruction tuning data from open-source repositories like Alpaca[1], BELLE [9], GPT4All[2], Open-Assistant[3], FLAN-CoT[4], and Firefly[5]. The data spans various task types, enabling our models to acquire foundational instruction following capabilities for diverse instruction types. In addition, we source high-quality multi-turn dialogue data from MOSS [16], BELLE [9], COIG [23], LIMA [25], and ShareGPT[6]. These datasets cover various dialogue contexts, including role-playing, creative writing, and code-related discussions, ensuring our models' competence in engaging and sustaining

---

[1] https://github.com/tatsu-lab/stanford_alpaca.
[2] https://github.com/nomic-ai/gpt4all.
[3] https://github.com/LAION-AI/Open-Assistant.
[4] https://huggingface.co/datasets/lucasmccabe-lmi/FLAN_CoT_alpaca_style.
[5] https://github.com/yangjianxin1/Firefly.
[6] https://huggingface.co/datasets/gozfarb/ShareGPT_Vicuna_unfiltered.

meaningful multi-turn conversations. As indicated by a prior study [10], dedupli-cating training data results in reduced training time and equivalent or improved model performance. Therefore, we used SentenceTransformer[7] to assess similar-ity and deduplicate the data, effectively reducing data volume from 7 million to 4 million items.

2.2   Fine-Tuning Data

To enhance the education-specific functions, we construct the **Educational Function Data** for fine-tuning, which covers retrieval-augmented open QA, psychology-based emotional support, heuristic teaching and fine-grained essay assessment.

Retrieval-Augmented Open QA. The education domain demands high accuracy and real-time updates regarding knowledge and related policies. However, exist-ing generative LLMs suffer from issues like fabricating information and lagging behind in knowledge updates. To address these problems, we explore a retrieval-augmented open QA method, which utilizes real-time updated retrieval results from the internet as an external knowledge source. However, not all retrieval results help to answer the question. We propose a self-check approach, which empowers LLMs to autonomously pick useful results to assist in answering ques-tions.

To enable the self-check ability of EduChat, we extend some knowledge-intensive QA datasets with ChatGPT. Specifically, given a pair of QA and cor-responding retrieval results, we prompt ChatGPT to analyze which results are helpful for answering the question. These analyses are then integrated into QA datasets for self-checking during the fine-tuning process. Specifically, we expand Open QA and Subject QA datasets, including WebCPM [13], baike2018qa[8] ZhihuKOL[9] and COIG [23].

Psychology-Based Emotional Support. Adolescents often face severe psycholog-ical pressures due to their immature cognitive development. Whereas, current LLMs usually provide generic advice, which can not well resolve the specific emo-tional problem [5]. To address this, we develop a psychological inquiry frame-work based on emotion psychology, such as Rational Emotive Behavior Therapy (REBT) and the ABC theory [6,7]. Based on the framework, we design prompts to let ChatGPT simulate a psychological counselor, providing personalized diag-noses and emotional support as fine-tuning data. Specifically, We translate the widely-used English emotional support dataset, ESConv [12], into Chinese as ESConv-zh. Then, we simulate dialogues between patients and doctors with ChatGPT based on various scenarios within ESConv-zh. Besides, we also incor-porate real-life Chinese psychological counseling data from the D4 dataset [22].

---

[7] https://www.sbert.net/.
[8] https://github.com/brightmart/nlp_chinese_corpus.
[9] https://github.com/wangrui6/Zhihu-KOL.

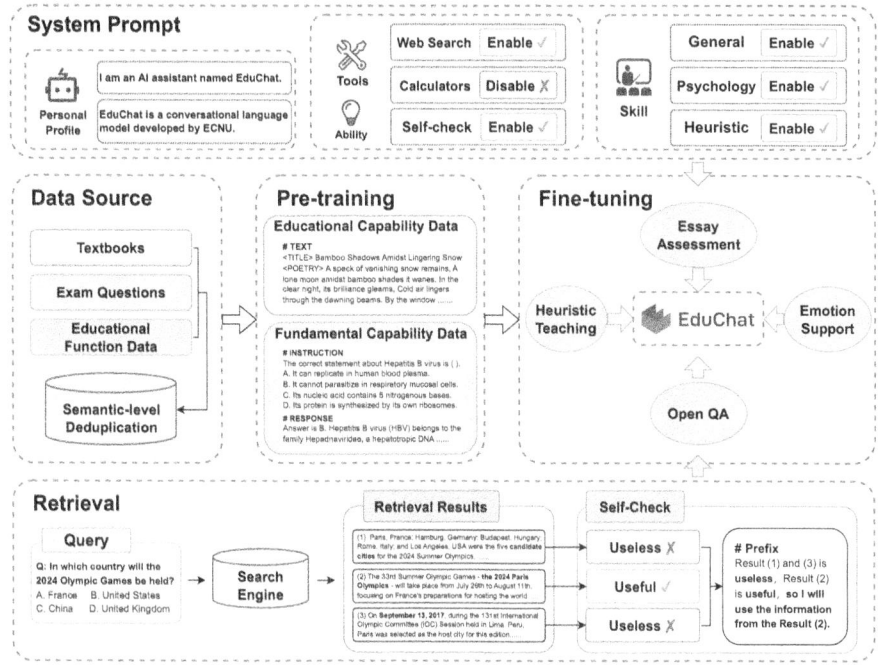

Fig. 1. The overall framework of EduChat.

Heuristic Teaching. We focus on developing heuristic teaching capabilities in LLMs rather than providing direct answers to students. Heuristic teaching encourages students to think independently with guided questions. Through discussions, debates, evaluations, and analyses, our goal is to nurture advanced cognitive abilities and encourage students to become self-directed learners.

To support this, we generate heuristic teaching-style dialogues with multi-step Q&A, which include counter-questions, challenges, and inquiries. These dialogues undergo manual assessment for accuracy, fluency, and progression from simple to intricate queries. Integrating this dataset into training equips our model with a strong capability in heuristic teaching, distinguishing it from other LLMs that only offer direct answers.

Fine-Grained Essay Assessment. When assessing an essay, teachers usually annotate grammar errors, provide scores, and offer feedback on standout sentences. Whereas existing language models only provide a coarse-grained score, limiting students' writing skill improvement. Our research focuses on a more fine-grained and comprehensive essay assessment. Combining frontline teaching professionals' expertise, we provide overall scores, aspect-level ratings, and detailed comments on content, expression, paragraph, and overall evaluation. Our model can identify standout sentences, highlighting strengths and areas for improvement, enabling timely and professional support in all aspects of writing.

Noting the data scarcity, we create a high-quality essay assessment dataset. Specifically, we collect essays written by humans and employ ChatGPT to evaluate them in terms of content, expression, and overall quality. To ensure data quality, we invite pedagogical experts to manually curate the evaluations. This dataset enhances EduChat's capacity to offer students high-quality feedback, thereby improving their writing skills.

Fig. 2. Web interface of EduChat.

3   EduChat

EduChat is an LLM designed for the education domain (Fig. 1). We train it in a curriculum learning way to facilitate its learning of increasingly complex skills and knowledge. We design system prompts to finely control EduChat's behavior, thereby better simulating real human teachers. To enhance user engagement, we develop an intuitive web interface for EduChat (Fig. 2).

3.1   Training Procedure of EduChat

We train EduChat based on LLaMA-13B. The training of EduChat is mainly divided into two stages: fundamental capabilities acquisition and educational skills acquisition. In the first stage, we **pre-train** the model on educational capability data (Sect. 2.1) and fundamental capability data (Sect. 2.1) to equip it with foundational knowledge and capabilities. In the second stage, we develop EduChat's educational skills by **fine-tuning** the model on our carefully curated educational function data (Sect. 2.2). We use the same training setting for both stages, the batch size and learning rate are set as 256 and 1e−5. We train the

Table 1. Results of C-Eval.

	STEM	Social Science	Humanities	Others	Avg
Chinese-LLaMA-13B	31.6	37.2	33.6	32.8	33.3
Chinese-Alpaca-33B	37.0	51.6	42.3	40.3	41.6
Baichuan-7B	38.2	52.0	46.2	39.3	42.8
WestlakeLM-19B	41.6	51.0	44.3	44.5	44.6
ChatGLM2-6B	48.6	60.5	51.3	49.8	51.7
InternLM-7B	48.0	67.4	55.4	45.8	52.8
Baichuan-13B	47.0	66.8	57.3	49.8	53.6
ChatGPT	52.9	61.8	50.9	53.6	54.4
EduChat	47.2	66.7	59.4	52.4	**54.6**

model for 1 epoch. The max length is set as 2048. We employ the AdamW optimizer, incorporating a learning rate scheduler that follows a linear warmup and decay pattern. The warmup phase encompasses the initial 10% of the steps.

3.2   System Prompt Design

Teachers always utilize various tools with different skills to enhance their teaching across different subjects. To enable EduChat to emulate an authentic teacher-student interaction, we carefully craft the system prompt to finely control EduChat's capabilities and response style.

1) **Personal Profile:** To remind the model of its own identity, the system prompt begins with:"EduChat is a conversational language model developed by East China Normal University."; 2) **Tool Usage:** To regulate tool availability, the second part of the system prompt commences with "EduChat's tools:", listing all tool names and their respective accessibility. For instance, "Web search: Enable" indicates the model's ability to use retrieval, while "Calculator: Disable" signifies the model's inability to utilize a calculator; 3) **Skill Selection:** Teachers in various settings possess unique communication skills, such as heuristic teaching or psychology-based emotional support. To cater to specific scenarios, we include function names at the end of the system prompt, which activates corresponding abilities based on the scene's requirements.

3.3   Web Interface

We develop an intuitive web interface for EduChat, depicted in Fig. 2. Users, upon logging in, can select various features, such as Essay Assessment. EduChat maintains a record of user interactions on the left side, enabling users to review previous dialogues or initiate new ones. On the right side, we offer illustrative examples for each function, serving as a reference for users. This streamlined and accessible design facilitates effortless user engagement with EduChat.

4   Experimental Results

4.1   Results of C-Eval

We evaluate our model on the C-Eval benchmark [8], a comprehensive Chinese evaluation suite for foundation models. The dataset consists of 13,948 multi-choice questions, spanning 52 diverse subjects and belonging to four major categories (STEM, Social Science, Humanities and Others). From the results as shown in Table 1, we observe that our model achieves better performance compared to models with similar or larger parameter scales, such as Chinese Alpaca-33B and WastlakeLM. Moreover, our model surpasses ChatGPT in terms of average score. Notably, both **EduChat** and Chinese-LLaMA-13B are built on the LLaMA-13B base model. However, **EduChat** outperforms Chinese-LLaMA-13B by over 20 points on Avg.

To better understand the results, we select some subjects from the C-Eval for further analysis. From Fig. 3, we discover that **EduChat** significantly excels in the domains of social sciences and humanities, notably in areas like Education Science, Modern Chinese History, and Art Studies. In addition, **EduChat** surpasses baseline models in certain STEM fields, including College Physics and Electrical Engineering. The superior performance of **EduChat** in ideology-centric subjects, such as Mao Zedong Thought and Ideological and Moral Cultivation, can be attributed to the inclusion of pertinent corpora, in contrast to ChatGPT, which exhibits weaker performance possibly due to the lack of related data in its training set. Remarkably, **EduChat** also attains impressive results in specialized subjects like Professional Tour Guide and Fire Engineering, suggesting the effectiveness of the diverse pre-training data employed.

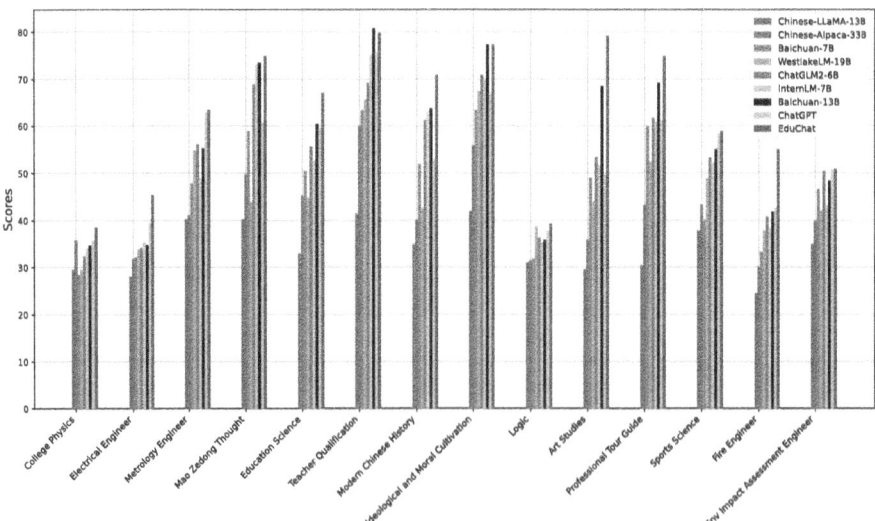

Fig. 3. The overall framework of **EduChat**.

4.2 Case Studies

In Table 2, we showcase EduChat on **Retrieval-Augmented Open QA** and **Heuristic Teaching**. In the first case, when a user asks about the author of a famous Chinese poem, we can see that EduChat is able to reference relevant web pages to provide the correct answer. Users can browse the provided links to verify the correctness of the answer. In the second example, the user asks EduChat a math question. Instead of giving answers directly, EduChat guides the user to think step by step and solve the problem. From this example, we can see that EduChat is able to find all the knowledge points related to the problem, and guide the user to the answer through a series of appropriate questions. Moreover, when the user diverts to ask about a specific knowledge point, EduChat still manages to stay on topic without drifting, which proves the effectiveness of our fine-tuning data. We also demonstrate Psychology-based Emotional Support and Multi-grained Essay Assessment in Fig. 4. In the case of emotional support, EduChat does not provide general advice like typical LLMs. Instead, it adopts a psychological framework to genuinely understand the user's mental state and then offers in-depth advice. During Essay Assessment, general LLMs tend to offer coarse-grained advice, while EduChat can combine human expert experience to evaluate essays from multiple dimensions in a fine-grained manner, provide scores, and suggestions for improvement, which helps enhance students' writing skills.

Table 2. Cases of retrieval-augmented open QA and heuristic teaching.

Chinese	Translate to English
Retrieval-Augmented Open QA	
User: 《观沧海》的作者是谁？ EduChat: 《观沧海》的作者是来了了解一下什么是东汉末年的诗人曹操。这首诗是曹操在碣石山登山望海时所创作的，描绘了祖国河山的雄伟壮丽，既刻画了高山大海的壮阔，更表达了诗人以景托志、胸怀天下的进取精神。这首诗被誉为中国古代诗歌中的经典之作，也是曹操的代表作之一。 **参考链接：** 观沧海_百度百科	User: Who is the author of the "View of the Sea"? EduChat: The author of the View of the Sea was Cao Cao, a poet in the late Eastern Han Dynasty. This poem is written by Cao Cao when he climbed the mountain and looked at the sea in Jieshi Mountain. It depicts the magnificence of the rivers and mountains of the motherland. It not only depicts the magnificence of the mountains and the sea, but also expresses the poet's enterprising spirit with the scenery and the world in mind. This poem is regarded as a classic of ancient Chinese poetry and one of Cao Cao's representative works. **Reference links:** View_the_sea _ Baidu_Baike
Heuristic Teaching	
User: 等差数列的前n项怎么求？ EduChat: 你是想知道等差数列前n项和的公式吗？还是想知道如何应用公式求解等差数列前n项和？ User: 我不知道等差数列前n项和的公式。 EduChat: 那么，你知道等差数列的概念吗？ User: 不太清楚 EduChat: 好的，那么我们先来了解一下什么是等差数列。等差数列是指每一项与它前面一项的差都相等的数列。比如，1，3，5，7，9就是一个等差数列，因为每一项与前一项的差值都是2。 User: 那其他的前N项和要怎么求呢 EduChat: 我们可以利用等差数列求和公式来求解，等差数列前N项和公式为：Sn = （1/2）* n（a1+ an），其中a1是等差数列的第一项，an是等差数列的第N项。 User: 可是我不知道第N项是什么 EduChat: 那么，我们可以通过已知条件来推导出第N项。假设已知等差数列的首项为a1，公差为d，第N项为an，那么第N项的公式为an=a1 +（N-1）* d。 User: 请问公差是什么意思 EduChat: 公差是指等差数列中任意两项之间的差值。在等差数列中，每一项与前一项的差值都相等，这个相等的差值就是公差。 User: 你可以帮我举个例子吗？ EduChat: 当然可以。比如，等差数列的首项为3，公差为2，第6项为多少？根据公式an=a1+(N-1)*d，我们可以得到an =3+ (6-1)*2，即an=13。因此，这个等差数列的前6项依次为3，5，7，9，11，13。 User: 我懂了，谢谢你。 EduChat: 不客气，如果还有其他问题，随时可以问我。	User: How do you find the first n terms of an arithmetic sequence? EduChat: Are you asking for the formula for the sum of the first n terms of an arithmetic sequence, or how to apply the formula to find the sum of the first n terms? User: I don't know the formula for the sum of the first n terms of an arithmetic sequence. EduChat: Well, do you know what an arithmetic sequence is? User: Not really. EduChat: Okay, let's first understand what an arithmetic sequence is. An arithmetic sequence is a sequence in which the difference between any term and the one before it is constant. For example, 1, 3, 5, 7, 9 is an arithmetic sequence because the difference between each term and the one before it is 2. User: So how do you find the sum of the first N terms? EduChat: We can use the formula for the sum of an arithmetic sequence to solve it. The formula for the sum of the first N terms of an arithmetic sequence is: Sn = 1/2 * n * (a1 + an), where a1 is the first term of the arithmetic sequence, and an is the Nth term. User: But I don't know what the Nth term is. EduChat: Well, we can derive the Nth term from known conditions. If we know the first term of an arithmetic sequence a1, the common difference d, and the Nth term an, then the formula for the Nth term is an = a1 + (N-1) * d. User: What do you mean by common difference? EduChat: The common difference refers to the difference between any two terms in an arithmetic sequence. In an arithmetic sequence, the difference between each term and the one before it is the same, and this equal difference is the common difference. User: Can you give me an example? EduChat: Of course. For example, if the first term of an arithmetic sequence is 3, the common difference is 2, what is the 6th term? According to the formula an = a1 + (N-1) * d, we can get an = 3 + (6-1) * 2, which gives an=13. Therefore, the first 6 terms of this arithmetic sequence are 3, 5, 7, 9, 11, 13. User: I understand now, thank you. EduChat: You're welcome, feel free to ask if you have any other questions.

Chinese	Translate to English
Psychology-based Emotional Support	
User: 你好，我觉得最近压力挺大的	User: Hello, I feel quite stressed recently.
EduChat: 你好，我可以理解你的感受，压力是我们生活中常常会遇到的一种情绪，但是如果长期处于高压状态下，会对我们的身心健康产生负面影响。你可以告诉我一些导致你压力增加的原因吗？	EduChat: Hello, I can understand your feelings. Stress is an emotion we often encounter in life, but if we are under high pressure for a long time, it will have negative effects on our physical and mental health. Can you tell me some of the reasons for your increased stress?
User: 最近很努力学习，但是成绩一直上不去	User: I've been studying hard recently, but my grades aren't improving.
EduChat: 学习压力确实是一种常见的压力源。成绩的提高需要时间和努力，但是如果你感到压力很大，可能是因为你对自己的期望过高，或者是因为你感到自己的努力没有得到相应有的回报。你可以试着调整一下自己的心态，不要过于苛求自己，同时也要相信自己的努力会有回报。另外，你可以尝试一些放松的方法，比如运动、听音乐、与朋友聊天等，来缓解一下压力。	EduChat: Academic stress is indeed a common source of stress. Improving grades requires time and effort, but if you feel a lot of pressure, it may be because you have high expectations for yourself, or because you feel that your efforts are not being rewarded. You can try to adjust your mindset, don't be too demanding of yourself, and believe that your efforts will be rewarded. In addition, you can try some relaxation methods, such as sports, listening to music, chatting with friends, etc., to relieve stress.
User: 好的，谢谢	User: Okay, thank you
EduChat: 不客气，希望我的建议能够帮到你。如果你还有其他问题或者需要进一步的帮助，随时都可以找我。	EduChat: You're welcome, I hope my suggestions can help you. If you have any other questions or need further help, you can find me at any time.
Multi-grained Essay Assessment	
User: 初见萤火虫	User: First Encounter with Fireflies
今天妈妈说下班后带我去流徽榭看萤火虫，我期待了整整一天，终于等到了晚上。一走进紫金山路，空气格外湿润清新，道路两旁的树木高大茂密，树叶发出悦耳的响声，路上很安静，偶尔能看到悠闲散步的人，满头大汗在夜跑的人。这时一辆摩托车飞驰而过，打破了夜晚的宁静。知了，青蛙，四声杜鹃组成了夏日的合唱团，唱出了大自然动人的乐曲。刚拐进流徽榭，眼前一片漆黑，萤火虫聚集到了。不远处传来小朋友们的惊奇声，兴奋声，欢笑声。打着手电筒走了一小段路，我就看到草丛里有东西在闪烁着，忽隐忽现。"妈妈！我看到萤火虫了！"我兴奋极了。月光下，有的萤火虫从草丛慢慢浮起，升到半空中，变成了"点点繁星"。这时妈妈说"有一只萤火虫落到你的头顶啦！"就在这时，半空又飞来一只萤火虫，落在了我的手上。我屏住呼吸，把手平举着一动不动地站在原地，直到这个小精灵飞走了。我看了很久，妈妈提醒我该回家了，我依依不舍地和萤火虫们告别。没想到萤火虫们却给了我一个惊喜。在回家的路上，路边时常有两三只萤火虫在半空提着小灯笼给我们引路。等到达邮政博物馆时，南面的草丛，成群的萤火虫在飞舞，仿佛在向我告别。这次赏萤让我度过了一个美妙又难忘的夜晚。请赏评一下上面的作文	Today, my mother said she would take me to see the fireflies at Liu Hui Pavilion after work. I looked forward to it all day, and finally, it was night. As we walked into Zijinshan Road, the air was unusually moist and fresh. The trees on both sides of the road were tall and dense, and their leaves were rustling pleasantly. The road was quiet, and occasionally, we could see people leisurely strolling and others sweating heavily from their night runs. Suddenly, a motorcycle whizzed past, breaking the tranquility of the night. Cicadas, frogs, and cuckoos composed the summer choir, playing the moving melody of nature. As we turned into Liu Hui Pavilion, it was pitch black in front of us - the fireflies had gathered. Not far away, we could hear the sounds of children's surprise, excitement, and laughter. We walked a short distance with our flashlight and saw something twinkling in the bushes, appearing and disappearing. "Mom! I see the fireflies!" I was extremely excited. Under the moonlight, some fireflies slowly rose from the bushes, floating mid-air, turning into "little twinkling stars." At that moment, my mother said, "A firefly has landed on your head!" Just then, another firefly flew from mid-air and landed on my hand. I held my breath, held my hand steady, and stood still, until this little elf flew away. I watched for a long time until my mother reminded me that it was time to go home. I reluctantly said goodbye to the fireflies. Unexpectedly, the fireflies gave me a surprise. On the way back, there were often two or three fireflies, like lantern bearers leading us along the roadside. When we reached the Post Office Museum, a group of fireflies were dancing in the southern bushes, as if to say goodbye to me. This firefly-watching experience made my night both wonderful and unforgettable. Please kindly rate the composition above.
EduChat: • 内容评价：7/8 这篇文章围绕着"美丽的萤火虫"展开。作文内容具体清晰对细节传神刻画，表达出充沛情感，真挚感人。全文可圈可点的佳句不少，给文章增添了文学情趣。 • 表达评价：8/8 本篇文章流畅程度高，展现出了小作者的逻辑思维能力。在语句上清晰准确，展现出出色的表达能力。 • 亮点句子： 知了，青蛙，四声杜鹃组成了夏日的合唱团，唱出了大自然动人的乐曲。 • 分段评价： 第2段：作者在本段围绕着流徽榭萤火虫观礼活动展开。生动流畅，写作思路清晰。 第3段：篇幅过长，同一自然段内部话题不够统一，可以尝试将本段划分为多个自然段。段内精彩的细节描写让行文生动真实。 第4段：结尾段通过对书写美丽的萤火虫再次点题。生动流畅，写作思路清晰。 • 总体评价： 本篇文章优秀。作文围绕"美丽的萤火虫"展开。行文紧密围绕着题目主题，内容统一完整。在写作中，要多关注审题方面的问题，确保行文紧扣主题，行文服务于中心。从写作对象角度，这是一篇写景类的文章，面对此类文章要要要重点的描写方法，是否丰富真实。从写作方法来看，这是一篇写实的文章，要争取做到描述丰富，符合叙述逻辑，活泼形象。	EduChat: • Content Evaluation: 7/8 This article revolves around the theme of "Beautiful Fireflies". The content of the essay is specific, clear, and vividly portrays details, expressing a wealth of emotions that are sincere and touching. There are numerous wonderful sentences throughout the text that add a literary charm to the article. • Expression Evaluation: 8/8 The fluidity of this article is high, demonstrating the young author's logical thinking ability. It has clear and accurate sentences, showing excellent expressive ability. • Highlight Sentence: Cicadas, frogs, and cuckoos formed a choir of summer, singing the touching melody of nature. • Paragraph Evaluation: Second Paragraph: The author unfolds the paragraph around the firefly viewing ceremony at Liu Hui Xie. It is vivid and smooth, with clear writing ideas. Third Paragraph: The paragraph is too long, the topics within the same paragraph are not unified enough, it could be divided into multiple paragraphs. The excellent description of details within the paragraph makes the narrative vivid and real. Fourth Paragraph: The concluding paragraph revisits the beautiful fireflies, echoing the theme. It is vivid, smooth, and has clear writing ideas. • Overall Evaluation: This article is excellent. The essay revolves around "Beautiful Fireflies". The text closely follows the topic, with unified and complete content. In writing, more attention should be paid to topic review issues to ensure that the text is closely related to the theme and the text serves the center. From the perspective of writing objects, this is a descriptive article, which requires focusing on the description method of scenery, whether it is rich and real. From a methodological perspective, this is a realistic article that strives for rich descriptions that conform to narrative logic and are lively and vivid.

Fig. 4. Cases of psychology-based emotional support and fine-grained essay assessment.

5   Conclusion

In this paper, we propose EduChat, an LLM-based chatbot system for intelligent education. By leveraging psychology and education theories, we enhance the model's educational functions like open QA, essay assessment, heuristic teaching, and emotional support. Through pre-training on educational corpus and fine-tuning with task-specific instructions, `EduChat` demonstrates great performance on the C-Eval benchmark. Overall, `EduChat` exhibits great potential towards revolutionizing intelligent education. In future work, we aim to expand `EduChat` on more functions, such as career planning, course guidance, question generation and so on.

Acknowledgments. This research is funded by the National Science and Technology Major Project (No. 2021ZD0114002), and the Science and Technology Commission of Shanghai Municipality Grant (No. 22511105901, No. 21511100402).

References

1. Baladn, A., Sastre, I., Chiruzzo, L., Ros, A.: RETUYT-InCo at BEA 2023 shared task: tuning open-source LLMs for generating teacher responses. In: Proceedings of the 18th Workshop on Innovative Use of NLP for Building Educational Applications (BEA 2023), pp. 756–765. Association for Computational Linguistics, Toronto, Canada, July 2023. https://aclanthology.org/2023.bea-1.61
2. Chen, K., Chen, Q., Zhou, J., He, Y., He, L.: DiaHalu: a dialogue-level hallucination evaluation benchmark for large language models. arXiv preprint arXiv:2403.00896 (2024)
3. Chen, M., et al.: Evaluating large language models trained on code. arXiv preprint arXiv:2107.03374 (2021)
4. Cui, J., Li, Z., Yan, Y., Chen, B., Yuan, L.: ChatLaw: open-source legal large language model with integrated external knowledge bases. ArXiv abs/2306.16092 (2023). https://api.semanticscholar.org/CorpusID:259274889
5. Dan, Y., Zhou, J., Chen, Q., Tian, J., He, L.: P-Tailor: customizing personality traits for language models via mixture of specialized LoRA experts. arXiv preprint arXiv:2406.12548 (2024)
6. Ellis, A.: The revised ABC's of rational-emotive therapy (RET). J. Rational-Emot. Cognitive-Behav. Ther. **9**(3), 139–172 (1991). https://doi.org/10.1007/BF01061227
7. Gu, Y., et al.: Enhancing depression-diagnosis-oriented chat with psychological state tracking. arXiv preprint arXiv:2403.09717 (2024)
8. Huang, Y., et al.: C-Eval: a multi-level multi-discipline Chinese evaluation suite for foundation models. arXiv preprint arXiv:2305.08322 (2023)
9. Ji, Y., et al.: Exploring the impact of instruction data scaling on large language models: an empirical study on real-world use cases. arXiv preprint arXiv:2303.14742 (2023)
10. Lee, K., et al.: Deduplicating training data makes language models better. arXiv preprint arXiv:2107.06499 (2021)
11. Li, Y., Li, Z., Zhang, K., Dan, R., Jiang, S., Zhang, Y.: ChatDoctor: a medical chat model fine-tuned on a large language model meta-AI (LLAMA) using medical domain knowledge. Cureus **15**(6) (2023)
12. Liu, S., et al.: Towards emotional support dialog systems. In: Proceedings of the 59th Annual Meeting of the Association for Computational Linguistics and the 11th International Joint Conference on Natural Language Processing (Volume 1: Long Papers), pp. 3469–3483. Association for Computational Linguistics, Online, August 2021. https://doi.org/10.18653/v1/2021.acl-long.269. https://aclanthology.org/2021.acl-long.269

13. Qin, Y., et al.: WebCPM: interactive web search for Chinese long-form question answering. arXiv preprint arXiv:2305.06849 (2023)
14. Schulman, J., et al.: ChatGPT: optimizing language models for dialogue. In: OpenAI blog (2022)
15. Stiennon, N., et al.: Learning to summarize with human feedback. In: Advances in Neural Information Processing Systems, vol. 33, pp. 3008–3021 (2020)
16. Sun, T., et al.: MOSS: training conversational language models from synthetic data (2023)
17. Tack, A., Kochmar, E., Yuan, Z., Bibauw, S., Piech, C.: The BEA 2023 shared task on generating AI teacher responses in educational dialogues. In: Proceedings of the 18th Workshop on Innovative Use of NLP for Building Educational Applications (BEA 2023), pp. 785–795. Association for Computational Linguistics, Toronto, Canada, July 2023. https://aclanthology.org/2023.bea-1.64
18. Touvron, H., et al.: LLaMA: open and efficient foundation language models. arXiv preprint arXiv:2302.13971 (2023)
19. Touvron, H., et al.: LLaMA 2: open foundation and fine-tuned chat models. ArXiv abs/2307.09288 (2023). https://api.semanticscholar.org/CorpusID:259950998
20. Wei, J., et al.: Finetuned language models are zero-shot learners. In: The Tenth International Conference on Learning Representations (2022)
21. Yang, H., Liu, X.Y., Wang, C.: FinGPT: open-source financial large language models. ArXiv abs/2306.06031 (2023). https://api.semanticscholar.org/CorpusID:259129734
22. Yao, B., et al.: D4: a Chinese dialogue dataset for depression-diagnosis-oriented chat. arXiv preprint arXiv:2205.11764 (2022)
23. Zhang, G., et al.: Chinese open instruction generalist: a preliminary release (2023)
24. Zhang, H., et al.: HuatuoGPT, towards taming language model to be a doctor. ArXiv abs/2305.15075 (2023). https://api.semanticscholar.org/CorpusID:258865566
25. Zhou, C., et al.: LIMA: less is more for alignment (2023)
26. Zhou, J., Ke, P., Qiu, X., Huang, M., Zhang, J.: ChatGPT: potential, prospects, and limitations. Front. Inf. Technol. Electron. Engineer., 1–6 (2023)

Manu-Eval: A Chinese Language Understanding Benchmark for Manufacturing Industry

Xiaoyi Liu$^{(\boxtimes)}$, Shuangtao Yang, Xiaozheng Dong, Honghui Rong, and Bo Fu

Lenovo Knowdee (Beijing) Intelligent Technology Co., Ltd., Beijing, China
{liuxy,yangst,dongxz,ronghh,fubo}@knowdee.com

Abstract. Large language models (LLMs) have demonstrated remarkable capabilities across various domains, fostering growing interest in their applications within the manufacturing industry. This paper presents a comprehensive evaluation of LLMs across 22 subcategories spanning 8 major sectors in the manufacturing industry, including machinery, automobile, electronics, chemical, light industry, pharmaceutical, transportation, and food manufacturing. To establish benchmark results on these tasks, we conduct a thorough evaluation of 20 top-performing LLMs. We anticipate that this evaluation will help analyze important strengths and shortcomings of both general domain models and domain-specific models, fostering their continued development and growth to better serve the Chinese manufacturing industry, thereby enabling AI-driven smart manufacturing. The dataset is available at https://github.com/KnowdeeAI/Manu-Eval.

Keywords: Manufacturing Industry · Large Language Models · Smart Manufacturing · LLM Benchmarking

1 Introduction

The manufacturing industry, a vital role in economic growth and technological innovation, is undergoing a transformative shift towards intelligent and automated systems. Artificial intelligence (AI), especially large language models (LLMs) which have demonstrated remarkable capabilities in understanding, reasoning, and generating natural languages, are being seen as potential powerful tools to drive this transformation.

The majority of existing LLMs are trained on broad, general-purpose data, lacking specialized and in-depth knowledge and reasoning abilities for specific industries and real-world scenarios. However, the manufacturing industry comprises a wide range of complex processes, operations, and domain-specific terminologies. Consequently, there remains a gap in comprehensively evaluating and benchmarking the performance of LLMs on manufacturing-specific

B. Xu et al. (Eds.): CCKS-IJCKG 2024, CCIS 2229, pp. 309–317, 2025.
https://doi.org/10.1007/978-981-96-1809-5_23

Table 1. Full list of tasks and Statistics of Manu-Eval

Supercategory	Task	# Question
机械制造业 Machinery	金属制品业 (Metal Products)	211
	通用设备制造业 (General Equipment)	228
	专用设备制造业 (Specialized Equipment)	179
	有色金属冶炼和压延加工业 (Non-Ferrous Metal Smelting and Rolling Processing)	153
	黑色金属冶炼和压延加工业 (Ferrous Metal Smelting and Rolling Processing)	188
汽车制造业 Automobile	汽车制造业 (Automobile)	182
电子制造业 Electronics	计算机、通信和其他电子设备制造业 (Computer, Communications and Other Electronic Equipment)	213
	仪器仪表制造业 (Instrument and Apparatus)	198
	电气机械和器材制造业 (Electrical Machinery and Equipment)	262
化工制造业 Chemical	化学原料和化学制品制造业 (Chemical Raw Materials and Chemical Products)	249
	化学纤维制造业 (Chemical Fiber)	88
	橡胶和塑料制品业 (Rubber and Plastic Products)	189
	石油、煤炭及其他燃料加工业 (Petroleum Coal and Other Fuel Processing)	188
轻工制造业 Light Industry	纺织业 (Textile)	130
	纺织服装、服饰业 (Textile Clothing and Apparel)	26
	造纸和纸制品业 (Papermaking and Paper Products)	24
	印刷和记录媒介复制业 (Printing and Record Media Reproduction)	43
	非金属矿物制品业 (Mon-Metallic Mineral Products)	198
医药制造业 Pharmaceutical	医药制造业 (Pharmaceutical)	219
运输设备制造业 Transportation	铁路、船舶、航空航天和其他运输设备制造业 (Railroads, Ships, Aerospace, and Other Transportation Equipment)	196
食品制造业 Food	农副食品加工业 (Agricultural and Sideline Food Processing)	71
	酒、饮料和精制茶制造业 (Wine, Beverage and Refined Tea)	22
Total		3457

tasks and knowledge areas. Existing general language understanding benchmarks [10,12,14,17–20], while valuable for assessing core language skills, fall short in capturing the nuances and complexities inherent to the manufacturing domain.

In this paper we introduce Manu-Eval, a Chinese benchmark specifically designed to evaluate the multi-task language understanding capabilities of LLMs within the manufacturing industry. Our benchmark comprises a diverse set of 22 subcategories spanning 8 major manufacturing sectors, including machinery, automobile, electronics, chemical, light industry, pharmaceutical, transportation, and food manufacturing. Through Manu-Eval, we conduct an extensive evaluation of 20 LLMs, covering top-performing commercial and open-source models, and multilingual and Chinese-oriented models, providing a comprehensive assessment of their strengths, limitations, and areas for improvement in handling manufacturing specific knowledge.

2   Related Work

2.1   Comprehensive Benchmarks

Large language models (LLMs) have demonstrated impressive general abilities, achieving state-of-the-art results across a wide range of natural language processing tasks. To evaluate and track their capabilities, various general language understanding benchmarks have been developed, such as GLUE [18], which combines a collection of natural language understanding (NLU) tasks, and Super-GLUE [17], which features a new set of more challenging NLU tasks styled after GLUE. For Chinese language evaluation, widely adopted benchmarks include CLUE [19], a benchmark for Chinese NLU tasks, and SuperCLUE [20], specifically designed to assess the performance of LLMs.

Asides from benchmarks for NLU capacities, the MMLU Benchmark [10] provides a multi-domain and multi-task test covering 57 subjects to evaluate models' world knowledge and problem solving ability. Several Chinese benchmarks have followed the MMLU style, such as C-Eval [12], designed to assess advanced knowledge and reasoning skills of models, and CMMLU [14], which includes more Chinese-specific topics.

2.2   Domain Specific Evaluation

Recently, SuperCLUE-Industry [1] provides a benchmark focus on the industry field. They evaluate models industrial knowledge and abilities using multiple tasks, question-answering, computation, and code generation, as well as applied abilities like data analysis, document question-answering, and acting as intelligent industrial agents.

China Academy of Industrial Internet released a comprehensive evaluation report on LLMs for industrial application [3]. They conducted evaluation on multiple LLMs across different tasks, including industrial knowledge question-answering, data analysis, engineering modeling, document generation, and code understanding. The report also publishes performance scores across different tasks, analyzing the strengths and limitations of each model. This provides enterprises with a valuable reference for selecting suitable industrial LLMs.

Compare to these benchmarks and evaluations, we focus exclusively on the manufacturing industry, aimed for a comprehensive evaluation of LLMs' domain-specific knowledge within manufacturing contexts. Moreover, our dataset will be publicly accessible, enabling broader research efforts towards developing robust industrial LLMs.

3 Manu-Eval

3.1 Overview

The goal of Manu-Eval is to provide a way for developers to quickly assess their models' knowledge in the manufacturing industry. By creating standardized test sets covering real-world manufacturing problems and an efficient evaluation system, we aim to drive the continuous improvement and industrial adoption of LLMs for smart manufacturing applications. We have created an extensive multi-task evaluation for the Chinese manufacturing industry, spanning 8 major sectors like machinery, automobiles, electronics, chemicals, light industries, pharmaceuticals, transportation equipment, and food manufacturing. And our goal is to cover the knowledge of the entire manufacturing pipeline for each category - research and development (R&D), production, supply chain, sales, and after-sales service, although R&D and production consist the major part of our current dataset.

3.2 Data Collection

Task Selection. We selected task subjects from the "Manufacturing Industry" category of the Industrial Classification for National Economic Activities [2] published by China's National Bureau of Statistics. From this, we chose 22 specific manufacturing industries and reorganized them into 8 broader sectors. The full list of tasks and their assigned categories is shown in Table 1.

Data Source. The core data for this benchmark is sourced from professional exams and evaluation materials specific to the manufacturing industry. Given the highly specialized and domain-specific nature of industrial applications, we primarily relied on questions and answers from professional qualification exams, vocational skills assessments, and national standards/guidelines related to manufacturing processes and operations.

Data Processing and Quality Check. To ensure high-quality and reliable evaluation data, we employed both human and automatic approach. We first use LLMs to convert non-multiple-choice QA pair into multiple-choice format. Then, an automated data cleaning pipeline filtered out invalid or malformed questions and choices. We then used LLMs gain to automatically classify the remaining questions into appropriate subject categories under our 22 chosen subject areas. To verify this categorization, we hired two annotators with college degrees or higher to manually review and confirm the classifications. To further ensure data quality and answer accuracy, we hired three additional annotators with direct manufacturing industry experience. These annotators used their own expert knowledge and online resources to validate the questions and verify the correctness of the provided answers.

4   Experiment

4.1   Setup

We evaluate LLMs only in zero-shot settings. This allows for an unbiased measurement of the models' out-of-the-box capabilities in handling industry-specific tasks. For all models, both commercial models and open-source models, we use the model response to get the answer, instead of the highest logits of the next predicted token of the four choices. Regular expressions is used to extract the answer choices from the model responses. For open-source models, we use vLLM [13] to create OpenAI compatible servers. This ensure us a consistent evaluation pipeline across commercial and open-source models. For prompt, we keep it simple and concise, following the format exemplified in Fig. 1.

请直接输出正确的答案选项。
(Please provide the correct answer choice directly.)
问题：将共析钢加热到 () 线温度以上，其珠光体组织将转变为奥氏体。
When hypoeutectoid steel is heated above the () line temperature, its pearlite structure will transform into austenite.
选项：
Options:
A. A1
B. A3
C. Acm
D. A4

Fig. 1. Zero-shot prompt example from Ferrous Metal Smelting and Rolling Processing Task.

4.2   Models

To provide a comprehensive overview of LLM performance in the Chinese manufacturing context, we evaluate Manu-Eval on 20 top-performing multilingual and Chinese-oriented LLMs from various organizations and with different model sizes, as listed in Table 2. For multilingual LLMs, we selected the Claude family [5], gpt-3.5-turbo-0125 and gpt-4-turbo from the OpenAI family, LLaMA3-8B/70B [16] by Meta, and Gemini-1.5 [15] introduced by Google. For Chinese-oriented model, we selected multiple sized model from the Qwen1.5 [4] family developed by Alibaba, ChatGLM3-6B [9], both SFT and DPO versions of MiniCPM [11], open-sourced deepseek-llm-67b-chat [7] and API DeepSeek-V2 [8] from DeepSeek-ai, Baichuan2-7B/13B [21] released by Baichuan-inc, and the 128k version of newly released ERNIE-speed [6] by Baidu. We also evaluated our own model, Zpoint-Mini which was trained on data covering multiple manufacturing areas. We only present chat version of models in the evaluation, since we found most base models struggle to follow instructions in a zero-shot setting, making it difficult to extract answers and resulting in poor performance. The access methods for models are listed in the table. "API" indicates that we evaluated the model by requesting the provider's API, while "Weight" denotes that we evaluated the model offline using its weights. Due to limited GPU resources, we evaluate some large models in their quantized format, indicated by an asterisk (*) in the table.

4.3 Results and Analysis

By Model. Table 3 presents the performance of all evaluated models under the zero-shot setting on our Manu-Eval benchmark. Several notable observations can be made.

Among open-source models, the Chinese-oriented Qwen1.5-72B-Chat and its smaller counterpart Qwen1.5-32B-Chat from Alibaba exhibit the strongest performance, achieving average accuracy of 88.4% and 81.65% respectively. These models outperform all other evaluated models, underscoring the effectiveness of their Chinese language understanding capabilities and manufacturing-specific knowledge.

For commercial models, the multilingual gemini-1.5-pro and the Chinese-oriented DeepSeek-V2 achieve an average accuracy of 78.15% and 77.63%. Unexpectedly, OpenAI's gpt-4-turbo, despite its impressive general capabilities, scores lower at 72.46% accuracy, ranking fifth across all models evaluated.

Contrary to the common assumption that larger models perform better, our results showcase that the Qwen family, with a maximum of 72B parameters, outperforms several larger commercial models. In addition, the 7B Qwen1.5-7B-Chat (69.22%) outperform the 70B LLaMA-3B-70B (64.45%) in accuracy, and smaller models like Zpoint-Mini (63.9%) also demonstrate comparable results. We attribute this to the quality and relevance of the training data used for these Chinese-oriented models, which likely aligns better with Chinese manufacturing domain.

By Task. At the last line of Table 3, we present the average accuracy of each supercategories. We observe a significant variation in model performance depending on the specific manufacturing domain. Food manufacturing emerges as the

Table 2. Models evaluated in this paper on Manu-Eval. "*" indicates INT4 version of the model is used.

Model	Organization	Access
claude-3-opus-20240229	Anthropic	API
deepseek-chat	deepseek	API
ernie-speed-128k	Baidu	API
gemini-1.5-pro	Google	API
gpt-3.5-turbo-0125	OpenAI	API
gpt-4-turbo	OpenAI	API
Baichuan2-7B-Chat	baichuan-inc	Weight
Baichuan2-13B-Chat	baichuan-inc	Weight
chatglm3-6b	Tsinghua	Weight
deepseek-llm-67b-chat*	deepseek	Weight
llama3-70b*	Meta	Weight
llama3-8b	Meta	Weight
MiniCPM-2B-DPO	ModelBest	Weight
MiniCPM-2B-SFT	ModelBest	Weight
Qwen1.5-1.8B-Chat	Alibaba	Weight
Qwen1.5-32B-Chat*	Alibaba	Weight
Qwen1.5-72B-Chat*	Alibaba	Weight
Qwen1.5-7B-Chat	Alibaba	Weight
Qwen1.5-4B-Chat	Alibaba	Weight
Zpoint-Mini	Knowdee	Weight

category where LLMs perform the best, achieving an impressive average accuracy of 74.64% across all models. This could be attributed to the relatively simpler and more common used in this supercategory, making it more accessible for current LLMs. Another factor could be that there are currently relatively fewer cases in this supercategory. The automobile and machinery manufacturing supercategories seem more challenging, with average accuracy of 54.31% and 56.57%, respectively. These two categories often involve intricate technical terminology, complex processes, and specialized knowledge, which LLMs may struggle to grasp without specific training or adaptation. These findings highlight the inherent complexity and diversity within the manufacturing industry, emphasizing the need of developing specialized LLMs that can effectively capture and leverage the domain-specific nuances of each manufacturing sub-industries.

Table 3. Zero-shot average accuracy on Manu-Eval.

Model	Mach.	Auto.	Elec.	Chem.	Light	Phar.	Trans.	Food	Avg.
Qwen1.5-72B-Chat	84.38	79.67	87.91	81.28	94.62	90.41	92.35	99.3	88.4
Qwen1.5-32B-Chat	75.69	75.82	82.38	75.01	88.95	81.74	82.65	92.8	81.65
gemini-1.5-pro	71.64	65.93	78.21	74.43	85.79	74.89	71.94	93.5	78.15
DeepSeek-V2	72.51	70.33	78.54	73.64	81.75	79.91	78.06	89.12	77.63
gpt4-turbo	65.52	67.58	73.41	68.55	80.31	70.32	66.33	83.16	72.46
claude-3-opus-20240229	65.04	58.24	72.89	67.38	73.84	68.49	65.82	78.94	69.68
ERNIE-speed-128k	61.28	59.89	68.24	67.95	77.91	64.38	62.24	84.73	69.48
Qwen1.5-7B-Chat	64.39	62.64	72.11	65.15	70.86	69.86	68.88	84.19	69.22
Meta-Llama-3-70B-Instruct	55.87	54.4	65.73	60.73	70.37	65.75	59.69	83.32	64.45
Zpoint-Mini	60.34	58.79	67.03	59.12	68.85	66.67	63.78	66.49	63.9
Qwen1.5-4B-Chat	58.9	51.1	66.4	57.16	71.39	57.08	57.65	71.57	63.1
deepseek-llm-67b-chat	47.42	48.35	56.64	51.73	64.7	52.51	52.55	68.6	55.82
Qwen1.5-1.8B-Chat	48.76	46.15	55.58	49.59	59.64	51.14	47.45	58.58	53.14
MiniCPM-2B-dpo-bf16	45.47	42.86	52.25	48.6	60.83	46.58	44.9	72.44	52.81
gpt-3.5-turbo-0125	45.88	46.15	56.11	49.47	57.72	45.66	50.51	67.19	52.77
Meta-Llama-3-8B-Instruct	46.33	45.6	53.61	47.64	60.39	47.49	48.47	65.24	52.59
MiniCPM-2B-sft-bf16	45.79	41.76	51.14	47.79	61.65	47.95	43.37	68.05	52.31
Baichuan2-7B-Chat	41.4	40.66	50.3	42.73	51.68	38.81	43.88	59.67	46.81
Baichuan2-13B-Chat	41.16	39.01	44.53	36.93	44.52	41.1	42.35	55.06	42.83
chatglm3-6b	33.58	31.32	39.19	36.78	50.71	36.53	35.71	50.83	40.52
AVG.	56.57	54.31	63.61	58.08	68.82	59.86	58.93	74.64	-

4.4 Conclusion

In this work, we introduce Manu-Eval, a Chinese benchmark designed to assess LLMs' multi-task capabilities within the manufacturing industry. Our extensive evaluation across 20 advanced LLMs reveals opportunities for improvement, underscoring the need for continued research in this domain. We anticipate our benchmark will catalyze advancements in AI-driven manufacturing and help developments of industrial LLMs made for the manufacturing sector.

While Manu-Eval covers a broad range of manufacturing areas, we consider it as first step since we only evaluate models' knowledge of this domain. There are many other critical aspects and abilities, such as mechanical design, troubleshooting, process optimization, and more, that require more in-depth domain knowledge for designing a comprehensive evaluation. We leave further exploration on evaluation of those aspects and applications of manufacturing domain knowledge in real-world scenarios for future work.

References

1. Superclue-industry: 中文原生工业大模型测评基准. https://www.cluebenchmarks.com/superclue_industry.html
2. 2017 年国民经济行业分类 (gb/t 4754-2017), July 2017. https://www.stats.gov.cn/xxgk/tjbz/gjtjbz/201710/t20171017_1758922.html
3. 重磅发布丨人工智能大模型工业应用准确性测评-中国工业互联网研究院_2024, March 2024. https://www.china-aii.com/newsinfo/6875620.html?templateId=1562263
4. Alibaba: Introducing qwen1.5, February 2024. https://qwenlm.github.io/blog/qwen1.5/
5. Anthropic: Introducing Claude. Anthropic blog (2022). https://www.anthropic.com/
6. Baidu. https://cloud.baidu.com/doc/WENXINWORKSHOP/s/6ltgkzya5
7. Deepseek. https://huggingface.co/deepseek-ai/deepseek-llm-67b-chat
8. DeepSeek. https://www.deepseek.com/
9. Du, Z., et al.: GLM: general language model pretraining with autoregressive blank infilling (2022)
10. Hendrycks, D., et al.: Measuring massive multitask language understanding (2021)
11. Hu, S., et al.: MiniCPM: unveiling the potential of small language models with scalable training strategies (2024)
12. Huang, Y., et al.: C-Eval: a multi-level multi-discipline Chinese evaluation suite for foundation models (2023)
13. Kwon, W., et al.: Efficient memory management for large language model serving with PagedAttention (2023)
14. Li, H., et al.: CMMLU: measuring massive multitask language understanding in Chinese (2024)
15. Team, G.: Gemini 1.5: unlocking multimodal understanding across millions of tokens of context (2024)
16. Touvron, H., et al.: LLaMA: open and efficient foundation language models (2023)
17. Wang, A., et al.: SuperGLUE: a stickier benchmark for general-purpose language understanding systems (2020)
18. Wang, A., Singh, A., Michael, J., Hill, F., Levy, O., Bowman, S.R.: GLUE: a multi-task benchmark and analysis platform for natural language understanding (2019)
19. Xu, L., et al.: CLUE: a Chinese language understanding evaluation benchmark. In: Scott, D., Bel, N., Zong, C. (eds.) Proceedings of the 28th International Conference on Computational Linguistics, pp. 4762–4772. International Committee on Computational Linguistics, Barcelona, Spain (Online), December 2020. https://doi.org/10.18653/v1/2020.coling-main.419. https://aclanthology.org/2020.coling-main.419
20. Xu, L., et al.: SuperCLUE: a comprehensive Chinese large language model benchmark (2023)
21. Yang, A.M., et al.: Baichuan 2: open large-scale language models. ArXiv abs/2309.10305 (2023). https://api.semanticscholar.org/CorpusID:261951743

A Comprehensive Ontology Knowledge Evaluation System for Large Language Models

Xiaotong Qin[1,2], Tong Zhou[1,2], Yubo Chen[1,2(✉)], Kang Liu[1,2], and Jun Zhao[1,2]

[1] The Laboratory of Cognition and Decision Intelligence for Complex Systems, Institute of Automation, Chinese Academy of Sciences, Beijing, China
{qinxiaotong2022,tong.zhou}@ia.ac.cn, {yubo.chen,kliu,jzhao}@nlpria.ac.cn
[2] School of Artificial Intelligence, University of Chinese Academy of Sciences, Beijing, China

Abstract. Large Language Models (LLMs) have acquired vast amounts of knowledge through extensive pre-training on large corpora. However, the depth of their understanding of ontology knowledge remains unclear. Ontology is about what types of entities exist, how they are grouped into categories, and how they are related to one another. Ontology knowledge provides a standardized approach to knowledge representation that aligns with human cognition. This study aims to analyze the comprehension and mastery of ontological knowledge within LLMs. In this paper, we proposes a comprehensive benchmark that includes generative ontology knowledge data, focusing both on classes and properties. We develop an extensive evaluation framework and design various tasks aiming to evaluate the memorization and utilization of ontological knowledge. The experimental results indicate that, while LLMs can memorize ontological knowledge, they struggle with truly comprehending and utilizing it effectively.

Keywords: Large language models · Ontology knowledge · Model evaluation

1 Introduction

Ontology is an abstract conceptual model derived from the objective world, encompassing classes, properties, and their relationships [9]. It offers a standardized method for describing knowledge that aligns with human cognition and facilitates effective communication among people [3,8,10,15].

Learning from their extensive pre-training corpora, Large Language Models (LLMs) have exhibited exceptional performance in encoding vast amounts of knowledge [7,12,17]. Many researches [12,14] has focused on analyzing the factual knowledge within LLMs. For instance, to extract the capital of France, we can query language models like BERT with *The capital of France is [MASK]*,

© The Author(s), under exclusive license to Springer Nature Singapore Pte Ltd. 2025
B. Xu et al. (Eds.): CCKS-IJCKG 2024, CCIS 2229, pp. 318–328, 2025.
https://doi.org/10.1007/978-981-96-1809-5_24

where France is the subject. However, factual knowledge is considered in isolation, because it typically involves discrete pieces of information [13], which lacks logical connections between individual pieces of information. From the perspective of human cognition [4,20], it is important to explore the relationships and organization of the knowledge these models possess. Thus, our research focuses on analyzing the understanding and mastery of ontological knowledge within LLMs, and further exploring the intrinsic connections and structure of knowledge in models.

Integrating ontology knowledge into large language models can substantially enhance their performance in specific tasks [2]. A comprehensive understanding of hierarchical structures and properties of classes (or concept) can improve the models' and inferential capabilities based on context [6,11]. As shown in Fig. 1, in a medical diagnosis system, understanding the hierarchical relationships between symptoms, diseases, and treatments can help the model provide more precise diagnoses and recommend appropriate treatments. This integration can result in more accurate and contextually relevant responses, thereby increasing the utility of these models in practical applications. Consequently, a systematic evaluation of ontology knowledge within large language models is essential. Such an evaluation allows us to understand how well these models internalize and utilize ontological structures and relationships. This insight can guide the development of more robust models and applications that leverage ontology knowledge more effectively, resulting in improved performance and utility.

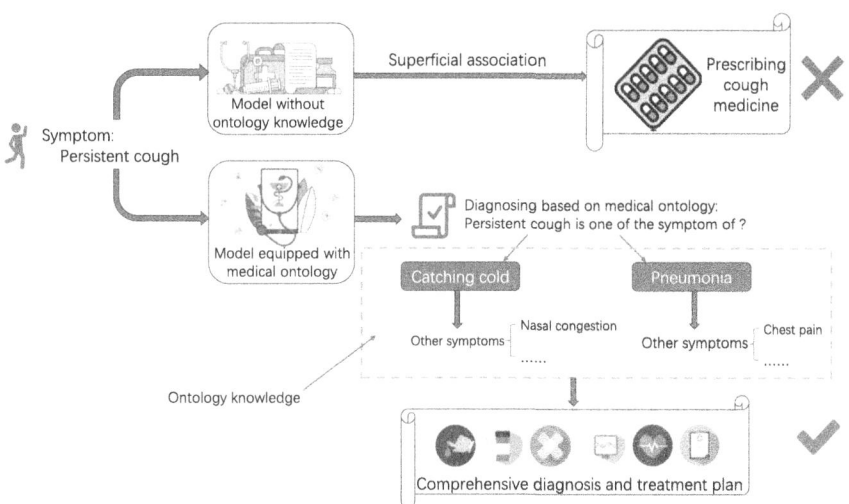

Fig. 1. Ontological knowledge is essential in practical applications. If a model knows that a persistent cough is a symptom of both the common cold and pneumonia but can differentiate based on additional context like fever and chest pain, it can give more accurate and useful responses.

Previous research [19] has explored ontology knowledge by designing memory and reasoning tasks. They have found that encoder-based PLMs, like BERT, can memorize certain ontology knowledge and follow ontology entailment rules for reasoning based on implicit knowledge. These findings indicate that PLMs possess a certain degree of understanding of ontology knowledge. However, there are two main limitations in their research. First, the reasoning tasks typically involve **superficial concept replacements** and do not reflect ontology-specific application scenarios. For example, given the premise *one has to be a person to be a player at a sports team, and [X] is a player at a sport team*, any *[X]* would be considered *person*. This does not reflect model's mastery of the properties in ontological knowledge. Second, there is **ambiguity in probing** the memorization of ontology knowledge using fill-in-the-blank methods. For instance, when presented with the prompt *Messi is [MASK]*, the model might struggle with the intended meaning, as words like *a football player*, *male* as well as *famous* could all fit the context. These confusing probings might misestimate the model's capabilities. Additionally, existing studies fail to take account in the **implicit application of ontology knowledge** across different levels of difficulty.

To address these issues, we aim to provide a more comprehensive and mensurable evaluation of ontology knowledge within LLMs. In this paper, we propose a comprehensive benchmark that includes general ontology knowledge data, focusing on the perspectives of classes hierarchy and properties. Classes hierarchy encompasses the subordination relationships between instances and their corresponding classes, as well as the hierarchical relationships among classes. On the other hand, properties encompass the definitions of each class, as well as other characteristics associated with entities. Meanwhile, we design various tasks related to ontology knowledge. Taking into account the objectivity of our evaluation tasks and the specification to the ontological knowledge, we carefully design the options in multiple-choice format to avoid ambiguity. Through knowledge evaluation on multiple LLMs, we aim to assess their understanding and application of ontology knowledge[1]. Our main contributions are summarized as follows:

1) Comprehensive Evaluation Framework: We develop an extensive evaluation framework for ontology knowledge, including classes hierarchy and properties.
2) Hierarchical Evaluation Tasks: We design 17 distinct evaluation tasks to hierarchically assess a model's understanding of ontology knowledge from both memory and utilization perspectives.
3) Model Evaluation and Insights: We evaluate 7 models covering 3 architectures and different parameter scales using our framework. Our evaluation find that LLMs can memorize certain ontology knowledge but cannot utilize it effectively.

---

[1] Our source codes with corresponding experimental datasets will be openly available at https://github.com/qinxt2022/OntologyKnowledgeProbing.

2   Benchmark

In this section, we introduce the benchmark for the ontology knowledge evaluation tasks. Considering the hierarchical class relationships and properties in ontology, we build ontology schema covering general classes and their corresponding properties. Meanwhile, inspired by Bloom's taxonomy [18], we design memorization and utilization tasks for class hierarchy and property respectively. Total of 4 types of tasks are included in our benchmark.

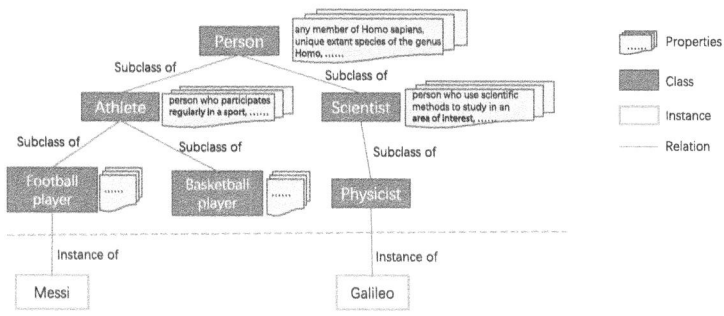

Fig. 2. An example of ontology schema.

2.1   Ontology Schema

The ontology schema is as Fig. 2 shows. We begin our project by leveraging DBpedia [1] to construct an initial concept schema, identifying a comprehensive set of 783 classes. Recognizing the need for clarity and organization, we manually curated the schema to ensure clear and logical hierarchical relationships, ultimately refining it to 370 well-defined classes.

To enrich the schema, we gather representative instances for each class. Utilizing SPARQL queries, we extracted 20 instances per class. In the final phase, we acquired descriptions of each class from Wikidata, which constitute the property data of the classes.

2.2   Evaluation Tasks

Our evaluation targets include investigating the relationships between instances and their corresponding classes, the hierarchical relationships among classes, the characteristics and the properties of each class. We address both the memorization and utilization of ontology knowledge through four distinct task types. To guarantee the model fully understands the intent of the input and to avoid ambiguity during model generating, we predominantly used multiple-choice formats in the tasks.

Table 1. Task examples of Class Hierarchy Memorization Tasks

Task type	Examples	Ground truth
IBJ	**What category does the entity 'Lionel Messi' belong to?** **A.** football player **B.** basketball player **C.** swimmer **D.** boxer	A
ITJ	**What category does the entity 'Lionel Messi' belong to?** **A.** person **B.** event **C.** device **D.** organization	A
BIJ	**Which of the following options is 'novel'?** **A.** Pride and Prejudice **B.** The Freewheelin' Bob Dylan **C.** Google Maps **D.** Lord's Prayer	A
TIJ	**Which of the following options is 'work'?** **A.** Pride and Prejudice **B.** Raphael **C.** All Saints' Day Flood of 1436 **D.** Hombolo Dam	A
BMJ	**What category does the entity 'bank' belong to?** **A.** sports league **B.** educational institution **C.** company **D.** performing group	C
BTJ	**What category does the entity 'bank' belong to?** **A.** person **B.** animal **C.** organisation **D.** work	C
MBJ	**Which of the following options can be classified as 'reptile'?** **A.** mosquito **B.** seagull **C.** skink **D.** snapper	C
TBJ	**Which of the following options is 'animal'?** **A.** tilia **B.** national basketball association **C.** ice cream **D.** skink	D

Table 2. Task examples of our four types of Class Hierarchy Utilization Tasks

Task type	Examples	Ground truth
NCJ	**Please identify the entity which one is not of the same type as others:** **A.** Kobe Bryant **B.** Lionel Messi **C.** Diego Maradona **D.** Johan Cruyff	A
SORT	**Order the subsequent items from the category that encompasses the most to the one that encompasses the least:** **A.** football player **B.** athlete **C.** Messi **D.** person	D B A C
NLJ	**Determine the option that differs in category level from the others:** **A.** athelete **B.** Lionel Messi **C.** Diego Maradona **D.** Johan Cruyff	A

Table 3. Task examples of our four types of Property Knowledge Memorization Tasks

Task type	Examples	Ground truth
CDJ	**Which of the following characteristics could 'song writer' have?** **A.** person who writes the words or music to songs **B.** facility built for racing of animals, vehicles, or athletes **C.** human-designed and -made structure **D.** fundamental facilities and systems serving a country, city, or other areas	A
DCJ	**If 'X' is defined as 'person who writes the words or music to songs', which category does 'X' fall into?** **A.** planet **B.** vegetable **C.** soccer player **D.** song writer	D
SRJO	**If[X] competes for [Y] and [X] is a(an) athlete, [Y] can be:** **A.** food **B.** building **C.** animal **D.** football team	D
ROJS	**If[X] competes for [Y] and [Y] is a(an) sports team, [X] can be:** **A.** food **B.** cat **C.** dam **D.** basketball player	D

Table 4. Task examples of our four types of Property Knowledge Utilization Tasks

Task type	Examples	Ground truth
NOJ	**If [X] competes for [Y] and [X] is a(an) athlete, [Y] cannot be:** **A.** football team **B.** track and field team **C.** basketball team **D.** food	D
NSJ	**If [X] competes for [Y] and [Y] is a(an) sports team, [X] cannot be:** **A.** football player **B.** swimmer **C.** basketball player **D.** food	D

Class Hierarchy Evaluation Tasks. These tasks evaluate hierarchical classes relationships in LLMs, and cover two scenarios: memorization and utilization.

For memorization, all of the inputs are of multiple choice format. On the one hand, given a query entity, this task requires LLMs to select the class that best matches this entity. On the other hand, given a class name, to select an entity that belongs to this class. The same group of options belongs to the same taxonomy layer, and we set option groups at different taxonomy layers to detect the model's ability to classify at various levels of abstraction. For example, for the input *What category does Messi belong to?* there are two sets of options: one includes *basketball player, football player, swimmer* and *boxer*, while the other includes *person, event, device* and *organization*. This task could evaluate not only whether LLMs can memorize entities and their corresponding class but also their ability to distinguish classes at different levels of abstraction. Task examples are shown in Table 1. The tasks includes Instance-Bottom-Class-Judgement (IBJ, choose bottom class for given instance), Instance-Top-Class-Judgement (ITJ, choose top class for given instance), Bottom-Class-Instance-Judgement (BIJ, choose instance for given bottom class), Top-Class-Instance-Judgement (TIJ, choose instance for given top class), Bottom-Class-Middle-Class-Judgement (BMJ, choose middle class for given bottom class), Bottom-Class-Top-Class-Judgement (BTJ, choose top class for given bottom class), Middle-Class-Bottom-Class Judgement (MBJ, choose bottom class for given middle class), and Top-Class-Bottom-Class-Judgement (TBJ, choose bottom class for given top class), where the bottom class is the most fine-grained class, the top class means the most abstract and fundamental class, and the middle class is between above two according general classification method.

For utilization, the input formats include multiple-choice questions and sorting tasks. The multiple-choice tasks require the LLMs to either select the option with different classes or hierarchical levels from others, while the sorting tasks ask LLMs to order given class names by their hierarchical ranking. These tasks help determine whether the models can handle tasks using different levels of abstraction of objects or entailment relationships between different classes. Task examples are shown in Table 2. The tasks includes Not-Same-Class-Judgement (NCJ, choose the entity of different type as others), SORT (sort the items according their layer in class hierarchy), and Not-Same-Layer-Judgement (NLJ, choose the entity of different layer in class hierarchy as others).

Property Evaluation Tasks. These tasks evaluate the definition and characteristics of certain class in LLMs, and also cover two scenarios: memorization and utilization. The properties considered include definitions and other characteristics specific to each class. The input format is multiple-choice questions.

For memorization, given a class name and a set of definitions, the models should select the most appropriate definition for that class. Conversely, given a definition and a set of class names, the models should select the most appropriate class. Task examples are shown in Table 3. The tasks includes Class-Definition-Judgement (CDJ, choose the definition of given class), Definition-Class-Judgement (DCJ, choose the class according given definition), Subject-Relation-Judge-Object (SRJO, choose the suitable object for given subject and relation), and Relation-Object-Judge-Subject (ORJS, choose the suitable subject for given object and relation).

For utilization, given a premise, the models need to select the possible subject or object. This task helps understand whether models can apply ontological knowledge to specific situations, which may be related to the rationality of the reasoning process. Task examples are shown in Table 4. The tasks includes Negative-Object-Judgement (NOJ, choose the unsuitable object for given subject and relation), and Negative-Subject-Judgement (NJS, choose the unsuitable subject for given object and relation).

3 Evaluation Setup

We introduce the various LLMs investigated in our experiments and the three evaluation methods we adopted.

3.1 Models

We primarily evaluate three types of LLMs: (1) Llama-2 series [16], including Llama-2-7b, Llama-2-7b-chat, Llama-2-13b, and Llama-2-13b-chat. (2) Mistral [5], including Mistral-7b-v0.1, and Mistral-7B-Instruct-v0.2. (3) GPT-3.5 (GPT-3.5-turbo API).

3.2 Evaluation Methods

We manually designed natural language inputs for each evaluation task, providing 5-shot examples for the models in each task. The examples will help the LLMs better grasp our intent and become more familiar with the multiple-choice format, improving the reliability of the evaluation results. Moreover, To better quantify the ontology knowledge inside the LLMs, we generate 5 different paraphrases for the diversity of input.

4 Experiments and Results

In this section, we present the performance of LLMs on all tasks and analyze the results to propose 5 main findings.

Table 5. Accuracy(%) of all models on each evaluation task. The average values of four types of tasks are shown respectively.

Scenario	Task type	Mistral-7B	Mistral-7B-instruct	llama-2-7b	llama-2-7b-chat	llama-2-13b	llama-2-13b-chat	GPT-3.5
Class Hierarchy Memorization	IBJ	81.3	81	71.1	77.3	81.9	80.2	87.0
	ITJ	78.9	75.3	46.8	64.5	64.8	67.6	80.4
	BIJ	80.2	80.6	53.6	64.3	78.2	73.8	85.3
	TIJ	75.0	87.5	45.0	52.5	45.0	87.5	100.0
	BMJ	92.0	90.8	70.0	86.5	89.8	90.2	94.6
	BTJ	92.4	91.1	74.5	84.4	84.7	86.7	93.5
	MBJ	94.5	90.6	75.5	87.1	88.9	87.1	97.0
	TBJ	88.0	86.1	62.2	73.9	76.3	78.3	91.0
	Average	85.3	85.4	62.3	73.8	76.2	81.4	91.1
Class Hierarchy Utilization	NCJ	55.5	64.1	26.7	40.2	29.1	43.7	78.7
	SORT	29.5	31.7	10.2	9.4	10.1	11.1	74
	NLJ	62.4	34.8	22.8	33.4	34.3	37.2	43.2
	Average	49.1	43.5	19.9	27.7	24.5	30.7	65.3
Property Memorization	CDJ	95.5	94.4	74.6	87.4	90.8	91.1	93.6
	DCJ	95.4	94.6	71.3	93.6	94.4	94.7	94.9
	SRJO	90.4	80.8	55.8	75	71.1	78.8	92.3
	ROJS	25	13.5	51.9	51.9	38.5	32.7	30.8
	Average	76.6	70.8	63.4	77.0	73.7	74.3	77.9
Property Utilization	NOJ	57.1	46.4	35.7	17.9	32.1	46.4	46.4
	NSJ	52.6	68.4	15.8	18.4	31.6	47.4	36.8
	Average	54.85	57.4	25.75	18.15	31.85	46.9	41.6

4.1 Ontology Evaluation Tasks Can Reflect the Models' Capabilities

The overall experimental results are shown in Table 2. In general, different model families exhibit varying levels of performance.

First of all, GPT-3.5 demonstrates the highest performance across all models. Meanwhile, among the open-source models, Mistral-7B stands out as the best performer. According to the technical report of Mistral-7B [5], which assert superior performance to LLaMA2-13B, we can conclude that our ontology evaluation tasks can reflect the models' capabilities.

4.2 LLMs Face Challenges in Utilizing Ontological Knowledge

Notably, all of the models struggle more with utilization tasks compared to memorization tasks. This suggests that while LLMs have acquired ontological knowledge, they still face significant challenges in utilizing this knowledge correctly.

These findings highlight a critical gap: although LLMs have successfully learned ontological relationships, effectively leveraging this knowledge in practical scenarios remains an area for improvement. This underscores the importance of developing techniques and approaches that enhance the utilization of learned ontological knowledge, thereby improving the overall performance of language models in complex tasks (Table 5).

4.3 Instruction Fine-Tuning Does Not Enhance Ontological Knowledge Mastery

Comparing the base models with their chat versions, we observe that the chat versions of the LLaMA-2 series show a comparably stable improvements. However, Mistral-7B-Chat does not consistently outperform its base model. This leads to a conclusion that instruction fine-tuning does not necessarily enhance a model's grasp of ontological knowledge. We propose that weaker models like LLaMA-2, instruction fine-tuning may improve the model's ability to understand the input and generate responses that follow the given format, which doesn't apply for all types of models. Yet, it may not necessarily improve the depth of comprehension or the effective utilization of knowledge. To more effectively handle ontological knowledge, we must explore alternative, more potent strategies such as knowledge retrieval or knowledge editing techniques.

4.4   LLMs Perform Worse with Abstract Classes in Ontology

When comparing IBJ task and ITJ task, we observe that models perform worse when dealing with abstract categories (the options of ITJ task include higher-level classes). That means, determining that *Messi is a person* is more challenging for the model than determining that *Messi is a football player*, even though the former is a more certain fact for humans. Similar phenomena are found when comparing BIJ task and TIJ task, BMJ task and BTJ task, along with MBJ task and BTJ task.

We attribute this to two main reasons. Firstly, entities with a large hierarchical gap are less likely to appear together in the pre-training corpus, making it difficult for the model to establish connections between them. Secondly, it is difficult for LLMs to understand more abstract ontological knowledge. The more abstract the concept, the harder it is for the model to understand and process it effectively.

4.5   LLMs Struggle with Distinguishing Hierarchical Relationships

The models perform particularly poorly on SORT task. While they can handle pairwise judgments of relationships between entities, they fail to distinguish the hierarchical relationships when entities from all nodes of a hierarchy chain are presented together. This indicates that the models might only be capable of judging the distance (associating with semantic similarity or co-occurrence frequency in the corpus) between two entities rather than understanding their hierarchical (superordinate-subordinate) relationships.

4.6   LLMs' Poor Reverse Thinking Ability Hinders Their Performance

LLMs exhibit relatively poor performance in NOJ task and NSJ task compared to SRJO task and ROJS task. This difficulty in retrieving knowledge in reverse order might stem from their left-to-right autoregressive training design. And when selecting the correct option, the models may rely more on the surface similarity of relationships between the correct option and the context rather than truly understanding their connections and differences. They might be aware of the most relevant answer but unable to differentiate the quality of other options.

5   Conclusion

In this paper, we systematically analyze the ontology knowledge in existing LLMs by constructing a comprehensive ontology knowledge evaluation benchmark. Extensive experiments show that certain amount of ontology knowledge can be find in LLMs. LLMs can recognize each class but is not particularly sensitive to their relationships, particularly hierarchical relationships. Though models can remember ontological knowledge, they struggle to truly understand and utilize it. In the future, we will explore ways to enhance the ontology knowledge in LLMs, either by integrating them with knowledge graphs or through ontology knowledge editing.

References

1. Auer, S., Bizer, C., Kobilarov, G., Lehmann, J., Cyganiak, R., Ives, Z.G.: Dbpedia: a nucleus for a web of open data. In: ISWC/ASWC (2007). https://api.semanticscholar.org/CorpusID:7278297
2. Goodwin, T., Demner-Fushman, D.: Enhancing question answering by injecting ontological knowledge through regularization. In: Agirre, E., Apidianaki, M., Vulić, I. (eds.) Proceedings of Deep Learning Inside Out (DeeLIO): The First Workshop on Knowledge Extraction and Integration for Deep Learning Architectures, pp. 56–63. Association for Computational Linguistics (2020). https://doi.org/10.18653/v1/2020.deelio-1.7, https://aclanthology.org/2020.deelio-1.7
3. Gruber, T.R.: A translation approach to portable ontology specifications. Knowl. Acquisition **5**, 199–220 (1993). https://api.semanticscholar.org/CorpusID:15709015
4. Holzinger, A., Saranti, A., Angerschmid, A., Finzel, B., Schmid, U., Mueller, H.: Toward human-level concept learning: pattern benchmarking for AI algorithms. Patterns **4** (2023). https://api.semanticscholar.org/CorpusID:259348460
5. Jiang, A.Q., et al.: Mistral 7b. ArXiv arXiv:2310.06825 (2023). https://api.semanticscholar.org/CorpusID:263830494
6. Jullien, M., Valentino, M., Freitas, A.: Do transformers encode a foundational ontology? probing abstract classes in natural language (2022)
7. Lin, Y., Tan, Y.C., Frank, R.: Open sesame: getting inside BERT's linguistic knowledge. In: BlackboxNLP@ACL (2019). https://api.semanticscholar.org/CorpusID:174799346
8. Neches, R., et al.: Enabling technology for knowledge sharing. AI Mag. **12**, 36–56 (1991). https://api.semanticscholar.org/CorpusID:7410387
9. Nilsson, J.F.: Ontological constitutions for classes and properties. In: International Conference on Conceptual Structures (2006). https://api.semanticscholar.org/CorpusID:206599406
10. Ogden, C.K., Richards, I.A.: The meaning of meaning: a study of the influence of language upon thought and of the science of symbolism. Nature **111**, 566–566. https://api.semanticscholar.org/CorpusID:4003776
11. Peng, H., et al.: COPEN: probing conceptual knowledge in pre-trained language models. In: Goldberg, Y., Kozareva, Z., Zhang, Y. (eds.) Proceedings of the 2022 Conference on Empirical Methods in Natural Language Processing, pp. 5015–5035. Association for Computational Linguistics, Abu Dhabi, United Arab Emirates (2022). https://doi.org/10.18653/v1/2022.emnlp-main.335, https://aclanthology.org/2022.emnlp-main.335

12. Petroni, F., et al.: Language models as knowledge bases? In: Inui, K., Jiang, J., Ng, V., Wan, X. (eds.) Proceedings of the 2019 Conference on Empirical Methods in Natural Language Processing and the 9th International Joint Conference on Natural Language Processing (EMNLP-IJCNLP), pp. 2463–2473. Association for Computational Linguistics, Hong Kong, China (2019). https://doi.org/10.18653/v1/D19-1250, https://aclanthology.org/D19-1250

13. Pinter, Y., Elhadad, M.: Emptying the ocean with a spoon: should we edit models? ArXiv arXiv:2310.11958 (2023). https://api.semanticscholar.org/CorpusID:264288692

14. Shin, T., Razeghi, Y., Logan IV, R.L., Wallace, E., Singh, S.: AutoPrompt: eliciting knowledge from language models with automatically generated prompts. In: Webber, B., Cohn, T., He, Y., Liu, Y. (eds.) Proceedings of the 2020 Conference on Empirical Methods in Natural Language Processing (EMNLP), pp. 4222–4235. Association for Computational Linguistics (2020). https://doi.org/10.18653/v1/2020.emnlp-main.346, https://aclanthology.org/2020.emnlp-main.346

15. Studer, R., Benjamins, R., Fensel, D.A.: Knowledge engineering: principles and methods. Data Knowl. Eng. **25**, 161–197 (1998). https://api.semanticscholar.org/CorpusID:2521178

16. Touvron, H., et al.: Llama 2: open foundation and fine-tuned chat models. ArXiv arXiv:2307.09288 (2023). https://api.semanticscholar.org/CorpusID:259950998

17. Wang, C., Liu, X., Song, D.X.: Language models are open knowledge graphs. ArXiv arXiv:2010.11967 (2020). https://api.semanticscholar.org/CorpusID:225062414

18. Wiggins, J.S.: Book reviews: taxonomy of educational objectives, the classification of educational goals, handbook ii: affective domain by David R. Krathwohl, Benjamin S. Bloom, and Bertram B. Masia. New York: David mckay company, 1964, pp. vii + 196. Educ. Psychol. Meas. **25**, 895 – 897 (1965). https://api.semanticscholar.org/CorpusID:86743286

19. Wu, W., Jiang, C., Jiang, Y., Xie, P., Tu, K.: Do PLMs know and understand ontological knowledge? In: Rogers, A., Boyd-Graber, J., Okazaki, N. (eds.) Proceedings of the 61st Annual Meeting of the Association for Computational Linguistics (Volume 1: Long Papers), pp. 3080–3101. Association for Computational Linguistics, Toronto, Canada (2023). https://doi.org/10.18653/v1/2023.acl-long.173, https://aclanthology.org/2023.acl-long.173

20. Zhao, B., Lucas, C.G., Bramley, N.R.: A model of conceptual bootstrapping in human cognition. Nat. Hum. Behav. **8**, 125–136 (2023). https://api.semanticscholar.org/CorpusID:264170103

Poster and Demo

Enhancing Traditional Chinese Medicine Information Extraction Using Instruction-Tuned Large Models

Jingyao Chen, Shuqi Xia[✉], Jinghua Li, and Tong Yu

Institute of Information on Traditional Chinese Medicine, China Academy of Chinese Medical Sciences, Beijing, China
ShakiraXiasq@163.com

Abstract. The task of information extraction is confronted with many difficulties in the field of traditional Chinese medicine. On top of that, existing methods, including traditional neural network and emerging large language model-assisted information extraction, show low accuracy and recall rates. In this paper, an instruction fine-tuning paradigm is proposed for traditional Chinese medicine text information extraction, which has achieved ideal results on data sets.

Keywords: information extraction · instruction fine-tuning · traditional Chinese medicine

1 Introduction

Traditional Chinese medicine (TCM), originated from ancient China, is a scientific study of Yin-Yang and the Five Elements. The four ways of diagnosis, namely, observing, listening, querying and feeling the pulse makes it possible for TCM experts to explore the changes of human meridians and body fluids. Following that, methods like Chinese herbal medicine and acupuncture could be applied to help human body regain yin-yang harmony. By expressing TCM in the form of knowledge graph, its scattered and complex concepts can be systematized, which paves the way for the smart applications of TCM in everyday life. Information Extraction (IE) can automatically identify and classify common medical terms such as disease names and prescriptions from medical records and literature. Therefore, it is not only the basic task of knowledge acquisition, but also a key step in constructing a TCM knowledge graph. Although relevant research in text information extraction is in full swing, very little is done to advance the work of TCM-IE. The possible reasons are as follows: firstly, the research of TCM-IE has a late start domestically, as such, there is a lack of high-quality supervision and training data of TCM, and the quality of labeled data is uneven due to technical person's insufficient of knowledge in TCM; Secondly, there are a large number of technical terms in Chinese medicine texts. These terms are not only complex in structure, but also vary significantly in expressions.

B. Xu et al. (Eds.): CCKS-IJCKG 2024, CCIS 2229, pp. 331–336, 2025.
https://doi.org/10.1007/978-981-96-1809-5_25

We propose an instruction fine-tuning paradigm for TCM text information extraction, which is characterized by guiding large language models (LLM) through customized instruction sets to improve model's accuracy and effectiveness of information extraction.

□Model fine-tuning: The Llama3 model are fine-tuned by the Low-Rank Adaptation (LoRA) method to ensure that the model can adapt to the complex terminology and relationships in TCM.

2 Related Work

2.1 Information Extraction

Deep learning models are able to learn rich information from large-scale datasets, making this data-driven approach well-received in IE tasks. The first model to apply deep learning to IE tasks is the LSTM + CRF model (Lample, 2016). The memory unit of LSTM compensates for CRF's limitations in processing context information to a certain extent, and greatly promotes the application of deep learning in IE. Nevertheless, although LSTM + CRF improves the processing power of TCM terms and complex relationships, its dependence on large amounts of labeled data makes it still perform poorly in the field of TCM where data is scarce. As a deep bidirectional Transformer model, BERT (Delvin, 2018) can read text from left to right and right to left at the same time, capture more contextual information, achieve deep understanding of text, and perform well in a variety of NLP tasks. The BERT-LSTM-CRF model is one of the most popular IE models at present, and it has been proven to be effective in many experiments. Qu et al. (2020) applied the Bert-BiLSTM-CRF model to TCM-IE, partially solving the challenge of identifying ambiguous entities in TCM texts. However, although BERT performs well in many tasks, its generalization ability is still limited when dealing with highly professional TCM terms and complex relationships.

2.2 Instruction Fine-Tuned Large Language Model

The generative large language model represented by GPT-4 (Achiam, 2023), which can generate natural language text through dialogue, has excellent performance in a variety of NLP tasks, demonstrating strong generation capabilities and context under-standing capabilities. However, recent studies have shown significant performance gaps in LLM regarding information extraction tasks (Chen, 2023). This is mainly because IE is a sequence labeling task, which requires assigning an entity type label to each word element in the sentence, while large models belong to text generation models.

Instruction fine-tuning, being a new paradigm, trains the LLM in a supervised manner by providing a data set consisting of (INSTRUCTION, OUTPUT) pairs. Instruction fine-tuning can bridge the gap between the next word prediction goal of the LLM and the user's goal of having the LLM comply with human instructions (Zhang, 2023). The key to instruction fine-tuning lies in fine-tuning the construction of the data set. As a method to effectively improving generative model's ability in understanding task instructions (Wei, 2021), instruction fine-tuning has gathered more and more attention. In existing studies, instruction fine-tuning models are rarely used for IE tasks, and mainly focus on general information extraction, such as InstructUIE (Wang, 2023), YAYI-UIE (Xiao, 2023), etc., ignoring its practice in TCM.

In response to these problems, this paper proposes a large language model paradigm based on instruction fine-tuning. Through the customized instruction set, our method can guide the model to identify and classify TCM entities more accurately in a more time-efficient manner. The experimental results show that this method has achieved significant performance improvement on Chinese datasets, providing an effective solution for the construction of TCM knowledge graph.

3 Methods

The essence of our method is to guide large language models to learn and identify information in TCM medical records during the fine-tuning and reasoning process through customized instruction sets, so as to fully and accurately identify TCM entities. The specific method is divided into the following steps:

3.1 Data Preparation

As the labeling data of TCM medical records is relatively scarce, in the fine-tuning stage, we used 1,800 TCM journal bibliography data and 200 TCM medical records data, which contain rich TCM terms as well as diagnosis and treatment records. To train and evaluate the model, we divide the data set into training set, test set at a ratio of 9:1.[1]

3.2 Instruction Data Set Design

In order to better guide the model, we designed a customized instruction set. To be more specific, we generated instructions for each piece of data by leveraging GPT4-o. Based on instruction engineering technology (Wang, 2023), the instruction set structurally defines tasks through pseudo-code brackets, guiding large language models to perform accurate information extraction. During LLM training period, plenty of high-quality codes have been inputted, therefore, the meaning of pseudo-code can be easily understood. The length of carefully designed pseudo-code is usually less than one page, which simplifies communication compared to actual code (Phuong, 2022). By using pseudo-code, the output results and execution logic of LLM can be accurately controlled and defined. Finally, the formatted json is generated, which will be directly used for knowledge graphs construction.

The pseudo-code we use in instruction fine-tuning and prompting engineering is as follows:

---

[1] The sample data is available at https://github.com/JYao-Chen/TCM-IE-LLM.

Algorithm 1 Basic Sample of the Instruction Set（指令集的基本样例）

这是中医医案命名实体识别任务，以下是一段需要抽取的实体定义和伪代码逻辑：
"""

² The sample data is available at https://github.com/JYao-Chen/TCM-IE-LLM.

Jingyao Chen and Shuqi Xia

```
def extract_entities(text):
    entities = {
        '方剂': "代表方剂类实体，如："健脾和胃，理气化积"等",
        '临床表现': "代表临床表现类实体，如：暖气、上腹饱胀等",
        '中医治疗': "代表中医治疗类实体，如：耳穴贴压、针刺等",
        '中药': "代表中药类实体，如：炒白术、陈皮、党参等",
    }
    output = {entity: [] for entity in entities}

    input_text = "采用胃康乐胶囊II号(香附、砂仁、山药、党参、白术、佛手等)治疗慢性
活动性胃炎120例..."

    output['方剂'].append("胃康乐胶囊II号")
    output['中医治疗'].append("胃康乐胶囊II号")
    output['中药'].extend(["白术", "党参", "佛手", "砂仁", "山药", "香附"])

    return output

extract_entities(input_text)
"""
```

3.3 Model Fine-Tuning

ChatGPT is not open source, luckily, the open source community provides multiple alternatives, such as LLaMa (Touvron, 2023), which is relatively inexpensive to train, so we adopt the Llama3-8B-Chinese model and used LoRA (Hu, 2021) method to perform Supervised Fine-Tuning tasks.

4 Experiment

In the experimental part, experiments in supervised settings and zero-shot settings are conducted respectively to evaluate the effect of TCM-IE with F1 value as the metrics.

4.1 Experiments on Supervised Settings

We used 180 topic data and 20 TCM medical record data to conduct supervised experiments. Unfine-tuned BERT-base-Chinese and Llama3-8B-Chinese are selected as baselines.

The experimental results, as can be observed in Table 1, indicate that on supervised datasets, the Llama3-8B-Chinese model fine-tuned by data instructions outperforms the baseline models in terms of F1 value. Specifically, the overall F1 value of the fine-tuned model reached 0.95, which is significantly higher than 0.79 of BERT and 0.71 of Llama.

Overall, fine-tuning through domain-specific instructions significantly improves the model's ability to extract accurate and relevant information from TCM texts.

Table 1. Overall F1 results on supervised datasets.

Dataset	BERT-base-Chinese	Llama3-8B-Chinese	TCM fine-tuned Llama3-8B-Chinese
Supervised Dataset	0.79	0.71	0.95

4.2   Experiments on Zero-Shot Settings

330 spleen and stomach TCM medical records are used to conduct zero-shot experiments. The unfine-tuned Llama3-8B-Chinese is selected as the baseline.

As presented in Table 2, when it comes to zero-shot data set, the Llama3-8B-Chinese model fine-tuned by TCM data instructions has more desired performance than the baseline model.

Overall, these results highlight the effectiveness of domain-specific instruction fine-tuning in enhancing the model's zero-shot learning ability, enabling it to accurately and efficiently extract relevant information from TCM texts without inputting similar data previously.

Table 2. Overall F1 results on zero-shot datasets.

Dataset	Llama3-8B-Chinese	TCM fine-tuned Llama3-8B-Chinese
Zero-Shot Dataset	0.45	0.82

5   Conclusion

In this paper, instruction fine-tuning model is proposed for information extraction from TCM texts. By fine-tuning the domain-specific instructions based on the Llama3 model, the performance of the model in supervised and zero-shot environments is significantly improved.

The experimental results show that the fine-tuned model outperforms the baseline model in terms of accuracy, recall ratio and F1 value, demonstrating its superior performance in complex domains. This is especially true in terms of zero-shot learning.

Considering the pros and cons of our work, what follows might inspire later academics: □

☐Expand the amount of data: Add standardized file data including TCM clinical diagnosis and treatment terms, which makes the model's judgment of categories more accurate and the recall more sufficient.

☐Expand categories: In instruction fine-tuning, expand more categories to enhance model's adaptability in different tasks.

Through these proposed improvements, the performance of the model will be further improved, providing a more accurate and efficient solution for the construction of TCM knowledge graphs.

Acknowledgments. This study was funded by

1. Development Research on Automatic Construction of TCM Knowledge Graph: Common Technologies and Platform (grant number ZZ150313), Task Leader: Tong Yu

2. Automatic construction and knowledge service of knowledge map of stomach diseases in Chinese medicine based on massive literature, Scientific and Technological Innovation Project of China Academy of Chinese Medical Sciences (grant number CI2021A05308), Task Leader: Tong Yu

3. Research on construction and integration of large-scale traditional Chinese medicine knowledgeable graph and effective prescription and medicine discovery, CACMS Innovation Fund (grant number CI2021B002), Task Leader: Jinghua Li.

References

Lample, G., Ballesteros, M., Subramanian, S., Kawakami, K., Dyer, C.: Neural architectures for named entity recognition. arXiv preprint arXiv:1603.01360 (2016)

Devlin, J., Chang, M.W., Lee, K., Toutanova, K.: Bert: pre-training of deep bidirectional transformers for language understanding. arXiv preprint arXiv:1810.04805 (2018)

Qu, Q., Kan, H., Wu, Y., Gao, Y.: Named entity recognition of TCM text based on Bert model. In: 2020 7th International Forum on Electrical Engineering and Automation (IFEEA), pp. 652–655. IEEE (2020)

Achiam, J., et al.: GPT-4 technical report. arXiv preprint arXiv:2303.08774 (2023)

Chen, X., et al.: How robust is GPT-3.5 to predecessors? A comprehensive study on language understanding tasks. arXiv preprint arXiv:2303.00293 (2023)

Zhang, S., et al.: Instruction tuning for large language models: a survey. arXiv preprint arXiv:2308.10792 (2023)

Wei, J., et al.: Finetuned language models are zero-shot learners. arXiv preprint arXiv:2109.01652 (2021)

Wang, X., et al.: Instructuie: multi-task instruction tuning for unified information extraction. arXiv preprint arXiv:2304.08085 (2023)

Xiao, X., et al.: YAYI-UIE: a chat-enhanced instruction tuning framework for universal information extraction. arXiv preprint arXiv:2312.15548 (2023)

Wang, J., et al.: Prompt engineering for healthcare: methodologies and applications. arXiv preprint arXiv:2304.14670 (2023)

Phuong, M., Hutter, M.: Formal algorithms for transformers. arXiv preprint arXiv:2207.09238 (2022)

Touvron, H., et al.: Llama: open and efficient foundation language models. arXiv preprint arXiv:2302.13971 (2023)

Hu, E.J., et al.: Lora: low-rank adaptation of large language models. arXiv preprint arXiv:2106.09685 (2021)

Development of an Intelligent Chinese Medicine Q&A System Based on Traditional Chinese Medicine Knowledge Graph and Large Language Models

Xu Yihan[✉], Yu Tong, Li Jinghua, Yu Qi, Zhang Zhulv, Zhang Xiehua, Xia Shuqi, and Li Ge

Institute of Information on Traditional Chinese Medicine, Beijing, China
1257154802@qq.com

Abstract. This paper briefly introduces the design and implementation of an intelligent traditional Chinese medicine Q&A system based on traditional Chinese medicine knowledge graph and large language models. By integrating the traditional Chinese medicine knowledge graph and natural language processing technology, the system aims to provide efficient traditional Chinese medicine Q&A services.

Keywords: traditional Chinese medicine knowledge graph · large language model · intelligent Q&A · natural language processing

1 Introduction

1.1 Background

Against the backdrop of continuous advancements in cutting-edge technologies such as big data and large language models, artificial intelligence is undergoing a rapid evolution. Traditional Chinese Medicine (TCM), as the essence of ancient Chinese wisdom, can still play a significant role in the development of national medical science through experience transmission and knowledge innovation. However, the transference of TCM experts' experience faces numerous challenges, such as how to achieve syndrome differentiation and treatment during TCM clinical practice, how to adjust expert prescriptions flexibly based on patients' clinical manifestations; as well as how to efficiently pass down TCM experts' experience using current cutting-edge technologies. This necessitates the flexible application of contemporary knowledge graph technologies [1] during the experience transmission process, leveraging the advantages of generative large language models in reinforcement learning, self-supervised learning, etc., to construct intelligent TCM consultation models so as to provide clinical decision support based on the foundation of experience transmission [2].

1.2 Objective

This paper aims to utilize Retrieval-Augmented Generation (RAG) technology to build a TCM expert knowledge base and develop an intelligent consultation system, serving the digital preservation and intelligent application of TCM experience [2].

2 Related Work

Based on deep learning, large language models are trained on a vast corpus of text, enabling them to generate seemingly reasonable responses. To enhance model's performance in specific domains, common methods include fine-tuning, prompt engineering, and RAG. Currently, the paradigm of fine-tuning is mainly adopted to improve model's performance in specific domains, like ChatDoctor [2] and DoctorGLM [2]. Despite fine-tuning can provide the model with more professional and accurate background knowledge, thus generating responses that are more in line with the characteristics of traditional Chinese medicine, a main drawback of fine-tuning lies in its failure to keep up with the latest knowledge. Given that theories and practice in traditional Chinese medicine are constantly evolving, with new research findings and treatment methods emerging continuously, RAG technology, by retrieving the latest materials and data, can ensure that the information provided by the Q&A system is up-to-date, meeting users' needs for the latest medical consultations, thus having a greater advantage in the field of traditional Chinese medicine Q&A.

The advantages of RAG combined with knowledge graphs are significant: First, it can well compensate for the hallucination problem of large models, enhancing the professionalism and reliability of the Q&A system. A case in point is OREOLM [2]. OREOLM has demonstrated through experiments on multiple Q&A datasets that using a knowledge graph can significantly improve performance compared to traditional large language models in a closed-book Q&A setting. Second, as the structure of the graph ensures that the retrieved information is not only relevant but also contextual, providing a richer background for response generation. Bert-MK [2] first used the Unified Medical Language System (UMLS) as a medical knowledge graph and constructed a subgraph of the medical knowledge graph to capture complex interactions between entities, thereby enhancing contextual understanding; then, the subgraph was modeled, and an attention mechanism was used to update node representations to ensure that the node representations reflect their contextual position in the graph. Experimental results show that Bert-MK has achieved better performance than existing state-of-the-art models in multiple medical NLP tasks, indicating its advantage in understanding the context of medical texts. Third, knowledge graphs are good at representing and storing heterogeneous and interconnected information in a structured way, and can easily capture complex relationships and attributes between different data types, making it capable of processing complex queries efficiently.

3 System Architecture

3.1 Overall Design

In order to build an intelligent knowledge graph interactive interface, this study used advanced technologies such as LangChain, Streamlit and Streamlit-Graph to build a Q&A system platform based on the original database of TCM expert knowledge graph constructed in the early stage.

3.2 Knowledge Graph

As for the construction of the knowledge graph, we choose Chinese medicine literature, classic works, renowned TCM experts' experience formulas, medical cases, and diagnostic rules as the data source. Based on the former data source, we use deep learning model combined with large language model to realize knowledge extraction, which are consist of entity recognition, relationship extraction and graph storage. During this period, the design of Schema is crucial. Schema defines the entity types and their attributes and relationships in the knowledge graph. Based on the characteristics of famous veteran doctors of traditional Chinese medicine, we design the following core entities and their relationships:

Entity: 1) famous TCM doctor; 2) medical case; 3) traditional Chinese medicine prescription; 4) scientific work; 5) scientific project.

Relation: 1) Be cured of; 2) Prescribe a prescription; 3) Be written by; 4) Participate; 5) Use.

The process of constructing a knowledge graph of renowned traditional Chinese medicine (TCM) doctors as the system's knowledge source mainly involves data preprocessing, entity and relationship extraction, triple formation, and finally knowledge graph visualization. In terms of data preprocessing, the collected data is first stored in Excel sheets, and then cleaned using Python's pandas module. Subsequently, GPT-4, a large language model, is employed to automatically extract and match triples from the cleaned data, such as relationships between medicine and disease treatment, formula composition, and associations between symptoms and syndromes. The extracted data is then subjected to knowledge integration, merging similar or identical data, with further deduplication to ensure extraction accuracy. Once entities and relationships are successfully extracted from the given text, we use Neo4j as the storage database to achieve knowledge visualization.

After successfully extracting entities and relationships from the given text, in order to achieve knowledge visualization, we use Neo4j as the storage database. We employ the corresponding Cypher language to import the triplets stored in the Excel spreadsheet into Neo4j, thereby enabling visual display (Figs. 1, 2, 3 and 5).

抽取下面文本的实体和关系，构建知识图谱，输出成neo4j代码。麻黄汤

《伤寒论》

【组成】麻

黄去节，三两（9g）

桂枝去皮，二两（6g）

杏仁去皮尖，七十个（9g）

　甘草炙，一两（3g）

【用法】

上四味，以水九升，先煮麻黄，减二升，去上沫，内诸药，煮取二升半，去滓，温服八合。覆取微似汗，不须啜粥，余如桂枝法将息（现代用法：水煎服，温覆取微汗）。

【功用】

发汗解表，宣肺平喘。

【主治】

外感风寒表实证。恶寒发热，头身疼痛，无汗而喘，舌苔薄白，脉浮紧。

【证治机理】

本证系由风寒束表，肺气失宣所致。风寒之邪侵袭肌表，营卫首当其冲，寒性收引凝滞，致使卫阳被遏，营阴郁滞，即卫闭营郁。卫气抗邪，正邪相争，则恶寒、发热；营卫不畅，腠理闭塞，经脉不通，则无汗、头痛、身痛、骨节疼痛；皮毛内合于肺，寒邪束表，肺气不宣，则上逆为喘；舌苔薄白，脉浮紧，皆是风寒束表之象。法当发汗解表，宣肺平喘。

【方解】

方中麻黄辛温，主入肺经，为发汗之峻剂，既开腠理，透毛窍，发汗以祛在表之风寒；又开

Fig. 1. Enter the extraction command and corresponding text into the large language model

10. 恶寒发热
11. 头身疼痛
12. 无汗而喘
13. 风寒束表
14. 肺气失宣

关系：

1. 麻黄汤 - 来源于 - 伤寒论
2. 麻黄汤 - 包含 - 麻黄
3. 麻黄汤 - 包含 - 桂枝
4. 麻黄汤 - 包含 - 杏仁
5. 麻黄汤 - 包含 - 甘草
6. 麻黄汤 - 功能 - 发汗解表
7. 麻黄汤 - 功能 - 宣肺平喘
8. 麻黄汤 - 主治 - 外感风寒表实证
9. 外感风寒表实证 - 症状 - 恶寒发热

Fig. 2. Entities and relationships extracted by the large language model

```cypher
CREATE
  (麻黄汤:Prescription {name: '麻黄汤'}),
  (伤寒论:Book {name: '伤寒论'}),
  (麻黄:Herb {name: '麻黄'}),
  (桂枝:Herb {name: '桂枝'}),
  (杏仁:Herb {name: '杏仁'}),
  (甘草:Herb {name: '甘草'}),
  (外感风寒表实证:Disease {name: '外感风寒表实证'}),
  (恶寒发热:Symptom {name: '恶寒发热'}),
  (头身疼痛:Symptom {name: '头身疼痛'}),
  (无汗而喘:Symptom {name: '无
```

Fig. 3. Import the triplets into Neo4j using Cypher language

Fig. 4. An overview of the knowledge map

3.3 Large Model Integration

In the selection of tools for deploying large language models locally, this paper comprehensively compares various large model inference tools and decides to adopt xInference, a distributed inference framework. In terms of Large Model Integration and

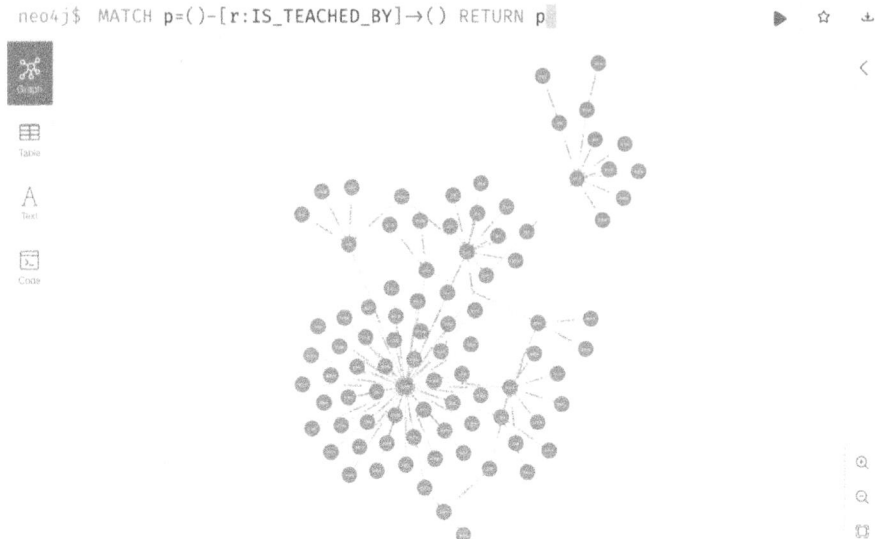

`neo4j$ MATCH p=()-[r:IS_TEACHED_BY]→() RETURN p`

Fig. 5. Characteristic medication relationship

RAG development, this paper is based on technologies such as Neo4j and Langchain to actualize knowledge graph text2cypher, namely, extracting relevant entities from natural language; automatically constructing related cypher statements based on entities, followed by returning results; generating embedded models or indexes for TCM knowledge text; and integrating large language models with inference frameworks to form the RAG system. This enables the TCM expert intelligent consultation system to have exceptional performance in information retrieval and query response. Based on large language models and knowledge graphs, with the TCM expert knowledge base and knowledge graph as knowledge sources, the specific operational process goes like this: what users input will be further transformed into keywords and vectors and then be inputted into blocks, where task processors include text analysis tools, OCR recognition tools, text layout analysis, table structure recognition as well as task data dispatched from the earlier knowledge base. After such primary processing, users' question will be recalled and re-ranked. Subsequently, after chatting with large language models, results will be returned to users. In addition, by seamlessly integrating the knowledge graph-based RAG system with the document vectorization similarity-based RAG system, we achieve hybrid search, which significantly enhances the results of question answering. This fusion approach allows the question-answering system to provide answers more accurately while effectively reducing the occurrence of AI hallucinations (Fig. 6).

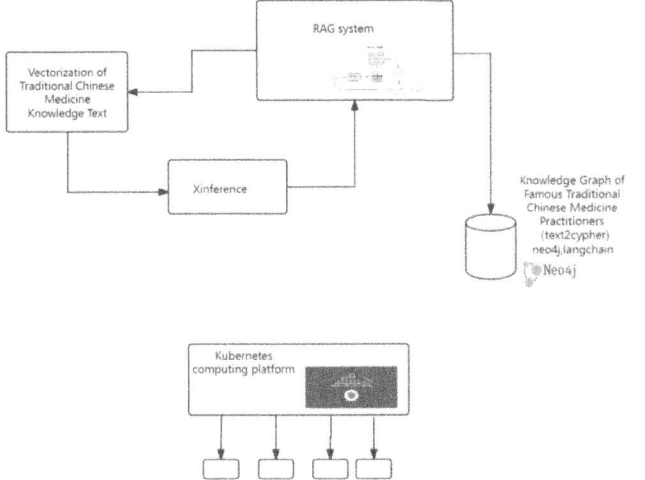

Fig. 6. Technical implementation diagram

4   Technical Implementation

4.1   User Interface

As illustrated in Fig. 4, the user interface is designed to be intuitive and user-friendly, allowing users to easily interact with the system and obtain the information they need.

- User Question: At the top of the content page, users can input their questions. For example, a user may ask, "What is Zhao Jianming's method for treating vitiligo? What are the ingredients of the related prescription?"
- System Response: Below the user query, the system provides a comprehensive answer. For instance, in the response to the query about Zhao Jianming's treatment for vitiligo, the system explains that Zhao's method includes the use of "Sanhuang Powder" and "Baixie Decoction." The response lists the ingredients of these prescriptions and their functions.
- Graphical Representation: At the bottom of the content page, a graphical representation of the relationships between different elements is displayed. For example, it shows connections between the practitioner (Zhao Jianming) and the prescriptions (Sanhuang Powder and Baixie Decoction) using labeled nodes and edges. This visual aid helps users understand the relationships and dependencies at a glance.

4.2   Front-End User Interface

- **User Selection and Problem-Raising:** Through the system, users can browse and select a specific question of interest. The system will load the relevant knowledge base based on the user's choice.
- **User Query Analysis:** Utilizing the natural language understanding and analysis capabilities of large language models, the system deeply parses the consultation information provided by users. The model can identify and extract users' key information,

such as name, age, and gender, as well as medically relevant information like chief complaints and past medical history. This extracted information is stored in databases to serve as references for subsequent diagnoses.

- **Deep Knowledge Analysis and Immediate Feedback:** Based on users' questions, the system first quickly provides answers in precise text form, and then visualizes the query results through dynamic graphical displays. This design not only enhances the efficiency of information transmission but also greatly promotes the reliability and transparency of the answers. It effectively suppresses misleading responses that AI might generate, ensuring that every interaction accurately meets the users' inquisitive needs.
- **Real-time Query and Display of Knowledge Graph:** The system can integrate existing knowledge graph databases, allowing for quick deployment and activation to provide users with an efficient Q&A experience based on knowledge graphs. Visualization components not only enhance the data presentation but also support users in tracing the sources of information. This deepens their understanding and eliminates potential misunderstandings.
- **Continuous Optimization of TEXT2CQL Technology:** The TEXT2CQL conversion process faces certain challenges in terms of stability. To overcome this problem, a Few-Shot Prompt Learning strategy is adopted to optimize this aspect. This approach will significantly enhance the system's adaptability and accuracy, ensuring that even complex or marginal queries can obtain high-quality answers.

4.3   Back-End Technologies

- **Technology Stack:** The system uses the Next.js framework [6] for both front-end and back-end interfaces. Next.js, an open-source JavaScript framework, offers server-side rendering and static site generation, enhancing page loading speed and optimizing search engine performance. Combined with TypeScript's robust type system and ChakraUI's attractive interface components, it builds an efficient and visually appealing application interface.
- **Data Storage:** A dual database architecture involving MongoDB [9] and PostgreSQL [10] (with Vector plugin) ensures data flexibility and security.
- **System Architecture Design:** The open-source Kubernetes (K8s) system [11] is utilized for its ability to automate container deployment, scaling, load balancing, and failure recovery. This paper leverages K8s' containerization [12] to build a highly available and scalable system. K8s' dynamic resource adjustment ensures optimal system operation [13]. The automated CD/CI deployment process speeds up software delivery while maintaining code quality. Distributed deployment strategies enable multi-replica operation, enhancing system availability and fault tolerance.
- **Scalability:** Horizontal scaling and shrinking mechanisms in the K8s platform automatically manage the number of Pod replicas to meet varying load demands. The platform adjusts the number of worker nodes based on the load, ensuring stable and efficient system operation. The system consists of six servers, including one master node, four worker nodes, and one harbor private repository, providing a solid foundation for stable application operation (Fig. 7).

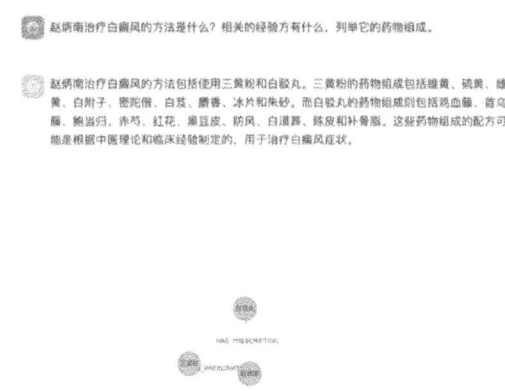

Fig. 7. Question answering system interface based on knowledge graph

5 Conclusion

Based on a TCM knowledge graph and a large language model, the Traditional Chinese Medicine (TCM) Q&A system, not only enhances the efficiency of TCM knowledge inheritance and application but also provides innovative solutions for the application of artificial intelligence in the field of medical consultation. Although this study has achieved significant results in the design and implementation of the TCM Q&A system, further optimization of the dynamic updating mechanism of the knowledge graph is still needed in the future to improve the system's adaptability and interaction experience. In addition, exploring the potential of the system in a broader range of TCM application scenarios, as well as how to better serve patients and medical professionals, will be the spotlight of future work.

Acknowledgments. This study was funded by: Research on Common Technologies and Platform Development for Automatic Construction of TCM KG (No. ZZ150313); Knowledge Service Research on Automatic Construction of TCM Spleen and Stomach Disease KG Based on Massive Literature (No. CI2021A05308); CACMS Science and Technology Innovation Project (No. CI2021B002): Research on Effective Prescription Discovery based on Large-Scale TCM KG Construction.

References

1. Walke, D., et al.: The importance of graph databases and graph learning for clinical applications. Database (Oxford) **2023**, baad045 (2024).
2. Zhao, W.X., et al.: A survey of large language models. arXiv:2303.18223 (2023)
3. Vaswani, A., et al.: Attention is all you need. In: Advances in Neural Information Processing Systems (2023)
4. Li, Y., Li, Z., Zhang, K., Dan, R., Jiang, S., Zhang, Y.: Chatdoctor: a medical chat model fine-tuned on a large language model meta-AI (llama) using medical domain knowledge. Cureus **15**(6) (2023)

5. Xiong, H., et al.: DoctorGLM: fine-tuning your Chinese doctor is not a herculean task. arXiv preprint arXiv:2304.01097 (2023)
6. Hu, Z., et al.: Empowering language models with knowledge graph reasoning for question answering. arXiv preprint arXiv:2211.08380 (2022)
7. He, B., Zhou, D., Xiao, J., Liu, Q., Yuan, N.J., Xu, T.: Integrating graph contextualized knowledge into pre-trained language models. arXiv preprint arXiv:1912.00147 (2019)
8. Fan, D.: Design and Research of a Front-end MVVM Framework Based on TypeScript. Beijing University of Posts and Telecommunications (2022)
9. Xianfu, B.: Research on data optimization patterns in MongoDB distributed cluster cloud. Wireless Internet Technol. **20**(22), 153–156 (2023)
10. Lirong, A., Kai, L.: Research on data management optimization technology based on postgres-XL. Comput. Technol. Dev. **28**(03), 11–14 (2018)
11. Liu, X., Cai, Z., Xu, J.: Container hybrid scaling method for kubernetes. Comput. Digit. Eng. **51**(10), 2219–2223+2241(2023)
12. Yan, S.: Research and Implementation of a Resource Scheduling Optimization Algorithm Based on Kubernetes. Beijing University of Posts and Telecommunications (2024)
13. Li, J., Dong, J., Li, K.: Research on cluster energy-saving strategies based on Kubernetes. Comput. Eng. **50**, 1–13 (2024)

KAOS: Large Model Multi-agent Operating System

Zhao Zhuo[1], Rongzhen Li[1(✉)], Kai Liu[1], Huhai Zou[1], KaiMao Li[1], Jie Yu[2], Tianhao Sun[1], and Qingbo Wu[2]

[1] Chongqing University, Chongqing, China
{lirongzhen,sthing}@cqu.edu.cn, leekaimao@stu.cqu.edu.cn
[2] National University of Defense Technology, Changsha, China
{yj,qingbo.wu}@nudt.edu.cn

Abstract. The intelligent interaction model based on large models reduces the differences in user experience across various system platforms but faces challenges in multi-agent collaboration and resource sharing. To demonstrate a uniform user experience across different foundational software platforms and address resource coordination management challenges, this paper proposes KAOS, a multi-agent operating system based on the open-source Kylin. The research method involves empowering agents with large models to serve applications. Firstly, the process involves the incorporation of management role agents and the establishment of a vertical collaborative framework among multiple agents, which serves to either build new or substitute existing application software. Secondly, the approach includes an in-depth examination of system-level strategies for the scheduling of shared resources, with the aim of improving the overall user experience and achieving an optimized allocation of resources. Lastly, the methodology concludes with the validation of the efficiency and the demonstrated superiority of a large model multi-agent operating system, accomplished through the application of real-world scenarios and the assessment of its intelligence capabilities. The feasibility of this system is demonstrated, providing a new perspective for the development of multi-agent operating systems. Experimental results show significant advantages of multi-agent collaboration in various application scenarios.

Keywords: large language model · multi-agent systems · resource scheduling policy

1 Introduction

In recent years, with the rapid advancement of artificial intelligence technologies, large language models (LLMs) have found widespread applications across various domains. The application of LLM-based agents in the field of operating systems (OS) is emerging as a popular research direction. Compared to traditional rule-based OS, integrating LLM agents into the OS allows for better understanding of user needs and offers a personalized, intelligent interactive experience. Users can engage in natural language conversations with the OS, ask various questions, and receive timely and accurate responses. Additionally, LLMs can assist users in routine office tasks such as drafting documents and emails, significantly enhancing work efficiency. Clearly, intelligent agents as a mode of functional software providing more natural and friendly interactions will be a future trend.

Users of mainstream operating systems currently exhibit significant differences in usage habits. Windows and Linux, as the two primary desktop operating systems, feature different user interfaces, command systems, and software ecosystems, which can cause inconvenience for users. There is a growing expectation for operating systems to provide more intelligent and personalized services to meet their increasing demands for information retrieval and task processing. By introducing LLMs, seamless switching and convenient user experiences between these two operating systems can be achieved, thereby improving user productivity and comfort. The LLM-based agent interaction model reduces the differences in using various operating systems; for example, organizing meetings on Linux through an agent is as straightforward as on Windows, making platform differences imperceptible to users.

However, LLM-based multi-agent systems face significant challenges in cooperative interaction and resource sharing. Presently, LLM-based OS models primarily involve single agents independently fulfilling user requirements, and the collaborative completion of tasks by multiple agents is still in its developmental stages. Similar to how humans solve complex problems, collaboration among multiple agents can provide better solutions. This requires not only considering resource sharing among multiple independent agents but also researching how to optimally schedule multiple agents for resource-efficient cooperative task completion.

To address these challenges, based on part of the framework of AIOS [35], this paper proposes KAOS, the first LLM-based multi-agent operating system built on the user-friendly open-source Kylin platform. The main innovations include:

1. A management role agent is proposed to offer more sophisticated functionalities, equipped with a desktop knowledge base and enhanced tool capabilities. This management agent assists in decision-making and state maintenance through a customized desktop knowledge base and tools. The desktop knowledge base integrates easily accessible information resources, including some with privacy rights, and tool enhancement provides a set of customized software tools such as Bing Search, Arxiv Search, and hotel recommendations, aiding the management role in more efficiently completing tasks.

2. A low-resource consumption vertical multi-agent collaboration model is introduced. In this vertical structure, agents have clear roles, with the management agent receiving user tasks and intelligently assigning them, ensuring each agent handles only its specialized part, thus reducing redundant computation and lowering communication overhead. Multiple agents collaborate using a shared resource pool to maximize resource utilization, such as shared computing power and data storage.

3. A timesharing resource scheduling strategy is researched and implemented to optimize resource utilization. This strategy divides resource usage time into multiple time slices, with different agents using resources in different slices to avoid conflicts. This approach maximizes resource utilization efficiency, ensuring the multi-agent system operates stably and efficiently under high load conditions.

The structure of this paper is as follows: Sect. 2 reviews related research on LLM-based agent operating systems; Sect. 3 introduces the system architecture and methods; Sect. 4 validates the effectiveness of the LLM-based multi-agent operating system; Sect. 5 provides conclusions and future outlook.

2   Related Work

Intelligent agents based on Large Language Models (LLMs) have garnered widespread attention in the field of artificial intelligence. These agents utilize large language models, trained on vast amounts of data, as their foundation for knowledge and capabilities, demonstrating language understanding, commonsense reasoning, and task execution abilities that closely resemble those of humans. Compared to traditional rule-based or machine learning-based agents, LLM-based agents possess broader and deeper knowledge, enabling them to flexibly address a variety of complex problems and tasks. The structure of these agents can be divided into four components: the large model, planning skills, experiential memory, and tool usage. Consequently, LLM-based agents exhibit a degree of autonomous learning and decision-making abilities, expanding their perception and action space through multimodal perception and tool usage. They continually refine their behavior strategies through interaction with their environment [1].

Research on LLM-based agents is primarily categorized into single-agent systems and multi-agent systems [2]. Single-agent systems rely on the language understanding and generation capabilities of LLMs, enabling smooth and natural interactions with humans or other systems. LLMs not only process information but also serve as the decision-making core, receiving natural language input, parsing it, and generating responses ranging from simple answers to complex action plans. Through iterative optimization, single-agent systems gradually enhance their perception and decision-making capabilities.

Currently, single-agent systems are capable of not only engaging in natural language interactions [3] but also operating virtual devices, utilizing configured tools [4–7], executing API calls [7–11], executing code [12], or managing web content [4,13–16]. For instance, Khot et al. [17] proposed a modular approach that decomposes complex tasks into sub-tasks and uses prompts to guide the model in generating outputs, thereby enhancing task performance and model interpretability.

Despite the strong natural language understanding and generation capabilities of single-agent systems [18], their independent operation limits their ability to acquire knowledge through teamwork. In contrast, multi-agent systems achieve common goals through cooperation and competition, offering greater functionality and becoming a focal point of research [19]. In multi-agent systems, multiple agents collaborate and compete to achieve common objectives [20]. Some agents are responsible for recognition, while others handle decision-making and planning, working together to solve complex tasks such as code development [21–24] and social simulation [25,26]. Competitive mechanisms are also necessary for certain tasks, such as solving debate problems [27–29].

Multi-agent systems leverage the unified language platform provided by LLMs for communication and collaboration [21], significantly enhancing the system's flexibility and scalability. Different agents communicate their intentions and needs through natural language, understanding feedback and suggestions. Hamilton et al. [30] proposed a decision-making method that simulates the behavior of the U.S. Supreme Court by following the majority opinion. Wang et al. [31] designed a system that generates final answers by integrating feedback from multiple agents, emphasizing the importance of effectively consolidating information.

Researchers are also exploring various methods to improve decision-making capabilities, such as having agents imitate human behavior. Many applications based on this idea have been investigated [21,22,27,32–34]. For example, MetaGPT [21] encodes standard operating procedures (SOPs) into prompt sequences, incorporating human workflows into LLM-based multi-agent collaboration. The advantage of this approach is that SOPs guide agents in generating structured intermediate outputs, such as high-quality requirement documents, design documents, flowcharts, and interface specifications. By using structured intermediate outputs, human workflows are seamlessly integrated into multi-agent collaboration, resulting in smoother workflows, fewer errors, and more coherent solutions.

In the system proposed by Du et al. [27], multiple agents can debate decisions. When different language agents are required to communicate with each other, they often change their views, leading to more accurate consensus answers. This study suggests that existing large language models are prone to errors and hallucinations, and the multi-agent debate approach can significantly enhance the mathematical and strategic reasoning capabilities of the models while reducing these issues.

In summary, although single-agent systems are powerful, their limitations in independent operation have prompted researchers to focus more on multi-agent collaborative systems. Exploring efficient collaboration and decision-making mechanisms is a major challenge with significant potential. Through collaboration and optimization, future agents will be better equipped to handle complex tasks, achieving higher levels of intelligence and automation.

3 System Architecture and Methodology

3.1 Basic Architecture

The KAOS architecture comprises three layers for an efficient, manageable agent system: The application layer for developing and deploying agent apps using the AIOS toolkit [35]. The kernel layer with an OS kernel for resource management and an LLM kernel for agent operation optimization. And the hardware layer with physical components managed for system security and stability.

3.2 Management Agents Enhanced with Desktop Knowledge Base and Tools

This paper proposes a management role agent equipped with a desktop knowledge base and tool enhancement capabilities. The desktop knowledge base provides an integrated and easily accessible information resource repository for the management role agent, containing both sensitive and non-sensitive business information. The tool enhancement offers a set of customized software tools to help the management agent efficiently complete decisions and daily tasks.

The desktop knowledge base interacts with the context manager and storage manager of the kernel layer to achieve data integration and quick access. Data is collected from the file system and external resources (such as web information), cleaned and formatted using data cleansing and integration tools, and stored in the knowledge base. Data in the knowledge base is classified according to sensitivity, with different access permissions set by the access manager to ensure that only authorized agents can access specific data.

To enhance management efficiency, the agent provides a set of customized software tools. These tools are integrated into the agent's work environment through the tool manager, such as Bing search, Arxiv query, and hotel recommendation.

3.3 Low Resource Consumption Vertical Multi-agent Collaboration

As shown in Fig. 1, the low resource consumption vertical multi-agent collaboration differs from traditional agent operating systems. Users interact only with the management agent, regardless of the task type. The management agent identifies the user's needs through LLM analysis and then delegates tasks to suitable agents. For example, when a user needs to write a document, the management agent uses LLM to identify the requirement and assigns the task to a writing agent.

For different types of agents, this paper designs two task decomposition methods: procedural and adaptive. The procedural method uses a workflow-like design model to organize limited functions logically. For agents with fixed tasks (such as writing agents and meeting agents), the procedural design is adopted. Researchers manually design sub-tasks. For instance, when a user inputs a task to generate a document with a given writing topic, the writing agent decomposes the task into steps: generating a title, outline, draft, checking grammar and logic, and finally saving the document. This decomposition method is suitable for agents with fixed task content.

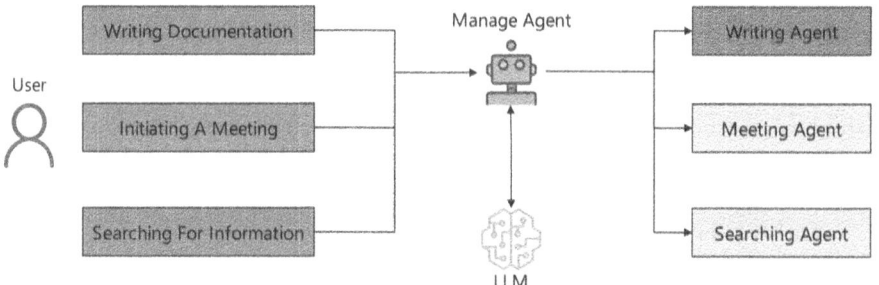

Fig. 1. Low Resource Consumption Vertical Multi-agent Collaboration

As shown in Fig. 2, a case of procedural sub-task decomposition based on intent disassembly for scheduling meetings is illustrated. The agent needs to understand the meeting's theme, time, and platform. Upon receiving the user's input, the meeting agent interacts with the LLM to utilize its natural language understanding capabilities to analyze the theme, time, and platform for the meeting.

The adaptive decomposition method is suitable for agents with high task uncertainty, such as search agents. For these agents, the LLM is used to automatically decompose tasks. For example, when searching "What are the differences between PyTorch and TensorFlow?", the search agent delegates the task to the LLM, which decomposes it into sub-questions like "What is PyTorch?", "What is TensorFlow?", "What are the main features of PyTorch and TensorFlow?", "What are the differences in programming models and API designs between PyTorch and TensorFlow?", etc. After the search agent answers all sub-questions, it summarizes and organizes the answers to provide a complete response. This decomposition method leverages the powerful natural language processing capabilities of the LLM to efficiently handle complex tasks, suitable for agents with high task uncertainty.

For complex user queries, the multi-agent operating system with a large model also has countermeasures. For instance, when a user needs an agent to search the web for information and write it into a document, the search agent collaborates with the document agent. The search agent gathers and summarizes information, while the document agent organizes and writes it into a document. This collaboration ensures efficient task completion and optimizes resource utilization.

Meeting Agent

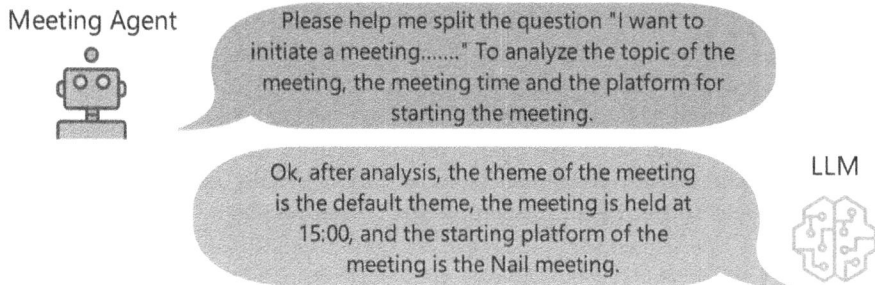

Fig. 2. Agent and Large Model Interaction

3.4   Time-Sharing Multi-agent Resource Scheduling Strategy

This paper develops and implements a time-sharing resource scheduling strategy based on time slices, aiming to optimize resource utilization efficiency in a multi-agent environment through refined time management and allocation strategies. The core of this mechanism is to divide continuous resource usage periods into multiple discrete time slices, allocating exclusive resource access periods for different agents.

This scheduling strategy effectively alleviates resource competition issues, avoiding potential resource conflicts and ensuring smooth system operation. By executing concurrently, the system's throughput is significantly improved, agents do not occupy each other's resources, thereby reducing the average waiting time and turnaround time for each agent, and allowing users to receive feedback from agents as soon as possible.

4   Experiments

4.1   Experimental Setup

The experiments in this paper were conducted using the following equipment configuration: Intel i9-9300H processor, GTX 4090 graphics card, 128GB RAM, running on the open-source Kirin 2.0 platform. The programming environment included Python 3.11, PyTorch 2.0.1, and CUDA 11.8. Five types of intelligent agents were used: (1) a management agent for task allocation and management, (2) a writing agent for text generation, (3) a search agent for web searches, (4) a meeting agent for initiating, recording, and forwarding meeting content, and (5) a text scoring agent for evaluating the texts generated by the agents.

Three sets of experiments were designed to test the collaborative efficiency of vertically integrated multi-agent systems, the resource scheduling strategy of time-shared multi-agent systems, and the performance of different large models. The experimental setup is detailed as follows.

Experiment 1: Vertical Multi-agent Collaboration. Experiment 1 includes an experimental group and a control group, using a text generation task as an example. The experimental group consisted of a management agent and a writing agent. The task was first sent to the management agent, which analyzed the task and then assigned it to the writing agent, providing several suggestions. The control group consisted of only a writing agent, and the task was directly sent to this agent for completion. Additionally, a text scoring agent evaluated the texts generated by both the experimental and control groups. Nine tasks were designed (see Table 1), and both groups completed these tasks to generate texts.

Table 1. The nine text generation tasks covering various dimensions from personal reflections, technological outlooks, literary creations to social issues, showcasing the comprehensiveness and diversity of the experiment.

Task Number	Task Description
1	Write a travel biography of Fujian.
2	Write your views on the future development of artificial intelligence.
3	Write an article on the importance of environmental protection.
4	Write a review of a book you recently read.
5	Write about an unforgettable travel experience.
6	Discuss the impact of modern technology on human life.
7	Write a letter to your future self.
8	Explain blockchain technology and its potential applications.
9	Write a short story about an unexpected adventure.

Experiment 2: Time-Shared Multi-agent Resource Scheduling Strategy. Experiment 2 tested the performance of three agents (writing agent, meeting agent, and search agent) under a time-shared resource scheduling strategy. Performance metrics included the waiting time, turnaround time, and response time of each agent, where the response time refers to the interval from the user submitting a request to the first response. The experiment also tested the performance of different large models running agents, using Qwen-14B-Chat-g4f32_1-MLC (referred to as Qwen-MLC) and Qwen-plus models, tested through local and Application Programming Interface (API) calls. The parameters of the large models are shown in Table 2.

Experiment 3: Testing the Text Generation Capability of Different Large Models. Experiment 3 evaluated the text generation performance of different large models within the system, with only the writing agent running.

Table 2. Parameters of the large models.

Model Name	Qwen-MLC	Qwen-plus	Gemini-pro	gpt-3.5-turbo
Open Source	14B	Not disclosed	100B	Not disclosed
Invocation Method	Local	API	API	API

Unlike Experiment 1, this experiment used three different large models (Qwen-plus, Gemini-pro, gpt-3.5-turbo) to generate texts based on the nine tasks in Table 1, and a text scoring agent evaluated the generated texts.

Through the above experimental setup, the effectiveness and performance of multi-agent collaboration and resource scheduling strategies can be comprehensively evaluated.

4.2 Analysis of Vertical Multi-agent Collaboration Experiments

This study designed and implemented experiments on vertical multi-agent collaboration to explore the impact of introducing a management agent on improving text quality and completion efficiency in text generation tasks. All tasks were meticulously designed, covering multiple dimensions from personal reflections, technological outlooks, literary creations to social issues, as shown in Table 1, ensuring the comprehensiveness and diversity of the experiment.

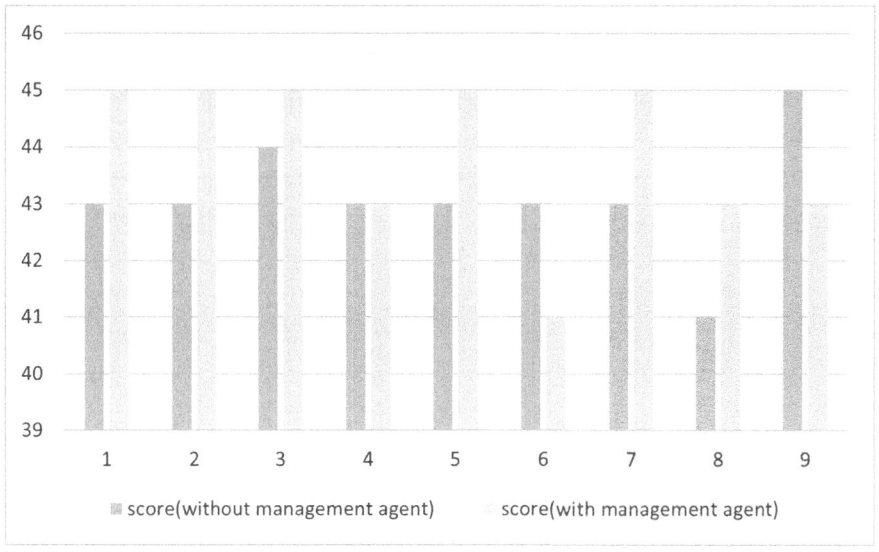

Fig. 3. Comparison of multi-agent collaboration and single-agent performance

Figure 3 demonstrates the efficacy of incorporating a management agent. The experimental group outperformed or matched the control group across five evaluation criteria: content quality, language fluency, structure and logic, topic relevance, and reader engagement. With an overall mean score of 43.89 compared to the control's 43.11, the experimental group achieved optimal or near-optimal scores in most tasks. These results underscore the management agent's critical function in task allocation, strategy refinement, and creative input, thereby elevating the overall quality of generated text.

4.3   Analysis of Time-Sharing Multi-Agent Resource Scheduling

The study compares time-slice, FIFO, and non-scheduling approaches in multi-agent systems, as depicted in Fig. 4, revealing that time-slice scheduling enhances overall system performance. Time-slice scheduling, while exhibiting a modestly elevated average waiting time compared to FIFO, notably diminishes the waiting time for the meeting agent, particularly in high-demand scenarios, thus optimizing system resource allocation.

The study's examination of turnaround time underscores the adaptability of time-slice scheduling to complex tasks. Despite a relative increase in turnaround time for the meeting agent, the strategy significantly outperforms non-scheduling, thereby mitigating the impact of high-load tasks. For writing and search agents, time-slice scheduling either slightly increases or mirrors FIFO's performance, yet markedly surpasses non-scheduling, ensuring multi-task equilibrium.

Scheduling Method	Evaluation Metrics	Qwen-MLC			Qwen-plus		
	Agent Type	Writing Agent	Searching Agent	Meeting Agent	Writing Agent	Searching Agent	Meeting Agent
Time-Sliced Scheduling	Waiting Time	3.07±2.86	1.19±1.79	3.14±1.79	2.42±3.63	1.03±1.00	6.67±1.50
	Turnaround Time	28.89±2.20	52.61±4.65	104.33±13.00	36.64±5.01	54.23±3.17	114.69±6.69
	Response Time	10.53±8.62	7.01±5.46	11.78±5.41	8.66±10.90	8.34±1.49	22.21±4.63
FIFO	Waiting Time	1.20±1.80	0.91±0.68	4.81±0.15	1.07±1.60	0.84±0.81	4.56±0.56
	Turnaround Time	27.55±3.10	56.96±2.44	118.55±7.85	33.17±4.31	51.57±5.30	106.00±12.18
	Response Time	4.94±5.36	6.19±2.02	16.78±0.37	4.58±5.15	5.71±2.52	16.20±1.65
No Scheduling	Waiting Time	0.00±0.00	20.75±4.60	60.20±9.56	0.00±0.00	28.03±6.80	64.64±10.21
	Turnaround Time	20.75±4.60	60.20±9.56	170.55±29.02	28.03±6.80	64.64±10.21	171.63±22.80
	Response Time	1.31±0.07	24.17±4.77	62.69+9.65	0.93±0.13	31.52±7.66	66.5+10.65

Fig. 4. Evaluation Metrics for Different Scheduling Methods

4.4   Performance Analysis of Large Models in Text Generation

In Experiment 3, the present study compares the three large models as depicted in Table 2. We evaluated three large models' text generation abilities through a single writing agent across nine tasks, with results in Fig. 5. Gpt-3.5-turbo emerged as the top performer with an average score of 44.67, demonstrating superior and consistent capabilities. Gemini-pro and Qwen-plus closely followed with scores of 43.33 and 43.11, respectively, showing stable performance, though Qwen-plus had a minor decline in Task 8. Tasks 1, 5, and 9 were less effective in distinguishing model capabilities. The study concludes that gpt-3.5-turbo is the preferred choice for high-quality text generation, while Qwen-plus and Gemini-pro are reliable alternatives, especially for tasks where model differentiation is less critical.

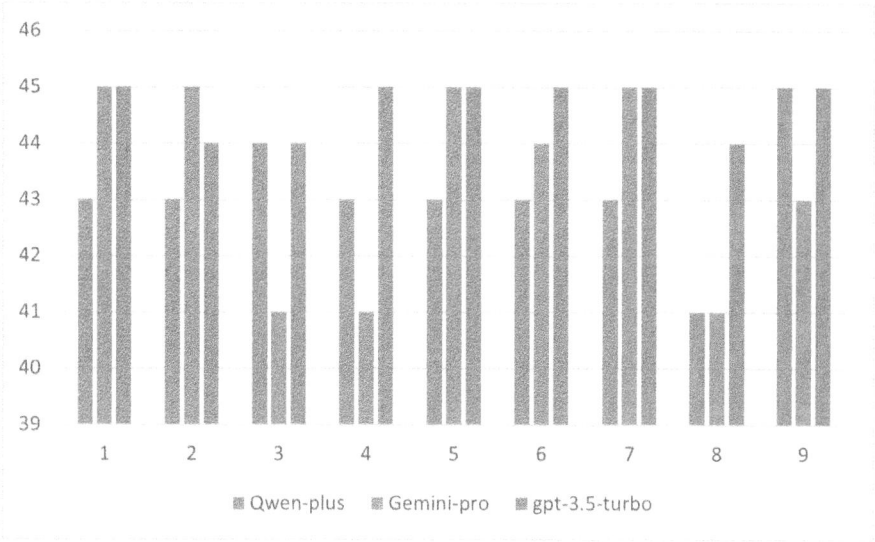

Fig. 5. Text generation capabilities for different large models within the framework. The scores in the chart represent the total score across five assessment dimensions: content quality, language fluency, structure and logic, topic relevance, and reader engagement.

5   Conclusion and Future Work

This paper explores the innovative applications of an LLM-based multi-agent operating system on the user-friendly, open-source Kirin 2.0 platform. We designed an intelligent, efficient, and resource-optimized multi-agent collaboration system featuring a management role agent to improve work efficiency. The study also investigates low-resource vertical collaboration, intelligent task allocation, and the reduction of redundant computation. Finally, we developed and implemented a time-sharing resource scheduling strategy to avoid resource conflicts, optimize resource utilization, and ensure system stability and efficiency under high-load conditions.

While this paper demonstrates the feasibility of multi-agent collaborative applications, there are some limitations. During the research, we identified several areas for improvement. For instance, the configuration of tool API parameters by the agents was adjusted by researchers. Future work can focus on enabling the large models to learn how to use various tool APIs independently. Scheduling methods will be refined for finer granularity, optimizing agent response times by incorporating real-time scheduling, shortest job first, and priority scheduling algorithms. Additionally, we plan to develop software development agents and system diagnostic agents on the open-source Kirin platform, leveraging the advantages of open-source code.

Our research and experiments indicate that LLM-based multi-agent operating systems have significant development potential from both theoretical and practical perspectives. Such systems can achieve seamless desktop operating system integration and lead the way in the intelligent development of operating systems. LLM-based multi-agent systems will realize greater potential in future advancements.

References

1. Xi, Z., et al.: The rise and potential of large language model based agents: a survey. arXiv preprint arXiv:2309.07864 (2023)
2. Guo, T., et al.: Large language model based multi-agents: a survey of progress and challenges. arXiv preprint arXiv:2402.01680 (2024)
3. Dambekodi, S., Frazier, S., Ammanabrolu, P., Riedl, M.O.: Playing text-based games with common sense. arXiv preprint arXiv:2012.02757 (2020)
4. Nakano, R., et al.: Webgpt: browser-assisted question-answering with human feedback. arXiv preprint arXiv:2112.09332 (2021)
5. Wu, C., Yin, S., Qi, W., Wang, X., Tang, Z., Duan, N.: Visual ChatGPT: talking, drawing and editing with visual foundation models. arXiv preprint arXiv:2303.04671 (2023)
6. Ahn, M., et al.: Do as i can, not as i say: grounding language in robotic affordances. arXiv preprint arXiv:2204.01691 (2022)
7. Parisi, A., Zhao, Y., Fiedel, N.: Talm: tool augmented language models. arXiv preprint arXiv:2205.12255 (2022)
8. Ge, Y., et al.: Openagi: when LLM meets domain experts. In: Advances in Neural Information Processing Systems, vol. 36 (2024)
9. Schick, T., et al.: Toolformer: language models can teach themselves to use tools. In: Advances in Neural Information Processing Systems, vol. 36 (2024)
10. Yao, S., Narasimhan, K.: Language agents in the digital world: opportunities and risks (2023). https://princeton-nlp.github.io/language-agent-impact/. Accessed 24 Jul 2023
11. Tang, Q., et al.: Toolalpaca: generalized tool learning for language models with 3000 simulated cases. arXiv preprint arXiv:2306.05301 (2023)
12. Zhang, K., et al.: Toolcoder: teach code generation models to use APIs with search tools. arXiv preprint arXiv:2305.04032 (2023)

13. Yao, S., Chen, H., Yang, J., Narasimhan, K.: Webshop: towards scalable real-world web interaction with grounded language agents. In: Advances in Neural Information Processing Systems, vol. 35, pp. 20744–20757 (2022)
14. Deng, X., et al.: Mind2web: towards a generalist agent for the web. In: Advances in Neural Information Processing Systems, vol. 36 (2024)
15. Furuta, H., et al.: Multimodal web navigation with instruction-finetuned foundation models. arXiv preprint arXiv:2305.11854 (2023)
16. Zhou, S., et al.: Webarena: a realistic web environment for building autonomous agents. arXiv preprint arXiv:2307.13854 (2023)
17. Khot, T., et al.: Decomposed prompting: a modular approach for solving complex tasks. arXiv preprint arXiv:2210.02406 (2022)
18. Sumers, T.R., Yao, S., Narasimhan, K., Griffiths, T.L.: Cognitive architectures for language agents. arXiv preprint arXiv:2309.02427 (2023)
19. Talebirad, Y., Nadiri, A.: Multi-agent collaboration: harnessing the power of intelligent LLM agents. arXiv preprint arXiv:2306.03314 (2023)
20. Ge, Y., et al.: LLM as OS, agents as apps: envisioning AIOS, agents and the AIOS-agent ecosystem. arXiv e-prints arXiv:2312.03815 (2023)
21. Hong, S., et al.: MetaGPT: meta programming for multi-agent collaborative framework. arXiv preprint arXiv:2308.00352 (2023)
22. Wu, Q., et al.: Autogen: enabling next-gen LLM applications via multi-agent conversation framework. arXiv preprint arXiv:2308.08155 (2023)
23. Qian, C., et al.: Communicative agents for software development. arXiv preprint arXiv:2307.07924 (2023)
24. Josifoski, M., et al.: Flows: building blocks of reasoning and collaborating AI. arXiv preprint arXiv:2308.01285 (2023)
25. Li, G., et al.: Camel: communicative agents for "mind" exploration of large language model society. In: Advances in Neural Information Processing Systems, vol. 36 (2024)
26. Park, J.S., et al.: Generative agents: Interactive simulacra of human behavior. In: Proceedings of the 36th Annual ACM Symposium on User Interface Software and Technology, pp. 1–22 (2023)
27. Du, Y., et al.: Improving factuality and reasoning in language models through multiagent debate. arXiv preprint arXiv:2305.14325 (2023)
28. Chan, C.-M., et al.: Chateval: towards better LLM-based evaluators through multi-agent debate. arXiv preprint arXiv:2308.07201 (2023)
29. Liang, J., et al.: Encouraging diverse and consistent behaviour in multi-agent systems with language models. arXiv preprint arXiv:2309.16834 (2023)
30. Hamilton, K., Wick, M., Chambers, N.: Blind debate: a contrastive learning framework for debater-agnostic verdict models. arXiv preprint arXiv:2310.12221 (2023)
31. Wang, D., et al.: Unleashing the power of LLM agents: challenges and opportunities. arXiv preprint arXiv:2311.02019 (2023)
32. Zhang, Z., et al.: Proagent: towards proactive large language model based autonomous agents. In: Advances in Neural Information Processing Systems, vol. 36 (2024)
33. Hassan, M., et al.: ChatGPT-eval: an open-source LLM evaluator with GPT-4 level performance. arXiv preprint arXiv:2310.13870 (2023)
34. Liu, Y., et al.: Training multi-agent decision making with machine-in-the-loop interaction. arXiv preprint arXiv:2311.10788 (2023)
35. Mei, K., et al.: AIOS: LLM agent operating system. arXiv preprint arXiv:2403.16971 (2024)

Integrating Large Language Models with Knowledge Graphs in Traditional Chinese Medicine Consultation: A Case Study

Heyi Zhang[1], Xin Wang[1(✉)], Zhaopeng Meng[2], Junhua Zhang[2], Zhe Chen[3], Pengwei Zhuang[4,5], Yongzhe Jia[6], Dawei Xu[6], and Wenbin Guo[1]

[1] College of Intelligence and Computing, Tianjin University, Tianjin 300354, China
wangx@tju.edu.cn
[2] Tianjin University of Traditional Chinese Medicine, Tianjin 301617, China
[3] Evidence-Based Medicine Center, Tianjin University of Traditional Chinese Medicine, Tianjin 301617, China
[4] First Teaching Hospital of Tianjin University of Traditional Chinese Medicine, Tianjin 300193, China
[5] National Clinical Research Center for Chinese Medicine Acupuncture and Moxibustion, Tianjin 300193, China
[6] Graph Intelligence (Tianjin) Co. Ltd., Tianjin, China

Abstract. In traditional Chinese medicine (TCM), Large Language Models (LLMs) face challenges due to theoretical differences from modern medicine and a scarcity of specialized data. We address these with a two-stage training: continuous pre-training followed by supervised fine-tuning. Our study organize a 2GB TCM corpus consisting of pre-trained and fine-tuning datasets. In addition, we have developed Qibo-Benchmark, a tool that evaluates the performance of LLM in the TCM on multiple dimensions, including objective, and three TCM NLP tasks. The medical LLM trained with our pipeline, named **Qibo**, exhibits significant performance boosts. Compared to the baselines, the average objective accuracy improved by 23% to 58%, and the Rouge-L scores for the three TCM NLP tasks are 0.72, 0.61, and 0.55. Finally, we propose a pipline to apply Knowledge Graphs and LLMs to TCM consultation and demonstrate the performance through a case.

Keywords: Large Language Models · Knowledge Graph · traditional Chinese medicine

1 Introduction and Related Works

Recently, significant advances have been made in Large Language Models (LLMs), from ChatGPT [1] to GPT-4o [2]. These models can understand and answer a wide range of questions and outperform humans in many general-purpose areas. To fill the gaps in the Chinese language processing capabilities of

B. Xu et al. (Eds.): CCKS-IJCKG 2024, CCIS 2229, pp. 360–366, 2025.
https://doi.org/10.1007/978-981-96-1809-5_28

these models, researchers have also introduced more powerful Chinese language models, such as Chinese LLaMA [3], GLM [4], Baichuan [5], and so on.

In the realm of traditional Chinese medicine (TCM), several medical LLMs have been proposed [6–8]. These LLMs are mainly trained through supervised fine-tuning (SFT). While SFT is crucial for acquiring domain-specific knowledge, it often results in limited knowledge infusion and can lead to overconfident generalizations [9]. As well as Reinforcement Learning from Human Feedback (RLHF) is a popular method to offset some of the limitations of SFT, but it is very complex and requires rigorous hyperparameter tuning [10]. [11,12] have shown that almost all knowledge is learned in pre-training, which is a crucial stage for accumulating basic knowledge.

Although many works have existed on LLMs in TCM, and these works have further advanced the development of large models in TCM, yet the characteristics of TCM domain are neglected by them. They have never considered the essential differences between TCM and modern medical theories.

The main contributions of this paper are as follows: Firstly, we provide a data processing scheme based on different granularity rules to further improve the quality of TCM training corpus. Then, we have trained a new TCM LLM based on Chinese-LLaMA in two stages from pre-training to SFT. Finally, a pipeline is designed to combine the model and knowledge graph to TCM consultation and syndrome differentiation to improve the interpretability and performance of the diagnosis process.

Related Works. Chinese medical domain studies include DoctorGLM [8], which used extensive Chinese medical dialogue data and an external medical knowledge base, and BenTsao [13], utilizing only a medical knowledge graph for dialogue construction. Zhang, et al. [7] created HuatuoGPT with a 25-million dialogue dataset, achieving better response quality through a blend of distilled and real data for SFT and ChatGPT for RLHF feedback ranking. Zhongjing [14], which is a Chinese medical LLaMA-based LLM that implements an entire training pipeline from pre-training, SFT, to RLHF and introduce a Chinese multi-turn medical dialogue dataset of 70,000 authentic doctor-patient dialogues, CMtMedQA, which significantly enhances the capability for complex dialogue and proactive inquiry initiation.

2 Method

This section explores the construction of Qibo, dividing into three stages: *Data Process, Training Phase,* and *TCM Consultation and Syndrome Differentiation.* The overall construction process of Qibo is shown in Fig. 1.

Data Process. There are fewer sources of TCM corpus expertise, mainly modern TCM textbooks, TCM ancient books, encyclopedia of TCM and so on. For

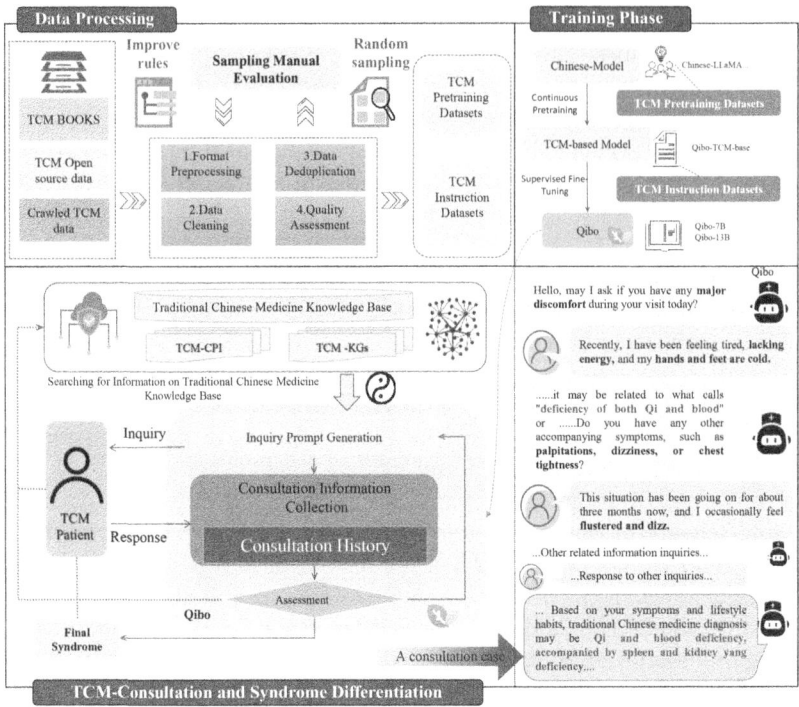

Fig. 1. The Overall Construction Process of Qibo.

the processing of raw data, we convert the raw data into a uniform JSON format, and then do the cleansing, de-duplication, and quality assessment of the data to obtain the training corpus of high-quality.

Training Phase. Qibo undergoes through two stages of training: *Continuous Pre-training* and *Supervised Instruction Fine-Tuning*. The former gained basic understanding of TCM, while the latter gained the ability to follow and answer TCM instructions.

Continuous Pre-training: We construct the TCM pre-training dataset and perform continuous pre-training on Chinese-LLaMA by predicting the probability of the next word to finally obtain a basic TCM model with TCM knowledge system, understanding, dialectical ability, and entity recognition ability. N indicates the numbers of sequences partitioned from TCM Pre-training Dataset, where each sequence $X_i = [x_{i,1}, x_{i,2}, ..., x_{i,T}]$ contains T tokens. We define the loss function as the sum of the negative log probabilities of the next token $x_{i,t+1}$ given the previous tokens $x_{i,1...t}$ in the sequence, where θ denotes the model parameters.

$$L = -\sum_{i=1}^{N}\sum_{t=1}^{T}\log[P(x_{i,t+1}|x_{i,1...t},\theta)] \tag{1}$$

Supervised Instruction Fine-Tuning: We fine-tune the model by translating data from multiple sources into a multi-turn conversation format[1] to enhance the capability of following and answering instructions. Considering each prompt $X_i = [X_{i,1}, X_{i,2}, ...]$ and its corresponding response $Y_i = [Y_{i,1}, Y_{i,2}, ...Y_{i,T_i}]$ contains T_i tokens, the loss function in the SFT stage can be defined as follows:

$$L = -\sum_{i=1}^{N}\sum_{t=1}^{T_i}\log[P(y_{i,t+1}|X_i, y_{i,1...t},\theta)] \tag{2}$$

where N represents the total number of training instances, and θ represents the model parameters.

TCM Consultation and Syndrome Differentiation. This method utilizes external TCM knowledge graph[2] to guide the model in asking patients, collecting consultation information, further explaining relevant consultation principles, increasing the interpretability and credibility of the consultation. On the other hand, by using multiple rounds of comprehensive information to judge the condition of patients, the accuracy of the consultation can be further improved. The specific case is shown in the bottom left part of Fig. 1, and some examples of a single consultation are shown in the bottom right part. Based on the initial judgment of the chief complaint, the model identified "deficiency of both Qi and blood syndrome" (气血两虚证). After multiple rounds of questioning with the patient to obtain more relevant information, it is determined that the "Qi and blood deficiency syndrome" is accompanied by "spleen and kidney yang deficiency syndrome" (脾肾阳虚证).

3   Experiments and Evaluation

In order to evaluate our model, we chose a series of LLMs with different parameter scales as benchmarks for comparison, including generalized LLMs and medical LLMs, including **ChatGPT, Chinese-LLaMA, BenTsao, DoctorGLM, HuatuoGPT, Zhongjing**.

3.1   Objective Evaluation

For the objective evaluation of models, referring to Med-PaLM [15], we used the model to select the correct option from multiple-choice questions for verification, and examined the knowledge and understanding ability related to TCM

---

[1] Convert script link, https://github.com/zhangheyi-1/ccks-IJCKG.
[2] Detail Link, https://github.com/zhangheyi-1/ccks-IJCKG/blob/main/KGs.

contained in the model. A total of 3,175 practice questions related to the 13 TCM practice exams are collected and organized as assessment data, which are measured by comparing the accuracy of responses across subjects. The experimental results are shown in Fig. 2, and our method achieves optimal performance

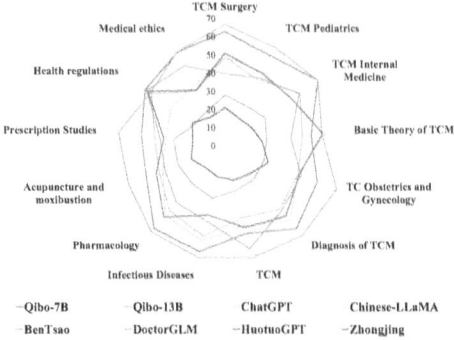

Fig. 2. The accuracy of different models in the 13 subject multiple-choice questions of the TCM Practicing Examination.

3.2   TCM NLP Tasks Evaluation

In order to further explore the comprehensive performance of the model and the generalization ability of TCM NLP tasks, we retained 517, 689, and 475 data from the prescription identification dataset, syndrome differentiation dataset, and reading comprehension quiz dataset, respectively, for assessment, and the assessment criteria are adopted from Rouge-L [16]. We compare the model results with the optimal results of traditional BERT-based proprietary methods [17–19] on the corresponding dataset to verify the performance of the model. The experimental results are shown in Table 1. Although not as optimal as traditional methods, our method achieves optimal results in LLMs.

Table 1. The results of three NLP tasks, where "* "indicates the best outcome of the method specifically designed for this task. TCM-NER refers to the entity recognition task of TCM prescriptions. TCM RP is a TCM reading comprehension quiz pair construction task. TCM-SD refers to the task of syndrome differentiation in TCM.

Task	ours	Zhongjing	BenTsao	HuatuoGPT	DoctorGLM	*
TCM-NER	0.72	0.54	0.23	0.63	0.21	**0.78**
TCM-RP	0.61	0.58	0.21	0.60	0.19	**0.63**
TCM-SD	0.54	0.34	0.12	0.31	0.10	**0.87**

4   Conclusion

We present Qibo, an LLM in the traditional Chinese medical domain, which is implemented from pre-training to SFT. Its performance exceeds that of other open-source Chinese medical models in TCM, and is comparable to models with significantly more parameters. We have collected high-quality training corpus data in TCM and constructed an evaluation benchmark, to address the evaluation gap. Furthermore, a pipeline is designed to apply the model and KG to TCM consultation, with the aim of improving the interpretability and performance of the diagnostic process. Finally, sufficient experiments have validated the excellent performance of our model in TCM.

References

1. OpenAI, T.: ChatGPT: optimizing language models for dialogue. OpenAI (2022)
2. OpenAI.: Hello GPT-4o (2024). https://openai.com/index/hello-gpt-4o/
3. Cui, Y., Yang, Z., Yao, X.: Efficient and effective text encoding for Chinese llama and alpaca. arXiv preprint arXiv:2304.08177 (2023)
4. Du, Z., et al.: GLM: general language model pretraining with autoregressive blank infilling. arXiv preprint arXiv:2103.10360 (2021)
5. Baichuan.: Baichuan 2: open large-scale language models. arXiv preprint arXiv:2309.10305 (2023)
6. Li, J., et al.: Huatuo-26m, a large-scale Chinese medical QA dataset. arXiv preprint arXiv:2305.01526 (2023)
7. Zhang, H., et al.: Huatuogpt, towards taming language model to be a doctor. arXiv preprint arXiv:2305.15075 (2023)
8. Xiong, H., et al.: Doctorglm: Fine-tuning your Chinese doctor is not a herculean task. arXiv preprint arXiv:2304.01097 (2023)
9. Luo, Y., et al.: An empirical study of catastrophic forgetting in large language models during continual fine-tuning. arXiv preprint arXiv:2308.08747 (2023)
10. Ramamurthy, R., et al.: Is reinforcement learning (not) for natural language processing?: benchmarks, baselines, and building blocks for natural language policy optimization. arXiv preprint arXiv:2210.01241 (2022)
11. Han, X., et al.: Pre-trained models: past, present and future. AI Open **2**, 225–250 (2021)
12. Zhou, C., et al.: Lima: less is more for alignment. arXiv preprint arXiv:2305.11206 (2023)
13. Wang, H., et al.: Huatuo: tuning llama model with Chinese medical knowledge. arXiv preprint arXiv:2304.06975 (2023)
14. Yang, S., et al.: Zhongjing: enhancing the Chinese medical capabilities of large language model through expert feedback and real-world multi-turn dialogue. arXiv preprint arXiv:2308.03549 (2023)
15. Singhal, K., et al.: Large language models encode clinical knowledge. arXiv preprint arXiv:2212.13138 (2022)
16. Lin, C.Y.: Rouge: a package for automatic evaluation of summaries. In: Text Summarization Branches Out, pp. 74–81 (2004)
17. Mucheng, R., et al.: TCM-SD: a benchmark for probing syndrome differentiation via natural language processing. In: Proceedings of the 21st Chinese National Conference on Computational Linguistics, pp. 908–920 (2022)

18. Alibaba: "Wanchuang Cup" challenge of TCM literature problem generation (2020). https://tianchi.aliyun.com/competition/entrance/531824/rankingList
19. Alibaba: "Wanchuang Cup" challenge of entity recognition for TCM instructions (2020). https://tianchi.aliyun.com/competition/entrance/531826/rankingList

Local Index File-Based Tool for Extracting Class Hierarchies from Wikidata

Kouji Kozaki[1,2(✉)] 🆔, Shusaku Egami[2] 🆔, and Ken Fukuda[2] 🆔

[1] Osaka Electro-Communication University, 18-8 Hatsucho, Neyagawa-shi,
Osaka 572-8530, Japan
`kozaki@osakaca.ac.jp`
[2] National Institute of Advanced Industrial Science and Technology (AIST),
Osaka, Japan

Abstract. Wikidata's class hierarchies provide crucial knowledge for building knowledge graphs. However, the hierarchy is so extensive that extracting the necessary class hierarchy is challenging. In this study, we develop a tool to efficiently extract class hierarchies by locally storing information on Wikidata class hierarchies in an index file.

Keywords: Class hierarchy · Wikidata · Knowledge graph · Ontology

1 Introduction

Wikidata is used by many knowledge systems as an open knowledge graph where a lot of factual information is registered [1]. Among these, the class hierarchy is fundamental for utilizing the semantics provided by the knowledge graph. For example, a search for the label "Osaka" (in English) in Wikidata returns 21 data entries. These entries belong to 23 classes, including "metropolis," "family name," "song," "ship," "Wikimedia disambiguation page," and so on. By checking these classes, the required data can be identified. In this study, we develop a method and tool to extract arbitrary class hierarchies from Wikidata in order to utilize Wikidata's class hierarchies more efficiently.

The paper is organized as follows: Sect. 2 describes a general method for extracting Wikidata class hierarchies and its challenges, Sect. 3 proposes the development of a tool to address these challenges, Sect. 4 presents the results of extracting class hierarchies using the proposed tool, Sect. 5 compares the results with related studies, and Sect. 6 concludes the paper.

2 Extracting Class Hierarchy from Wikidata

Wikidata defines about 110 million data items, which are classified into 4 million classes using the instance-of property (wdt:P31). The classes are hierarchically classified using the subclass-of property (wdt:P279). Therefore, to extract the

B. Xu et al. (Eds.): CCKS-IJCKG 2024, CCIS 2229, pp. 367–372, 2025.
https://doi.org/10.1007/978-981-96-1809-5_29

class hierarchy, we can recursively retrieve the triple "subclass wdt:P279 super-class". Fig. 1 shows an example of a class hierarchy obtained from Wikidata with "entity (wd:Q35120)" and "product (wd:Q2424752)" as root[1] However, with this method, the number of SPARQL queries increases rapidly as the class hierarchy gets deeper, which increases the load on the server and the time required for retrieval. To mitigate this, the tool used in Fig. 1 limits the depth and width of the class hierarchy to be expanded, and expands it as needed.

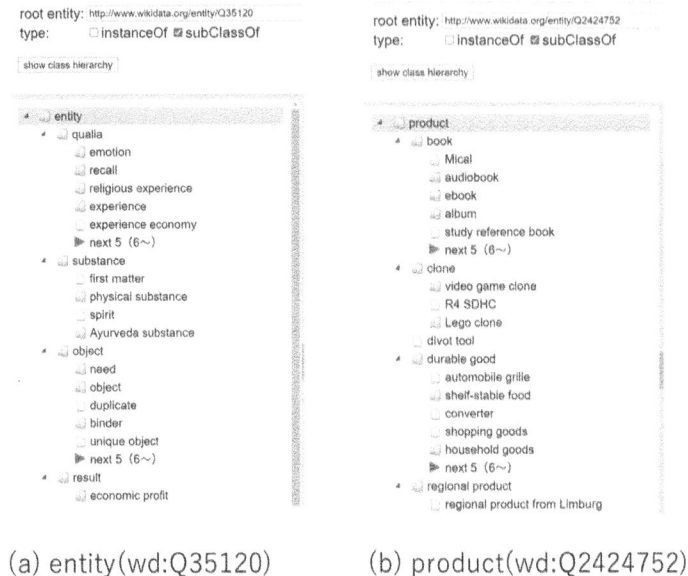

(a) entity(wd:Q35120)   (b) product(wd:Q2424752)

Fig. 1. Examples of class hierarchies in Wikidata.

Therefore, we consider a SPARQL query that can retrieve all the information in a class hierarchy in a single query. Figure 2 shows an example of a query to obtain the class hierarchy for three levels, using "entity (wd:Q35120)" as the root. In this query, "wdt:P279/wdt:P279/wdt:P279" represents the condition to repeat subclass-of property (wdt:P279) three times, thus obtaining the third level subclasses. By increasing the number of iterations, a class hierarchy of arbitrary depth can be obtained. However, the deeper the hierarchy in this query, the larger the number of combinations to be searched, and a timeout occurs on the server, making it impossible to retrieve the class hierarchy. In the case of "entity (wd:Q35120)", the timeout occurred when the depth of the class hierarchy to be retrieved was set to 5 levels.

Based on the above discussion, we concluded that to extract arbitrary class hierarchies from Wikidata, it is better to use a tool that can be operated on

---

[1] They are available at https://hozo.jp/wikidata_tree/index.html.

```
SELECT DISTINCT *
WHERE {
    ?s wdt:P279|wdt:P279/wdt:P279|wdt:P279/wdt:P279/wdt:P279  wd:Q35120 .
    ?s ?p ?o.
    FILTER((?p = wdt:P279)||((((?p = rdfs:label)||(?p = skos:altLabel))
                            &&((lang(?o)="en")||(lang(?o)="ja"))))
}ORDER BY ?s
```

Fig. 2. An example of a SPARQL query to obtain the class hierarchy for three levels from Wikidata, using "entity (wd:Q35120)" as the root. It also obtains labels and aliases of each classes in English and Japanese.

a local PC instead of using a SPARQL query to the Wikidata Query Service. Therefore, this study aims to develop a local index file-based tool for extracting class hierarchies from Wikidata.

3   Local Index File-Based Tool for Extracting Class Hierarchy from Wikidata

3.1   Local Index File for Class Hierarchy

In developing a tool to use Wikidata information on a local PC, the tool does not target Wikidata as a whole, but only the data necessary to extract class hierarchies, since Wikidata is a huge database. However, using SPARQL queries to extract the necessary data is difficult due to server load issues such as timeouts, as described in the previous section. Therefore, the necessary data is extracted from Wikidata dump files and used.

Specifically, triples containing "subclass-of (wdt:P279)" and "instance-of (wdt:P31)" as predicates are extracted from the N-triples format dump file, where the conditions for data extraction can be easily specified. In addition, "label (rdfs:label)", "alias (soks:altLabel)" and "Description (schema:description)" of each data item are extracted as language information to understand the contents of the extracted data.

In order to make these data easily accessible independent of the computer environment, we do not use specific database software, but manage them using index files with file names corresponding to the ID of each data item. Index files are prepared for each property type. For example, the subclass list for data with ID Q35120 is stored in the file "/Q35/120.dat". For "subclass-of (wdt:P279)" and "instance-of (wdt:P31)", the index file should be created separately in both directions, since it requires bidirectional processing of specifying a subject and obtaining an object/specifying an object and obtaining a subject. On the other hand, for language information, an index file is created only in one direction for the language information for the data[2].

---

[2] Searching for IDs from strings such as labels requires indexing in the reverse direction. Although not used in this study, we have implemented a tool using indexing by Apache Lucene for string searches.

3.2   Functionalities for Extracting Class Hierarchies

Basic Function: Extraction of Class Hierarchy Specifying Root Class.
To extract the class hierarchy, specify the class to be the root class and retrieve subclasses recursively, as described in Sect. 2. To retrieve instances classified into each class along with the class hierarchy, retrieve instances recursively along with the subclasses. If necessary, language information (label, alias, description) of each data item is acquired at the same time. This is the basic function of this tool.

Optional Setting: Restrictions on Subclass Search. The class hierarchy of Wikidata is so large that it may be redundant to extract all subclasses and instances. For this reason, the tool provides the following options to limit the search for subclasses.

– Depth of class hierarchy to search : 1 or more integers
– Explore instances? : yes or no
– Skip searching for duplicate subclasses? : yes or no
– Skip searching for subclasses that do not have instances? : yes or no
– Skip searching for subclasses that do not have labels? : yes or no

These settings can be combined to obtain a range of class hierarchies as needed.

Bottom-Up Determination of the Root of the Class Hierarchy. When extracting the class hierarchy of a domain, it is sometimes difficult to identify the root class. In such cases, there is a method to identify the root class by collecting multiple terms (entities) in the target domain and searching for the upper classes common to them in a bottom-up manner [2].

To support this method, this tool provides a function that extracts the upper classes common to multiple entities and designates them as the root class when the number of common lower classes exceeds a certain threshold. By applying the root class obtained here to the extraction of class hierarchies described above, a class hierarchy for a specific domain can be obtained collectively.

4   Implementation and Discussion

We implemented the local index file-based tool for extracting class hierarchies from Wikidata described in Sect. 3 using Java. A Wikidata N-Triples format dump file dated June 9, 2024, was used as the source data for index generation for this tool. The developed tool is available at https://github.com/oecu-kozaki-lab/WD-HierarchyTool.

To verify the performance of the developed tool, we tested the extraction of class hierarchies. The extraction conditions were "entity (wd:Q35120)" and "product (wd:Q2424752)" as root, and the restrictions for subclass exploration set to a depth of 1 to 5 and "no" for all other conditions. Tables 1-(a) shows the size of the obtained class hierarchies and the time required for extraction.

As a comparison, Table 1-(b) shows the results of SPARQL queries against the Wikidata query service to obtain the class hierarchies under similar conditions. The SPARQL query was the one shown in Fig. 2, with a modified specification of "the number of layers of class hierarchy" and "root class."

All the above processes were performed on the same Windows PC (Windows 11 pro 64bit, Xeon processor, 32 GB memory).

Comparing Table 1-(a) and (b), we can see that the SPARQL query is faster in terms of processing time, but the SPARQL query cannot retrieve class hierarchies higher than 5 levels due to server timeout. Considering that the number of instances is much larger than the number of classes in this comparison, the range of hierarchies that can be obtained by SPARQL queries is limited, and the proposed tool is useful for obtaining larger hierarchies. Additionally, the proposed tool allows the specification of extraction conditions for various class hierarchies as described in Sect. 3.2, whereas it is difficult to specify such details in a SPARQL query. Thus, the SPARQL query is suitable for quickly retrieving class hierarchies of small depth, while the proposed tool seems to be suitable for retrieving larger class hierarchies, considering detailed conditions.

Table 1. Table captions should be placed above the tables.

(a) Class Hierachies extracted using the proposed tool

Root Class	Layers	1	2	3	4	5
wd:Q35120 (entity)	time(s)	0.33	1.34	22.00	565.00	2208.94
	triples	256	3,175	26,567	167,751	4,549,912
	(subclass-of)	(39)	(809)	(7,646)	(52,515)	(1,847,465)
wd:Q2424752 (product)	time(s)	1.42	7.47	34.58	176.57	737.18
	triples	1,591	19,922	104,370	329,792	2,855,032
	(subclass-of)	(501)	(6,570)	(35,849)	(115,613)	(1,045,876)

(b) Class Hierachies extracted using SPARQL queries to Wikidata

Root Class	Layers	1	2	3	4	5
wd:Q35120 (entity)	time(s)	1.13	2.14	7.88	28.70	timeout
	triples	249	3,765	30,611	172,662	–
	(subclass-of)	(36)	(1,235)	(10,911)	(65,300)	
wd:Q2424752 (product)	time(s)	2.02	3.99	11.17	49.43	timeout
	triples	2035	24,581	111,194	266,818	–
	(subclass-of)	(825)	(9,765)	(44,337)	(105,136)	

5 Related Works

Various tools have been developed using Wikidata and are listed on the Wikidata Tools site[3]. Similar to this study, some tools have been developed for the

---

[3] https://www.wikidata.org/wiki/Wikidata:Tools.

Wikidata class hierarchy. Wikidata Graph Builder makes it easy to visualize a variety of data, including class hierarchies, using the Wikidata query service [3]. Wikidata Class Browser makes it easy to search for information about class definitions, such as the specified class hierarchy and properties common to that class [4]. The Wikidata Class Browser allows the users to quickly view a list of subclasses of a given class with statistics about those subclasses [5].

However, since all of these tools are implemented using the Wikidata query service, they may not be usable due to server timeouts when the size of the target class hierarchy increases, as described in Sect. 4. The unique feature of this research is that it does not use the Wikidata query service but works only with locally created index files. Therefore, class hierarchies can be extracted without timeouts, even if it takes a long time to extract the hierarchy.

6   Conclusion and Future Works

In this study, we developed a tool to extract class hierarchies from Wikidata. The tool is characterized by the fact that it extracts information on class hierarchies from Wikidata dumps in advance and operates based on data stored as indexed local files. This feature allows the system to handle huge class hierarchies that cannot be obtained using the Wikidata query service. Future tasks include improving the efficiency of indexing and class hierarchy extraction, visualizing the extracted class hierarchies, and extending the indexing to include structures and string searches other than the target class hierarchy.

Acknowledgments. This work was supported in part by JSPS KAKENHI Grant Number 23K28152 and the Environment Research and Technology Development Fund (JPMEERF20241M01) of ERCA and the New Energy and Industrial Technology Development Organization (NEDO) under Project JPNP20006.

Disclosure of Interests. The authors have no competing interests to declare that are relevant to the content of this article.

References

1. Vrandečić, D., Krötzsch, M.: Wikidata: a free collaborative knowledgebase. Commun. ACM **57**(10), 78–85 (2014)
2. Kume, S., Kozaki, K.: Extracting domain-specific concepts from largescale linked open data. In: IJCKG 2021: The 10th International Joint Conference on Knowledge Graphs, December 2021, Virtual Event, Thailand, pp. 28–37 (2022)
3. Wikidata graph builder. https://angryloki.github.io/wikidata-graph-builder/, Accessed 17 Aug 2024
4. Wikidata ontology explorer, https://lucaswerkmeister.github.io/wikidata-ontology-explorer/, Accessed 17 Aug 2024
5. Wikidata Class Browser, https://bambots.brucemyers.com/WikidataClasses.php, Accessed 17 Aug 2024

A Study on the Metadata System and the Construction of Knowledge Graph of *the Classic of Mountains and Rivers*-- Taking *the Classic of the Southern Mountains* as an Example

Yu Qin[✉] and Enbo Jiang

National Science Library (Chengdu), Chinese Academy of Sciences, Chengdu 610041, China
qinyu23@mails.ucas.ac.cn

Abstract. As one of the cutting-edge technologies for knowledge organization and management, knowledge graph is gradually becoming an important force driving the in-depth development of digital humanities research. Its unique semantic correlation ability is conducive to revealing the complex connections and deep knowledge structures hidden in textual information, opening up new perspectives and paths for cultural inheritance and innovation. As one of the three mysterious books of ancient times, *the Classic of Mountains and Rivers* contains profound cultural and research values. Taking *the Classic of the Southern Mountains* in *the Classic of Mountains and Rivers* as a specific example, this paper systematically constructs a set of metadata description system for *the Classic of Mountains and Rivers*, and introduces Protege and Neo4j to build domain ontology and knowledge graph, revealing its internal knowledge structure and logical relationship. This provides a useful exploration and practice for the application of knowledge graph in the field of digital humanities.

Keywords: *the Classic of Mountains and Rivers* · Metadata · Ontology · Knowledge Graph · Digital Humanities

1 Introduction

As a representative work of ancient Chinese books, *the Classic of Mountains and Rivers* has had a profound impact on later generations with its rich knowledge of natural geography, biology, and mythology.

However, due to the complexity of the content of *the Classic of Mountains and Rivers*, it is difficult to comprehensively and systematically explore its intrinsic value by using traditional methods. As an advanced knowledge organization method, the knowledge graph is conducive to revealing the complex connections and deep knowledge structures hidden in text information due to its unique semantic association ability.

B. Xu et al. (Eds.): CCKS-IJCKG 2024, CCIS 2229, pp. 373–383, 2025.
https://doi.org/10.1007/978-981-96-1809-5_30

2 Related Work

2.1 Research on Antiquities

This paper uses CNKI as the source data of search with "ancient books" and "digital humanities" as the search terms to conduct a combined search, and a total of 194 journal articles were retrieved. The topics of which involve the digitization of ancient book texts, ancient book data technology, and research on the value-added of ancient book research.

Through screening by paper titles and abstracts, a total of 35 articles involve ancient book research in specific fields, mainly focusing on traditional Chinese medicine, ancient Chinese history, and Chinese literature.

The digital humanities project of ancient books is different from the traditional digitization of ancient books, focusing on mining the textual content of ancient books, paying attention to the word and semantic analysis of ancient books, such as grammatical features, syntactic features and semantic recognition [1]. It aims to use machine learning, text mining, natural language processing and other technologies to extract the elements of ancient book texts and realize the association and interaction of elements.

2.2 Research on *the Classic of Mountains and Rivers*

Since there are fewer foreign studies on *the Classic of Mountains and Rivers* abroad, this paper mainly focuses on domestic research. At present, the domestic research on *the Classic of Mountains and Rivers* mainly focuses on its version characteristics, mythology and legends, clan lineage and other aspects, with less research in digital humanities [2].

In addition, the National Library has constructed a knowledge base platform for *the Classic of Mountains and Rivers*, which achieves multi-dimensional and multi-directional connections in browsing, retrieval and data analysis of texts, images and geographic information of the 92 kinds of ancient books of *the Classic of Mountains and Rivers* that have been included [3]. However, the national map has not yet fully explored and applied the association between knowledge elements and further implemented the retrieval function based on the knowledge elements. Therefore, the paper will achieve the association of knowledge elements on the basis of extracting knowledge elements, constructing a knowledge map, and realizing scenario-based retrieval based on the knowledge graph.

3 Construction of the Metadata System

3.1 Metadata Definitions and Classifications

The Classic of Mountains and Rivers contains a large number of entities, but its description of entities is significantly different from the general description method, and the existing metadata standards cannot be well reused. Therefore, this paper takes *the Classic of the Southern Mountains* as an example to try to construct a metadata description system for *the Classic of Mountains and Rivers*.

In the construction of metadata description system, this paper divides it into basic metadata and content metadata, in order to meet the information retrieval and analysis of different levels and different needs. Among them, basic metadata is mainly used to describe the basic information of the whole text. Combining with Dublin Core Metadata and specific needs, this paper stipulates that the descriptive metadata of the text are the main title and the responsible person; content metadata describes various entities in the book more specifically. In animal metadata, Darwin Core is reused and adjusted according to the text content.

3.2 Metadata Description Regularization

A clear, unified and standardized metadata description can ensure the accuracy, completeness and consistency of the data, which is conducive to data governance; at the same time, in-depth analysis of metadata is conducive to digging and discovering the association and potential value between the data.

The metadata description specification constructed in this paper aims to provide a systematic framework for *the Classic of Mountains and Rivers* about different categories of natural and humanistic elements outside books as well as inside books, and to record and describe in detail the basic attributes, characteristics, locations, resources and other relevant information of the physical elements through standardized fields.

Specifically, the specification covers six entity categories: Chapter (a collection of ancient classical texts), Mountain, River, Botany, Animal and Mineral, with multiple metadata fields under each category to ensure the comprehensiveness and accuracy of the information.

For example, in the category of Chapter, the unique name of the sutra and the compiler are identified through the fields of "title proper" and "author", which is conducive to quickly locating the entity's position in the book; the category of Mountains, which is a collection of books, has several metadata fields to ensure the comprehensiveness and accuracy of the information. The Mountains category records the names, locations, characteristics, vegetation and resources of mountain ranges in detail, providing an important basis for research in the fields of geography, ecology and resource exploration.

Similarly, other entity categories such as River, Botany, Animal, and Mineral are also recorded through their respective metadata fields. This is shown in Table 1 of the appendix.

4 Ontological Construction

In this paper, we use Ontology101 seven-step method to construct the ontology library of *the Classic of the Southern Mountains*, and choose to select Protégé software as the ontology modelling tool.

4.1 Entity Definitions and Attribute Descriptions

To construct the ontology model of *the Classic of the Southern Mountains*, on the one hand, it is necessary to define the core concepts (classes) of *the Classic of the Southern Mountains*, identify the entities in the content and classify them; on the other hand, by extracting the attribute features of *the Classic of the Southern Mountains*, the attributes of the concepts and the relationship between the concepts are summarised inductively. It is noteworthy that due to the emergence of the concept of mountain ranges in t*he Classic of Mountains and Seas*, "Mountains" are utilized to denote a series of mountains. The details are shown in Tables 1 and 2.

Table 1. Entity and Data Attributes of *the Classic of the Southern Mountains*

Class	Entity		DataProperty	range
Geography	Mountains	Mountain	mountain_name	string
			mountain_feature	string
	River		river_name	string
			river_feature	string
Biology	Botany		botany_name	string
			botany_feature	string
	Animal		animal_name	string
			animal_feature	string
			animal_habit	string
			animal_sound	string
	God		god_name	string
			god_feature	string
Mineral			mineral_name	string
			mineral_feature	string
			mineral_founction	string
Chapter			chapter_name	string
			chapter_author	string

4.2 Ontology

In this paper, the above ontology model of *the Classic of the Southern Mountains* is implemented in Protégé 5.6.3 software, and uses the OntoGraft function to display the overall structure of the domain ontology, shown in Fig. 1.

Table 2. Attributes of Entity Objects of the Classic of the Southern Mountains

ObjectProperty	Domain	Range	Relationship Description	Example
contain	Chapter	Geography, Biology, Mineral	Warp section contains rivers, mountains, organisms, minerals	<the *the Classic of the Southern Mountains*-I, contains, Zhaoyao Mountain>
produce	Geography	Mineral	The mountains produce minerals	<Zhaoyao Mountain,produces, Jade>
belongs_to	Mountain	Mountains	The peaks belong to the mountain system	<Zhaoyao Mountain, belongs to, the Magpie Mountain system>
stem_from	River	Mountain	Rivers originate in mountain ranges	<Yongshui River, stems from, Daoguo Mountain>
flow_to	River	River	Riverine inflows to water bodies	< Zeshui River, flows to, Pangshui River>
grow_in	Biology	Mountain	Creatures grow in mountains	<Huahui, grows in, Yaoguang Mountain>
is_far_from	Geography	Geography	Relative position of rivers to rivers or mountains to mountains	<Cabinet Mountain, 450 miles southeast, Changyou Mountain>

5   Construction of a Knowledge Graph

5.1   Entity Recognition and Ternary Extraction

Recently, BERT has been widely used in entity recognition across various fields. However, due to the unique annotation of the training set of *the Classic of Mountains and Rivers*, the annotation effect of using the public training set is not ideal, and manual annotation requires a lot of time. Therefore, this paper utilizes GPT and ERNIE Bot, and then combines with manual entity recognition, the identified entities are marked with different colors.

Structured information can be extracted from the rich content of the Classic of the Southern Mountains. This paper determines the relationships between entities based on rules and semantic associations. Specifically, it first applies the entity object attributes defined above to set the semantic associations between entities and connect them. Then, by reading and summarizing the characteristic vocabulary related to entity relationships

Fig. 1. Protégé Realisation of the ontology *the Classic of the Southern Mountains* —OntoGraft Presentation

in the Classic of the Southern Mountains, the accuracy of the relationships is further verified based on rules. For example, "The Fang River originates from here (Jitaishan) and flows south into the Yu River". The "Fang River"originates from "Jitaishan" and flows into the "Yu River". On the one hand, there is an entity relationship type of "… Originates from…" between rivers and mountains, and a relationship type of "… Flows into…" between rivers. On the other hand, characteristic words such as "originate" and "flow into" appear. After extraction, 295 relationships of 7 relationship types are formed.

5.2 Knowledge Graph Visualization

In Neo4j, different entity types are distinguished by different colors, and the colors can be customized. The graph constructed in this article includes the direction and distance relationships between mountains, the origin relationships between rivers and mountains, the flow direction and inflow relationships between rivers, the output relationships between mountains and minerals, etc.

Neo4j provides a query function that can display different contents separately, which helps to view and understand a certain type of entity in a focused manner. Figure 2 shows the relationship between Zhaoyao Mountain and other entities such as animals, rivers, plants, etc.

5.3 Semantic Query

This paper realizes semantic retrieval based on some examples. By entering keywords in the text box, relevant retrieval results can be obtained. For example, entering "Feature of Guanguan" in the search box will return the results as shown in Fig. 3.

Based on the knowledge graph constructed above, it can be further improved to achieve the following scenario applications:

Cultural Dissemination and Education. Through the visual representation of the knowledge graph, complex information in the Classic of the Southern Mountains can be

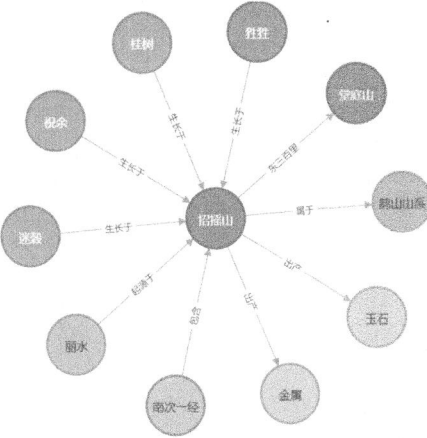

Fig. 2. The picture demonstrates the relationships between Zhaoyao Mountain and animals, rivers, minerals, and so on.

Fig. 3. Search results for "Feature of Guanguan"

presented to the public in an intuitive and comprehensible manner, thereby enhancing the dissemination and influence of cultural heritage. This approach aids learners in better understanding and studying the Classic of Mountains and Seas as well as the cultural connotations embedded within it.

Intelligent Applications. Within the knowledge graph, algorithms such as graph mining computation and ontology reasoning can be employed to uncover hidden knowledge, patterns, or rules within the Classic of the Southern Mountains. This enables the construction of intelligent question-answering systems that provide users with precise and rapid information retrieval services.

6   Summary

Taking *the Classic of the Southern Mountains* as an example, this paper constructs a metadata description system and knowledge graph for *the Classic of Mountains and Rivers*.

However, due to the constraints of professional backgrounds, this paper acknowledges that there remain limitations in the level of cognition and research depth regarding the Classic of Mountains and Seas. Further exploration and excavation are necessary to enhance our understanding. In terms of the metadata system and knowledge graph construction, refinements are also required. Additionally, this paper ought to improve upon entity recognition and annotation, establishing a suitable corpus and exploring the application of intelligent automated entity recognition in the Classic of Mountains and Rivers. Moving forward, there is potential to delve deeper into and leverage the knowledge embodied in the Classic of Mountains and Rivers, thereby promoting the development of new forms of domain-specific knowledge services in this field.

Appendix

Table 3. Metadata Description System and Specification

Object	Metadata	Explanation	Description Specification	Example
Chapter	title proper	A unique name for a section or chapter that uniquely identifies a section of scripture	String type, must be the name of an actual section and unique within the same system	the Classic of the Southern Mountains-I
	author	The compiler of the scriptures or the author of the apocryphal texts	String type, may contain "anonymous" or specific historical period, compiler's name, etc	compilers of the pre-Qin period
Mountain	name	The official name of a mountain range, the unique identifier of a mountain range	String type, must be the name of an actual mountain and unique within the same system	Kuaiji Mountain
	placement	Location of mountains in relation to other geographical features	String type, can contain information such as direction, distance, reference, etc	Qiwu Mountain: Five hundred miles to the east of Luwu Mountain
	feature	Physical characteristics of mountains, i.e. the shape, height and other natural features of the mountain range	String type describing the natural form of the mountain range	rectangle

(*continued*)

Table 3. (*continued*)

Object	Metadata	Explanation	Description Specification	Example
	plant cover	Species and distribution of plants growing on the mountains	String type with plant species	laurel tree
	resource	Natural resources that can be extracted or utilized in the mountains	String list type, with each resource separated by a comma	the mountain is rich in metallic minerals and jade, and the lower part of the mountain is full of crystal clear agate stones
River	name	Official name of the river, unique identifier of the river	String type, must be the name of a river that actually exists and is unique within the same system	rushing water
	birthplace	Where the river begins to flow	String type, usually the name of a mountain range or a specific location	Hushao Mountain
	flow	Direction of river flow	String type with direction and end point information	eastward into the sea
	inflow	The body of water into which the river ultimately flows	String type for the name of the water body	flow into Chishui River, Inner Mongolia
	resource	Natural resources in rivers		There's a lot of purple snails in the water
Botany	name	Official name of the plant, unique identifier of the plant	String type, must be the name of a plant that actually exists and is unique within the same system	laurel tree
	growing place	Where plants grow	String type, can contain name of mountain range, geographic location, etc	Hushao Mountain

(*continued*)

Table 3. (*continued*)

Object	Metadata	Explanation	Description Specification	Example
	feature	Describe the shape and other characteristics of the plant	String type, specific description according to element type	black texture
Animal	name	Official name of the animal, unique identifier of the animal	String type, must be the name of an animal that actually exists and is unique within the same system	Shengsheng
	feature	Describe the natural forms of animals	string type	It's shaped like a normal sheep without the mouth
	habit	The habits of the animal, including its diet, activity patterns, reproduction methods, etc	string type	You can live without eating and not die
	habitat	The metadata details the habitat of animals, encompassing their living environments	string type , typically representing specific mountains or rivers	Changyou lives in Changyou Mountain
	sound	Characteristics of sounds made by animals, including grunts, chirps, etc	String types, including volume, pitch, tempo, etc	The roar sounds like people singing
	affect	Impact of the animal on its surroundings or on humans, including positive and negative impacts	string type	Wherever there's a long right, there's a big flood

(*continued*)

Table 3. (*continued*)

Object	Metadata	Explanation	Description Specification	Example
Mineral	name	Official name of the mineral, unique identifier of the mineral	String type, must be the name of a mineral that actually exists, possibly the mineral type	metal minerals
	Place of origin	Location of mineral output	String type, can contain name of mountain range, geographic location, etc	Aikisan, Shunsan (southern sunny side of the mountain), etc
	feature	Distinguishing features of the mineral, e.g. colour, texture, etc	string type	the size of a grain of millet
	function	Use or value of minerals	string type	used as chess pieces

References

1. Wang, Q.Y., Long, H.: Analysis of the current situation of digital humanities research and practice in the field of ancient books in China. Library World **4**, 19–25 (2023)
2. Wang, M.L.: Research on information visualization design of the book of classics of the mountains and Sea A, Master's thesis, Yanshan University, Hebei, China (2022)
3. The National Library opens the Classic of Mountains and Seasknowledge base online for free, https://news.bjd.com.cn/2024/06/20/10810759.shtml, Accessed 28 July 2024

Evaluations

A Two-Stage Approach for Knowledge Editing in LLM

Hao Xiong, Wenbiao Shao, Tailai Han, and Wenliang Chen[✉]

Soochow University, SuZhou, China
{hxiongxionghao,wbshao,tlhan0812}@stu.suda.edu.cn, wlchen@suda.edu.cn

Abstract. Knowledge Editing (KE) aims to modify specific knowledge within large models and address issues of knowledge misinformation or inaccuracies. Most previous studies were mainly based on structured facts, whereas real-world knowledge updates commonly emerge in unstructured texts. In this work, we focus on the CCKS-IJCKG 2024 Competition of Knowledge Editing, which features a Chinese dataset containing both structured and unstructured knowledge. We propose a two-stage approach for KE, using LoRA based method for model editing and incorporating the In-Context Learning Knowledge Editing (IKE) method to refine the editing process. In the first stage, we demonstrate that standard fine-tuning based on LoRA can yield superior performance with some key modifications. We align the inference process with the training process and implement template-based training, specifically without masking special tokens. To enhance portability, we also incorporate reverse masked loss during training. In the second stage, the IKE method leverages in-context learning to improve the model's understanding and application of the edited knowledge in various contexts. Our method takes the first place in the CCKS-IJCKG 2024 Knowledge Editing Evaluation Task Competition, achieving a score of 52.6388%.

Keywords: Large Language Model · Knowledge Editing · Parameter Efficient Tuning · In-Context Learning

1 Introduction

As the world continually changes, the need to update Large Language Models (LLMs) to correct outdated information and integrate new knowledge is becoming increasingly crucial. However, retraining these models from scratch for updates is both costly and impractical. Therefore, many researchers have turned their attention to Knowledge Editing (KE) as an efficient method for refining and modifying specific knowledge within LLMs, addressing inaccuracies or misinformation [16]. Traditional KE methods predominantly concentrate on structured datasets, where information is presented in a well-organized format. However, real-world knowledge updates often come from unstructured texts,

B. Xu et al. (Eds.): CCKS-IJCKG 2024, CCIS 2229, pp. 387–397, 2025.
https://doi.org/10.1007/978-981-96-1809-5_31

which pose unique challenges and opportunities for effective knowledge modification [2,14]. While the CCKS-IJCKG 2024 Knowledge Editing Competition provides an opportunity to explore KE methodologies within the realm of dataset containing structured and unstructured knowledge. This competition highlights the necessity for robust approaches to handle and edit knowledge embedded in these diverse formats, reflecting more realistic scenarios where knowledge is frequently updated and needs to be integrated seamlessly into existing models. In our current research, we concentrate on the application of KE to LLMs, particularly as demonstrated in the CCKS-IJCKG 2024 Knowledge Editing challenge. The example shown in Fig. 1 clearly illustrates the process of knowledge editing.

To address the challenges in knowledge editing, we first choose the baseline model using the LoRA method. We then combine this approach with a template-based training strategy, which leads to significant improvements in performance. Specifically, we do not mask special tokens within the templates and calculate the cross-entropy loss to further refine the model's predictions. For the unique challenge of ancient poem, we integrate reverse masked loss into the loss function, which helps to further enhance the model's ability to generate accurate and contextually appropriate responses. In addition to these techniques, we employ the IKE method to address more complex scenarios in model editing. The IKE approach leverages in-context learning by retrieving context examples during the editing process, dynamically guiding the model's understanding and application of newly edited knowledge.

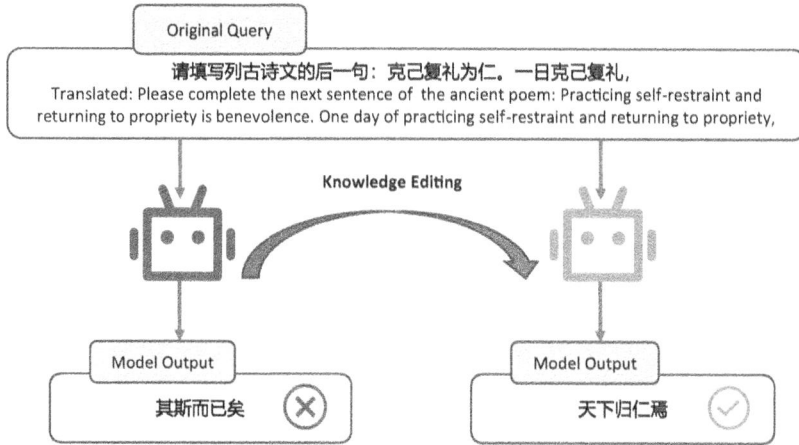

Fig. 1. Demonstration of an example from the evaluation task for knowledge editing.

2   Related Work

2.1   Knowledge Editing

Knowledge editing has attracted great attention in recent years [16]. Existing methods mainly focus on edit accuracy, locality, portability and fluency. Researchers diligently classify existing knowledge editing approaches into two primary paradigms [15].

Preserve Models' Parameters. The additional parameters approach involves integrating extra trainable components into the existing model architecture. CaliNet [3] and T-Patcher [6] fix neurons for each edit. And memory-based methods store corrective examples in a memory module and use a retrieval mechanism to select relevant updates for new inputs. SERAC [11] adopts a scope classifier and decides whether to use the original model or the counterfactual model. While WISE [12] proposes a dual parametric memory scheme to separate pre-trained knowledge from edited knowledge and uses a knowledge-sharding mechanism to manage continual edits without conflicts which enables reliability, generalization, and locality simultaneously achieved. IKE [18] refines the model's output to align with given knowledge through the use of specialized in-context prompts.

Modify Models' Parameters. Meta-learning methods teach a hypernetwork to learn the change of gradient for editing. MEND [10] employs auxiliary networks and facilitates scalable editing through the decomposition of gradients, which in turn enables efficient and effective modifications for knowledge editing. Regarding the approach locate-then-edit, they identify the specific parameters related to the target knowledge and then directly modify these parameters to achieve the desired knowledge editing. ROME [8] and MEMIT [9] use causal tracing to identify the relevant knowledge and then propose modifying multiple layers parameters to facilitate extensive edits. UNKE [2] addresses the limitations of previous methods by extending assumptions in the layer and token dimensions, demonstrating superior performance on both unstructured and traditional structured knowledge editing datasets.

2.2   Parameter-Efficient Tuning

Parameter-efficient tuning emerges as a key technique for adapting large pre-trained models to specific tasks without the need for extensive retraining. Among the most representative methods in this category are Adapter, Prompt Tuning, and LoRA. The Adapter [4] introduces a trainable, bottleneck-shaped neural network that is appended to the output of the transformer block. Prompt Tuning [7] takes a different approach by optimizing task-specific prompts, which are either discrete tokens or continuous vectors, appended to the input. LoRA [5] further refines this concept by keeping the pre-trained model entirely frozen and updating only the rank decomposition matrices that are introduced into the target modules. This significantly reduces the number of trainable parameters while still allowing for effective task adaptation. Building upon LoRA, AdaLoRA [17] introduces an adaptive mechanism that dynamically adjusts the rank according to the specific needs of each task. This not only enhances flexibility and adaptability but also maintains the parameter efficiency that is central to these approaches.

3   Methodology

As shown in Fig. 2, our proposed methodology consists of two key components: 1) parameter-efficient tuning, 2) in-context learning for knowledge editing. Detailed descriptions of the specific implementations for each component are provided in subsequent sections.

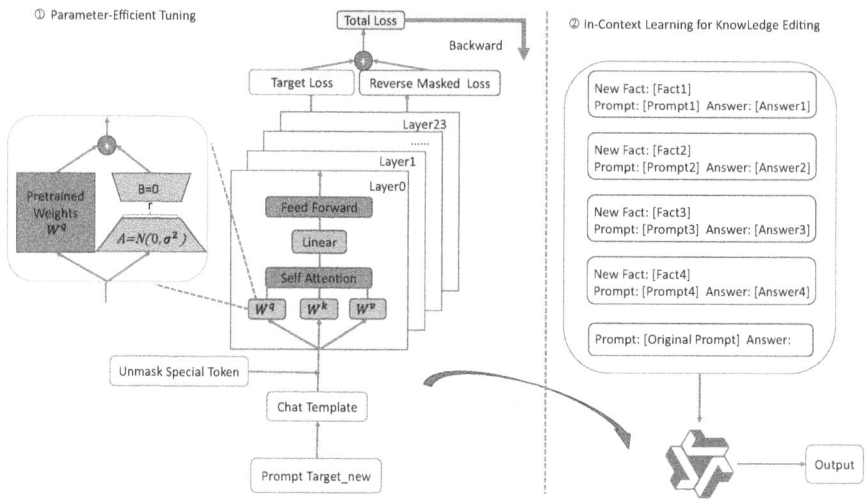

Fig. 2. The architecture of our proposed LoIKE method.

3.1   Parameter Efficient Tuning

Training Template. According to the method UNKE [2], unstructured knowledge has a significantly higher knowledge density compared to structured knowledge. This higher density makes unstructured knowledge even less prone to localization. Consequently, conventional methods for editing previous knowledge are inadequate when it comes to handling tasks related to editing unstructured knowledge. To address this, we use LoRA fine-tuning, which allows for training more layers while requiring fewer parameters. As shown in Fig. 3. The modules we trained not only the MLP layers but also the attention layers.

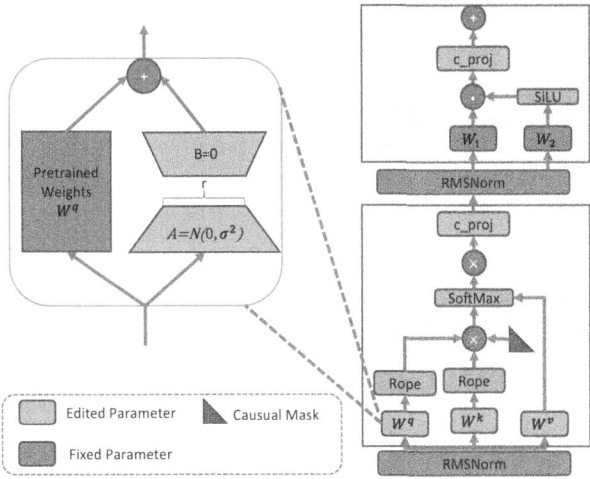

Fig. 3. The left part shows the LoRA training process, and the right part shows a block of Qwen, which is trained using LoRA. In the Qwen block, W^q, W^k, W^v are the weight matrices that project input embeddings into the query, key, and value spaces. The Rope is the Rotational Positional Encoding mechanism for encoding positional information. The c_proj layer serves as a projection layer that maps the attended values back into the original feature space. Additionally, W_1, W_2 represent the linear transformations in the feed-forward network and SiLU is the Sigmoid Linear Unit activation function.

Chat models are trained with specific formats that convert conversations into a single tokenizable string. Deviating from the format used during training can lead to severe performance degradation. Therefore, it is crucial to match the evaluation format to the training format. Since the base model used in this task is a chat model, we align the evaluation process with the appropriate chat template. This alignment results in a significant improvement over the baseline performance.

During the evaluation process, we observe that the edited model often repeats itself when generating output. To address this issue, we draw inspiration from the fine-tuning code provided in the Qwen [1] official repository and implement a method called **U**nmask **S**pecial **T**okens (UST) during the model training process. In this approach, special tokens such as the end token, which indicates the conclusion of a sequence, are kept unmasked. This is crucial for ensuring that the model accurately learns when to end a sequence, thereby producing complete and coherent text. By keeping special tokens unmasked, we maintain consistency between the training and inference phases. This consistency allows the model to apply the patterns and rules it learns during training directly to the inference phase, reducing the need for further adjustments and enhancing its overall performance.

Objective Function. The target loss used during the fine-tuning of GPT-based models is a masked cross-entropy loss. This loss function evaluates the model's performance in generating text based on a given prompt by measuring its ability to predict the correct continuation. In this setting, each input sequence consists of a prompt and an expected answer. During training, the prompt is masked out so that the model does not consider it when calculating the loss. The masked cross-entropy loss is then computed between the model's predictions and the actual answers, with only the answer part contributing to the loss.

To encourage the model to pay attention to the prompt and better understand the input context, we introduce an additional token loss by masking the answer part. This compensates for the loss in the prompt section. With this approach, the model gains a degree of bidirectional understanding, as it learns not only to predict subsequent words but also to reconstruct preceding words. This process is illustrated in Fig. 4.

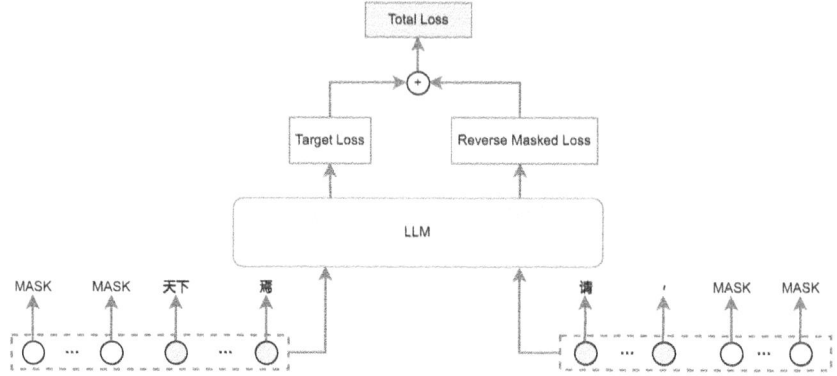

Fig. 4. The total loss is calculated in two steps: first by adding the Target Loss, and then by incorporating the Reverse Masked Loss.

The target loss is represented by a vector x, where each element corresponds to a token in the input sequence. For the answer tokens, the values in x are set to 1, while for the prompt tokens, the values are set to 0. This setup ensures that the cross-entropy loss is calculated only for the answer tokens. To incorporate the **R**everse **M**asked **L**oss (RML), we introduce another vector m, which is constructed as the complement of x. In vector m, the values are set to 1 for the prompt tokens and 0 for the answer tokens, allowing the model to compute the loss for the prompt tokens instead. The reverse masked loss is then calculated by multiplying m with a coefficient λ. This coefficient λ controls the weight of the reverse masked loss in the overall loss function, enabling fine-tuning of the model's attention to both the prompt and answer tokens during training.

$$Loss = (x + \lambda m) \times \sum_{t=1}^{T} \sum_{i=1}^{V} y_{t,i} log p_{t,i} \tag{1}$$

3.2 In-Context Learning for Knowledge Editing

In order to better guide the model output, we combine the IKE method for knowledge editing. This method uses a pre-trained sentence encoder to encode both the prompt of a new fact and its original answer. Similarly, the data in the training set is encoded in the same way. We then use cosine similarity between these encoded vectors to retrieve the top k facts that are most similar to the query. These k facts serve as In-Context Learning (ICL) samples, where the ranking of these in-context examples is also determined by cosine similarity. The ICL samples are then combined with the prompt and input into the previously trained, edited model for evaluation. Since the primary task is to correct the internal knowledge errors of the model, the demonstration used here mainly focuses on the "copy" type in IKE. This approach involves teaching the model to replicate the prediction of the target prompt when presented with new facts, which is essential for effectively injecting new knowledge into language models.

4 Experiments

4.1 Datasets

The Experimental dataset CCKS-CKnowEdit used in this paper is all derived from the brand-new Chinese knowledge editing dataset CKnowEdit provided by the CCKS-IJCKG 2024 Conference's Large Model Knowledge Editing Evaluation Task. The dataset CCKS-CKnowEdit used in this competition covers all 7 types of CKnowEdit in terms of categories, and consists of 700 data items. The statistics of the CCKS-CKnowEdit dataset are shown in Table 1.

Table 1. The statistics of CCKS-CKnowEdit.

Data Type	Quantity
古诗(Ancient Poem)	133
谚语(Proverb)	80
成语(Idiom)	40
拼音注音(Pinyin Annotation)	50
文言文(Classical Chinese Text)	70
中国地理(Chinese Geography)	50
ruozhi吧(Ruozhi Bar)	277
Total	700

4.2 Experiment Setup

Our experiment is primarily based on the official code framework Easyedit[1] [13]. According to the competition requirements, the Qwen-1.8B-Chat[2] is selected as

---

[1] https://github.com/zjunlp/EasyEdit.
[2] https://huggingface.co/Qwen/Qwen-1_8B-Chat.

the target model for knowledge editing. For the LoRA fine-tuning, the rank is set to 8, lora_alpha is set to 32, and lora_dropout is set to 0.1. The target module is configured as attn.c_attn, attn.c_proj, mlp.c_proj. The model is trained with a learning rate of $5e^{-5}$ for 70 epochs, using float32 precision to maintain numerical stability and precision during training. Due to limited computing resources, the number of ICL examples is set to 4.

Baselines. We compare our method LoIKE against several baselines including ROME, MEMIT, FT-M, LoRA, and UNKE.

Evaluation. The competition releases evaluation methods including **Edit Accuracy**, **Portability**, **Locality**, and **Fluency**. The final ranking is based on the **Score**, which combines these metrics. The goal of knowledge editing is to alleviate the problem of knowledge errors by modifying specific knowledge in a large model. The difficulty lies in the fact that there is only one piece of knowledge, and the model needs to be edited successfully to answer questions related to this piece of knowledge and minimize the impact on other performance of the model.

Edit Accuracy measures the ability of the model to produce the edited response for a query, **Portability** assesses how well the model generalizes the knowledge for rephrased or logically related queries within the edit scope, **Locality** evaluates if the model maintains original predictions for queries outside the edit scope. Due to unstructured text, the above three indicators are not easy to match directly. We use the following two key metrics, word-level overlap **ROUGE-L** and semantic similarity **BERT-Score**, they are employed to evaluate the surface-level accuracy and the deeper semantic understanding of the generated text. We leverage a pre-trained multilingual sentence encoder, specifically the sentence-transformers[3] to obtain text embeddings. The **BERT-Score** is calculated by measuring the cosine similarity between these embeddings. Finally, the **Accuracy** for each generated answer is computed as a weighted average of the **ROUGE-L** and **BERT-Score** values:

$$\text{Accuracy} = 0.5 \times \text{ROUGE-L} + 0.5 \times \text{BERT-Score} \tag{2}$$

Fluency assesses the linguistic quality of the model's output after editing, using a weighted sum of bi-gram and tri-gram entropies:

$$\text{Fluency} = -\sum_{n=2}^{3} w_n \sum_x f_n(x) \log f_n(x) \tag{3}$$

where f_n is the n-gram distribution. The overall **Score**, which determines the final ranking, is computed as:

---

[3] https://huggingface.co/sentence-transformers/paraphrase-multilingual-MiniLM-L12-v2.

Score $= 0.2 \times$ Edit Accuracy $+ 0.35 \times$ Portability

$$+ 0.35 \times \text{Locality} + 0.1 \times \text{Fluency} \quad (4)$$

4.3 Main Results

According to the official final evaluation results, the final rating of this experiment is shown in Table 2. Our method LoIKE demonstrates significant enhancements in Edit Accuracy, Portability, Fluency and overall Score compared to the baselines. Among the baseline methods, UNKE performs the best, followed by LoRA and FT-M, while ROME and MEMIT exhibit relatively poor performance, indicating their limited effectiveness on Chinese dataset containing structured and unstructured knowledge. Notably, LoIKE achieves the highest Edit Accuracy at 96.7107%, representing a substantial improvement over the second-best method UNKE. This notable enhancement underscores LoIKE's superior capacity for accurately updating the model's knowledge. The considerable improvement in Edit Accuracy further suggests that the integration of LoRA and IKE within our approach effectively enhances the model's precision in executing accurate updates. In terms of Portability, LoIKE records a score of 40.1028%, outperforming all other methods. This result highlights our approach's effectiveness in transferring and applying edited knowledge across different contexts and scenarios. Regarding Fluency, LoIKE attains a score of 60.6419%, which is markedly higher than MEMIT by 12.8740%. This improvement in Fluency emphasizes that LoIKE not only excels in accurately updating knowledge but also maintains the naturalness and fluency of the generated text, a crucial attribute for producing high-quality outputs.

Table 2. The performance of our method and the baselines.

Method	Edit Accuracy (%)	Portability (%)	Locality (%)	Fluency (%)	Score (%)
ROME	36.4479	32.1940	36.3099	38.4830	35.1143
MEMIT	34.5220	31.5927	36.5680	47.7949	35.5401
FT-M	55.8094	33.5613	36.2363	37.5716	39.3482
LoRA	62.3686	34.1357	36.1436	36.2289	40.6944
UNKE	75.9009	34.1618	**38.1743**	40.0398	44.5018
LoIKE (ours)	**96.7107**	**40.1028**	37.7043	**60.6419**	**52.6388**

4.4 Abalation Study

To evaluate the individual contributions of the components in our proposed method LoIKE, we conducted an ablation study. with the results presented in

Table 3. The analysis highlights the importance of each module in our approach. Removing the IKE module significantly impacts Fluency, leading to a decrease of 25.4482%, which in turn causes the overall Score to decline by 1.7890%. This result underscores the crucial role that the IKE module plays in maintaining high fluency and overall performance. The removal of the RML leads to an improvement in Fluency, which increases by 8.2683%. However, this improvement in Fluency comes at the cost of Portability, which decreases by 2.6193%. This suggests that while RML significantly boosts Portability, it negatively impacts the Fluency. Omitting the UST results in a slight decrease in Edit Accuracy and Locality. While the Fluency decreases by 6.5011%, and the overall Score decreases by 0.7817%. These results emphasize the importance of maintaining special tokens for enhanced contextual understanding and fluency. Finally, the removal of template-based training leads to a drastic reduction in Edit Accuracy by 37.3957%, with Portability declines by 2.9276% and the overall score falls by 9.3518%. These results demonstrate the critical role of template-based training in achieving high accuracy and maintaining robust performance across multiple metrics.

Table 3. Abalation study, gradually removing each module. Note that the UST indicate that keep unmasking special token, the RML means the reverse masked loss.

Method	Edit Accuracy (%)	Portability (%)	Locality (%)	Fluency (%)	Score(%)
LoIKE (ours)	96.7107	**40.1028**	37.7043	**60.6419**	**52.6388**
- IKE	99.9666	39.5482	38.5578	35.1937	50.8498
- RML	**100.0000**	36.9288	**38.7245**	43.4620	50.8249
- UST	99.7643	37.0633	38.3489	36.9609	50.0432
- Template	62.3686	34.1357	36.1436	36.2289	40.6944

5   Conclusion

In this paper, we present a two-stage approach to enhance the LoRA framework by utilizing a training template, incorporating the reverse maked loss, and integrating the IKE method. The IKE method is essential for maintaining high fluency, while RML enhances portability at the cost of slight reductions in other metrics. Additionally, the UST module is essential for contextual understanding, and template-based training is fundamental for achieving high accuracy. This approach is developed to complete the CCKS-IJCKG 2024 Knowledge Editing Evaluation Task Competition, where it demonstrates remarkable effectiveness. Specifically, our method achieve a score of 52.6388% and won the first prize in the competition.

Acknowledgments. This work is supported by the National Natural Science Foundation of China (Grant No. 62036004) and Project Funded by the Priority Academic Program Development of Jiangsu Higher Education Institutions.

References

1. Bai, J., et al.: Qwen technical report. arXiv preprint arXiv:2309.16609 (2023)
2. Deng, J., Wei, Z., Pang, L., Ding, H., Shen, H., Cheng, X.: Unke: unstructured knowledge editing in large language models. arXiv preprint arXiv:2405.15349 (2024)
3. Dong, Q., Dai, D., Song, Y., Xu, J., Sui, Z., Li, L.: Calibrating factual knowledge in pretrained language models. arXiv preprint arXiv:2210.03329 (2022)
4. Houlsby, N., et al.: Parameter-efficient transfer learning for nlp. In: International Conference on Machine Learning, pp. 2790–2799. PMLR (2019)
5. Hu, E.J., et al.: Lora: low-rank adaptation of large language models. arXiv preprint arXiv:2106.09685 (2021)
6. Huang, Z., Shen, Y., Zhang, X., Zhou, J., Rong, W., Xiong, Z.: Transformer-patcher: one mistake worth one neuron. arXiv preprint arXiv:2301.09785 (2023)
7. Li, X.L., Liang, P.: Prefix-tuning: optimizing continuous prompts for generation. arXiv preprint arXiv:2101.00190 (2021)
8. Meng, K., Bau, D., Andonian, A., Belinkov, Y.: Locating and editing factual associations in GPT. Adv. Neural. Inf. Process. Syst. **35**, 17359–17372 (2022)
9. Meng, K., Sharma, A.S., Andonian, A., Belinkov, Y., Bau, D.: Mass-editing memory in a transformer. arXiv preprint arXiv:2210.07229 (2022)
10. Mitchell, E., Lin, C., Bosselut, A., Finn, C., Manning, C.D.: Fast model editing at scale. arXiv preprint arXiv:2110.11309 (2021)
11. Mitchell, E., Lin, C., Bosselut, A., Manning, C.D., Finn, C.: Memory-based model editing at scale. In: International Conference on Machine Learning, pp. 15817–15831. PMLR (2022)
12. Wang, P., et al.: Wise: rethinking the knowledge memory for lifelong model editing of large language models. arXiv preprint arXiv:2405.14768 (2024)
13. Wang, P., et al.: Easyedit: an easy-to-use knowledge editing framework for large language models. arXiv preprint arXiv:2308.07269 (2023)
14. Wu, X., Pan, L., Wang, W.Y., Luu, A.T.: Updating language models with unstructured facts: towards practical knowledge editing. arXiv preprint arXiv:2402.18909 (2024)
15. Yao, Y., et al.: Editing large language models: problems, methods, and opportunities. arXiv preprint arXiv:2305.13172 (2023)
16. Zhang, N., et al.: A comprehensive study of knowledge editing for large language models. arXiv preprint arXiv:2401.01286 (2024)
17. Zhang, Q., et al.: Adalora: adaptive budget allocation for parameter-efficient fine-tuning. arXiv preprint arXiv:2303.10512 (2023)
18. Zheng, C., et al.: Can we edit factual knowledge by in-context learning? arXiv preprint arXiv:2305.12740 (2023)

LLM-Based Functional Query Generation with Multi-relation Alignment for Complex Knowledge Based Question Answering

He Dong, Wang Song, Li Hao, Liu Jianzhu, He Ji$^{(\boxtimes)}$, Li Peiyao, Tao Jiang, and Xu Bingyu

PICC Information Technology Company Limited, Shanghai, Songjiang District, China
{hedong,wangsong,lihao01,liujianzhu,heji,lipeiyao01,taojiang, xubingyu}@picc.com.cn

Abstract. Complex Knowledge Based Question Answering (C-KBQA) is inherently more difficult than traditional KBQA tasks, as it typically cannot be resolved solely through direct queries, necessitating relational reasoning to derive answers. To address this challenge, this paper focuses on generating Large Language Model (LLM)-friendly functional queries from natural language questions based on LLMs. In addition, to effectively generate promising functional queries, entities and relations alignment methods as well as multi-LLM fusion techniques are utilized to extract accurate answers from complex textual questions. In the CCKS2024 CGQA competition, the method proposed in this paper achieved the second place with an F1 score of 0.8186, clearly demonstrating the high effectiveness and efficiency of the method.

Keywords: Complex KBQA · NL2Query

1 Introduction

A Knowledge based question answering (KBQA) system is utilized to retrieve answers from natural language questions by extracting relevant entities and relations from a pre-built knowledge graph [1]. In recent times, KBQA has gained significant attention as it provides a more effective and efficient way of fetching knowledge, which makes it smarter and more accurate in responding to user inquiries. A KBQA system typically consists of several main components, including question parsing, knowledge graph querying, answer extraction, and answer generation. Question parsing refers to the process of understanding the natural language questions and transforming them into a format that can be processed by computers. Knowledge graph querying refers to extract information from entities and relations using the parsing information. Information in the knowledge graphs is often demonstrated using triples, which contains head, relation and tail [2]. A

B. Xu et al. (Eds.): CCKS-IJCKG 2024, CCIS 2229, pp. 398–408, 2025.
https://doi.org/10.1007/978-981-96-1809-5_32

knowledge graph represents information in a structured mode, typically connecting various pieces of information through nodes and edges, where nodes signify entities and edges represent relations. Answer extraction and answer generation refer to the process of retrieving the most proper answer from the knowledge graph and demonstrating the answer in a human-understandable format, respectively.

The three main methods of KBQA are retrieval-based methods, semantic parsing-based methods, and template-based methods [3]. Retrieval-based methods [4] recognize the core entity from input questions and utilize ranking to determine the most proper answer. The process often necessitates focusing on entities relevant to the queries, leading to efficient performance with simple queries but struggling with complex questions. Semantic parsing-based methods [5–7] mainly comprehend the semantics of natural language questions by converting them into the logical forms which can be executed and handled by knowledge graphs. A limitation of this approach is its heavy reliance on logical formats, where parsing errors can lead to process failure. Template-based methods response to questions by aligning them with pre-established templates, potentially lead to faster response times, but the requirement for a large number of templates reduces the efficiency.

Large Language Models (LLMs) are complicated artificial intelligence systems which make use of deep learning techniques, particularly neural networks, to comprehend and produce human languages. These LLMs are trained on immense quantities of text data, ranging from various articles and books to a wide scope of web pages and social media content. Through this extensive training, LLMs learn the statistical patterns and structures of languages. LLMs have attracted significant attention due to the outstanding performance over a broad spectrum of natural language processing tasks [8], such as text classification, machine translation, text generation, and question answering. Notable examples like GPT-4 [9], BERT [10], and PaLM [11] have demonstrated the effectiveness in enhancing the performance of these natural language tasks.

Compared with ordinary KBQA, complex Knowledge Base Question Answering (C-KBQA) faces many challenges: 1) Multi-hop reasoning, where C-KBQA tasks typically involve multiple mentions with complex logical and semantic aspects, requiring multi-hop querying to get the answer; 2) Incomplete knowledge base, where knowledge bases often lack completeness, entities, or relations to solving C-KBQA tasks; 3) Complex reasoning challenge, where some questions cannot be directly retrieved from a given knowledge base, and on the contrary, complex reasoning is necessary to derive the final answer from intermediate results, such as numerical and set operations, keyword matching, and comparisons. To address multi-hop reasoning challenge, a Chain-of-Thought(CoT) [12] prompting technique is employed, which enables LLMs to decompose complex questions into step-by-step sub-questions for sequential resolution, explicitly revealing intermediate reasoning steps. For incomplete knowledge base problem, a multi-relation alignment method is introduced to combine and generate multiple queries sorted by their rank. This method of generating multiple queries, particularly in cases of incomplete knowledge bases, significantly increases the probability of obtaining the correct answers. For the challenge of complex reasoning, the method innovatively leverages a functional query framework based on the CG graph (conditional graph) [13], which can flexibly handle C-KBQA tasks through custom query functions.

The main method proposed in this paper to solve the complex KBQA consists of four core modules, including Prompt module, LLM module, Multi-relation alignment module, and Query module. We combined the Prompt module and LLM module to derive basic queries from natural language questions in the first place, followed by Multi-relation alignment module which identifies multiple candidate relations to generate several queries regarding to a single question. These multiple queries for a single question generated from potential candidate relations can be able to effectively enhance the opportunity of discovering the correct answer. Finally, the Query module transforms these queries into executable formats, enabling the search for the definitive answer. The main contributions of this paper is to propose an effective method based on LLM, entity alignment and multi-model fusion, which performs well in handling complex KBQA tasks. In the CCKS2024 CGQA competition, the method proposed in this paper achieved an outstanding performance with an F1 score of 0.8186. In addition, sufficient experiments are conducted to demonstrate the importance of each module within the method.

This paper begins with Related work chapter which gives a review of some literature related to this paper. The Methods chapter will elaborate the method step by step. Afterwards, the Results chapter will evaluate the method through experimental demonstrations of statistical data, followed by the Conclusion chapter summarizing this paper.

2   Related Work

This paper is closely related to the studies on Complex KBQA and LLM-based query generation and reasoning. The common approaches of KBQA include semantic parsing methods and information retrieval methods.

Semantic Parsing Based (SP-Based). SP-based approach converts natural language questions into corresponding symbolic logical forms and executes it against the knowledge base to obtain the final answers. For Complex KBQA, many SP-based approaches utilize syntactic parsing information, such as dependency syntax trees [14] [15] and Abstract Meaning Representation (AMR) [16], to better align question components with logical forms.

Information Retrieval Based (IR-Based). IR-based approach constructs a question-specific subgraph from the knowledge base, and ranks all entities extracted from the subgraph based on their relevance to the question. The top-ranked entities are predicted as the answers to the question [3]. In Complex

KBQA tasks, IR-based methods suffer from the incompleteness of KBs [17]. To address this issue, some researchers are attempting to use additional information for knowledge base completion [18–20].

Large Language Model Based (LLM-Based). Luo, Haoran, et al. [21] employs a generate-then-retrieve framework for KBQA tasks. Zhang et al. [13] proposes a novel method which utilizes a Condition Graph (CG) for knowledge representation and employs an LLM-based functional query for CG querying.

3   Methods

As described in Sect. 1, C-KBQA is a sophisticated task, facing challenges such as multi-hop reasoning, incomplete knowledge base and complex reasoning. To address these challenges, a LLM-based NL2Query method is proposed, which flexibly handles complex natural language questions and dynamically generates queries based on the complexity and variation of the question. Specifically, a Chain-of-Thought(CoT) prompting technique is employed for LLMs to improve the multi-hop reasoning. A multi-relation alignment method is introduced to generate multiple functional queries for a single question, which mitigating the impact of incomplete knowledge bases and increasing the probability of obtaining the correct answer. For complex reasoning, a functional query framework based on the CG graph (conditional graph) is utilized, which enables programming custom query functions to flexibly handle C-KBQA tasks.

As shown in Fig. 1, the method proposed in this paper comprises four modules: 1) Prompt module, 2) LLM module, 3) Multi-relation Alignment module, and 4) Query module. When a natural language question input to our model, the Prompt module will merge it with a predefined set of prompts to get the final prompt. Then the final prompt is sent to an LLM to generate the basic functional query. The functional query is defined as a sequence of query functions as shown in the lower part of Fig. 1. Once been generated, the functional query will be input to the Multi-relation Alignment module, which will link entities and relations within the functional query to the specific knowledge base. Besides, the Multi-relation Alignment module will also generate candidate relations for each individual query generated by the LLM module and rank them based on word similarity. The Multi-relation Alignment module will generate multiple copies of original query by replacing the original relations to candidate relations. Finally, the queries generated by the Multi-relation alignment module will be input to the Query module, which will transform the queries into executable queries to obtain the final answers. The detailed description of each module will be expand in the following subsections.

3.1   Prompt Module

The Prompt module contains the system definition, the description of custom query functions, and few-shot prompting (few-shot examples of how to use these

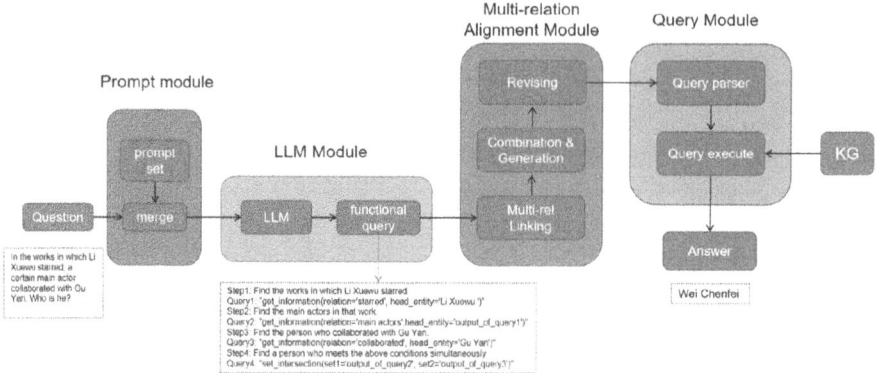

Fig. 1. The overall architecture diagram of this method. The method is composed of four modules: the Prompt module, the LLM module, the Multi-relation Alignment module, and the Query module.

custom query functions to retrieve information and reason to get the answer). For example, in the system definition, the LLM is required to learn how to create conditional graph queries for natural language questions. The descriptions of custom functions and few-shot examples on how to use them are added. Those functions are word_in, get_relation, max, min, count, etc.

Upon receiving a natural language question as input, the Prompt module will merge the predefined system definition, the description of custom query function, and few-shot prompting with the question together to get the final prompt. The final prompt will then be input to the LLM module to generate the basic queries.

3.2 LLM Module

In the LLM module, two kinds of LLMs are used: Qwen-2-7b-Instruct and gpt-4o. The fine-tuned Qwen2-7b-Instruct is utilized as the main model, while gpt-4o as the auxiliary model. Using different sets of super parameters, such as temperature, will generate several different queries. Then, these queries are input to the Multi-relation Alignment module and the Query module, resulting in several different answers.

To select the most promising answer, two fusion strategies were employed and tested: the voting strategy and the fallback strategy. The voting strategy involves selecting the answer that appears most frequently among the answers. The fallback strategy necessitates sorting the answers by priority and then selecting an answer based on priority. If a higher-priority answer is available, it is returned; otherwise, the next highest priority answer is selected, continuing until no candidate answers remain.

Qwen2-7B-Instruct, as the main model, has been fine-tuned through the low-rank adaptation (LoRA) for functional query generation. The parameter settings of LoRA are shown in Table 1.

Table 1. LoRA parameters.

Parameter	Quantity
Lora_alpha	16
Lora_rank	64
Lora_dropout	0.05

3.3   Multi-relation Alignment Module

The Multi-relation Alignment module contains three sub-modules: Linking & Ranking, Combination & Generation, and Revising.

Linking and Ranking. In ranking procedure, multiple relations (three) are selected as candidates for each single query, as shown in Fig. 2.

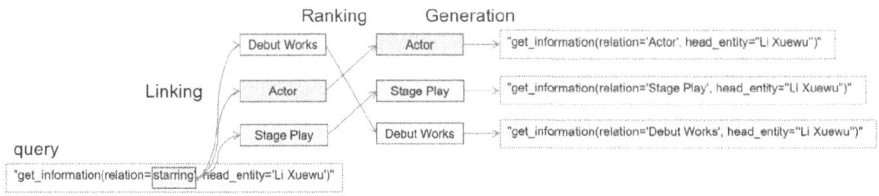

Fig. 2. Linking & Ranking module

When top-3 relations are selected, they will be ranked based on the similarity score between the original relation (for instance, "starring" in Fig. 2) and the selected relations. The similarity score is defined in Eq. (1)

$$score(a, b) = cos\_sim(emb(a), \quad emb(b)) \tag{1}$$

where a and b are two arbitrary words and cos_sim is cosine similarity. Given two word vector q and k, cos_sim can be written as Eq. (2)

$$cos\_sim(q, k) = \frac{q \cdot k}{||q|| \times ||k|| + \epsilon} \tag{2}$$

In Eq. (1), emb represents an embedding model. In the experiment, various embedding models are tested, including GTE [22] (a BERT based embedding model), text-embedding-ada-v2, and text-embedding-ada-v3 (black-box embedding API provided by Open AI).

Combination and Generation. After Linking & Ranking procedure, the top relations for each individual query of the LLM-generated functional queries will be selected and ranked. In Combination & Generation procedure, the top relations from different queries are combined to obtain multiple relation combinations. As shown in Fig. 3, for simplicity, it is assumed that two functional queries are generated by an LLM with the top 3 and top 2 relations, respectively, which will results in $3 \times 2 = 6$ relational combinations.

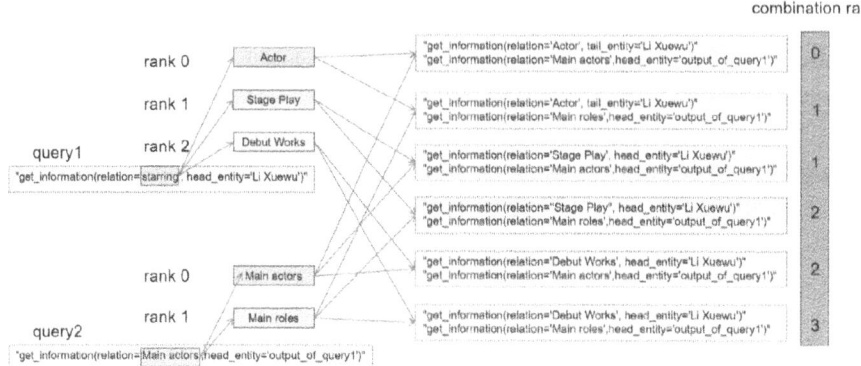

Fig. 3. Combination & Generation module

As shown in Fig. 3, these relational combinations are then sorted based on a combination score, which is calculated by adding all rank IDs of all relations in a combination. For instance, the ranking ID of the relation "Actor" in query 1 is 0, and the ranking ID of the relation "Main roles" in query 2 is 1. So, the combination score of "Actor" and "Main roles" will be $0 + 1 = 1$.

Once the relational combinations are obtained, query combinations can be generated by replacing original relations with relation combinations accordingly. Then, these queries are sent to the Query module one by one to obtain the result. If the result is not empty, stop sending anymore queries and return the result.

Revising. After the Combination & Generation module, multiple query combinations are generated and sent to the Revising module. Revising module modifies the head and tail parameters in query function "get_information" based on the head and tail information in the knowledge base (marked in red in Fig. 4). Besides, considering the existence of mutual relations in the knowledge base, such as "mate" and "friend", a custom query function "get_information_mutual" is defined to retrieve information from knowledge bases with these relations. A set of mutual relations are maintained. When a query involves one of these relations, the function "get_information" is replaced with the function "get_information_mutual" (marked in blue in Fig. 4).

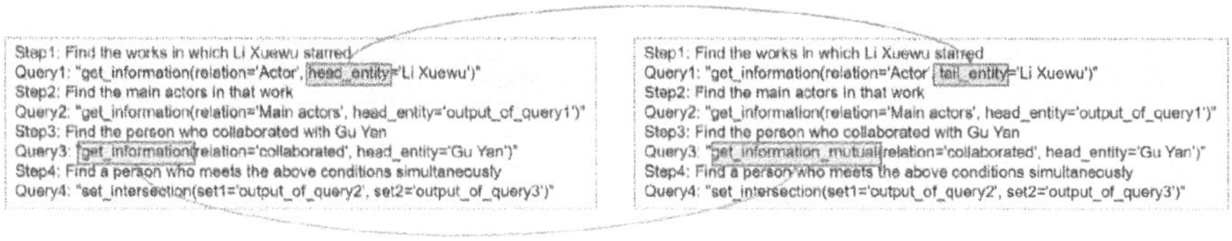

Fig. 4. Query Revising

3.4 Query Module

The Query module includes a Query Parser and a Query Execution module. The Query Parser model parses the LLM-generated queries into executable queries, and the parser method can be found in paper TrustUQA [13]. In the Query Execution module, the parsed queries are executed on the answer.

4 Results

4.1 Datasets

This section evaluates the method using the Evaluation of Complex Question Answering Reasoning in Person Knowledge Graph Task test dataset in CCKS2024. The person knowledge graph contains 785,553 triplets in total, including 44,406 head entities, 252,184 tail entities, and 240 relations. Table 2 presents some vital statics of the knowledge graph and testing dataset.

Table 2. Knowledge Graph and Data Information.

Item	Quantity
Triplet	785,553
Head Entity	44,406
Tail Entity	252,184
Relation	240
Test Case	2,000

4.2 Multi-relation Alignment Module Results

On the baseline, text-embedding-ada-002, text-embedding-ada-003 and GTE [22] for Multi-relation Alignment module are compared. As shown in Table 3, the F1 score of text-embedding-ada-003 and GTE are relatively higher, at 95.72%, while the F1 score of the text-embedding-ada-002 is 95.12%. Due to the local deployment of the GTE model, the response time for a single request is 0.003 s. The text-embedding-ada-003 works through OpenAI's API and is completed within 0.4 s. This indicates that GTE is more effective for the Multi-relation Alignment module.

Table 3. Relation Alignment Results.

Model	F1 Performance (100%)	Time Cost
text-embedding-ada-002	95.12%	0.4 s
text-embedding-ada-003	95.72%	0.4 s
GTE	95.72%	0.003 s

4.3 LLM Module Results

LLM Evaluation with Different Tempetature. Firstly, the performance of fin-tuned Qwen2-7B-Instruct and GPT4o-mini at different temperatures are compared. As shown in Table 4, for complex KBQA tasks, the effectiveness (F1-score) of the two models is inversely proportional to the temperature. This is because as the temperature decreases, the generalization capability of the LLMs decreases, while the precision increases, thereby generating functional queries that are more in line with the prompt specifications. This reduces errors in functional queries and improves the overall query performance. Moreover, the experiment also indicates that at the same temperature, the performance of fin-tuned Qwen2-7B-Instruct is superior to that of GPT4o-mini.

Table 4. Model Performance at Different Temperature

Model	F1		
	Temperature		
	1	0.7	0.1
GPT4o-mini	0.6850	0.6921	0.7035
Fin-tuned Qwen2-7b	0.742	0.7567	0.7628

Multi-LLM Results Fusion. In order to fully utilize the generalization ability of the model, this paper conducted experiments on the complementarity of various quantitative models at the same temperature of 1. It should be noted that as the number of models increases, the complementarity also increases. The inflection point occurs when the quantity reaches 5. However, as the number continues to increase, there is a diminishing return in terms of improving complementarity. Considering accuracy and efficiency, this paper chooses to utilize five models to enhance the complementarity of query statements.

Table 5. Fusion Strategy Comparison.

Strategy	F1
Fallback	81.20%
Voting	81.86%

In addition, experiments indicate that although a single low-temperature model performs better individually, recalling candidates from multiple high-temperature models (3 to 5 models) and then determining the final answer based on the ranked returned results is superior. This is because the latter method achieves a better balance between precision and recall across the models.

To fully leverage the complementarity of models at various temperatures and enhance the overall accuracy, a model fusion framework is developed, involving a fin-tuned model (temperature = 0.1), five fin-tuned models (temperature = 1), and GPT4o. Table 5 illustrates the impact of different fusion strategies on the accuracy within this framework. As shown in Table 5, the voting strategy performs better than the fallback strategy.

5   Conclusion

This paper proposes an LLM-based method that combines functional query generation and multi-relation alignment to address the complex KBQA tasks. By integrating the LLM with multi-relation alignment steps, multiple queries are generated for a single question, reducing the reliances of relation extraction and significantly increasing the likelihood of obtaining the correct answer. In addition, the voting strategy is utilized to perform multi-model fusion, making use of the precision and generalization capabilities at different temperatures of LLM, to obtain the optimal results through multi-path recall. The method ultimately achieved an outstanding performance in CCKS2024 CGQA with an F1 score of 0.8186, indicating that the method has excellent performance in dealing with complex KBQA tasks.

References

1. Berant, J., et al.: Semantic parsing on freebase from question-answer pairs. In: proceedings of emnlp (2013)
2. Tian, L., Zhou, X., Wu, Y.P., et al.: Knowledge graph and knowledge reasoning: a systematic review. J. Electron. Sci. Technol. **20**(2), 100159 (2022)
3. Lan, Y., et al.: A survey on complex knowledge base question answering: methods, challenges and solutions. arXiv preprint arXiv:2105.11644 (2021)
4. Zhang, J., et al.: Subgraph retrieval enhanced model for multi-hop knowledge base question answering. arXiv preprint arXiv:2202.13296 (2022)
5. Nie, L., et al.: GraphQ IR: unifying the semantic parsing of graph query languages with one intermediate representation. arXiv preprint arXiv:2205.12078 (2022)
6. Lan, Y., Jiang, J.: Query graph generation for answering multi-hop complex questions from knowledge bases. Assoc. Comput. Linguist. (2020). https://doi.org/10.18653/V1/2020.ACL-MAIN.91
7. Sun, Y., Zhang, L., Cheng, G., et al.: SPARQA: skeleton-based semantic parsing for complex questions over knowledge bases. In: Proceedings of the AAAI Conference on Artificial Intelligence, vol. 34, no. 5, pp. 8952–8959 (2020)
8. Min, B., Ross, H., Sulem, E., et al.: Recent advances in natural language processing via large pre-trained language models: a survey. ACM Comput. Surv. **56**(2), 1–40 (2023)

9. OpenAI: GPT-4 technical report. arXiv preprint arXiv:2303.08774 (2023)
10. Devlin, J., et al.: BERT: pre-training of deep bidirectional transformers for language understanding (2018). https://doi.org/10.48550/arXiv.1810.04805
11. Chowdhery, A., Narang, S., Devlin, J., et al.: PaLM: scaling language modeling with pathways. J. Mach. Learn. Res. **24**(240), 1–113 (2023)
12. Wei, J., et al.: Chain-of-thought prompting elicits reasoning in large language models. In: Advances in Neural Information Processing Systems (2022)
13. Zhang, W., et al.: TrustUQA: a trustful framework for unified structured data question answering. arXiv preprint arXiv:2406.18916 (2024)
14. Abujabal, A., et al.: Never-ending learning for open-domain question answering over knowledge bases. In: Proceedings of the 2018 World Wide Web Conference, pp. 1053–1062 (2018)
15. Luo, K., et al.: Knowledge base question answering via encoding of complex query graphs. In: Proceedings of the 2018 Conference on Empirical Methods in Natural Language Processing, pp. 2185-2194 (2018)
16. Kapanipathi, P., et al.: Question answering over knowledge bases by leveraging semantic parsing and neuro-symbolic reasoning. arXiv preprint arXiv:2012.01707 (2020). https://doi.org/10.48550/arXiv.2012.01707
17. Min, B., Grishman, R., Wan, L., et al.: Distant supervision for relation extraction with an incomplete knowledge base. Assoc. Comput. Linguist. (ACL) (2013)
18. Sun, H., Tania, B.-W., William, W.C.: Pullnet: open domain question answering with iterative retrieval on knowledge bases and text. arXiv preprint arXiv:1904.09537 (2019)
19. Saxena, A., et al.: Improving multi-hop question answering over knowledge graphs using knowledge base embeddings. In: Proceedings of the 58th Annual Meeting of the Association for Computational Linguistics, pp. 4498–4507 (2020). https://doi.org/10.18653/v1/2020.acl-main.412
20. Han, J., et al.: Open domain question answering based on text enhanced knowledge graph with hyperedge infusion. In: Findings of the Association for Computational Linguistics: EMNLP, pp. 1475-1481 (2020)
21. Luo, H., et al.: Chatkbqa: a generate-then-retrieve framework for knowledge base question answering with fine-tuned large language models. arXiv preprint arXiv:2310.08975 (2023)
22. Li, Z., et al.: Towards general text embeddings with multi-stage contrastive learning. arXiv preprint arXiv:2308.03281 (2023)

A Person Attribute Knowledge-Based Question Answering Method Leveraging Large Language Models

Jiabei Chen[1,2(✉)], Yao Xu[1,2], Shizhu He[1,2], Jun Zhao[1,2], and Kang Liu[1,2(✉)]

[1] The Laboratory of Cognition and Decision Intelligence for Complex Systems
Institute of Automation, Chinese Academy of Sciences, Beijing, China
`chenjiabei2024@ia.ac.cn`, `{yao.xu,shizhu.he,jzhao,kliu}@nlpr.ia.ac.cn`
[2] School of Artificial Intelligence, University of Chinese Academy of Sciences,
Beijing, China

Abstract. Recent advancements in large language models (LLMs) have been significantly propelled by pre-trained language models, enabling them to excel in various natural language processing tasks, including complex reasoning and mathematical problem-solving. Despite their impressive capabilities, LLMs still encounter challenges with factual knowledge judgment, often leading to inaccuracies or "hallucinations". To address this, integration of knowledge graphs with LLMs has been explored to enhance Knowledge Base Question Answering (KBQA). In the context of person attribute knowledge graph question answering, which often involves numerical data and complex relationship reasoning, we introduce an iterative reasoning framework that leverages LLMs to interact with the knowledge graph effectively. Additionally, we propose a dynamic demonstrations retrieval method to augment LLMs' reasoning capabilities with high-quality prompts. Experimental results demonstrate that our approach not only achieves superior performance but also won the first prize in the CCKS2024 Person Attribute KBQA task leaderboard.

Keywords: Question Answering · Knowledge Graph · LLM

1 Introduction

In recent years, the emergence of pre-trained language models has led to remarkable advancements in the performance of large language models (LLMs) across various natural language processing tasks [1]. Thanks to extensive pre-training on large-scale text corpus, LLMs are capable of addressing complex reasoning questions that involve intricate knowledge relationships [14], as well as solving mathematical problems that require sophisticated derivation processes [7].

However, despite their promising performance in many tasks, LLMs still face challenges due to their lack of robust factual knowledge judgment, which

B. Xu et al. (Eds.): CCKS-IJCKG 2024, CCIS 2229, pp. 409–418, 2025.
https://doi.org/10.1007/978-981-96-1809-5_33

can result in the generation of inaccurate or misleading information—a phenomenon often referred to as "hallucination" [3]. To mitigate this issue, various approaches that integrate knowledge graphs with large language models have been proposed [4,9] to meet the challenge of Knowledge Base Question Answering (KBQA). These methods capitalize on the factual knowledge provided by knowledge graphs while simultaneously harnessing the powerful language understanding and reasoning capabilities of large language models.

Previous works for KBQA reasoning can be classified into two groups: information retrieval (IR)-based methods and semantic parsing (SP)-based methods [5]. Semantic parsing-based approaches typically involve converting natural language questions into a symbolic logical form, which can be interpreted as an executable query by machines. This transformation allows the system to retrieve answers directly from the knowledge base by mapping the semantics of the question to the structured data contained within the knowledge graph [6,8]. Conversely, information retrieval-based methods start from an entity and iteratively explore the graph until finding the answer [4,11,13].

For the person attribute knowledge graph question answering task, the questions related to individuals exhibit two distinct characteristics: (1) They often involve numerical statistics and calculations, such as querying the number of albums released by a singer in 2024, and (2) They frequently require intricate reasoning about character relationships, such as identifying the nephew of an individual when this information is not explicitly stored but must be inferred based on sibling and parental relationships. To tackle this issue, we propose an **iterative reasoning framework leveraging LLMs**, which interacts with the knowledge graph and performs reasoning processes faithfully and reliably, as shown in Fig. 1. Additionally, we propose a **dynamic demonstrations retrieval** method to enhance LLMs reasoning ability with high-quality prompts. Moreover, experiments show that our method achieves great performance and effectively meet the challenge of the person Attribute knowledge graph question answering for **getting the 1st prize** on the final leaderboard of CCKS2024 Person Attribute KBQA task.

2   Method

2.1   Dynamic Demonstrations Retrieval

Large language models have demonstrated the remarkable ability of few-shot in-context learning (ICL), which allows LLMs to quickly adapt to new downstream tasks when a few examples, or *demonstrations*, are provided with the specific task input. However, it is precisely due to the introduction of demonstrations that the selection and design of examples have a significant impact on the model's ability of ICL [10]. In order to mitigate LLM's sensitivity to the choice of few-shot demonstrations, instead of static, fixed demonstrations, a dynamic, context-sensitive approach is used to retrieve query-related demonstrations which concatenated with the target question.

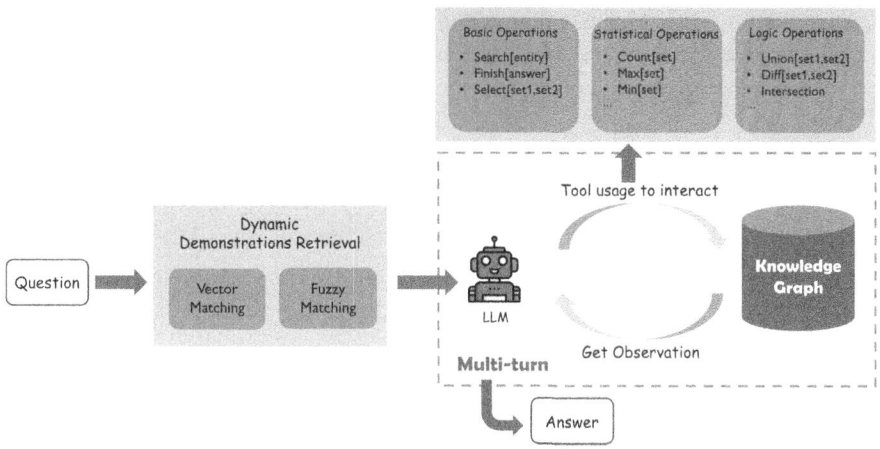

Fig. 1. The overall pipeline of our KBQA framework

Retrieval Demonstrations Construction. Since the training set only contains question-answer pairs omitting the detailed reasoning steps, we need to construct examples that include solving steps to obtain high-quality prompts. In order to save costs, we randomly select 500 questions from the training set $\mathcal{D}_{train} = \{(q_{train}, a_{train})\}$. For each question q_i from training questions $\mathcal{S}$, we use an LLM to generate the trajectory e_i and get the the final answer a_i' from e_i. If the result a_i' is equal to the labeled answer a_i, we regard (q_i, e_i) as a question-trajectory demonstration pair. Moreover, we manually annotated 60 samples adapted from the training set questions to address the challenges of the final test set for more complex queries. Finally, we obtain the demonstration corpus $\mathcal{D}_{retrieve} = \{(q_{retrieve}, e_{retrieve})\}$ for retrieval.

Retrieval Approach. To enhance the diversity and accuracy of retrieval, we have designed the following retrieval methods:

Vector-based Retrieval. Semantic search aims to enhance search accuracy by comprehensively understanding the semantic meaning of both the search query and the corpus. The fundamental concept of semantic search involves embedding all entries within a corpus—whether they are sentences, paragraphs, or documents-into a vector space. Given a question q, we apply a dense text encoder E to encode both q and each question $q_{retrieve}$ from retrieval corpus into vectors. Then, we compute the similarity between these question vectors and select the k most similar questions in the retrieval corpus. Then we will utilize the selected questions with their trajectories to build the prompt.

Fuzzy Matching-based Retrieval. Compared to vector-based retrieval, the primary advantage of fuzzy matching lies in its robustness to lexical variations and typographical errors. Moreover, fuzzy matching can enhance the recall of similar questions because of keywords matching. Specifically, we use Levenshtein Distance to measure the similarity between two sequences, which is defined as follows:

$$
lev(s,t) = \begin{cases} |s| & \text{if } |t| = 0, \\ |t| & \text{if } |s| = 0, \\ lev(s[1:], t[1:]) & \text{if } s[0] = t[0], \\ 1 + \min \begin{cases} lev(s[1:], t), \\ lev(s, t[1:]), & \text{otherwise.} \\ lev(s[1:], t[1:]) \end{cases} \end{cases} \quad (1)
$$

where s and t are strings, $|s|$ denotes the length of string s, $s[0]$ denotes the first character of string s and $s[1:]$ denotes the substring of s starting from the second character to the end.

Hybrid Retrieval. Finally, we combine the vector-based and fuzzy matching-based retrieval methods to produce a hybrid retriever. Our hybrid retriever uses Reciprocal Rank Fusion (RRF) score to combine multiple ranked lists into a single, more comprehensive ranking. Given the vector-based and fuzzy matching-based ranked lists, the RRF score for an item d is calculated as:

$$
RRF(d) = \frac{1}{rank_{vec}(d) + C} + \frac{1}{rank_{fuzzy}(d) + C} \quad (2)
$$

where $rank_{vec}(d)$ and $rank_{fuzzy}(d)$ is the rank of item d in the list. If the item d is not included in a particular list, the $rank(d)$ should be infinite. And C is a constant which helps to dampen the influence of lower-ranked items.

2.2   Iterative KG Reasoning Framework

In this work, motivated by ReAct [15], we regard LLMs as an agent interacting with the knowledge graph and use the ReAct-style trajectory to conduct the reasoning process which additionally generates Chain-of-Thought rationales prior to taking actions. Given a question q, the LLM first generates the initial thought t_0 to decompose original question into the first sub-questions. At each round except for the first round, the model should generate a thought t that concludes the result of the previous observation obtained from last round and plans the sub-question for the next round. Then, the LLM is required to generate an action a from the action space which provides tools to assist large models in searching knowledge graphs, performing logical operations, or conducting statistical counting. After executing the action a, the returned result serves as the current round's observation o. When encountering *Finish* in the action a_T, we parse the result within it as the final answer. Thus, every trajectory e generated from the LLM is defined as below:

$$
e = \{q, t_0, a_0, o_0, t_1, ..., t_n, a_n\} \quad (3)
$$

Action Space. Although LLMs performs well across a variety of downstream tasks, it struggles to manipulate structured data with precision and efficiency [4]. To address this difficulty, we draw inspiration from the use of specialized and predefined tools to solve complex tasks and design the following three categories of operations to assist LLMs:

- **Basic Operations** This category primarily aids LLMs in interacting with knowledge graphs and filtering existing data. It includes operations for retrieving candidate triples through interactions with the knowledge graph (*Search[entity]*), selecting intermediate entities based on the second hops (*Select[set1,set2]*), and obtaining final answers (*Finish[answer]*).
- **Logic Operations** This category is designed to assist LLMs in performing logical operations on the extracted entity sets, which consist of entity set intersection (*Intersection[set1,set2]*), union (*Union[set1,set2]*) and difference (*Difference[set1,set2]*).
- **Statistical Operations** This category focuses on enabling the model to perform statistical analyses and counting operations on the extracted data. It includes functions for calculating the number of elements in a set (*Count[set]*), determining the maximum and minimum elements within a set (*Max[set]*, *Min[set]*), comparing the sizes of two sets (*Subtract[num1,num2]*), and checking for the presence of specific elements within a set (*Contain[set,element]*).

2.3 Interaction with KG

Exploration in the Search Stage. In the Search stage, it primarily consists of three processes: key entity recognition, entity linking and relation exploration, ultimately resulting in a set of entities most relevant to the current thought.

Key Entity recognition. We use LLMs to extract key entity from the current thought in the process. LLMs just follow the demonstrations in the context and returns through the parameters in the *Search* action. As shown in Fig. 2(a), when the sub-question is "需要知道王建杰的音乐作品有哪些(*Need to know what* 王建杰*'s musical works are*)" (Thought 1), the LLM generates the action "*Search[*王建杰*]*" and the key entity, "王建杰" , is extracted for future use.

Entity Linking. The extracted key entity will be matched and linked to the entity in the knowledge graph. However, not all extracted entities can be grounded in the knowledge graph, possibly because the extracted entities are incomplete or are other aliases. If the corresponding entity cannot be matched, we perform semantic similarity matching with all entities in the knowledge graph and use the longest string matching to obtain the final extracted entity.

Relation Exploration. After identifying the entity most relevant to the current thought, we conduct relation exploration for the entity. Initially, we search out all relations linked to the entity. Subsequently, we compute the semantic similarity between these relations and the current thought, thereby identifying the top-k

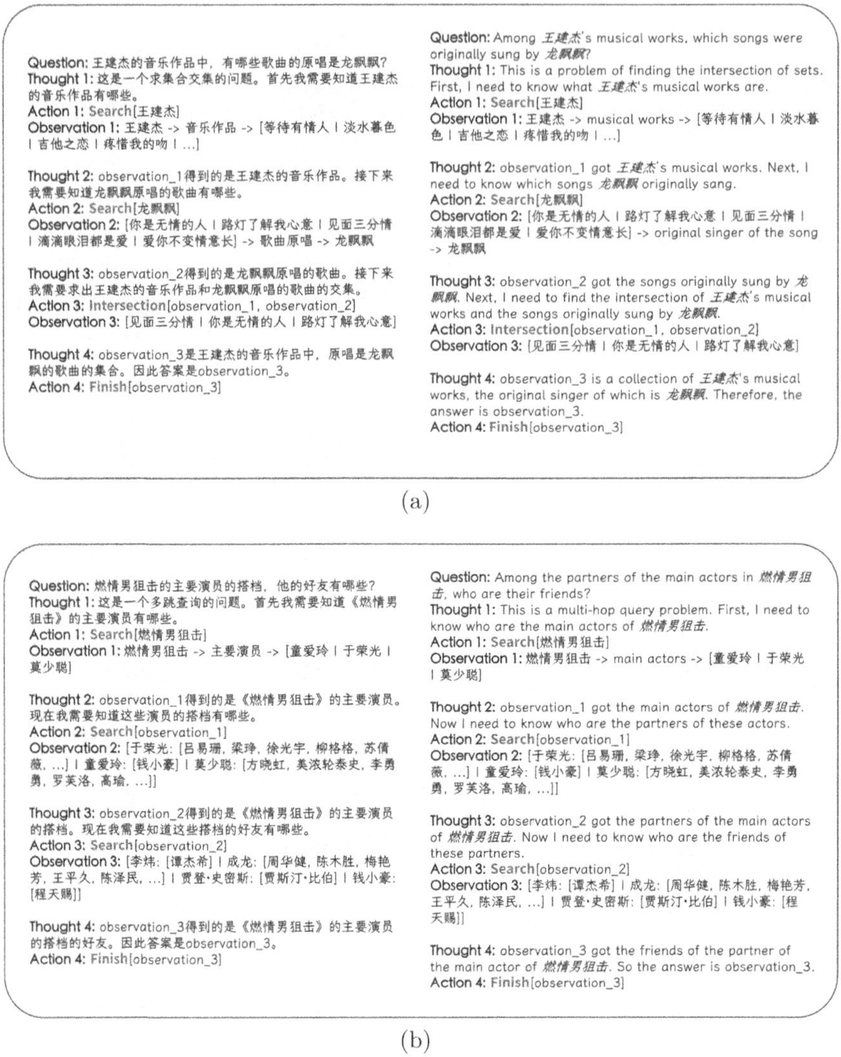

(a)

(b)

Fig. 2. Iterative reasoning trajectory examples, the right side is the translated version. (a) is an example of using logic tool, **_Intersection_**. And (b) is an example of **Recursive Exploration**.

relationships ranked by similarity. These top-k relationships, along with their respective directions, are then presented to LLMs for ultimately selecting the particular relation with the highest relevance. Specially, some relations, like "搭档(Partner)" "好友(Friend)", are bidirectional relations so we will use the extracted entity as either the head entity or the tail entity to search and obtain the final set of target entities.

After executing the three process described above, we will obtain the target entity set that is most related to the current thought. To balance precision and recall, and to avoid issues related to the excessive context length of LLMs, we replace portions of the target entity set that exceed k elements with "...", as shown in Fig. 2(a) (Observation 1). All target entity sets are retained for use in subsequent stages and can be referenced as parameters in action operations, such as "*observation_1*".

Recursive Search for Multi Candidates. For multi-hop queries, we typically need to conduct further exploration of the previously obtained entity set. At this point, exploring each candidate entity effectively corresponds to querying a sub-question for that entity. Therefore, we employ a recursive approach to explore the candidate entity set using the same question-and-answer framework. Specifically, for each candidate entity, we rewrite the current thought to generate new corresponding sub-questions for recursive calls. We then merge the final target entity set based on the answers returned from each question-and-answer instance, which will include both the candidate entities and the results from the most recent exploration, as shown in Fig. 2(b) (Observation 2 & 3).

However, when the number of candidate entities in the Observation is excessive, the efficiency of recursively exploring each entity becomes significantly low. To address this issue, we directly utilize the semantic similarity between the current thought and all the relations in the knowledge graph when the referenced observation containing more than K elements. We then employ LLMs to filter and select the top-1 relation. Finally, we efficiently retrieve the corresponding target entities using this relationship with predefined SPARQL.

3   Experiments and Results

3.1   Dataset

Our method is evaluated on the CCKS2024 Person Attribute Knowledge-Based Question Answering Competition dataset. The given knowledge graph provides 256,378 entities, 240 relations and 785,553 triples related to personal relationships. The training set contains 1238 question-answer pairs. And the performance is evaluated on two different test sets with 1,000 samples in the preliminary test set and 2,000 in the final test set.

3.2   Evaluation Metric

In this task, the F1 score is employed to evaluate the prediction quality for each individual sample, and the final test result is determined by averaging the F1 scores across all test samples, as Averaged F1, which is defined below:

$$P_i = \frac{|A_i \cap G_i|}{|A_i|}, R_i = \frac{|A_i \cup G_i|}{|A_i|} \tag{4}$$

$$F1 = \frac{1}{|Q|} \sum_{i=1}^{|Q|} \frac{2P_i R_i}{P_i + R_i} \tag{5}$$

where $|Q|$ denotes the number of samples in the test set, A_i, G_i denotes the prediction set and the ground truth set of the i-th sample respectively.

Table 1. The Averaged F1 in different settings for the preliminary and final stage.

Methods		Averaged F1 (Preliminary)	Averaged F1 (Final)
Prompting w/DeepSeek-V2-Chat	Fixed 16-shot	0.7769	\
	Dynamic 5-shot(vector)	\	0.7749
	Dynamic 5-shot(vector)+SC	**0.8177**	0.7871
	Dynamic 5-shot(hybrid)	\	0.8052
	Dynamic 5-shot(hybrid)+SC	\	**0.8213**

3.3   Implementation

We use DeepSeek-V2-Chat as the base model for our experiments. The maximum token length for each generation is set to 1024. The temperature parameter is set to 0.7. We also use the technique of Self-Consistency (SC) [12] and the sampling number is set to 5. The embedding model for computing semantic similarity is BGE-M3, which can work across multiple languages [2].

3.4   Results

Table 1 shows the Averaged F1 scores in different settings for the preliminary and final stage. As the table shows, our methods achieves **0.8177** in the preliminary test set which is ranked 1st on the leaderboard and gets 1st prize in the final with F1 score of **0.8213**. From the results, it can also be seen that by integrating dynamic demonstrations retriever and the technique of self-consistency, our iterative knowledge graph reasoning framework can achieve great performance and effectively meet the challenge of the person Attribute knowledge graph question answering.

4   Conclusion

In this paper, we propose an iterative reasoning framework leveraging large language models for the Person Attribute Knowledge Graph Question Answering task. Through the integration of key techniques such as dynamic demonstrations retrieval, agent-based tool usage and iterative interaction with KG, we have built an effective and reliable QA system. Experimental results show the system's excellent performance in dealing with complex questions involving person relationships, achieving a remarkable first-place rank in the CCKS2024 Person Attribute KBQA task leaderboard.

Acknowledgments. This work was supported by the Beijing Natural Science Foundation (L243006), the National Natural Science Foundation of China (No. 62376270) and the Youth Innovation Promotion Association CAS.

References

1. Brown, T.B., et al.: Language models are few-shot learners (2020). https://arxiv.org/abs/2005.14165
2. Chen, J., Xiao, S., Zhang, P., Luo, K., Lian, D., Liu, Z.: Bge m3-embedding: multi-lingual, multi-functionality, multi-granularity text embeddings through self-knowledge distillation (2024). https://arxiv.org/abs/2402.03216
3. Huang, L., et al.: A survey on hallucination in large language models: principles, taxonomy, challenges, and open questions (2023). https://arxiv.org/abs/2311.05232
4. Jiang, J., Zhou, K., Dong, Z., Ye, K., Zhao, X., Wen, J.R.: StructGPT: a general framework for large language model to reason over structured data. In: Bouamor, H., Pino, J., Bali, K. (eds.) Proceedings of the 2023 Conference on Empirical Methods in Natural Language Processing, pp. 9237–9251. Association for Computational Linguistics, Singapore, December 2023. https://doi.org/10.18653/v1/2023.emnlp-main.574, https://aclanthology.org/2023.emnlp-main.574
5. Lan, Y., He, G., Jiang, J., Jiang, J., Zhao, W.X., Wen, J.R.: Complex knowledge base question answering: a survey. IEEE Trans. Knowl. Data Eng. **35**(11), 11196–11215 (2023). https://doi.org/10.1109/TKDE.2022.3223858
6. Li, T., Ma, X., Zhuang, A., Gu, Y., Su, Y., Chen, W.: Few-shot in-context learning on knowledge base question answering. In: Rogers, A., Boyd-Graber, J., Okazaki, N. (eds.) Proceedings of the 61st Annual Meeting of the Association for Computational Linguistics (Volume 1: Long Papers), pp. 6966–6980. Association for Computational Linguistics, Toronto, Canada, July 2023. https://doi.org/10.18653/v1/2023.acl-long.385, https://aclanthology.org/2023.acl-long.385
7. Luo, H., et al.: Wizardmath: empowering mathematical reasoning for large language models via reinforced evol-instruct (2023). https://arxiv.org/abs/2308.09583
8. Luo, H., et al.: Chatkbqa: a generate-then-retrieve framework for knowledge base question answering with fine-tuned large language models (2024). https://arxiv.org/abs/2310.08975
9. LUO, L., Li, Y.F., Haf, R., Pan, S.: Reasoning on graphs: faithful and interpretable large language model reasoning. In: The Twelfth International Conference on Learning Representations (2024). https://openreview.net/forum?id=ZGNWW7xZ6Q
10. Luo, M., Xu, X., Liu, Y., Pasupat, P., Kazemi, M.: In-context learning with retrieved demonstrations for language models: a survey (2024). https://arxiv.org/abs/2401.11624
11. Sun, J., et al.: Think-on-graph: Deep and responsible reasoning of large language model on knowledge graph. In: The Twelfth International Conference on Learning Representations (2024). https://openreview.net/forum?id=nnVO1PvbTv

12. Wang, X., et al.: Self-consistency improves chain of thought reasoning in language models (2023). https://arxiv.org/abs/2203.11171
13. Xu, Y., et al.: Generate-on-graph: treat llm as both agent and kg in incomplete knowledge graph question answering (2024). https://arxiv.org/abs/2404.14741
14. Yang, Z., et al.: Hotpotqa: a dataset for diverse, explainable multi-hop question answering (2018). https://arxiv.org/abs/1809.09600
15. Yao, S., et al.: React: synergizing reasoning and acting in language models (2023). https://arxiv.org/abs/2210.03629

Instruction Fine-Tuning of Large Language Models for Traditional Chinese Medicine

Juntao Li, Ling Luo$^{(\boxtimes)}$, Tengxiao Lv, Chao Liu, Jiewei Qi, Zhihao Yang, Jian Wang, and Hongfei Lin

School of Computer Science and Technology, Dalian University of Technology, Dalian, China
lingluo@dlut.edu.cn

Abstract. Traditional Chinese Medicine (TCM) is a holistic healthcare system that encompasses a rich body of medical theories, extensive clinical experiences, and a vast pharmacopoeia, which plays a significant role in medical practices. To promote the development of large language models (LLMs) for TCM, the CCKS 2024 challenge organized the TCMBench track, including TCM knowledge comprehension evaluation and TCM natural language inference tasks. In this paper, we present our instruction fine-tuning method for this track. We first constructed a rich training set by collecting existing Chinese medical exam datasets and TCM-related internet sources. We then designed various instructional prompts tailored to different tasks, enabling the model to fully exploit the knowledge acquired during fine-tuning to answer questions accurately and contextually. We also conducted extensive tests on different LLMs. Experimental results demonstrate the effectiveness and robustness of our method, achieving the first place in this challenge.

Keywords: Large Language Model · Traditional Chinese Medicine · Instruction Fine-Tuning

1 Introduction

Traditional Chinese Medicine (TCM) [1], with a history spanning thousands of years, remains integral to global healthcare, grounded in the theories of Yin-Yang balance [2], the Five Elements [3], and Qi. These foundational concepts emphasize holistic and syndrome differentiation approaches, forming the basis of TCM's diagnostic and therapeutic practices. Although TCM is gaining international recognition for its effectiveness in treating chronic and neurological diseases [4, 5], its complexity and reliance on practitioner experience present challenges for standardization and modern research [6].

The advent of Large Language Models (LLMs) like ChatGPT and GPT-4 has opened new possibilities for the preservation and innovation of TCM [7]. However, significant differences between TCM and Western medicine in theory and practice mean that traditional Western benchmarks are inadequate for evaluating LLMs in the TCM domain. To promote the development and evaluation of LLMs for TCM, Yue et al. [8] developed the TCM evaluation benchmark and organized the TCMBench track at the 2024 China

© The Author(s), under exclusive license to Springer Nature Singapore Pte Ltd. 2025
B. Xu et al. (Eds.): CCKS-IJCKG 2024, CCIS 2229, pp. 419–430, 2025.
https://doi.org/10.1007/978-981-96-1809-5_34

Conference on Knowledge Graph and Semantic Computing (CCKS 2024). This track is divided into two subtracks: efficient fine-tuning and non-fine-tuning. We participated in the efficient fine-tuning track, where participants are restricted to fine-tuning LLMs with parameters not exceeding 1% of the model's total.

The efficient fine-tuning track consists of two core subtasks: the TCM knowledge comprehension evaluation and the TCM natural language inference. The first task includes an evaluation dataset, TCM-ED, composed of real practice questions from the TCM Licensing Exam (TCMLE), which reflect the fundamental medical knowledge and reasoning required to obtain a TCM license in China. However, no training data is provided. The second task is a TCM-specific natural language inference task to determine if there is semantic consistency between two sentences, using the TMNLI dataset. Unlike the first task, the TMNLI dataset includes a substantial amount of training data. Therefore, we focus more on the first subtask, which presents a more significant challenge.

In this paper, we present our instruction fine-tuning method of LLMs for this track. Our main contributions are as follows:

– We collected data from multiple sources, including available Chinese medical exam datasets, TCMLE-related web resources, and TCM LLMs training data. To enhance the dataset, we used advanced LLMs to generate answer analysis, particularly for questions lacking medical interpretations. After standardizing the format, we ultimately obtained over 100,000 TCM-related data instances.
– We explored various instructional prompt methods for generating instruction-tuning data. Experimental results indicate that the model achieves better performance by analyzing both correct and incorrect options, compared to analyzing only the correct answers. Furthermore, providing question option information in multi-round dialogues significantly improves the accuracy on B1-type questions.
– We conducted comprehensive tests on different LLM bases. The results show that InternLM series outperforms other models.

Owing to the above contributions, our method achieves the first place in the efficient fine-tuning track of the TCMBench Challenge at CCKS 2024, with the accuracy scores of 89.01 and 97.65% in the TCM-ED and TMNLI tasks, respectively.

2   Related Work

Recently, LLMs (such as GPT-4 [9] and LLaMA [10]) have shown promising results in various natural language processing (NLP) tasks and have received widespread attention around the world. However, theseLLMs often perform sub-optimal and struggle to follow instructions when applied to specific domains like biomedicine. Training LLMs from scratch typically requires much resource. To address these challenges, Instruction Fine-tuning (IFT) methods [11–13] have been proposed to enhance the performance of LLMs, particularly in the medical field. The primary objective of IFT is to improve the model's ability to follow various human instructions and execute specific tasksIFT achieves this goal by utilizing specially constructed training datasets that typically contain instruction-input-output triplets. These approaches enable the development of LLMs that can more accurately interpret and respond to medical queries, follow complex medical instructions, and perform specialized medical tasks, thereby potentially improving the application of LLMs in healthcare and medical research.

In the field of TCM question-answering (QA), the introduction of LLMs has brought new vitality to the integration of traditional medicine with modern technology. Models like ShenNong-TCM-LLM [14] and CMLM-ZhongJing [15] represent significant advancements in TCM-focused LLMs. ShenNong-TCM-LLM is built upon a TCM knowledge graph and utilizes a vast amount of TCM-related instructional data generated with the assistance of ChatGPT and similar models. Subsequently, it undergoes instruction fine-tuning. In clinical TCM diagnosis and treatment, ShenNong-TCM-LLM show exceptional capabilities in prescription recommendation. CMLM-ZhongJing employs specialized tables and specific prompt templates to achieve precise simulation and reasoning of TCM prescription data and complex diagnostic logic. It can also generate detailed diagnostic and treatment plans. This provides comprehensive intelligent assistance for TCM clinical diagnosis and treatment, from diagnosis to therapy, thus further advancing the modernization of TCM practice.

3   Method

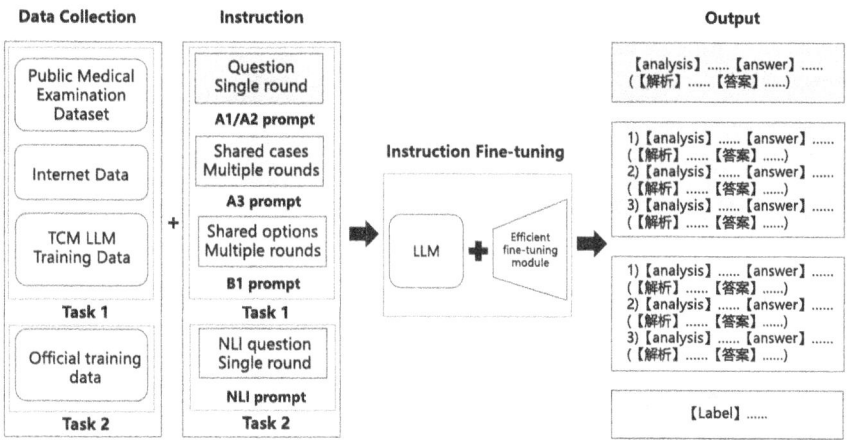

Fig. 1. Overview of our instruction fine-tuning method

To enhance the model's performance on TCMBench, we focused on two main aspects: data quality and instructional prompts design. An overview of our fine-tuning method for TCMBench is shown in Fig. 1. We first collected relevant data about TCM licensing exams from multiple sources for the TCM-ED task. To improve data quality, we employed GLM-4 [16] to generate medical interpretations for questions that lacked explanations. Next, we developed and tested specialized instructional prompts for different question types. We continuously optimized and adjusted our prompting strategies based on the structural characteristics of each question type. Through these methods, we aim to optimize training data and design targeted prompting strategies to enhance the model's overall performance on TCMBench.

3.1 Training Data Collection

Since official TCMBench only provides training data for the TMNLI task, data for the TCM-ED task must be collected and processed extensively. This section details our data collection and processing methods. The distribution of the final dataset across different question types is presented in Table 1. It is important to note that while some data for A1 and A2 question types lack explanations, the A3 and B1 question types include complete explanations.

Table 1. Information of dataset for fine-tuning

Type	Task1				Task2	Other
	A1/A2 (analysis)	A1/A2 (no analysis)	A3	B1	NLI	QA
Official training data	–	–	–	–	25,581	–
CMExam	44,199	7,912	–	–	–	–
CMB	–	31,561	–	–	–	–
Internet Data	7,940	–	2,779	4,641	–	–
TCM LLM Training Data	–	–	–	–	–	10,000
Total	52,139	39,473	2,779	4,641	25,581	10,000

To enhance the model's performance on Task1 (i.e., the TCM-ED task), we aggregated comprehensive sets of related data from three sources: public medical examination datasets, internet data and TCM LLM training data.

(1) **Public Medical Examination Dataset**

CMExam [17]. The questions are sourced from China's National Medical Licensing Examination. CMExam consists of over 60,000 multiple-choice questions designed for standardized and objective assessment, with 85.24% of the questions including medical analysises.

CMB [18]. This is a comprehensive, multi-level Chinese medical evaluation dataset. It contains 280,839 multiple-choice questions and 74 complex case consultation questions, covering all clinical medical specialties and different professional levels. However, CMB's answers do not include any medical analysis. It aims to comprehensively assess models' medical knowledge and clinical consultation abilities. We selected TCM-related single-choice questions, cleaned the data, and used them to improve the model's reasoning accuracy during fine-tuning.

(2) **Internet Data**

The datasets mentioned above mainly cover the A1/A2 type questions but lack other question types present in TCMBench. Therefore, we collected a large number of TCM licensing exam simulation questions and practice problems from the internet, such as Weipu[1] and Huanqiuwangxiao[2]. These questions are carefully screened and organized, primarily covering A3 and B1 type questions, which are relatively scarce in public datasets. We employed OCR technology to extract the question content and used regular expressions to standardize the data, ensuring all data conforms to a unified format. For some questions lacking standard analyses, we employed an advanced LLM, GLM-4, to generate relevant medical analyses. This data collection and processing not only addressed the scarcity of A3 and B1 type questions but also provided more diverse training materials, enabling the model to perform more stably and accurately on these scarce question types.

(3) **Traditional Chinese Medicine LLM Training Data**

ShenNong_TCM_Dataset[3]. The dataset serves as the training corpus for the ShenNong-TCM-LLM, aiming to enhance the LLM's knowledge in TCM and its ability to respond to medical inquiries. However, the dataset also contains some queries that cannot be answered with certainty, as well as questions unrelated to TCM.. Through a rigorous keyword matching and filtering strategy, we eliminated question-answer pairs that lack direct, clear answers or are not directly relevant to TCM topics, ensuring the dataset's purity and high relevance. Ultimately, we randomly extracted 10,000 dialogue entries closely from the filtered dataset, s. These entries serve as a fine-tuning training corpus, supporting the model's specialized optimization.

For Task2 (i.e., the TMNLI task), we used the training dataset provided by CCKS2024 TCMBench. In this dataset, each piece of data includes a premise, hypothesis, and a label indicating the entailment relationship. The data is constructed using questions, answers, and standard explanations from the TCM licensing exam. To eliminate bias caused by significant length differences between question-answer generated sentences and standard explanations, the premise and hypothesis were swapped for half of the data pairs, resulting in 25,581 instances of training data.

Through the collection and processing of these data, we have constructed a comprehensive and optimized training dataset, providing strong support for the model's excellent performance in the TCMBench task. This process has not only enhanced the model's ability to handle different types of questions but also promoted the digital inheritance and modernization research of TCM knowledge.

3.2 Instruction-Tuning Data Construction

To enable the LLMs to understand TCM task instructions for multitasking, we constructed and optimized instructional prompts for fine-tuning, covering the tasks described in the previous section. First, we utilized the Chain of Thought (CoT) prompt templates

---

[1] https://oldvers.cqvip.com/view/professional/index.aspx.

[2] https://www.hqwx.com/zyzyys-kaoshi/moniti/.

[3] https://huggingface.co/datasets/michaelwzhu/ShenNong_TCM_Dataset.

provided by the CCKS-2024 TCMBench organizers, which included A1/A2, A3, and B1 question types. For the TMNLI task that do not have predefined prompts, we referred to prompt templates from other question types and created instructional prompts that matched the characteristics of the TMNLI task. To maximize the use of A1/A2 question type data that lack explanations, we designed new instruction templates modeled after the prompt templates for questions with explanations. These customized instructional prompts not only provide the model with clearer problem-solving guidance but also enable it to demonstrate strong understanding and reasoning abilities even when faced with unannotated data. For ShenNong dialogue data, we developed specific instruction templates. These templates help the model distinguish between different tasks and significantly enhance its performance on TCM tasks, particularly by aiding the model in learning a large amount of TCM knowledge.

Notably, due to the unique structure of B1 question types—where options are extracted as shared content—we designed two different forms of instructional prompts. As shown in Fig. 2, Prompt1 places the shared options only before the first sub-question, with subsequent questions providing only the question content. This format requires the model to have strong memory capabilities to effectively utilize the previously provided information in subsequent answers. In contrast, Prompt2 places the options after each sub-question, making the structure of each sub-question more similar to A1/A2 types. This design leverages the model's learned TCM knowledge from other question types, thereby improving performance on B1 questions.

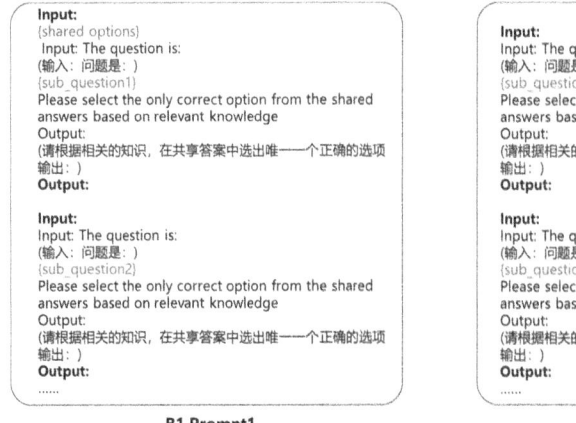 

B1 Prompt1   B1 Prompt2

Fig. 2. Two methods of prompts for B1 type questions

Furthermore, we explored the effect of two different types of answer analyses for the models. The first is the original Long Analysis provided by the CMExam dataset. This analysis provides a comprehensive explanation of the answers, including explanations for the correct options as well as the reasons for all incorrect options. We also tested the second type, i.e., Shot Analysis, where the data was cleaned and processed to focus on key points, making the explanations shorter and more consistent with the length and format of standard explanations in the TCMBench validation set.. An example of different analyses is shown in Fig. 3.

Question

Question:食管疾病的胸痛特点是（ ）．
The chest pain characteristics of esophageal diseases are ().
A. 精神紧张诱发 Psychological tension induced
B. 含化硝酸甘油减轻 Reduction of nitroglycerin content
C. 呼吸时加重，屏气时消失 Exacerbating during breathing, disappearing during breath hold
D. 压迫加剧 Intensifying oppression
E. 进食加剧 Increased eating intensity

Long analysis

Answer: "E"
Analysis: 食管疾病导致的胸痛是指由食管疾病或食管功能障碍引起的胸痛，表现为烧心伴胸骨后或胸骨下发作性疼痛，进食时发作或加剧，服用抗酸剂或促动力药物可减轻或消失（E对）．精神紧张诱发是指精神紧张状态导致的脏器血供相对不足，多见于心绞痛，也可见于食管疾病（A错）．含化硝酸甘油减轻是心绞痛的胸痛特点（B错）．呼吸时加重，屏气时消失是指呼吸时胸膜摩擦而疼痛加重，是结核性胸膜炎以及心包炎的特点（C错）．压迫加剧见于多种可致胸痛的疾病（D错）．
Chest pain caused by esophageal disease refers to chest pain caused by esophageal disease or esophageal dysfunction, manifested as heartburn accompanied by episodic pain behind or below the sternum, which occurs or worsens during eating. Taking antacids or prokinetic drugs can alleviate or disappear (E correct). Mental stress induction refers to the relative lack of blood supply to organs caused by mental stress, which is more common in angina pectoris and can also be seen in esophageal diseases (A false). Containing nitroglycerin to alleviate chest pain is a characteristic of angina pectoris (B false). Exacerbation during breathing and disappearance during breath holding refers to the increased pain caused by pleural friction during breathing, which is a characteristic of tuberculous pleurisy and pericarditis (C false). Exacerbation of compression is seen in various diseases that can cause chest pain (D false).

Short analysis

Answer: "E"
Analysis: 食管疾病导致的胸痛是指由食管疾病或食管功能障碍引起的胸痛，表现为烧心伴胸骨后或胸骨下发作性疼痛，进食时发作或加剧，服用抗酸剂或促动力药物可减轻或消失。
Chest pain caused by esophageal disease refers to chest pain caused by esophageal disease or esophageal dysfunction, manifested as heartburn accompanied by episodic pain behind or below the sternum, which occurs or worsens during eating. Taking antacids or prokinetic drugs can alleviate or disappear.

Fig. 3. An Example of different analyses for TCM questions

3.3 Model Fine-Tuning

After completing the preliminary data preparation, we constructed various training sets by combining different instructional prompts and dataset. Participants are required to choose one of the following open-source models in the challenge: Qwen series[4], InternLM series[5], ChatGLM3-6B [16] and ShenNong-TCM-LLM [14]. During the model fine-tuning process, we employed the Firefly[6] fine-tuning framework in conjunction with QLoRA [19] technology for efficient fine-tuning. This strategy allowed us to develop an efficient and accurate model fine-tuning process while ensuring that the

---

[4] https://huggingface.co/qwen.

[5] https://huggingface.co/internlm.

[6] https://github.com/yangjianxin1/Firefly.

model maintains a high level of understanding and reasoning ability with complex TCM tasks.

4 Experiments

4.1 Datasets and Evaluation Metrics

The official final test set includes TCM-ED and the TMNLI test sets. In the TCM-ED test set, there are 1,598 combined samples of types A1 and A2, and 198 samples of type A3, which include 642 sub-questions. For type B1, there are 1,481 test samples with 3,231 sub-questions. The TCM-NLI test set consists of 3,916 samples.

For the two evaluation tasks, CCKS2024 TCMBench uses different evaluation metrics as follows: For the TCM-ED task, the accuracy metric is first used to compare the correct options with the model's responses. The other evaluation metrics are for generating question analyses, using Rouge1 [20], RougeL [20], BertScore [21], and SARI [22] to comprehensively evaluate the quality of model-generated analyses. For the TMNLI task, the accuracy is used as the evaluation metric. The model's ability to judge the semantic similarity between two sentences is assessed by comparing the labels generated by the model with the correct labels. The average of all scores is used as the final score.

4.2 Implementation Details

We used QLoRA to efficiently fine-tune the base model. During the training process, we set the learning rate to 0.0001, the maximum sequence length to 1024 and the number of training epochs to 4. We adopted a constant learning rate strategy with warm-up, setting the warm-up steps to 100, and selected paged_adamw_32bit as the optimizer. In the QLoRA parameter configuration, we set the rank to 32, the alpha value to 32 and the dropout rate for LoRA layers to 0.05, without fine-tuning the bias terms.

During the efficient fine-tuning process, we used three Nvidia A5000 GPUs for training and two separate GPUs for inference testing. During inference, we set the temperature coefficient to 0.35, top_p to 0.9, and the maximum number of generated tokens to 500.

4.3 Performance of Different LLMs Without Fine-Tuning

To determine the base model for fine-tuning, we first tested different chat models using zero-shot prompt without fine-tuning. The accuracies of various models on different tasks are shown in Table 2. Compared to other models, ChatGLM3-6B shows lower accuracy across all tasks. For the Qwen series, we found that Qwen2-7B-Instruct performed better than Qwen1.5-14B-Chat on Task1 (the TCM-ED task), but its accuracy on Task2 (the TMNLI task) was significantly lower. This was because when Qwen2-7B-Instruct was instructed to judge whether there was an entailment relationship between the hypothesis and premise sentences, with label values of 0 and 2 (where 0 indicates an entailment relationship between the premise and hypothesis, and 2 indicates no relation or contradiction), it outputted a large number of cases with label value 1.

Table 2. Accuracy and final score of each model without fine-tuning on the final test set

Model	Task1	Task2	A1/A2	A3	B1	Score
ChatGLM3-6B	40.09	37.22	40.68	51.02	37.77	33.15
Qwen1.5-14B-Chat	70.48	73.77	74.84	75.17	67.49	48.61
Qwen2-7B-Instruct	71.66	56.35	**77.97**	75.17	67.93	47.32
InternLM2-20B-Chat	67.60	**84.85**	66.62	73.13	67.05	50.96
InternLM2_5-7B-Chat	**74.08**	76.45	**77.97**	**77.89**	**71.48**	**52.28**

For the InternLM series, we found that InternLM2_5-7B-Chat outperformed other models overall. Although its accuracy on Task2 is lower than that of InternLM2-20B-Chat from the same series, it achieves higher accuracy on other tasks compared to other models. Based on these findings, we conducted fine-tuning experiments using a small amount of data on Qwen2-7B-Instruct, InternLM2_5-7B-Chat, and InternLM2_5-7B. Ultimately, we decided to use InternLM2_5-7B as the base model for subsequent experiments.

4.4 Main Results

To evaluate the performance of different datasets and instructional prompts, we submitted multiple runs with various settings. After efficient fine-tuning, the model's accuracy on different tasks is shown in Table 3. Our best submission (i.e., Run5) achieves the highest final score, securing the first place in the efficient fine-tuning track of the challenge. The results of the top 2 and 3 teams are also shown for comparison. We further analyzed the results of different runs, and our main findings are as follows.

(1) **The TCM training data collected from various sources contributes to the model's performance in TCM knowledge comprehension.** The results from Run1, 4 and 5 show that incorporating internet data into the public medical examination dataset notably improves the accuracy in A1/A2 and A3 question types. This suggests that the internet data effectively help the model acquire the medical knowledge necessary for completing Task1. Furthermore, the results of fine-tuning with additional TCM LLM training data show an improvement in Task1 accuracy compared to Run4, suggesting that the dialogue data help the model supplement some TCM-related medical knowledge. By comparing the effects of different data sources, we found that the internet data contributed the most to improving the model's accuracy on Task1.

(2) **Providing question option information in multi-round dialogues significantly improves the model's performance on B1-type questions.** Comparing Run2 and Run3, we observed that prompt optimization significantly improves the model's accuracy on specific question types. The accuracy for type B1 increases from 83.26 to 90.03. This is because in Run2, we only provided shared answer options for the first sub-question, while in Run3, we added shared answer options after each sub-question's prompt, making the prompt for each B1 sub-question more similar to

Table 3. Accuracy and final score of different methods on the final test set

Method	Setting	Task1	Task2	A1/A2	A3	B1	Score
sf_cloud (Top2)	–	84.90	**98.42**	86.55	85.51	83.97	59.39
ZZUNLP (Top3)	–	85.98	96.99	**88.30**	**88.47**	84.34	58.97
Run1	Data: Exam; Prompt:1; Analysis: Long	77.98	97.20	83.44	81.29	74.71	57.67
Run2	Data: Exam + Internet; Prompt:1; Analysis: Short	85.19	97.27	88.05	87.85	83.26	58.53
Run3	Data: Exam + Internet; Prompt:2; Analysis: Short	88.83	97.14	87.30	86.60	90.03	59.15
Run4	Data: Exam + Internet; Prompt:2; Analysis: Long	88.52	97.45	86.36	87.69	89.76	59.42
Run5 (Top1)	Data: Exam + Internet + LLM; Prompt:2; Analysis: Long	**89.01**	97.65	86.98	85.98	**90.62**	**59.75**

the A1/A2 question types. This improvement allows the model to better leverage knowledge and experience gained from A1/A2 type training data during reasoning.

(3) **The models trained with long analysises achieve higher analysis scores than those trained with short analysises.** As shown in Table 4, a comparison between Run3 and Run4 reveals that using longer explanations resulted in slightly lower RougeL and BertScore scores but notably improve Rouge1 and SARI scores. Overall, the average score when using longer analysises is slightly higher than that of using shorter analysises. We also found that Rouge1 and SARI were positively correlated, as were RougeL and BertScore, while there was a negative correlation between these two groups of metrics. Additionally, the incorporation of these data sources effectively improved the model's average scores in generating analyses.

Table 4. The analysis scores of the different methods on the final test set

Method	Setting	Rouge1	RougeL	BertScore	SARI	*AVE*
sf_cloud (Top2)	–	42.35	28.54	**73.54**	28.60	**43.26**
ZZUNLP (Top3)	–	**45.65**	23.68	71.14	30.40	42.72
Run3	Data: Exam + Internet; Prompt:2; Analysis: Short	39.15	**28.99**	73.33	27.10	42.14
Run4	Data: Exam + Internet; Prompt:2; Analysis: Long	42.47	26.33	72.35	29.42	42.64
Run5 (Top1)	Data: Exam + Internet + LLM; Prompt:2; Analysis: Long	43.48	25.78	72.05	**30.54**	42.96

5 Conclusion

In this paper, we present our instruction fine-tuning method for LLMs in the TCM-Bench track. We successfully constructed training data using web scraping techniques and advanced GLM-4, while fully utilizing open-source TCM-related datasets. Experimental results show that incorporating long analysis data and additional TCM dialogue data can enhance models' reasoning performance during the fine-tuning process. Our method achieves the state-of-the-art performances on TCMBench test sets, demonstrating its effectiveness for TCM. In future work, we plan to further explore additional TCM capabilities of LLMs, such as assisting in disease diagnosis and treatment.

Acknowledgments. This research was supported by the CIPSC-SMP-Zhipu.AI Large Model Cross-Disciplinary Fund (NO. ZPCG2024010204), the National Natural Science Foundation of China (No. 62302076), and the Fundamental Research Funds for the Central Universities (No. DUT23RC (3)014).

References

1. Nestler, G.: Traditional chinese medicine. Med. Clin. **86**(1), 63–73 (2002)
2. Li, P.P.: The unique value of yin-yang balancing: a critical response. Manag. Organ. Rev. **10**(2), 321–332 (2014)
3. Pachuta, D.M.: Chinese medicine: the law of five elements. India Int. Centre Q. **18**(2/3), 41–68 (1991)
4. Jiang, M., Zhang, C., Cao, H., Chan, K., Lu, A.: The role of Chinese medicine in the treatment of chronic diseases in china. Planta Med. **77**(09), 873–881 (2011)
5. Ren, Z.L., Zuo, P.P.: Neural regeneration: role of traditional Chinese medicine in neurological diseases treatment. J. Pharmacol. Sci. **120**(3), 139–145 (2012)
6. Fung, F.Y., Linn, Y.C.: Developing traditional chinese medicine in the era of evidence-based medicine: current evidences and challenges. Evid.-Based Complement. Altern. Med. **2015**(1), 425037 (2015)

7. Thirunavukarasu, A.J., Ting, D.S.J., Elangovan, K., Gutierrez, L., Tan, T.F., Ting, D.S.W.: Large language models in medicine. Nat. Med. **29**(8), 1930–1940 (2023)
8. Yue, W., et al.: Tcmbench: a comprehensive benchmark for evaluating large language models in traditional chinese medicine. arXiv preprint arXiv:2406.01126 (2024)
9. Achiam, J., et al.: Gpt-4 technical report. arXiv preprint arXiv:2303.08774 (2023)
10. Touvron, H., et al.: Llama: open and efficient foundation language models. arXiv preprint arXiv:2302.13971 (2023)
11. Luo, L., et al.: Taiyi: a bilingual fine-tuned large language model for diverse biomedical tasks. J. Am. Med. Inf. Assoc. ocae037 (2024)
12. Tran, H., Yang, Z., Yao, Z., Yu, H.: Bioinstruct: instruction tuning of large language models for biomedical natural language processing. J. Am. Med. Inf. Assoc. ocae122 (2024)
13. Singhal, K., et al.: Large language models encode clinical knowledge. Nature **620**(7972), 172–180 (2023)
14. Wei Zhu, W., Wang, X.: Shennong-tcm: A traditional chinese medicine large language model (2023)
15. Yang, S., et al.: Zhongjing: Enhancing the Chinese medical capabilities of large language model through expert feedback and real-world multi-turn dialogue. In: Proceedings of the AAAI Conference on Artificial Intelligence, vol. 38, pp. 19368–19376 (2024)
16. Glm, T., et al.: Chatglm: a family of large language models from glm-130b to glm-4 all tools. arXiv preprint arXiv:2406.12793 (2024)
17. Liu, J., et al.: Benchmarking large language models on cmexam-a comprehensive chinese medical exam dataset. In: Advances in Neural Information Processing Systems, vol. 36 (2024)
18. Wang, X., et al.: Cmb: a comprehensive medical benchmark in Chinese. In: Proceedings of the 2024 Conference of the North American Chapter of the Association for Computational Linguistics: Human Language Technologies (Volume 1: Long Papers), pp. 6184–6205 (2024)
19. Dettmers, T., Pagnoni, A., Holtzman, A., Zettlemoyer, L.: Qlora: efficient finetuning of quantized llms. In: Advances in Neural Information Processing Systems, vol. 36 (2024)
20. Lin, C.Y.: Rouge: A package for automatic evaluation of summaries. In: Text Summarization Branches Out, pp. 74–81 (2004)
21. Zhang, T., Kishore, V., Wu, F., Weinberger, K.Q., Artzi, Y.: Bertscore: evaluating text generation with bert. arXiv preprint arXiv:1904.09675 (2019)
22. Xu, W., Napoles, C., Pavlick, E., Chen, Q., Callison-Burch, C.: Optimizing statistical machine translation for text simplification. Trans. Assoc. Comput. Linguist. **4**, 401–415 (2016)

Enhancing Traditional Chinese Medicine Question Answering and Semantic Reasoning via Historical Exam Retrieval and Sentence Similarity

Qin Fang, Yifan Wang, Pan Yuan, Zheng Zhang, and Xian Peng[✉]

National Engineering Research Center of Educational Big Data, Central China
Normal University, Wuhan, China
{fangqin,wangyifan0122,yuanpan,zhangz}@mails.ccnu.edu.cn,
pengxian@ccnu.edu.cn

Abstract. In this paper, we present our system for the CCKS2024-TCMBench. TCMBench is a comprehensive benchmark designed to evaluate the performance of large language models (LLMs) in the traditional Chinese medicine (TCM) domain. Specifically, TCMBench defines two tasks: (1) TCM question answering with explanations, and (2) TCM natural language inference. For Task 1, we first supplemented the knowledge points associated with questions in the CMExam dataset. During the question-answering process, we retrieved relevant knowledge points by matching similar historical questions, thereby enhancing the LLMs' answering performance. For Task 2, we calculated sentence similarity between premises and hypotheses, determining the levels of similarity based on the computed values, which served as supplementary input to enhance the LLMs' reasoning capabilities. By applying these methods, our system achieved second place in the CCKS2024-TCMBench (Non-Finetuning Track), validating the effectiveness of the proposed approach.

Keywords: Large Language Models · Question Answering · Knowledge Retrieval · Sentence Similarity

1 Introduction

In recent years, large language models (LLMs) such as GPT-4 [1] and LLaMA [2] have brought about profound transformations in the field of natural language processing (NLP) [3]. These models, which are built on the Transformer architecture and trained on diverse, large-scale datasets, have demonstrated exceptional capabilities in tasks such as natural language understanding, generation, and reasoning [4]. The applications of LLMs have evolved from basic text generation to more sophisticated tasks, including complex dialogue systems and reasoning-based challenges, leading to significant improvements in both the accuracy of natural language understanding and the quality of generated text [5].

B. Xu et al. (Eds.): CCKS-IJCKG 2024, CCIS 2229, pp. 431–439, 2025.
https://doi.org/10.1007/978-981-96-1809-5_35

Although LLMs have demonstrated outstanding performance in natural language processing tasks within Western medicine, their evaluation and application in the domain of Traditional Chinese Medicine (TCM) remain underexplored [6]. Unlike Western medicine, TCM is based on distinct theoretical foundations, diagnostic methods, and treatment approaches, with language that is often philosophical and symbolic [7]. These characteristics pose unique challenges for LLMs, making existing benchmarks designed for Western medicine insufficient for accurately assessing language models in the TCM domain [6].

TCMBench is a comprehensive benchmark designed to evaluate the performance of LLMs in the domain of traditional Chinese medicine [8]. It is derived from the Traditional Chinese Medicine Licensing Examination, specifically tailored for the TCM domain. The benchmark includes two tasks: 1) TCM Knowledge Understanding: Given a question, select the correct answer and provide an explanation; 2) TCM Semantic Reasoning: Given a premise and a hypothesis, determine whether they have an entailment relationship.

The main contributions of this paper are as follows:

(1) For Task 1, we considered that the questions in the benchmark are similar to those in the historical TCM exam question, and that the historical questions contain potentially valuable knowledge. Therefore, we extracted the knowledge points related to the questions from the CMExam dataset[1]. During the question-answering process, we retrieved relevant knowledge points by matching similar historical questions and used them as context to enhance the LLMs' performance.

(2) For Task 2, we observed that there are differences in sentence similarity between premises and hypotheses depending on whether an entailment relationship exists. Therefore, we determined the levels of similarity based on the similarity values and used these levels as supplementary input to improve the reasoning capabilities of LLMs.

(3) We conducted experiments on the dataset provided by the competition, and our method achieved second place in the CCKS2024-TCMBench (Non-Finetuning Track), validating the effectiveness of our approach.

2   Methodology

2.1   TCM Question Answering with Explanations

Knowledge Extraction and Refinement in the CMExam Dataset.
CMExam is the first Chinese medical exam dataset to provide comprehensive medical annotations, comprising over 60,000 multiple-choice questions [9]. This dataset includes not only the questions but also their corresponding answers and explanations, which serve as valuable knowledge resources. To effectively utilize this information, we constructed prompt templates using the questions, answers, and explanations from CMExam. These templates were then input into DeepSeek-V2 [10], which summarized and refined the knowledge points. Through

---

[1] https://github.com/williamliujl/CMExam/tree/main/data.

this approach, we achieved a comprehensive mapping of each historical question to its associated knowledge points.

Vectorized Storage of Historical Exam Questions. We first embedded each question from the CMExam dataset using an embedding model. The resulting vectorized data was then stored in a vector database, a specialized system designed for the storage, indexing, and retrieval of high-dimensional data. In this study, we employed FAISS [11], an efficient approximate nearest neighbor search library, for data storage and retrieval. FAISS is capable of handling large volumes of vector data, offering advantages such as fast retrieval and precise ranking.

The entire process, from knowledge extraction to the vectorized storage of historical exam questions within the CMExam dataset, is summarized in Fig. 1.

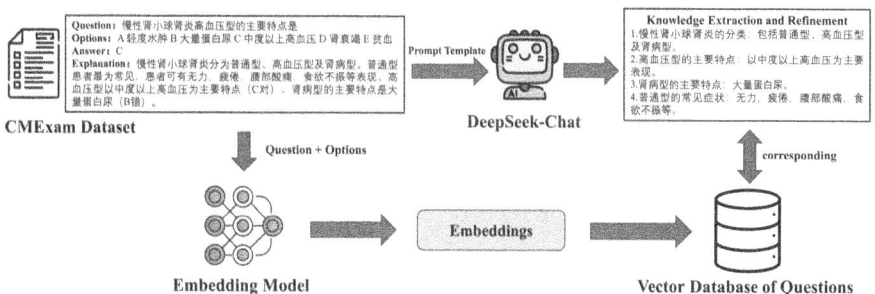

Fig. 1. Example Overview of Knowledge Extraction and Vectorized Storage Processes in the CMExam Dataset.

Euclidean Distance. Euclidean distance [12] is a mathematical method used to measure the linear distance between two points in a multidimensional space. It quantifies the absolute distance between points by reflecting the differences in numerical characteristics across dimensions. For two points $A(x_1, y_1, z_1, \ldots, n_1)$ and $B(x_2, y_2, z_2, \ldots, n_2)$ in an n-dimensional space, their Euclidean distance is calculated using the formula:

$$D(A, B) = \sqrt{(x_2 - x_1)^2 + (y_2 - y_1)^2 + (z_2 - z_1)^2 + \cdots + (n_2 - n_1)^2}$$

In essence, Euclidean distance represents the straight-line distance between two points in Euclidean space, calculated as the square root of the sum of squared differences in coordinates across dimensions. In this study, we compute the Euclidean distance between two vectors to assess their similarity, and based on this metric, we construct the FAISS IndexFlatL2 index.

TCM Question Answering with Explanations. In the question-answering process of the LLMs, the workflow is illustrated as shown in Fig. 2. Specifically, the process begins with the vectorization of the question using an embedding model. Subsequently, semantic retrieval is employed to identify the most similar historical questions. We set a similarity threshold of 0.25, and if the similarity score between the retrieved question and the current query is below 0.25, the corresponding knowledge point information from the retrieved question is obtained. Finally, depending on the availability of knowledge point information, different prompts are constructed and provided to the LLM to generate the final answer.

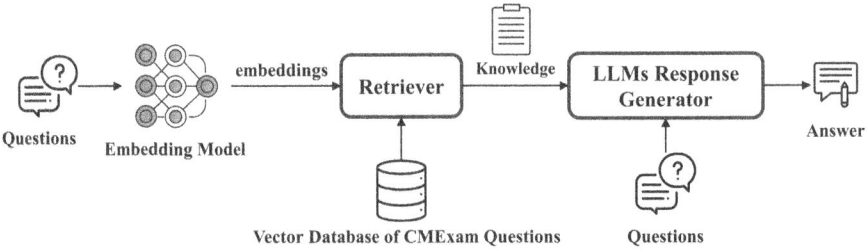

Fig. 2. Workflow of TCM Question Answering with Explanations.

2.2   TCM Natural Language Inference

Cosine Similarity. Cosine similarity assesses the degree of similarity between entities in a multidimensional space by calculating the cosine of the angle between two vectors [13]. Unlike Euclidean distance, which measures the magnitude of difference, cosine similarity focuses on the directional difference between vectors. For two non-zero vectors $A = (a_1, a_2, \ldots, a_n)$ and $B = (b_1, b_2, \ldots, b_n)$, their cosine similarity can be calculated using the following formula:

$$cosine\ similarity(A, B) = \cos(\theta) = \frac{A \cdot B}{\|A\|\|B\|}$$

where $A \cdot B$ denotes the dot product of vectors A and B, and $\|A\|$ and $\|B\|$ represent the magnitudes (norms) of the vectors. Values closer to 1 indicate smaller angles and higher similarity between the vectors, whereas values closer to 0 indicate lower similarity.

Analysis and Classification of Sentence Similarity in Premises and Hypotheses. We computed the cosine similarity for all sentence pairs in the training set and performed a classification analysis based on their entailment labels. The results indicated that the average similarity for sentence pairs with entailment was 0.810, while for those without entailment it was 0.484. This finding suggests that sentence similarity could be a valuable indicator for detecting entailment. Consequently, we classified sentence similarity into three levels (low, medium, and high) based on the average similarity, with the ranges defined as [0, 0.484], (0.484, 0.810], and (0.810, 1], respectively.

TCM Natural Language Inference. In the TCM natural language inference process, the workflow is illustrated in Fig. 3. First, we convert the premise and hypothesis into embeddings using an embedding model. Next, we compute the cosine similarity between these embeddings and classify the sentence similarity based on the similarity score. Subsequently, we incorporate this similarity level information into a prompt template, which is then provided to the LLM. Finally, the LLM generates the inference result, determining whether an entailment relationship exists between the premise and the hypothesis.

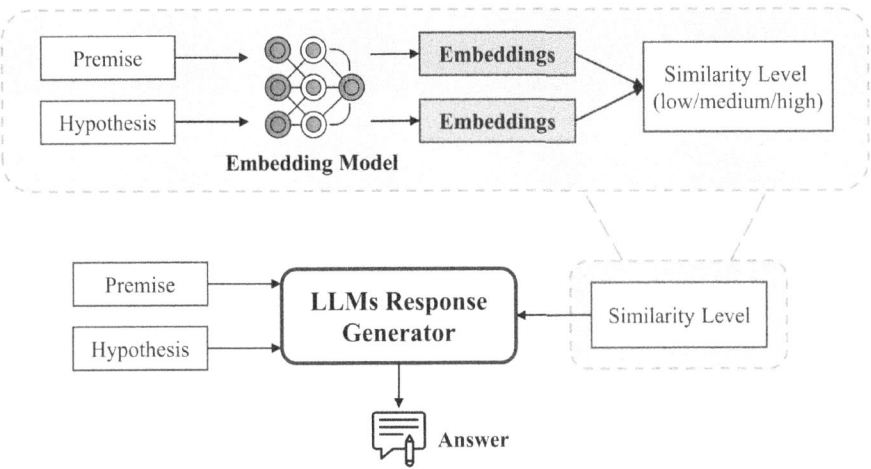

Fig. 3. Workflow of TCM natural language inference.

3 Experiments

3.1 Datasets

Our experiments were conducted on the dataset provided by the CCKS 2024-TCMBench competition. For Task 1, the dataset includes three types of questions: the single-sentence best-choice questions (A1) and the case summary best-choice questions (A2), the best-choice questions for case groups (A3), and the standard compatibility questions (B1). For Task 2, the training set contains 17,907 samples, the validation set contains 2,558 samples, and the test set contains 3,916 samples. The detailed statistical information is presented in Table 1.

Table 1. The statistical information of the Task1(upper) and Task2(lower).

Question Type	#Qustions	#Sub-Question	#All
A1/A2	1,598	1,598	
A3	198	642	5,471
B1	1,481	3,231	

Dataset	#Entailment	#Contradiction	#All
Train	4,489	13,418	17,907
Validation	1,775	783	2,558
Test	/	/	3,916

3.2 Baselines

Following the competition's evaluation guidelines, for Task 1, we employ the official prompt templates provided by the organizers to generate both the options and the corresponding analysis from the LLMs, serving as the baseline.

For Task 2, we adopt a few-shot learning approach to guide the model's output generation, establishing this as the baseline.

3.3 Implementation Details

In accordance with the competition's constraints, we selected stella-base-zh-v3-1792d[2] as the embedding model and internlm2.5-7b[3] [14] as the large language model (LLM). All models were run on a single A800-80G GPU, and we utilized the vLLM library [15]. For the generation process, we initially set the model temperature to 0 to remove the effect of random sampling. If the model's output did not meet the format requirements, we adjusted the temperature to 0.5. The maximum number of tokens for generation was set to 512.

3.4 Evaluation Metrics

In this study, for Task 1, the evaluation system uses accuracy as the evaluation metric to compare the options generated by LLMs with the correct options, thereby assessing the models' understanding and application of TCM knowledge. Additionally, traditional evaluation metrics for text generation tasks are employed, including word overlap-based methods such as ROUGE [16], and deep learning-based semantic similarity measures like BertScore [17]. These are used to evaluate the semantic similarity between the analysis generated by the LLMs and the standard analysis. Moreover, the evaluation system incorporates the

---

[2] https://huggingface.co/infgrad/stella-base-zh-v3-1792d.
[3] https://huggingface.co/internlm/internlm2_5-7b.

SARI [18] metric to evaluate text simplification by assessing lexical modifications.

For Task 2, the evaluation system employs accuracy to evaluate the performance of LLMs in predicting semantic entailment relations.

3.5   Experimental Results

For Task 1, we evaluated the impact of incorporating historical exam question retrieval into the baseline model. For Task 2, we extended the zero-shot approach by adding sentence similarity level prompts to assess the LLMs' performance. The results of both experiments are presented in Tables 2 and 3, respectively.

Table 2. The Detailed Data of Experimental Results for Task 1.

Method	Acc	Rouge-1	Rouge-L	BertScore	SARI
Baseline	0.7670	0.4680	0.1922	0.6886	0.3108
Ours(with retrieval)	0.8309	0.4611	0.2236	0.7013	0.3109

Table 3. The Detailed Data of Experimental Results for Task 2.

Method	Acc
Baseline	0.8636
Ours(with similarity level)	0.9091

3.6   Additional Analyses

For Task 1, the experimental results demonstrate that incorporating historical exam question retrieval led to a 6.39% increase in the accuracy of the LLM's answers compared to the baseline. Both Rouge-L and BertScore showed improvement, indicating that the retrieval mechanism enabled the model to generate analysis that was more semantically and structurally aligned with the standard analysis. The increase in Rouge-L suggests better sequence alignment, while the higher BertScore reflects enhanced semantic similarity. However, Rouge-1 and SARI showed only minor changes, likely because these metrics focus on word-level matching and text simplification. The retrieved knowledge may already be present in the LLMs but was not effectively utilized to generate more coherent analyses. This observation implies that improving the semantic coherence and relevance of the analysis could lead to more accurate model responses.

For task 2, the results indicate that incorporating sentence similarity prompts in the zero-shot setting improved accuracy on the NLI task by 4.55% compared to the baseline. This enhancement is likely due to the prompts guiding the LLM to consider not only surface-level features but also deeper semantic similarities between sentences, providing clear direction for semantic matching. In contrast, the few-shot approach lacks this explicit semantic focus, which limits its ability to generalize to unseen sentence pairs. This observation highlights the significant impact of prompt design on LLM performance, where similarity-based prompts direct the LLM to better align semantic meanings during inference, thereby improving its ability to detect implicit relationships and achieve better generalization compared to task-specific few-shot examples.

Due to the limited number of result submission attempts and time constraints during the competition, we could not conduct more extensive comparative experiments for Task 1 and Task 2. As a result, the depth of the experimental results and analyses was somewhat constrained. In future work, we plan to build on the insights gained from this experience to broaden the scope of experiments, particularly by testing LLM performance across other domains. We will also adopt more detailed experimental designs and evaluation methods with the goal of further enhancing the overall performance of LLMs.

4   Conclusion

In this paper, we present a method for enhancing the performance of large language models (LLMs) on TCM question answering and TCM natural language inference tasks without requiring fine-tuning. For the first task, we leveraged a comprehensive historical question database (CMExam) to extract knowledge points from each historical exam question. Then we employed a retrieval-based approach to enhance the performance of LLMs by obtaining relevant knowledge points for the evaluation questions. For the second task, we improved inference performance by providing additional similarity level information for premises and hypotheses to the LLMs. Experimental results demonstrate the effectiveness of our approach, and we achieved second place in the CCKS2024-TCMBench (Non-Finetuning Track).

Acknowledgments. This work was supported by the Research Funds from National Natural Science Foundation of China (Grant No. 62107016, 62077024, 62293555, 62293550), Hubei Provincial Natural Science Foundation of China (2023AFA020), Humanities and Social Sciences Foundation of the Ministry of Education (Grant No. 21YJC880057).

References

1. Achiam, J., et al.: GPT-4 technical report. arXiv preprint arXiv:2303.08774 (2023)
2. Dubey, A., et al.: The Llama 3 herd of models. arXiv preprint arXiv:2407.21783 (2024)
3. Qiu, J., et al.: Large AI models in health informatics: applications, challenges, and the future. IEEE J. Biomedical Health Inform. (2023)

4. Peng, B., et al.: Graph retrieval-augmented generation: a survey. arXiv preprint arXiv:2408.08921 (2024)
5. Pan, S., Luo, L., Wang, Y., Chen, C., Wang, J., Wu, X.: Unifying large language models and knowledge graphs: a roadmap. IEEE Trans. Knowl. Data Eng. (2024)
6. Cai, Y., et al.: MedBench: a large-scale Chinese benchmark for evaluating medical large language models. In: Proceedings of the AAAI Conference on Artificial Intelligence, vol. 38, pp. 17709–17717 (2024)
7. Zhang, Q., Zhou, J., Zhang, B.: Computational traditional Chinese medicine diagnosis: a literature survey. Comput. Biol. Med. **133**, 104358 (2021)
8. Yue, W., et al.: TCMBench: a comprehensive benchmark for evaluating large language models in traditional Chinese medicine. arXiv preprint arXiv:2406.01126 (2024)
9. Liu, J., et al.: Benchmarking large language models on CMExam-a comprehensive Chinese medical exam dataset. In: Advances in Neural Information Processing Systems, vol. 36 (2024)
10. Liu, A., et al.: DeepSeek-V2: a strong, economical, and efficient mixture-of-experts language model. arXiv preprint arXiv:2405.04434 (2024)
11. Douze, M., et al.: The Faiss library. arXiv preprint arXiv:2401.08281 (2024)
12. Cardarilli, G.C., Di Nunzio, L., Fazzolari, R., Nannarelli, A., Re, M., Spanò, S.: N-dimensional approximation of Euclidean distance. IEEE Trans. Circuits Syst. II Express Briefs **67**(3), 565–569 (2019)
13. Zhu, S., Wu, J., Xiong, H., Xia, G.: Scaling up top-K cosine similarity search. Data Knowl. Eng. **70**(1), 60–83 (2011)
14. Cai, Z., et al.: InternLM2 technical report. arXiv preprint arXiv:2403.17297 (2024)
15. Kwon, W., et al.: Efficient memory management for large language model serving with paged attention. In: Proceedings of the 29th Symposium on Operating Systems Principles, pp. 611–626 (2023)
16. Lin, C.Y.: ROUGE: a package for automatic evaluation of summaries. In: Text Summarization Branches Out, pp. 74–81 (2004)
17. Zhang, T., Kishore, V., Wu, F., Weinberger, K.Q., Artzi, Y.: BERTscore: evaluating text generation with BERT. arXiv preprint arXiv:1904.09675 (2019)
18. Xu, W., Napoles, C., Pavlick, E., Chen, Q., Callison-Burch, C.: Optimizing statistical machine translation for text simplification. Trans. Assoc. Comput. Linguist. **4**, 401–415 (2016)

Chinese Knowledge Base Question Answering System with Retrieval Augmented Generation

Ruihan Zhang, Jiayang Li[(⊠)], and Ziyun Chen

Nari Group Corporation (State Grid Electirc Power Research Institute),
Nanjing 210061, Jiangsu, China
{lijiayang,chenziyun}@sgepri.sgcc.com.cn

Abstract. Knowledge base question answering (KBQA) aim to answer given question based on facts from a structured knowledge base. Chinese KBQA (CKBQA) limited the task with Chinese. Previous approaches including semantic parsing-based or retrieval-based approaches have limited performance in large scale CKBQA tasks. In this paper, we proposed a Retrieval Augmented Generation (RAG) based CKBQA system including facts retrieval and semantic parsing (SPARQL generation) with large language models (LLM). Our approach aims to maximize the use of the LLM's information filtering capabilities, thereby reducing the logical demands of SPARQL query generation. Our approach possesses strong scalability, allowing it to adapt to new questions related to newly added facts without requiring additional supervised fine-tuning (SFT) of the LLM. We achieve 1st in CCKS2024 CKBQA competition with F1 scores of 85.15%.

Keywords: CKBQA · LLM · RAG

1 Introduction

Knowledge Base Question Answering (KBQA), as a subdomain task in Question Answering (QA) in Natural Language Processing (NLP), focuses on answering given questions based on facts in the existing Knowledge Base (KB). KB contains collections of facts, which are mostly in the form of alias triplets (subjects, relations, objects). The task of KBQA requires the answer to the question through facts in the KB. Chinese KBQA (CKBQA) specifies that the language used in KBQA is Chinese. In recent years, the continuous development of Large Language Models (LLMs) [22] has significantly improved the performance of NLP-related tasks, while also driving the adaptation of traditional approaches in subdomain tasks to better align with the strengths of LLMs [13]. The two categories of KBQA methods are semantic parsing-based (SP-based) and information retrieval-based (IR-based) methods [8]. The former generates symbolic

© The Author(s), under exclusive license to Springer Nature Singapore Pte Ltd. 2025
B. Xu et al. (Eds.): CCKS-IJCKG 2024, CCIS 2229, pp. 440–452, 2025.
https://doi.org/10.1007/978-981-96-1809-5_36

logic forms to execute against the KB, and the latter constructs a question-specific subgraph of the full KB to simplify the difficulties of answer generation. Recent approaches tend to use powerful LLM to better solve the problem.

In CCKS2023 CKBQA competition, the KB contains over 60 million facts with over 20 million entities (subjects and objects) and nearly half a million relations. Relations can be complex for different queries. For instance, the question "代表食物是烤包子的民族所信奉的宗教? (What is the religion believed in by the ethnic group whose representative food is baked buns?)". Here, the main difficulty is the entity linking for the relation "believe". In the KB, for ethnic groups, the relation name of "faith" is not the word mentioned in the question "信奉" but the word "信奉". Although these words have similar meanings in Chinese, facts with different subjects require different relations.

Our CKBQA system is designed to maximize the potential of Large Language Models (LLMs). Current LLM models possess strong retrieval capabilities, but they are somewhat lacking in logical reasoning skills. Therefore, we opted to use the Retrieval-Augmented Generation (RAG) [4] approach with LLMs as SP-based methods for SPARQL generation.

RAG enhances accuracy by combining retrieval and generation. It first retrieves relevant documents from a database and then uses these documents to generate more precise responses. SPARQL generation requires high model accuracy. Even the wrong words or order of triplet components may result in a failed query. With retrieval information, generated answers are grounded in up-to-date, contextually relevant information, improving overall accuracy. Also, with the recent improvements in LLMs to handle long-context processing, we tend to provide highly precise but low-recall information to the LLM for SPARQL generation. In other words, we supply as much relevant information as possible to the LLM model, leveraging its capacity to sift through large amounts of data and identify the pertinent facts necessary for correctly generating SPARQL queries.

Potentially, our CKBQA system also reduces the cost of updating new data. There is no need for retraining (SFT) the LLM again when generating SPARQL queries for new questions and facts based on RAG retrieval results; we only need to update the information retrieval database to quickly add new triplet data. As for the computationally intensive LLM component, with relevant or even accurate facts provided for given questions, we aim for the LLM to focus on generating SPARQL with correct grammar, entities, and relations.

2   Related Works

Commonly used KBQA approaches can be separated into Information Retrieval-based (IR-based) methods and Semantic Parsing-based (SP-based) methods [8, 10]. Some more recent approaches tried to include LLMs in the pipeline.

IR-based KBQA methods [11,14,20] aim to retrieve relevant facts (triplets) based on given questions to a subgraph. Then determine answers based on generated subgraph. SP-base KBQA methods aim to use models to generate logical forms from the questions and then excute them against KBs. Commonly used logical forms include SPARQL which used in this paper. Some SP-based methods [3,7,9] designed a pipeline, generate SPARQL step-by-step. Some other methods [15,19,21] tend to solve the problem directly with models.

Some recent works [5,6,16] tried to introduce powerful LLM tools in the pipeline to improve the overall performance. Mostly, they tend to ultilize the thinking capabilities of LLM to seek for answers by retrieving from the graph in step-wise manner. LLMs like ChatGPT [12] has shown to contain the ability to simplify many traditional downstream NLP tasks [13]. With some open source pretrained LLMs like "LLaMA" [17] and "Qwen" [1], supervised finetuning can further improve the performance in downstream tasks. In this paper, as we focus on Chinese questions, "Qwen" series LLMs are our best choice.

3   Methods

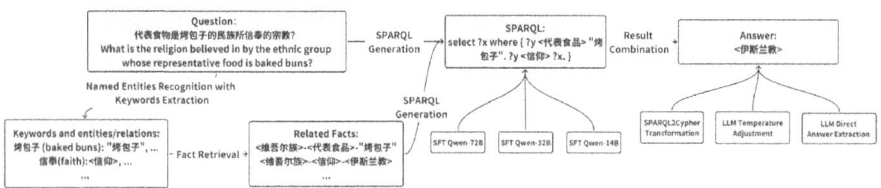

Fig. 1. Overview of our system. It's a RAG-based approach. The LLM take the question and retrieved related facts as input in the prompts and generate the SPARQL queries. We trained several LLMs and use results combination to provide the final answer.

Our approach contains the pipeline with: 1, named entity recognition (NER); 2, facts retrieval (FR); 3, symbolic logic forms of SPARQL generation (SG); and 4, results combination (ensemble methods, RC). We selected the RAG-based approach for utilizing LLMs, aiming to maximize the retrieval capabilities while minimizing their involvement in logical reasoning tasks. And, given the current advancements in LLMs' long-context processing abilities, we incorporate as much relevant information as possible, relying on the model's capacity to retrieve the final result from the extended context. Figure 1 shows the overall processing progress of our system.

In detail, we extract keywords from the given question and transfer them to entities and relations in KB in NER, and retrieve question-related facts based on these keywords in FR. These question-related facts are filled as context information for RAG in LLM inference. For SG, we finetune (SFT) the LLM model to enhance RAG-based SPARQL generation ability. In the final stage of RC, we combine results from different SFT models with different temperatures. Additionally, we introduced a rule-based SPARQL-to-Cypher query conversion, followed by result querying in Neo4j. The purpose of this approach is to correct certain SPARQL queries with incorrect sequence order by leveraging the direction-agnostic nature of Cypher queries. We also SFT LLMs with abilities of directly extract answers from the FR facts for the best performance.

3.1   Named Entity Recognition (NER) with LLM and ES

The goal of Named Entity Recognition (NER) is to extract relevant entities and relationships from the Knowledge Base (KB) based on the given question. This process is divided into two steps. First, extracting keywords from the question; and then querying the relevant entities or relations based on the extracted keywords. For each question, NER will provide a set of entities and relations as candidates for the next steps. The keyword extraction is achieved through a pre-trained LLM with few-shot inference. We use the offline "Qwen1.5-110B-Chat" model here. In the prompt, only questions are provided.

Next, given keywords extracted from the question, we use Elasticsearch (ES) to retrieve potential entities and relations. We established separate ES databases for the triplets for subjects, relations, and objects. For each keyword, we consider the possibility of it appearing in any of these positions, making requests to each database and returning the top-5 relevant entities or relations.

For ES search, we utilize 3 search tools, the exact match search, fuzzy search based on IK-analysis, and semantics search based on K-nearest neighbor (KNN) using BERT vectors. Exact match and fuzzy search with IK-analysis can respond in real-time. For semantics search, we only create the search index for relations as the size of the relation set is only about half a million.

3.2   Facts Retrieval (FR) for RAG

Based on the results from the previous subsection, for each question, we have a series of potentially relevant entities and relations. In this step, we need to retrieve the facts related to the question from these potential entities and relations to facilitate SPARQL generation. The target of this part is to directly get the accurate facts for SPARQL generation. Here the accurate fact means the fact in the chain to answer the question. In other words, in the SPARQL answers. For instance, for the question "代表食物是烤包子的民族所信奉的宗教? (What is the religion believed in by the ethnic group whose representative food is baked buns?)", the triplets "< 维吾尔族 >-< 代表食品 >-" 烤包子" (<Uyghur>-<representative_food>-"Baked Buns")" and "< 维吾尔族 >-< 信仰 >-< 伊斯兰教 > (<Uyghur>-<faith>-<Islam>)" are the accurate facts as the SPARQL query that answer the question is "select ?x where ?y < 代表食品 > "烤包子". ?y < 信仰 > ?x. ". The accurate facts can be considered as the facts used in the SPARQL query.

The NER module may miss the correct entities or relations. To address this, we considered different scenarios and proposed four fact retrieval strategies: "single", "combine", "ES", and "shortest".

Single Strategy. "Single" is a simple strategy for fact searching. We consider each entity or relation individually. Using the neo4j, we extract facts that contain target entities and relations and randomly select 25 of them. The logic is that most entities are contained in only a few facts (fewer than 25). So querying the facts based on a single entity could return the accurate facts in some questions.

However, most relations may have facts more than 25. The strategy of "single" may have limited performances.

Combine Strategy. "Combine" is the strategy seeking facts with two extracted entities or relations contained. This is the simplest strategy with combinations of extracted entities or relations. It's efficient but mostly no results returned. The logic behind this is that for most questions, each retrieved fact is based on querying two of the entities or relations in the fact triplets.

ES Strategy. "ES" means to query via Elasticsearch with the IK-analysis tool. We use combinations of extracted entities to query the ES engine. That is, for each subset of keywords (not the extracted entities and relations), we combine these keywords and query the ES with fuzzy search. In this context, fuzzy search can be understood as returning the facts that share the highest number of common sub-tokens with the query. To limit the query time cost, we set the maximum size of the keyword combination to 5.

Shortest Strategy. "Shortest" refers to shortest path. For each pair of keywords, we retrieve if there exists a path connecting entities or relations of selected keywords with a max length of 2 (connected within 2 facts, multi hops). NER may fail to connect keywords, entities, and relations to questions. Here, we assume that at most one keyword extraction error may occur. The shortest-path strategy bypasses this erroneous keyword to retrieve the fact relevant to the answer. Results are limited to 5 for each keyword combination.

Measurements. To evaluate these results, we proposed the measurement of accuracy (precision) for entities and relations. That is, for the training set with the SPARQL query, we could measure the percentage of all entities/relations in the SPARQL queries that can be found in the retrieved facts.

$$ACC\_FR = \frac{1}{N} \sum_{i=1}^{N} I(SPARQL_i, Facts_i)$$

where

$$I(SPARQL, F) = \begin{cases} 1, & \text{if } |SPARQL| \subseteq |Facts| \\ 0, & \text{otherwise} \end{cases}$$

Here N is the size of the dataset, $|SPARQL|$ refers to the set of entities and relations that appear in the given SPARQL query, and $|Facts|$ refers to the set of all entities and relations that appear in the retrieval facts of FR step. This is a rigorous measurement; even if only a part of accurate facts are retrieved, the LLM still has a probability of generating the correct answer. Also, other than ACC_FR, the size of facts is an important measurement as the context window is limited.

In conclusion, we design 4 different related fact retrieval strategies. Using RAG, we aim to get accurate facts for each question. The combination of facts from these methods is used as retrieval information for RAG.

3.3   RAG-Based SPARQL Generation with LLM

At this stage, we have extracted the facts relevant to the question. Using RAG-related techniques, we need to generate a precise SPARQL query corresponding to the question.

Current LLMs excel at retrieval but lack strong logical reasoning capabilities. Unlike previous works [18], we directly include all related facts in the forms of triplets in the prompt without any process. This reduces the logical demands on the LLM, allowing it to directly identify relevant triplets without needing to extract fasts from a processed graph.

We chose the "Qwen1.5" LLM series for SFT. We choose "Qwen1.5-72B-Chat", "Qwen1.5-32B-Chat", and "Qwen1.5-14B-Chat" as candidates. The retrieved facts (triplets) and questions are included in the prompt. Querying SPARQL will provide you with the answer entities set.

As described below, we combined results from different models with different temperatures. Thus we proposed two different measurements for the CKBQA answers. Commonly used average F1 scores are calculated through averaged precisions and recall, measuring the accuracy of SPARQL.

$$\text{Precision:} \quad P_i = \frac{|A_i \cap G_i|}{|A_i|}$$

$$\text{Recall:} \quad R_i = \frac{|A_i \cap G_i|}{|G_i|}$$

$$\text{Averaged F1:} \quad F1 = \frac{1}{|Q|} \sum_{i=1}^{|Q|} \frac{2 P_i R_i}{P_i + R_i}$$

Due to the need to merge results from different models, the correctness of each non-empty answer is more critical than the overall F1 score. Therefore, we proposed a rescaled F1 score for this method, where the traditional F1 score is adjusted based on the proportion of non-empty answers.

$$\text{Rescaled F1:} \quad F1_{\text{rescaled}} = F1 * \frac{|G|}{|\{A_i \mid A_i \neq \emptyset \text{ for } A_i \in A\}|}$$

Higher F1 scores with more empty answers refer to confident prediction (high precision) from LLM, benefiting the performance of different results' combinations.

3.4   Post-Hoc Improvement for CKBQA System

Having SPARQL queries and results against KB, this part aims to apply some post-hoc methods to improve performance after the LLM inference. The basic method in this part involves answer merging, where answers obtained through different approaches are combined. The correct answer is then identified by querying the LLM.

Results Combination with LLM. Pass@k is a commonly used measurement in the code generation task [2]. For each question with k generated samples from LLM, it measures the pass rate if any of these samples pass the test. Probability-based LLMs can provide different responses to the same question when the temperature is set above zero. Thus for this task, to maximize the performance, we could query different models with different temperatures multiple times, and then combine the results.

The result combination is done through LLM. We designed a prompt asking LLM which answer refers to the question best. LLMs can handle most irrelevant responses, effectively filtering out answers that are clearly unable to address the question. E.g. For the question, "代表食物是烤包子的民族所信奉的宗教? (What is the religion believed in by the ethnic group whose representative food is baked buns?)", "< 维吾尔族 >"(<Uyghur>) is the semantically wrong answer.

We combine the results from different inferences, temperatures, and models many times. In scenarios where the LLM cannot identify the correct answer, we conservatively choose to leave the question blank to improve the $F1_{rescaled}$ scores. The questions and the answer candidates are included in the prompt. We use few-shots techniques in this prompt.

Transformation from SPARQL to Cypher. In contrast, Cypher allows for undirected queries, meaning that unlike SPARQL, which requires the exact order of triplets in each query, Cypher can retrieve results where the subject and object are swapped when the specific syntax is used. This enables querying undirected triples. For instance, for the relation "<a>-<b>-<c>", in cypher, querying with "<c>-<b>-?" will also return the result with Cypher. But in SPARQL, sequences are strictly defined.

We designed a rule-based transformation function from SPARQL to Cypher, querying with neo4j. This transformation function cannot handle complex queries, e.g. including conditions.

LLM Directly Extract Answers. LLMs possess strong retrieval capabilities, and since the answers to most questions can be directly retrieved during the FR process, we also tested the ability of LLMs to replace the SPARQL query process and return results directly.

We SFT two models for direct answer extraction by the LLM. The first is an aggressive model (LLM-direct-extraction-aggressive), used to fill in cases where the SPARQL query returns no results. If the answer is not found within the facts retrieved by the FR process, the LLM will attempt to generate an answer based on its own knowledge. The second model is a conservative approach (LLM-direct-extraction-conservative). For questions where an answer cannot be extracted from the FR facts, it will return"result unknown". This approach is aimed at achieving the highest $F1_{rescaled}$ score by avoiding incorrect answers when the available information is insufficient.

4 Evaluation

Table 1. Facts retrieval methods with ACC_FR and facts size performance.

Single	Combine	ES	Shortest	ACC_FR	Facts Size
✓				58.45%	148.59
	✓			14.75%	1.69
		✓		51.10%	28.54
			✓	32.84%	21.15
✓	✓			58.75%	150.28
✓		✓		73.86%	177.13
✓			✓	62.71%	169.74
	✓	✓		52.42%	30.23
	✓		✓	33.73%	22.84
		✓	✓	61.11%	49.68
✓	✓	✓		73.89%	178.82
✓	✓		✓	62.81%	171.43
✓		✓	✓	76.07%	198.28
	✓	✓	✓	61.21%	51.38
✓	✓	✓	✓	**76.07%**	199.97

Our RAG-based CKBQA system is primarily divided into two components: the information retrieval component and the SPARQL generation component. Our evaluation is also conducted based on these two components separately.

After our prepossessing phase, the provided knowledge base contains 68567459 facts, with 11303474 subjects, 408806 relations, and 15316753 objects. The training set contains 7625 questions with correlated SPARQL queries and answers. The eval set (phase 1 data) contains 1292 questions with answers (no correlated SPARQL queries). The test set (phase 2 data) has 440 questions. Our evaluation for FR will focus on the training set as SPARQLs are provided. For the CKBQA accuracy part, we will use the eval set as training set is used in the SFT.

4.1 Facts Retrieval Accuracy

For the FR part, recall that the measurements mentioned in the previous section are the ACC_FR and the facts size. Selected LLM "Qwen1.5" series provide the context size of 32k, allowing long context. Table 1 shows the results of startegies combinations. When generating SPARQLs with LLM, we checked the input size dynamtically. Based on the ACC_FR, once the facts size is too long for the prompt, we will reduce the chosen strategies to fit the prompt size limit.

4.2 CKBQA Accuracy

This part focus on evaluating CKBQA answers. Mostly, they are from SPARQL queries' results. However, in our pipeline, Cypher and LLM-direct-extraction will also provide CKBQA answers. We will evaluate each method individually and then followed by an ablations experiments with results combinations.

We finetuned 6 different models. Four of them are SPARQL generating models and 2 are LLM-direct-extraction models. We choose "Qwen1.5-14B-Chat" (14B), "Qwen1.5-32B-Chat" (32B), and "Qwen1.5-72B-Chat" (72B) LLMs for SFT. All of them belongs to "Qwen1.5-Chat" model series. Due to time cost, 14B is our default choice. For the SPARQL generation, 3 models of 14B, 32B, and 72B are trained on the training set, named SPARQL-14B, SPARQL-32B, and SPARQL-72B respectively. For better performance, we trained another SPARQL-14B-V2 models on the combination of training and eval set (with some manual annotations). We cannot evaluate it on the eval set so this model will be ignored in this evaluation part. For LLM-direct-extraction, we trained two models based on 14B, named " LLM-direct-extraction, 14B-aggressive" and "LLM-direct-extraction, 14B-conservative". These two models are used to extract the answers from the FR facts.

Evaluation on Single Methods. For SPARQL-14B, 32B, and 72B, follow the idea of pass@k in code generation task, we query the LLM multiple times with different temperatures and combine these results. Table 2 shows the performance of individual LLMs and methods. Without SFT, even with few-shot methods, the $F1$ score is around 45% with powerful 110B LLM. After SFT, small 14B models could provide good performance with around 65%. Combining results from multiple inferences, the $F1$ slightly increased while $F1_{resacled}$ drops. As we focus on results combination, $F1_{rescaled}$ will be the most important measurement. Overall, larger LLM provides slightly worse performance. This might due to the lack of training epochs. Larger LLM may requires more epochs for better performance. LLM-direct-extraction-conservative provide the best $F1_{rescaled}$ performance. The SPARQL2Cypher approach is primarily used for cases where the order of SPARQL statements is incorrect.

Table 2. Performance of all individual LLMs and methods. Details about these models can be found in the Sect. 4.2. Performance of results combination, LLM-direct-extractions are included in the table.

Model Parameters		Performance	
Model Name	Inference Times	$F1$	$F1_{rescaled}$
No SFT, with few shots, temperature 0.2			
Qwen1.5-110B-Chat	1	44.95%	84.91%
SFT, no few shots, temperature 0.2			
SPARQL-14B	1	64.61%	85.26%
SPARQL-14B	5	68.33%	83.79%
SPARQL-32B	3	67.27%	82.26%
SPARQL-72B	3	66.29%	81.07%
14B+32B+72B	$5 + 3 + 3$	**75.32%**	84.27%
SPARQL2Cypher, rule-based transformation			
14B+32B+72B	$5 + 3 + 3$	69.54%	82.44%
SFT, LLM-direct-extraction temperature 0.2			
14B-aggressive	1	57.37%	57.37%
14B-conservative	1	29.83%	**96.72%**

Table 3. Performance with different temperature and their combinations. Evaluated with SPARQL-14B. Combination of five inferences for each temperature.

Temperature			Performance	
0.2	0.5	0.8	$F1$	$F1_{rescaled}$
✓			68.33%	**83.79%**
	✓		66.44%	81.25%
		✓	64.97%	80.14%
✓	✓		**69.43%**	83.01%
✓		✓	68.66%	81.63%
	✓	✓	67.38%	79.89%
✓	✓	✓	68.73%	80.82%

Table 3 shows the results with different temperatures. A higher temperature introduces more instability, and an excessively high temperature can even reduce performance after merging the results. For the best performance, for the final result, we exclude the results with temperature 0.8.

Table 4. Ablation experiments. Note that last performance with SPARQL-14B-V2 is evaluated on the test set, in CCKS2024 CKBQA task phase 2. This SPARQL-14B-V2 is finetuned on combination of training and eval set. It should not be evaluated on the eval set. 85.15% is our final performance in the task.

Methods	$F1$	$F1_{rescaled}$
LLM-direct-extraction-conservative	29.83%	96.72%
+ SPARQL-14B+32B+72B temperature 0.2+0.5	77.17%	85.56%
+ SPARQL2Cypher	81.72%	87.41%
+ LLM-direct-extraction-aggressive	83.15%	83.15%
+ SPARQL-14B-V2	85.15%	85.15%

Ablation with Result Combination. We combine results from different LLMs and methods for the best performance. Table 4 shows the results from ablation experiments. Our final score for this task is 85.15% in the test set.

5   Conclusion and Future Works

We proposed our RAG-based CKBQA system with SPARQL generate LLM. Our system include two steps of related facts (triplets) retrieval and SPARQL generation with LLM. Our system exhibits exceptional scalability. Since the SPARQL generation module based on LLMs focuses on generating SPARQL queries with correct SPARQL grammar, entities, and relations based on retrieved relevant facts. Therefore, the LLM does not require additional SFT for facts newly added to the KB. Our system achieve the 1st place in CCKS2024 CKBQA competition with F1 scores of 85.15%.

SPARQL generation based on RAG still holds significant untapped potential. More precise fuzzy search techniques, an increased volume of training data, and improved LLMs can all enhance system performance. We can even leverage the agent capabilities of LLMs by providing appropriate query tools, enable LLMs to perform CKBQA tasks with the added capability of self-correction.

References

1. Bai, J., et al.: Qwen technical report. arXiv preprint arXiv:2309.16609 (2023)
2. Chen, M., et al.: Evaluating large language models trained on code. arXiv preprint arXiv:2107.03374 (2021)
3. Chen, Z.Y., Chang, C.H., Chen, Y.P., Nayak, J., Ku, L.W.: UHop: an unrestricted-hop relation extraction framework for knowledge-based question answering. In: Proceedings of the 2019 Conference of the North American Chapter of the Association for Computational Linguistics: Human Language Technologies, Volume 1 (Long and Short Papers), pp. 345–356 (2019)
4. Gao, Y., et al.: Retrieval-augmented generation for large language models: a survey. arXiv preprint arXiv:2312.10997 (2023)

5. Gu, Y., Deng, X., Su, Y.: Don't generate, discriminate: a proposal for grounding language models to real-world environments. In: Proceedings of the 61st Annual Meeting of the Association for Computational Linguistics (Volume 1: Long Papers), pp. 4928–4949 (2023)

6. Jiang, J., Zhou, K., Dong, Z., Ye, K., Zhao, W.X., Wen, J.R.: StructGPT: a general framework for large language model to reason over structured data. In: Proceedings of the 2023 Conference on Empirical Methods in Natural Language Processing, pp. 9237–9251 (2023)

7. Jiang, J., Zhou, K., Zhao, W.X., Wen, J.R.: UniKGQA: unified retrieval and reasoning for solving multi-hop question answering over knowledge graph. arXiv preprint arXiv:2212.00959 (2022)

8. Lan, Y., He, G., Jiang, J., Jiang, J., Zhao, W.X., Wen, J.R.: Complex knowledge base question answering: a survey. IEEE Trans. Knowl. Data Eng. **35**(11), 11196–11215 (2022)

9. Lan, Y., Jiang, J.: Query graph generation for answering multi-hop complex questions from knowledge bases. In: Proceedings of the 58th Annual Meeting of the Association for Computational Linguistics, pp. 969–974 (2020)

10. Luo, H., et al.: ChatKBQA: a generate-then-retrieve framework for knowledge base question answering with fine-tuned large language models. arXiv preprint arXiv:2310.08975 (2023)

11. Miller, A., Fisch, A., Dodge, J., Karimi, A.H., Bordes, A., Weston, J.: Key-value memory networks for directly reading documents. In: Proceedings of the 2016 Conference on Empirical Methods in Natural Language Processing, pp. 1400–1409 (2016)

12. OpenAI, R.: GPT-4 technical report. arxiv 2303.08774. View in Article **2**(5) (2023)

13. Pan, S., Luo, L., Wang, Y., Chen, C., Wang, J., Wu, X.: Unifying large language models and knowledge graphs: a roadmap. IEEE Trans. Knowl. Data Eng. (2024)

14. Shi, J., Cao, S., Hou, L., Li, J., Zhang, H.: TransferNet: an effective and transparent framework for multi-hop question answering over relation graph. In: Proceedings of the 2021 Conference on Empirical Methods in Natural Language Processing, pp. 4149–4158 (2021)

15. Shu, Y., et al.: TIARA: multi-grained retrieval for robust question answering over large knowledge base. In: Proceedings of the 2022 Conference on Empirical Methods in Natural Language Processing, pp. 8108–8121 (2022)

16. Sun, J., et al.: Think-on-graph: deep and responsible reasoning of large language model on knowledge graph. In: The Twelfth International Conference on Learning Representations (2024)

17. Touvron, H., et al.: Llama: open and efficient foundation language models. arXiv preprint arXiv:2302.13971 (2023)

18. Yang, S., Teng, M., Dong, X., Bo, F.: LLM-based SPARQL generation with selected schema from large scale knowledge base. In: China Conference on Knowledge Graph and Semantic Computing, pp. 304–316. Springer (2023). https://doi.org/10.1007/978-981-99-7224-1_24

19. Yu, D., et al.: DecAF: joint decoding of answers and logical forms for question answering over knowledge bases. In: The Eleventh International Conference on Learning Representations (2022)

20. Zhang, J., et al.: Subgraph retrieval enhanced model for multi-hop knowledge base question answering. In: Proceedings of the 60th Annual Meeting of the Association for Computational Linguistics (Volume 1: Long Papers), pp. 5773–5784 (2022)

21. Zhang, L., et al.: FC-KBQA: a fine-to-coarse composition framework for knowledge base question answering. In: The 61st Annual Meeting Of The Association For Computational Linguistics (2023)
22. Zhao, W.X., et al.: A survey of large language models. arXiv preprint arXiv:2303.18223 (2023)

Fast Assortativity Coefficient Calculation in Large-Scale Social Networks

Liru Cao[1,2], Wangyang Liu[1,2(✉)], Lican Zhang[1,2], Wei Liao[1,2], and Yi Shen[1,2]

[1] Shenzhen CyberAray Network Technology Co., Ltd., Shenzhen 518000, Guangdong, China
liuwy06@foxmail.com, shenyi_wlar@cetc.com.cn
[2] Cyberspace Cognitive Domain Engineering Research Center of Guangdong Province,
Shenzhen 518000, Guangdong, China

Abstract. Assortativity coefficient, a metric utilized to quantify the tendency of nodes to connect with others sharing similar attributes, is important for analyzing social network structures, elucidating user behavioral preferences and predicting the evolution of networks. However, real-world social networks usually involve immense amounts of data and complex network structures, posing a challenge to rapidly and efficiently computing this coefficient within large-scale social networks. This paper proposes a Fast Assortativity Calculation (FAC) method tailored for large-scale social networks. FAC leverages a hybrid approach that combines breadth first search with random walks to achieve incomplete yet effective traversal of the graph structure, concurrently updating degree correlation coefficient during the traversal. Experimental results from four large-scale social network datasets, which contain millions of nodes and tens of millions of edges, demonstrate that FAC achieves a minimum improvement of 43.62% and a maximum of 66.91% in computational efficiency compared with the standard calculation modules provided by classical graph analysis tools. Notably, the absolute value of precision loss remains below 0.0068, maintaining an acceptable level of accuracy.

Keywords: Assortativity Coefficient · Degree Correlation · Graph Computation · Social Networks · Large-scale Networks

1 Introduction

The concept of assortativity was first articulated by Newman in 2002, referring to the tendency of nodes to connect with other similar nodes [1]. Assortativity coefficient, a crucial metric for quantifying network assortativity, is commonly defined as the Pearson correlation coefficient between the degrees of all node pairs in a graph.

Initially, the assortativity coefficient measured degree assortativity for undirected and unweighted graphs. Subsequent researches [2–5] extended this concept to directed and weighted graphs by introducing corresponding degree correlation formulas. [6] proposed a generalization of Newman's assortativity coefficient, providing a unified approach to assortativity that considers pairs of nodes that may not be directly adjacent but are connected through paths. These generalizations enriched the theoretical framework for

B. Xu et al. (Eds.): CCKS-IJCKG 2024, CCIS 2229, pp. 453–462, 2025.
https://doi.org/10.1007/978-981-96-1809-5_37

analyzing assortativity across various network types and structures. Some researchers [7–9] put their attention on the local structure of networks, investigating the assortativity and disassortativity of subcomponents such as individual nodes, groups and communities. This local-level analysis helps in identifying which substructures pose threats to the stability of the network and which contribute positively. Furthermore, Peel et al. contributed to a deeper understanding of the relationship between local and global assortativity, revealing that for many real-world networks, the distribution of assortativity is skewed, overdispersed, and multimodal [10].

As assortativity encompasses not only information about network structures but also helps the study of dynamic network development stages and network robustness, and assortative mixing is a common phenomenon in many real-world networks [11], the analysis of assortativity is an important task in social network research. For instance, studies such as [12] employed the assortativity as a primary indicator to investigate the relationship between commuting and social network integration, while [13] focused on understanding centralized and decentralized networks through assortativity and hierarchical structures. Furthermore, [14] discovered that degree assortativity initially increases and then declines to a long-term stable level in evolving social networks. Consequently, the analysis of assortativity serves as a significant reference for studying social network structures and evolutionary stages, understanding information dissemination within these networks as well as analyzing user behaviors and influence.

Currently, online social networks have emerged as crucial tools for studying social behaviors, information dissemination and group interactions. As of April 2024, the number of active users on online social media has exceeded 5 billion [15], with the most popular applications such as WhatsApp boasting a user base of at least 2 billion and Facebook reaching 3.065 billion [16]. These platforms typically exhibit vast data scales, intricate network architectures, rapid dynamic changes and significant computational resource consumption. However, traditional approaches for calculating assortativity coefficients often encompass all nodes and edges within the entire network, which is both computationally expensive and time-consuming. Given that real-world social network analysis tasks often focus on specific nodes and communities, traditional assortativity coefficient calculation approaches often struggle in these situations.

In this paper, we propose an approximate calculation method for assortativity coefficients in large-scale social networks. This approach estimates the assortativity coefficients based on a subset of edges and nodes within the network and the performance of the algorithm is evaluated by real world social network datasets. The remainder of this paper is structured as follows: Sect. 2 gives a brief introduction of related concepts and techniques. In Sect. 3, the method we designed is presented, and in Sect. 4, the experimental results are presented, and the computational efficiency and accuracy are compared with a benchmark model. Finally, the conclusions and the suggestions for further research are given in Sect. 5.

2 Related Concepts and Technologies

2.1 Breadth First Search and Random Walk

Breadth First Search
Breadth first search is an algorithm employed for traversing or searching a graph. It starts from a single node and explores all its immediate neighbors before proceeding to the next level of neighbors, following the order in which they were discovered. This process is repeated until all reachable nodes have been visited or a certain condition is met. Breadth first search is the cornerstone for improving the performance of many iterative graph algorithms.

Random Walk
In network analysis, the fundamental concept of the random walk involves starting from one or multiple nodes, where at each step, a neighboring node is selected based on a probability parameter to form a random walk path. Specifically, at any given node, with a probability of $1 - \alpha$, the walk proceeds to one of its neighboring nodes, while with a probability of α, the walk jumps randomly to any node in the graph, and α represents the jump probability. After each step of the walk, a new probability distribution is computed, which serves as the input for the subsequent step. By iteratively repeating this process, a stable probability distribution is eventually obtained.

2.2 Degree Correlation

Degree correlation is a metric that measures the relationship between the degrees of nodes in a graph, commonly used to describe whether nodes in a graph tend to connect nodes with similar degrees. Common methods for calculating degree correlation include the Pearson correlation coefficient, Kendall's tau rank correlation coefficient and Spearman's rank correlation coefficient. In this paper, we adopt the Pearson correlation coefficient to compute the degree correlation of networks.

For a graph $G(V, E)$, $V = \{v_1, v_2, \cdots\}$ is the set of nodes, and $E = \{e_1, e_2, \cdots\}$ is the set of edges. For each edge $e_i = (x, y)$, $1 \leq i \leq |E|$ in the graph G, x_i and y_i respectively to represent the degrees of the source node and the target node of e_i. The global degree correlation coefficient of the graph is

$$\rho_{degree} = \frac{\sum (x_i - \bar{x})(y_i - \bar{y})}{\sqrt{\sum (x_i - \bar{x})^2} \sqrt{\sum (y_i - \bar{y})^2}} \tag{1}$$

where $\bar{x} = \frac{1}{|E|} \sum_{i=1}^{|E|} x_i$ represent the mean degree of source nodes and $\bar{y} = \frac{1}{|E|} \sum_{i=1}^{|E|} y_i$ represent the mean degree of target nodes, respectively, across all edges in the graph.

3 Method

This section introduces the methodology we propose. Initially, multiple nodes are randomly sampled from the entire graph to serve as starting nodes for traversal. Subsequently, a hybrid strategy combining breadth first search and random walk is employed to incompletely traverse the nodes and edges of the graph. Throughout this traversal process, the degree correlation coefficient is updated in real-time.

3.1 Traversal Strategy

A random sampling approach is employed to select nodes from the entire graph based on a sampling rate δ, which constitute the initial set of nodes for the subsequent partial traversal process. The traversal methodology incorporates breadth first search with random walk, aiming to maintain a certain level of search breadth while enhancing the randomness of exploration. Specifically, upon encountering a node during the traversal, a probability $(1 - \alpha)$ dictates whether to proceed with visiting its adjacent nodes, while a complementary probability α controls the decision to potentially jump to other unvisited nodes that are k-hops away from the current node, implying a shortest path distance of k between them.

When opting to explore adjacent nodes with probability $(1 - \alpha)$, a BFS strategy is adopted to systematically access all unvisited neighbors of the current node, facilitating the batch computation of newly discovered edges. Furthermore, a maximum walk depth d is imposed as a constraint to limit the extent of the traversal.

Algorithm 1 Random Traversal Strategy

Input: an undirected and unweighted graph $G = (V, E)$.

δ is the sampling ratio.

α is the jump probability during the random walk.

d is maximum walk length of random walk.

Output: ρ is the approximate result of degree correlation of graph G.

begin

1.   SAMPLE $S = \{s_1, s_2, \cdots\}$ from V by ρ
2.   for s_i in S do
3.   visitednodes, calculatenodes←{s_j}, { }, step ← 0, node ←s_j, queue.pop (s_j)
4.   while step < d and queue not empty do
5.   n←queue.top() and step ++
6.   If node n not in calculatenodes do
7.   SKIP to another node with probability α or
8.   Neighbors←GetNeighbors(n, visitednodes)
9.   UpdateDegreeCorrelation(n, neighbors)
10.   calculatenodes.update(n), visitednodes.update(neighbors),
11.   queue.pop(n), queue.push(neighbors)

end

3.2 Update of Degree Correlation Coefficient

The calculation of the degree correlation coefficient is updated during traversal, rather than computed after traversal. Two sets are utilized to keep track of visited nodes and processed edges, respectively, ensuring that each edge contributes exactly once to the final result.

Reviewing Eq. (1), it can be decomposed into three subcomponents, each of which will be updated individually during the traversal process.

$$Cov(X, Y) = \sum (x_i - \bar{x})(y_i - \bar{y}) \tag{2}$$

$$VarX = \sum (x_i - \bar{x})^2 \tag{3}$$

$$VarY = \sum (y_i - \bar{y})^2 \tag{4}$$

When accessing all the unvisited neighbors of the current node with probability $1-\alpha$ during the traversal phase, the newly acquired edges are denoted as $\{\varepsilon_1, \varepsilon_2, \cdots, \varepsilon_{\Delta l}\}$, with the corresponding degree lists of the two endpoint nodes being $\{x_1, x_2, \cdots, x_{\Delta l}\}$ and $\{y_1, y_2, \cdots, y_{\Delta l}\}$, respectively, where Δl represents the number of newly discovered edges. Assuming the number of visited edges is l, the update of the degree correlation coefficient is as follows

$$\bar{x}_{new} = \frac{\bar{x}_{old} \times l + \sum_{i=1}^{\Delta l} x_i}{l + \Delta l} \tag{5}$$

$$\bar{y}_{new} = \frac{\bar{y}_{old} \times l + \sum_{i=1}^{\Delta l} y_i}{l + \Delta l} \tag{6}$$

$$\Delta Cov(X, Y) = \sum_{i=1}^{\Delta l} (x_i - \bar{x}_{new}) \times (y_i - \bar{y}_{new}) + l \times (\bar{x}_{new} - \bar{x}_{old})(\bar{y}_{new} - \bar{y}_{old}) \tag{7}$$

$$\Delta VarX = \sum_{i=1}^{\Delta l} (x_i - \bar{x}_{new})^2 + l \times (\bar{x}_{new} - \bar{x}_{old})^2 \tag{8}$$

$$\Delta VarY = \sum_{i=1}^{\Delta l} (y_i - \bar{y}_{new})^2 + l \times (\bar{y}_{new} - \bar{y}_{old})^2 \tag{9}$$

where $\bar{x}_{new}$ and $\bar{x}_{old}$ represent the mean degrees of the source nodes for all visited edges before and after this walk, respectively, and $\bar{y}_{new}$ and $\bar{y}_{old}$ represent the mean degrees of the target nodes for all visited edges before and after the update, respectively.

Finally, upon completion of the travel phase, an estimated degree correlation coefficient can be obtained by Eq. (10).

$$\rho_{est} = \frac{Cov(X,Y)}{\sqrt{VarX}\sqrt{VarY}} \tag{10}$$

4   Experiments and Results

We conducted tests on four real-world social network datasets with millions of nodes and tens of millions of edges to evaluate the stability, computational efficiency and accuracy of our designed model, FAC. The benchmark model used for comparison is the standard NetworkX module. In Sect. 4.1, we introduce the basic information of the datasets and the benchmark model for comparison. In Sect. 4.2, we present the metrics used to evaluate the performance of the model. Finally, in Sect. 4.3, we show the experimental parameter settings, experimental results, and provide an analysis of the results.

4.1 Datasets and Benchmark Model

The datasets employed in this paper are exclusively sourced from the open-source experimental dataset, SNAP (Stanford Large Network Dataset Collection), which is provided by Stanford University. This comprehensive collection encompasses over 50 large-scale networks, ranging from tens of thousands to tens of millions of nodes and edges, covering various domains such as social networks, biological structures, road networks, citation networks, communication networks and more.

For this experiment, four real-world social network datasets with millions of nodes were selectively utilized. Table 1 presents the fundamental information of these datasets.

Table 1. Information of 4 large social network datasets from SNAP

Name	Number of nodes	Number of edges	Number of communities
com-LiveJournal	3997962	34681189	287512
com-Youtube	1134890	2987624	8385
soc-Pokec	1632803	30622564	NA
wiki-topcats	1791489	28511807	17364

The benchmark model utilized in this paper is the standard computational module provided by NetworkX, which is an algorithmic package written in Python for graph analysis and complex network modeling. It serves as a classic tool in the realm of graph theory research and network data analysis, offering capabilities for graph analysis, creation, manipulation as well as the exploration of the structure, dynamics and functionalities of complex networks. Characterized by its simplicity, flexibility and an extensive algorithm library, NetworkX also supports visualization and boasts comprehensive functionalities, making it one of the most widely adopted graph analysis tools.

4.2 Evaluation Metrics

To evaluate the stability of FAC, multiple experiments would be conducted on the same dataset, and the coefficient of variation (CV) would be calculated among these output results, which is expressed as the ratio of the standard deviation to the mean value of the data. In practical applications, CV is commonly used to assess the stability and variability of data, providing insights into the consistency and reliability of the model's performance across repeated trials.

We measure the algorithm's accuracy by the loss between the mean value of multiple experiments results from the model and the true value, while the average running time is used to evaluate the computational efficiency of the model.

4.3 Results and Analysis

We conducted 10 experiments on each of the 4 datasets to observe the stability of the model's output. Subsequently, we applied trimmed mean to each set of data, effectively mitigating the influence of outlier values, thereby achieving a more robust statistical estimation. In terms of parameter settings, the sampling rate δ during the sampling phase was set to 0.001. In the traversal phase, we configured the maximum access step length d to 10, the probability parameter α to 0.2 and the jump step length k to 6.

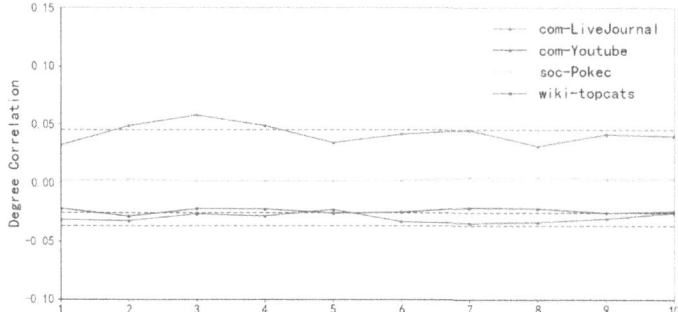

Fig. 1. Degree correlation coefficients of 4 datasets, calculated by both FAC and NetworkX

Figure 1 presents the results of the FAC model on four real-world social network datasets after 10 runs. The various datasets are distinguished by different colored lines. The solid lines in the figure represent the estimated degree correlation coefficients from FAC, while the dashed lines depict the output results from the benchmark model. Table 2 displays the coefficients of variation (CV) of the output results from FAC after running 10 times on each of the different datasets.

Table 2. The coefficients of variation for the estimated values of the FAC across 4 datasets.

Dataset	CV
com-LiveJournal	0.001518814
com-Youtube	0.000569434
soc-Pokec	0.000524396
wiki-topcats	0.000237467

Table 3 presents the trimmed mean values of the estimated degree correlation coefficients obtained from 10 calculations of the FAC model on these social networks, the degree correlation coefficients of the networks output by the benchmark model and the losses of the FAC model.

Combining the results from Table 2 and Table 3, it can be concluded that the coefficient of variation for FAC's multiple runs on all tested datasets is less than 0.002, indicating a high degree of stability for the FAC model. Moreover, when compared to the output results of the benchmark model, the maximum loss incurred by FAC is less

Table 3. Mean estimated values of FAC, degree correlation calculated by the benchmark model

Dataset	Degree correlation	Mean estimated value of FAC	Loss
com-LiveJournal	0.04515	0.04195	0.00321
com-Youtube	−0.03691	−0.03012	0.00679
soc-Pokec	0.00164	0.00247	0.00084
wiki-topcats	−0.02580	−0.02409	0.00171

than 0.0068, with an average error of 0.003153 (typically, degree correlation coefficients range between −1 and 1). This demonstrates that FAC maintains a certain level of stability and accuracy when estimating degree correlation coefficients for large datasets.

Table 4. The average runtime of the benchmark model and FAC on test datasets (Unit: Sec)

Dataset	Model		Improvement
	NetworkX	FAC	
com-LiveJournal	686.5261	387.0521	43.62%
com-Youtube	56.3120	26.9106	52.21%
soc-Pokec	366.6424	121.2934	66.91%
wiki-topcats	379.6482	178.6616	52.94%

Table 4 compares the computational efficiency of FAC with the benchmark model. Both were run 10 times on each dataset, and the runtime duration was recorded to calculate the mean runtime for these runs. Across all four datasets, FAC demonstrates a minimum improvement of 43.62% and a maximum improvement of 66.91% in runtime compared to the benchmark model. Clearly, FAC exhibits an advantage in computational efficiency over the benchmark model.

Overall, the fast correlation coefficient approximation method proposed in this paper is capable of handling the estimation task of correlation coefficients in social networks with millions of nodes. FAC outperforms traditional computational methods in terms of computational efficiency while maintaining a certain level of stability and accuracy.

5   Conclusions and Future Work

Experiments conducted on 4 large-scale real-world social network datasets have demonstrated that FAC can enhance the algorithm's operational efficiency while maintaining stability and precision. This can be attributed to two main reasons.

Firstly, FAC's graph traversal strategy combines breadth first search with random walk, allowing it to retain structural information of the graph while selecting edges minimally, rather than computing on the entire graph. By selecting unvisited neighbors of nodes through breadth first search, FAC preserves local structural information. Meanwhile, the random walk strategy introduces randomness in accessing the entire graph, enabling jumps to other nodes within k-hop distances with a certain probability, thus avoiding confinement to the local structural information of sampled nodes.

Secondly, unlike traditional approaches, FAC updates the degree correlation coefficient during the access process, rather than computing it after acquiring all subgraphs, thereby improving the algorithm's computational efficiency.

Moving forward, we summarize the following aspects that still require further exploration and optimization. Firstly, the degree correlation coefficient update method designed in this paper is derived based on the assumption that the graph is unweighted. Next, we will consider studying the update formula for degree correlation coefficients that is applicable to weighted graphs. Secondly, explore more efficient incomplete traversal strategies and design a parallel computing scheme.

Acknowledgments. This study was funded by National Key R&D Program of China (grant number 2022YFB3105400).

References

1. Newman, M.E.: Assortative mixing in networks. Phys. Rev. Lett. **89**(20), 208701–208705 (2002). https://doi.org/10.1103/PhysRevLett.89.208701
2. Mahendra, P., Prokopenko, M., Zomaya, A.: Assortative mixing in directed biological networks. IEEE/ACM Trans. Comput. Biol. Bioinf. **9**, 66–78 (2012). https://doi.org/10.1109/TCBB.2010.80
3. Foster, J., Foster, D., Grassberger, P., et al.: Edge direction and the structure of networks. In: Proceedings of the National Academy of Sciences, issue (107), pp. 10815–10820 (2009)
4. Pigorsch, U., Sabek, U.: Assortative mixing in weighted directed networks. Phys. A: Stat. Mech. Appl. (604) (2022). https://doi.org/10.1016/j.physa.2022.127850
5. Arcagni, A., Grassi, R., Stefani, S., et al.: Extending assortativity: An application to weighted social networks. J. Bus. Res. (129), 774–783 (2021). https://doi.org/10.1016/j.jbusres.2019.10.008
6. Arcagni, A., Grassi, R., Stefani, S., et al.: Higher order assortativity in complex networks. Eur. J. Oper. Res. (262), 708–719 (2017). https://doi.org/10.1016/j.ejor.2017.04.028
7. Piraveenan, M., Piraveenan, M., Prokopenko, M., et al.: Local assortativeness in scale-free networks. EPL (Europhysics Letters)(84), 28002 (2008). https://doi.org/10.1209/0295-5075/84/28002
8. Zhang, G., Cheng, S., Zhang, G.: A universal assortativity measure for network analsis. ArXiv (2012)
9. Sabek, M., Pigorsch, U.: Local assortativity in weighted and directed complex networks. Phys. A: Stat. Mech. Appl. (630) (2023). https://doi.org/10.1016/j.physa.2023.129231
10. Peel, L., Delvenne, J., Lambiotte, R.: Multiscale mixing patterns in networks. In: Proceedings of the National Academy of Sciences of the United States of America, issue (115), pp. 4057–4062 (2017). https://doi.org/10.1073/pnas.1713019115
11. Newman, M.: Mixing patterns in networks. Phys. Rev. E, Stat. Nonlinear Soft Matter Phys. (67 2 pt 2), 026126 (2002). https://doi.org/10.1103/PhysRevE.67.026126
12. Bokányi, E., Juhász, S., Karsai, M., et al:. Universal patterns of long-distance commuting and social assortativity in cities. Sci. Rep. (11), 20829 (2021). https://doi.org/10.1038/s41598-021-00416-1

13. Zhang, D.Y., He, B.F., Chen, P.: Accessing the hierarchical structure and assortativity coefficient in social netwrok service. J. Syst. Sci. Complexity **10**(01), 45–52 (2013)
14. Zhou, B., Lu, X., Holme, P.: Universal evolution patterns of degree assortativity in social networks. Soc. Netw. (63), 47–55 (2020). https://doi.org/10.1016/j.socnet.2020.04.004
15. Digital 2024: Global Overview Report. https://datareportal.com/reports/digital-2024-global-overview-report
16. Most popular social networks worldwide as of April 2024, by number of monthly active users. https://www.statista.com/statistics/272014/global-social-networks-ranked-by-number-of-users

MQATG: An Automatic Military Equipment Question-Answer Test Case Generation Framework Using Large Language Models

Tongtong Bai[1], Song Huang[1(✉)], Yunhuan Wu[1], Binsheng Hong[1], Jiangtao Lu[1], Zhen Yang[1], Jinchang Hu[2], Min Gu[1], and Yubin Qu[1]

[1] College of Command and Control Engineering, Army Engineering University of PLA, Nanjing, China
{huangsong,yangzhen}@aeu.edu.cn, yubinqu@icloud.com
[2] School of Software and Internet of Things Engineering, Jiangxi University of Finance And Economics, Nanchang, China
hujinchang@aeu.edu.cn

Abstract. The military equipment question-answer system (MEQAS) is designed to provide information related to military equipment and has made significant progress with the rapid advancement of artificial intelligence. The quality of MEQAS is particularly important, as errors in responses can lead to substantial losses. To evaluate the quality of MEQAS, high-quality test cases need to be generated. However, due to the specialized, closed, and difficult-to-access nature of data in the military equipment domain, the cost of relying solely on domain experts to manually construct datasets is prohibitively high and not effectively feasible. To improve the efficiency of generating test cases for MEQAS, this paper proposes MQATG, an automatic military equipment question-answer test case generation framework, which is based on large language models. MQATG encompasses the generation, evaluation, and selection of question-answer test cases. The question-answer test cases generation utilizes a large language model, guided by a generation prompt, to analyze the document and produce initial question-answer test cases. The question-answer test cases evaluation employs an evaluation prompt to engage the large language model in scoring each generated test case. The question-answer test cases selection sorts the question-answer test cases according to the scoring results, and allowing for the identification of the high-ranked cases as the final test cases. We evaluated the question-answer test cases for military equipment generated by MQATG using five metrics and calculated the comprehensive score. The experiment results demonstrate that MQATG can produce high-quality military equipment question-answer test cases, and is more effective than other method MQATG_b.

Keywords: Military equipment question-answer system · Question-Answer test cases · Large language models

B. Xu et al. (Eds.): CCKS-IJCKG 2024, CCIS 2229, pp. 463–474, 2025.
https://doi.org/10.1007/978-981-96-1809-5_38

1   Introduction

The advent of artificial intelligence has revolutionized the field of military technology, leading to the development of sophisticated systems that cater to the unique informational needs of military personnel. Among these systems, the military equipment question-answer system (MEQAS) stands out as a critical tool designed to provide detailed and accurate information related to military equipment [1]. Compared to general question-answer systems, MEQAS must function within the context of high-stakes, security-sensitive environments. Therefore, the accuracy and reliability of MEQAS are paramount, as any misinformation can have severe consequences, potentially leading to substantial losses in operational efficiency and strategic planning [2].

In order to assess the efficacy of question-answer systems, numerous evaluation methodologies have been proposed [3–7]. These methodologies either utilize pre-existing datasets or generate new datasets for the purpose of evaluation. A wide array of datasets [8–11] has been established to facilitate a thorough evaluation of question-answer systems. However, it is crucial to recognize that these datasets are predominantly designed for general applications and thus are not appropriate for the direct evaluation of MEQAS. The evaluation of MEQAS necessitates the development of specialized test data that is specifically tailored to the unique context of military equipment. This requirement arises from the need to ensure that the MEQAS is capable of handling the specific types of queries and information pertinent to military operations. Nevertheless, the specialized, closed, and difficult-to-access nature of data within the military equipment domain presents significant challenges [12]. Relying on domain experts to manually construct datasets is not only prohibitively expensive but also impractically time-consuming, thereby limiting the feasibility of creating sufficiently high-quality datasets for MEQAS evaluation.

To reduce the cost of generating high quality test data for MEQAS, we propose an automatic military equipment question-answer test case generation framework, named MQATG. This innovative framework leverages the capabilities of large language models to enhance the efficiency and effectiveness of test case generation. MQATG is structured around three core components: question-answer test cases generation, question-answer test cases evaluation, and question-answer test cases selection. The generation phase employs a large language model, which is directed by a generation prompt, to examine pertinent documents and produce preliminary question-answer test cases. Subsequently, the evaluation phase utilizes an evaluation prompt, prompting the large language model to assess each generated test case according to five established metrics. Ultimately, the selection phase organizes the test cases based on their scores, facilitating the identification and application of those with higher rankings.

In order to determine the effectiveness of our approach, we evaluated MQATC using five different evaluation metrics and a comprehensive score. The experimental results show that MQATC can generate four types of question-answer test cases with high quality, including fill-in-the-blank, multiple-choice, true-or-false, and complex question-answer pairs. Moreover, the quality of the questions generated by MQATC is high and closely related to the core of the document. In addition, we also compare MQATC with the other method MQATG_b. The results show that MQATG is more effective than MQATG_b, and has better performance in generating test cases for true-or-false question-answer test cases.

In summary, we make the following contributions in this work.

- We employ large language models to produce question-answer test cases and develop high-quality prompts to facilitate the generation process of these models.
- We propose an automatic framework for generating question-answer test cases for military equipment, named MQATG. MQATG encompasses the generation, evaluation, and selection of question-answer test cases.
- We conduct multiple experiments to evaluate MQATG, and the results demonstrate that MQATG is effective in generating high-quality question-answer test cases.

2   Background

2.1   Large Language Models

Large language models represent a significant advancement in natural language processing, characterized by their vast parameter counts and ability to perform a multitude of tasks without task-specific training [13,14]. These models are trained on massive amounts of text data, which enables them to capture intricate patterns and dependencies within human language. They serve as foundational tools for a variety of applications, including translation, summarization, question answering, and content generation.

One of the pioneering large language models is Google's transformer-based model BERT (Bidirectional Encoder Representations from Transformers) [15], which introduced bidirectional training of attention mechanisms and achieved state-of-the-art results on a wide range of natural language processing tasks. However, BERT's reliance on masked language modeling leads to inefficiencies during inference due to the need for sequential predictions [16]. Another notable model is GPT-3 (Generative Pre-trained Transformer 3) developed by OpenAI [17]. GPT-3 is known for its unprecedented scale, boasting over 175 billion parameters, which allows it to generate human-like text with remarkable fluency. Despite its impressive capabilities, GPT-3 has been criticized for its lack of common sense reasoning and potential biases inherited from the training data [18]. The most recent iteration of OpenAI's GPT series, GPT-4 [19], has taken the advancements made by its predecessors to new heights. GPT-4 is anticipated to have better understanding and handling of complex instructions and queries, along with improvements in areas such as code generation and multilingual support. However, like its predecessors, GPT-4 likely still faces challenges related to factual accuracy and ethical considerations, although ongoing research aims to mitigate these issues. In addition, open-source large language models have emerged (such as DeepSeek-V2-0628 [23], Qwen2-72B-Instruct [22],

Llama 3 [24]), offering similar capabilities while being accessible to a broader audience. Due to the low barrier to access associated with open-source large language models and their comparable performance to proprietary models like GPT, MQATG utilizes open-source large language models to generate and evaluate question-answer test cases.

3   Approach

3.1   Overview

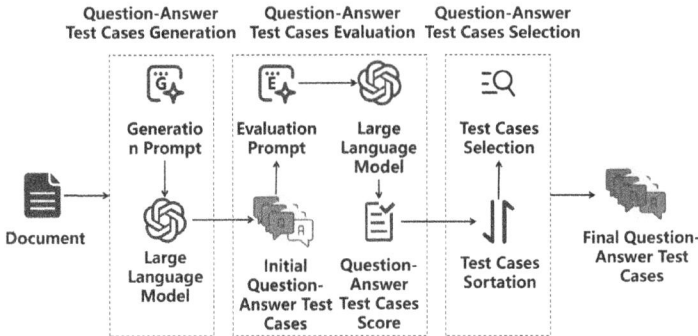

Fig. 1. Overview of MQATG

Figure 1 shows the overview of MQATC, which comprises question-answer test cases generation, question-answer test cases evaluation and question-answer test cases selection. The question-answer test cases generation employs a large language model, which is directed by a generation prompt, to analyze the document and create initial question-answer test cases. Then, the evaluation of the question-answer test cases utilizes an evaluation prompt to prompt the large language model to score each generated test case. Finally, The question-answer test cases selection process sorts the generated test cases based on their scores, enabling the identification of the high-ranked cases as the final test cases.

3.2   Question-Answer Test Cases Generation

Large language models have excellent understanding and generation capabilities in natural language [20]. To generating high quality question-answer test cases, MQATG employs the large language model to read the document and generate the initial question-answer test cases. Therefore, the performance of large language model determines the quality of question-answer test cases. The quality of the prompt is a critical factor influencing the effectiveness of large language model. To generate high-quality test cases, it is essential to construct high-quality prompts beforehand. Thus, we introduce the generation prompt, which includes

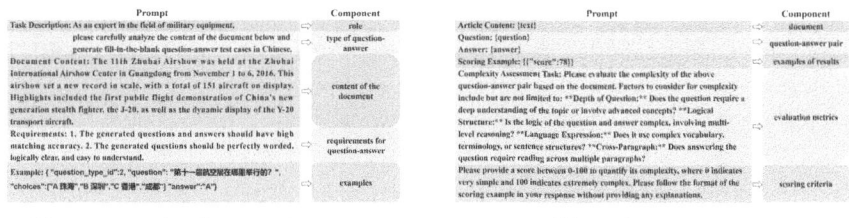

(a) an example of generation prompt   (b) evaluation prompt

Fig. 2. Generation prompt and evaluation prompt

the role, type of question-answer, content of the document, requirements for question-answer, and examples. As the example shown in Fig. 2a, the components of the prompt contain role, type of question-answer, content of the document, requirements for question-answer, examples. For large language model, different types of question-answer pairs require distinct prompts. Therefore, we designed specific prompts for fill-in-the-blank, multiple-choice, true-or-false, and complex question-answer pairs, denoted as p_f, p_s, p_t, p_c:

- p_f: The components of the prompt for fill-in-the-blank question-answer pairs is the same as Fig. 2a, as shown in Fig. 3a. In order to obtain high-scoring fill-in-the-blank for subsequent evaluation, we asked the large language model to generate 10 question-answer pairs. For the requirements of fill-in-the-blank, we first emphasize the feature requirements and the format requirements of fill-in-the-blank, such as the need to leave blank in the question and the output format needs to be a string. Then propose the content requirements of fill-in-the-blank, mainly requires that the question and answer pair must be closely related to the topic of the document. The example of fill-in-the-blank is from real military equipment question-answer pairs.

- p_s: Different fill-in-the-blank question-answer pairs, multiple-choice question-answer pairs necessitate the inclusion of a question, an answer, and a set of options. Therefore, we emphasize the choices type in the first requirement of the question-answer pair requirements, which is shown in Fig. 3b. The structure of multiple-choice questions can be either declarative or interrogative. To elucidate this distinction, we present two examples of question-answer pairs.

- p_t: The prompt for the true-or-false question-answer pairs is presented in Fig. 3c, From the example given, it can be observed that the format of the question-answer pair is relatively simple. The question is presented as a declarative, and the response is binary, consisting of either 'correct' or 'incorrect'.

- p_c: The requirements of complex question-answer pairs are higher than other types of question-answer pairs, encompassing not only quality and format specifications but also complexity requirments, as shown in Fig. 3d. The complexity requirments necessitates that the problem exhibit depth, possess a sophisticated logical structure, and utilize a more specialized vocabulary. To facilitate the generation of increasingly intricate question-answer pairs, we generate complex question-answer pairs in two steps. Firstly, we call prompt with ID 1 to generate the initial complex question answer pair and let the large language model list the thinking steps. Then we call prompt with ID 2 to generate more complex question-answer pairs according to the thinking step than the initial complex question-answer pair.

Task Description: As an expert in the field of military equipment, please think step-by-step and carefully analyze the content of the article below to generate ten high-quality, varied fill-in-the-blank question-answer pairs in Chinese. Output the results only, without any explanations.
Article Content: (text)
Requirements for Generating Fill-in-the-Blank Questions: 1.Blank Setting: Carefully select a key piece of information in the question to leave blank (indicated by), ensuring that the answer to the blank is directly from the original text or can be inferred from it. 2.The type of question_type_id is an integer with a fixed value of 1, the question is a string, and the answer is also a string. 3.The generated questions must be directly and closely related to the core information of the article. 4.The generated questions and answers should have a high matching accuracy. 5.The generated questions should be perfectly worded, logically clear, and easy to understand.
Example Generation (for reference only, actual content should comply with the above requirements) { 'question_type_id': 1, 'question': '战斗机，轰炸机和侦察机的留空时间取决于空中加油路径与的 接近程度。', 'answer': '战斗空域'}

(a) prompt for fill-in-the-blank question-answer pairs

Task Description: As an expert in the field of military equipment, please think step-by-step and carefully analyze the content of the article below to generate ten high-quality, varied multiple-choice question-answer pairs in Chinese. Output the results only, without any explanations.
Article Content: (text)
Requirements for Generating Multiple-Choice Questions: 1. Multiple-Choice Question Format: Strictly follow the example format below, using Chinese characters. The type of question_type_id is an integer with a fixed value of 2, the question is a string, choices is an array type containing four options such as {A, B, C, D}, and the answer is a string that can only be A, B, C, or D. 2.The generated questions must be directly and closely related to the core information of the article. 3.The generated questions and answers should have a high matching accuracy. 4.The generated questions should be perfectly worded, logically clear, and easy to understand.
Example Generation (for reference only, actual content should comply with the above requirements) { 'question_type_id': 2, 'question': '为增强美空中加油机战备态势，并确保更全面的作战范围，美军寻求在未来的全球部队管理计划中增加中队的数量。', 'choices': ['A 战斗机', 'B 加氢', 'C 飞行员', 'D 加油机'], 'answer': 'D'}. { 'question_type_id': 2, 'question': '美国跨与印太区域合作伙伴签订新的空中加油机基地协议，不包括以下哪个地区？', 'choices': ['A 东亚', 'B 东南亚', 'C 大洋洲', 'D 南美洲'], 'answer': 'D'}

(b) prompt for multiple-choice question-answer pairs

Task Description: As an expert in the field of military equipment, please think step-by-step and carefully analyze the content of the article below to generate ten high-quality, varied true-or-false question-answer pairs in Chinese. Output the results only, without any explanations.
Article Content:(text)
Requirements for Generating True-or-False Questions: 1. True-or-False Question Format: Strictly follow the example format below, using Chinese characters. The type of question_type_id is an integer with a fixed value of 3, the question is a string, and the answer is a string. 2. The generated questions must be directly and closely related to the core information of the article. 3. The generated questions and answers should have a high matching accuracy. 4. The generated questions should be perfectly worded, logically clear, and easy to understand.
Example Generation (for reference only, actual content should comply with the above requirements): { 'question_type_id': 3, 'question': '自适应基地是一个旨在提升空中资产生存能力的概念。', 'answer': '对'}

(c) prompt for true-or-false question-answer pairs

ID	Prompt
1	**Task Description:** You are an expert with extensive professional knowledge in the field of military equipment. Carefully read and analyze the content of the military equipment article provided below, and generate two high-quality complex question-answer pairs in Chinese. This must be done step-by-step according to the complexity, quality, and formatting requirements. List the steps of thinking one by one. **Article Content:**(text) **Requirements for Generating Complex Question-Answer Pairs:** Complexity **Requirements:** 1. Depth of Question: The question requires a deep understanding of the main topic or involves advanced concepts. 2. Logical Structure: The logic of the question and answer must be very complex and involve multi-level reasoning. 3. Language Expression: Complex vocabulary, terminology, or sentence structures need to be used. 4. Answers must be derived from reading multiple paragraphs and should include summarized and reasoned text. **Quality Requirements:** 1. The generated questions must be directly and closely related to the core information of the article. 2. The generated questions and answers should have a high matching accuracy. 3. The generated questions should be stated very clearly, with perfect grammar and logical coherence. **Formatting Requirements:** 1. Complex Question-Answer Format: Strictly follow the example format below, using Chinese characters. The type of question_type_id is an integer with a fixed value of 4, the question is a string, and the answer is a string. **Example Generation** (for reference only, actual content should comply with the above requirements): { 'question_type_id': 4, 'question': '美国空军加油机的"自适应基地"构想是什么？', 'answer': '这一概念设想在面临空中威胁的情况下，部队能够从多个规模较小、更具弹性的基地进行部署、战场生存、作战、机动并遂施战力。'}
2	Based on the feedback from the thinking steps, regenerate new high-complexity question-answer pairs.

(d) prompt for complex question-answer pairs

Fig. 3. Prompt for four question-answer pairs

3.3 Question-Answer Test Cases Evaluation

The large language model is directed by generation prompts to produce multiple question-answer test cases for each type. In order to select suitable question-answer pairs for the final output, a evaluation process is necessary. To improve the evaluation efficiency, we used the large language model to score the question-answer test case. In order to evaluate question-answer test cases more comprehensively, we select five evaluation metrics [21]:

- **Question Relevance:** The question-answer test cases are constructed based on the content of the document, hence the first step involves assessing the relevance of the questions to the document. If the question is directly and closely related to the core information of the document, the question relevance score is high.

- **Question-Answer Alignment:** In addition to evaluating the relevance of questions and document, we also evaluate the relevance between questions and answers. Different types of questions lead to different styles of answers. Therefore, three criteria were given for different style answers. Factual questions should be answered with direct and accurate references to information in the text. Explanatory questions require answers that can reasonably analyze the relevant content of the text. Reasoning questions should be answered by reasonable inference based on the content of the text.
- **Question Fluency:** To generate more understandable questions, we evaluate the fluency of the questions. The assessment ranges from clarity of question formulation, grammatical correctness, logical coherence, and suitability to the target audience.
- **Answer Fluency:** In addition to evaluating question fluency, it is imperative to assess answer fluency as well. The fluency of words is consistent for questions and answers. Therefore, the evaluation range of answer fluency is the same as question fluency.
- **Question-Answer Complexity:** The considerations for question-answer complexity include depth of the question, complexity of the logical structure, complexity of language expression, and depth of the answer. The depth of the question pertains to its relevance to the article's main theme. The complexity of the logical structure refers to the need for multi-level reasoning to arrive at the answer. The complexity of language expression indicates the requirement for complex vocabulary and structures in both the question and the answer. The depth of the answer signifies that it must be derived through summarization across multiple paragraphs.

MQATG constructs four evaluation prompts based on the evaluation metrics, where the same prompt was used to evaluate the fluency of question and answer. As the evaluation prompt for question-answer complexity shown in Fig. 2b, evaluation prompts require input of the document, question-answer pair, examples of results, evaluation metrics and scoring criteria. The main difference between the four evaluation prompts is that the evaluation metrics are different.

3.4   Question-Answer Test Cases Selection

To comprehensively evaluate the quality of a question-answer test case, MQATG utilizes multiple evaluation metrics. For the fill-in-the-blank, multiple-choice, and true-or-false question-answer test cases, MQATG selects question relevance, question-answer alignment, and question fluency metrics. For complex question-answer test cases, additional answer fluency and question-answer complexity metrics are required. The answers of complex question-answer test cases are more complex than other question-answer pairs, so the fluency of the answers needs to be evaluated. The comprehensive score for a question-answer test case is calculated as follows:

$$S_q = \begin{cases} 0.2*S_1 + 0.2*S_2 + 0.1*S_3 + 0.1*S_4 + 0.4*S_5, \ complex \ question - answer \\ 0.4*S_1 + 0.4*S_2 + 0.2*S_3, \qquad\qquad\qquad\qquad otherwise \end{cases} \quad (1)$$

where S_1, S_2, S_3, S_4, and S_5 represent question relevance, question-answer alignment, question fluency, answer fluency, and question-answer complexity, respectively. The complexity of the question and answer for complex question-answer test cases necessitates further examination. Additionally, there is a need to focus on the relevance of the questions and the alignment between questions and answers in other types of question-answer test cases.

MQATG calculates comprehensive score for each test case within the respective categories utilizing formula 1, subsequently ranks the test cases and ultimately selects the top-ranked test cases.

4   Evaluation

4.1   Research Questions

RQ1: Do MQATG generates each type of question-answer test cases with high scores? In this RQ, we aim to explore the ability of MQATG to generate each type of question-answer test cases.

RQ2: How effective is MQATG in generating question-answer test cases? We show the scores of question-answer test cases generated by MQATG and compare it with other approach.

4.2   Large Language Models and Dataset

We selected two high-performing open-source models, DeepSeek-V2-0628 [23] and Qwen2-72B-Instruct [22]. DeepSeek-V2-0628 is employed to create three categories question-answer test cases that include fill-in-the-blank, multiple-choice, and true-or-false questions and answers. Qwen2-72B-Instruct is utilized for the generation of complex question-answer test cases. To assess the quality of these question-answer test cases, DeepSeek-V2-0628 serves as the evaluation large language model. The document dataset is provided by the evaluation task of question and answer generation technology in military equipment field based on large model [21].

5   Result Analysis

5.1   Answer to RQ1

In this RQ, we use MQATG to generate four distinct categories of question-answer test cases and record the top-ranked test cases. To verify the performance of MQATG, we random picked 20 documents from the dataset. The scores of the question-answer test cases generated by MQATG are shown in Table 1, which includes the scores for each evaluation metric and comprehensive score. The type id denotes the category of question-answer test cases, where 1 indicates fill-in-the-blank, 2 signifies multiple-choice, 3 represents true-or-false, and 4 corresponds to complex question-answer. S_1, S_2, S_3, S_4, and S_5 represent question relevance, question-answer alignment, question fluency, answer fluency, and question-answer complexity, respectively. S_q denotes the comprehensive score for a question-answer test case, as indicated in formula 1. The scores represent the average scores obtained from generating each type of question-answer test case across 20 articles. Moreover, for each document, we did not select just one instance per type of question-answer test case. Specifically, for fill-in-the-blank and complex question-answer test cases, we selected two instances each, while for multiple-choice and true-or-false, we chose three instances each. All the question-answer test cases generated by MQATG achieved scores exceeding 80 for each evaluation metric. Although the complex question-answer test cases generated by MQATG scored low, the scores for the remaining three types test cases were close to 100, indicating a strong performance of MQATG in generating question-answer test cases. For different evaluation metric, MQATG performed particularly well on question relevance, question-answer alignment, and question fluency. While the answer fluency and question-answer complexity criteria had somewhat lower scores for complex question-answer test cases compared to other criteria, MQATG achieved scores above 80. Note that the average score for question relevance was 100, suggesting that the quality of questions generated by MQATG is very high.

Table 1. Average scores for four categories of question-answer pair test cases generated by MQATG.

Type ID	S_1	S_2	S_3	S_4	S_5	S_q
1	100.00	100.00	92.25	-	-	98.45
2	100.00	100.00	94.92	-	-	98.98
3	100.00	100.00	93.92	-	-	98.78
4	100.00	97.63	85.88	85.08	80.00	88.62
Avg.	100.00	99.41	91.74	85.08	80.00	96.21

5.2   Answer to RQ2

To evaluate the effectiveness of MQATG, we introduce a new method for generating question-answer test cases, named MQATG_b, which only generates the required number of test cases when generating the initial question-answer test cases, instead of generating 10 for subsequent selection like MQATG. For example, for a fill-in-the-blank question-answer test case, MQATG_b specifies in the generation prompt that the number of generation is 2 instead of 10. In addition, MQATG_b replaces the selection module of MQATG with the optimization module. When the comprehensive score S_q of the question-answer test case is lower than 90, the optimization module regenerates a new question-answer test case, and selects the test case with higher comprehensive score from the old and new test cases. Moreover, to study the impact of large language models on question-answer test case generation, MQATG_b uses Llama 3 [24] for generating the initial question-answer test cases and DeepSeek-V2-0628 for both the optimization module and the scoring module.

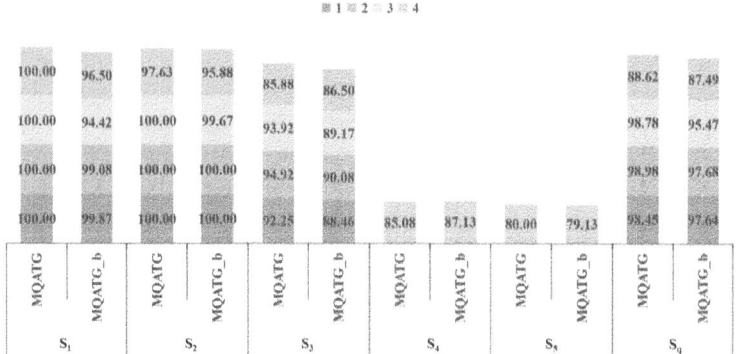

Fig. 4. Average scores for four categories of question-answer pair test cases generated by MQATG and MQATG_b.

We conducted a comparative analysis of the performance of MQATG and MQATG_b in generating four question-answer test cases across a set of 20 documents. The experimental results are shown in Fig. 4. For most evaluation metrics and comprehensive score, the total score of all test cases generated by MQATG surpasses that of MQATG_b. For instance, MQATG achieves a total score of 500 in metric S_1, while MQATG_b obtains a total score below 500. Although MQATG exhibited lower performance than MQATG_b in metric S_4, the difference in scores was minimal, at only 2.05 points. Furthermore, MQATG consistently outperformed MQATG_b in every category of question-answer test cases. For example, the S_q score for the fill-in-the-blank question represented by ID 1 is 98.45 for MQATG, compared to 97.64 for MQATG_b. In addition, for the true-or-false question-answer represented by ID 3, the total score gap between MQATG and MQATG_b is more than that of the other question-answer test cases.

6   Conclusion

In this paper, we propose an automatic military equipment question-answer test case generation framework, MQATG, which encompasses the generation, evaluation, and selection of question-answer test cases. The generation of question-answer test cases employs a large language model, which is directed by a generation prompt to examine the document and create preliminary question-answer test cases. The evaluation of these test cases utilizes an evaluation prompt to prompt the large language model to assess and assign scores to each generated test case. Subsequently, the selection process organizes the question-answer test cases based on the scoring outcomes, facilitating the identification of the highest-ranked cases to be designated as the final test cases.

In order to assess the effectiveness of MQATG, we analyze the question-answer test cases for military equipment produced by MQATG through the application of five evaluation metrics and a comprehensive score. The results of the experiments indicate that MQATG is capable of generating high-quality question-answer test cases related to military equipment and outperforms a baseline method referred to as MQATG_b.

References

1. Hou, X., Zhu, C., Li, Y., et al.: Question answering system based on military knowledge graph. In: International Conference on Electronic Information Engineering and Computer Communication (EIECC 2021), vol. 12172, pp. 33–39. SPIE (2022)
2. Szabadföldi, I.: Artificial intelligence in military application-opportunities and challenges. Land Forces Acad. Rev. **26**(2), 157–165 (2021)
3. Lowe, R., Noseworthy, M., Serban, I.V., Angelard-Gontier, N., Bengio, Y., Pineau, J.: Towards an automatic Turing test: learning to evaluate dialogue responses. In: Proceedings of the 55th Annual Meeting of the Association for Computational Linguistics (Volume 1: Long Papers), pp. 1116–1126, Vancouver, Canada. Association for Computational Linguistics (2017)
4. Kannan, A., Vinyals, O.: Adversarial evaluation of dialogue models. arxiv preprint arxiv:1701.08198 (2017)
5. Ma, Q., Wei, J.T.Z., Bojar, O., et al.: Results of the WMT19 metrics shared task: segment-level and strong MT systems pose big challenges. In: Association for Computational Linguistics (2019)
6. Ghazarian, S., Wei, J.T.-Z., Galstyan, A., Peng, N.: Better automatic evaluation of open-domain dialogue systems with contextualized embeddings. CoRR, abs/1904.10635 (2019)
7. Tan, Y., et al.: Can ChatGPT replace traditional KBQA models? An in-depth analysis of the question answering performance of the GPT LLM family. arxiv preprint arxiv:2303.07992 (2023)
8. Rajpurkar, P., Jia, R., Liang, P.: Know what you don't know: unanswerable questions for SQuAD. arxiv preprint arxiv:1806.03822 (2018)
9. Yang, Y., Yih, W.-T., Meek, C.: WikiQA: a challenge dataset for open-domain question answering. In: Proceedings of the 2015 Conference on Empirical Methods in Natural Language Processing (2015)
10. Kwiatkowski, T., Palomaki, J., Redfield, O., et al.: Natural questions: a benchmark for question answering research. Trans. Assoc. Comput. Linguist. **7**, 453–466 (2019)
11. Nguyen, T., Rosenberg, M., Song, X., et al.: MS MARCO: a human-generated machine reading comprehension dataset (2016)
12. Rettore, P.H., Zißner, P., Alkhowaiter, M., Zou, C., Sevenich, P.: Military data space: challenges, opportunities, and use cases. IEEE Commun. Mag. (2023)
13. Mann, B., et al.: Language models are few-shot learners. arxiv preprint arxiv:2005.14165 (2020)
14. Vaswani, A.: Attention is all you need. In: Advances in Neural Information Processing Systems (2017)

15. Devlin, J.: BERT: pre-training of deep bidirectional transformers for language understanding. arxiv preprint arxiv:1810.04805 (2018)
16. Radford, A.: Improving language understanding by generative pre-training (2018)
17. Floridi, L., Chiriatti, M.: GPT-3: Its nature, scope, limits, and consequences. Mind. Mach. **30**, 681–694 (2020)
18. Bender, E. M., Gebru, T., McMillan-Major, A., Shmitchell, S.: On the dangers of stochastic parrots: can language models be too big?. In: Proceedings of the 2021 ACM Conference on Fairness, Accountability, and Transparency, pp. 610–623 (2021)
19. Achiam, J., et al.: GPT-4 technical report. arxiv preprint arxiv:2303.08774 (2023)
20. Roumeliotis, K.I., Tselikas, N.D.: ChatGPT and Open-AI models: a preliminary review. Future Internet **15**(6), 192 (2023)
21. Evaluation Task of Question and Answer Generation Technology in Military Equipment Field Based on Large Model. China Conference on Knowledge Graph and Semantic Computing (2024). https://www.osredm.com/competition/zstp2024/fingerpost
22. Yang, A., et al.: Qwen2 technical report. arxiv preprint arxiv:2407.10671 (2024)
23. Liu, A., et al.: DeepSeek-V2: a strong, economical, and efficient mixture-of-experts language model. arxiv preprint arxiv:2405.04434 (2024)
24. Meta, A.I.: Introducing meta llama 3: the most capable openly available LLM to date (2024). https://aimeta.com/blog/meta-llama-3/. Accessed 26 Apr 2024

Boosting Q&A Generation for Military Equipment via Example Selection and Automated Prompt Engineering

Kunli Zhang[(⊠)], Yongqi Zhu, Yu Song, Bohan Yu, Guangyu Zhou,
Chenkang Zhu, and Pengcheng Wu

School of Computer Science and Artificial Intelligence, Zhengzhou University,
Zhengzhou 450001, China
ieklzhang@zzu.edu.cn, zyq44827@gs.zzu.edu.cn

Abstract. Question and Answer Generation (QAG) is a text generation task that aims to generate contextually relevant questions and answers based on a given text. Due to the closed nature and difficulty in obtaining data in the military equipment field, relying entirely on domain experts to manually build datasets is too costly and difficult to implement effectively. In response, this paper proposes a QAG method based on example selection and automatic prompting engineering. This method utilizes the generative capabilities of large language models (LLMs), enhanced by high-quality examples and optimized prompts to improve reasoning and adaptability to problems. Initially, high-quality examples are selected from the training set documents based on a question-answer pair scoring strategy, ensuring coverage of different question types and complexities. Furthermore, the highest-scoring examples from each category are selected as prompt examples to further optimize the accuracy and relevance of question-answer generation. Additionally, by testing various LLMs, the most suitable model is chosen for task execution. Finally, we employ automatic prompting engineering to generate prompts, using the reasoning ability of LLMs to create an initial set of prompts and iteratively optimize to select the best prompts. Our method ranked second in the CCKS 2024 evaluation task 10, achieving a score of 91.5%.

Keywords: Question answer generation · Prompt engineering · Large language model

1 Introduction

Question and Answer Generation (QAG) [4] is a task that produces relevant questions and answers from text, useful in dialogue, information retrieval, and extraction. It includes rule-based, machine learning, and deep learning methods, facing issues like high resource use, limited generalization, and uneven quality.

Large language models (LLMs) [3] and prompting methods present new approaches to QAG. LLMs, built on the Transformer architecture like GPT-4 [1] and LLaMA [9], are trained on extensive data to handle complex language. Prompting enhances LLMs' generative abilities by supplying specific cues, thereby increasing the relevance and quality of outputs.

In the military equipment sector [6], where data is highly guarded and expensive to access, utilizing this data to generate question-answer pairs poses challenges due to its specialized nature and inaccessibility. This paper introduces a new method for QAG that integrates example selection with automatic prompting engineering (APE). It employs a meticulous strategy based on scoring question-answer pairs to select high-quality examples from the training set. This approach evaluates the completeness and linguistic quality of examples, ensuring a broad spectrum of question types and complexity levels are covered, thereby improving the QAG's applicability and educational value.

This method emphasises the use of a scoring system in the selection of QAG examples. It establishes detailed scoring systems for different problem categories and selects the highest scoring examples as prompts after analysing their key features. The method also tests a variety of LLMs to determine the most appropriate LLM for the task. Once selected, it applies automated cue engineering techniques to generate an initial cue word that is appropriate for the model and task requirements. This results in an efficient, high-quality QAG.

Altogether, the following are the primary contributions of our work:

- We proposed a method combining example selection and APE, where high-quality examples are selected from the training set documents through a question-answer pair scoring strategy to serve as prompt examples. Subsequently, the most suitable model for executing this task was selected by testing various models.
- We utilized APE to generate an initial set of prompt words and iteratively optimized them to select the best prompts. Leveraging the reasoning capabilities of the selected LLMs, we achieved high-quality QAG.
- Our method ranked second in the CCKS 2024 evaluation task 10, achieving a score of 91.5%, validating the effectiveness of our approach and its practical value in real-world applications.

2   Related Works

2.1   Question Answer Generation

The task of QAG is one of the important tasks in the field of natural language processing, aiming to learn and understand given text and automatically generate high-quality question-answer pairs from it. In recent years, research on QAG has mainly focused on automatically generating answers to questions based on given text or knowledge bases. Deepak Gupta [5] proposed the NQG-Knowledge model, which encodes entities and relevant information within paragraphs, aiding the generation of questions. Zhang [11] proposed the BERT-QG-QAP model, which enhances semantic understanding to standardize the generation of semantically effective questions, thereby controlling question quality. Zhou [13] proposed a QAG method based on the Seq2Seq framework and text data processing (Seq2Seq-TDP-QAG).

In recent years, LLMs have garnered significant attention due to their superior natural language understanding and generation capabilities. Gupta [5] proposed constructing high-quality question-answer datasets using LLMs combined with retrieval-augmented generation (RAG) frameworks. Tan [8] introduced a hierarchical conditional variational autoencoder (HCVAE) designed to generate question-answer pairs given unstructured text as context while maximizing mutual information between QAG to ensure consistency.

2.2 Prompt Engineering

Prompt engineering, as an emerging research direction in generative AI and natural language processing, has received increasing attention in recent years. Recent work, such as that of Arora and Simran [2] introduced the Ask-Me-Anything framework, suggesting a method to aggregate multiple imperfect prompts to enhance model performance, particularly in question-and-answer formats. Wei [10] proposed chain-of-thought prompting, which improves the performance of complex tasks by generating intermediate reasoning steps. Zhou [12] introduced a novel prompting strategy, Least-to-most prompting, whose core idea is to break down a complex problem into a series of simpler sub-problems, solving them sequentially. Mitra Chancharik [7] proposed a novel zero-shot chain-of-thought prompting method called Compositional Chain-of-Thought, which utilizes LLMs to generate a scene graph and then uses this graph in prompts to produce responses.

Based on the aforementioned related research on QAG, most methods can achieve satisfactory results. However, most traditional approaches fail to understand and handle complex contexts, generating highly targeted questions and answers, which is crucial in fields involving deep knowledge or technical details. Therefore, this paper employs prompt engineering technology by incorporating domain knowledge in prompts to obtain more accurate question-answer pairs.

3 Preliminaries

In this evaluation task, different types of question-answer pairs need to be generated based on the given articles. Given a document $\mathcal{D} = \{t_1, t_2, ..., t_n\}$, where t_i represents the i-th word or phrase in the text, and N is the total length of the document. The text $\mathcal{D}$ contains all the contextual information required for QAG. The type of question is denoted by $\mathcal{Q}_{Type}$, which indicates the type of the generated question, chosen from a set of predefined types, formally represented as $\mathcal{Q}_{Type} \in \{\mathcal{Q}_{fill}, \mathcal{Q}_{choice}, \mathcal{Q}_{bool}, \mathcal{Q}_{complex}\}$, $\mathcal{Q}_{fill}, \mathcal{Q}_{choice}, \mathcal{Q}_{bool}, \mathcal{Q}_{complex}$ represent fill-in-the-blank questions, multiple choice questions, true or false questions, and complex questions, respectively. Below are the definitions of these four types of questions(Since the original data are in Chinese, most of the examples in this paper are expressed in Chinese.):

Document: 《超越美国空军加油机的"自适应基地"构想:改善印太司令部空中加油战备态势、扩大作战范围的替代性方案》……战斗机、轰炸机和侦察机的留空时间完全取决于空中加油路径与战斗空域的接近程度,尤其是在对抗性的反介入/区域拒止环境中,受油飞机为了应对来自陆地、海上和空中的威胁而做出规避动作时具备更高的燃油消耗率……	
Fill in the blanks	**Question**: "战斗机、轰炸机和侦察机的留空时间取决于空中加油路径与的_____接近程度。" **Answer**:"战斗空域"
Multiple choice	**Question**: "美国将与印太区域合作伙伴签订新的空中加油机基地协议,不包括以下哪个地区?" **Choices**:['A东亚', 'B东南亚', 'C大洋洲', 'D南美洲'] **Answer**:"D"
True or False	**Question**: "自适应基地是一个旨在提升空中资产生存能力的概念。" **Answer**:"错误"
Complex Question	**Question**: "国空军加油机的"自适应基地"构想是什么?" **Answer**:"这一概念设想在面临空中威胁的情况下,部队能够从多个规模更小、更分散、更具弹性的基地进行部署、战场生存、作战、机动并迅速恢复战力。"

Fig. 1. Examples of the four question types.

- $\mathcal{Q}_{fill}$ represents fill-in-the-blank questions, where sentences with blanks need to be generated from the text, requiring the respondent to fill in the correct word or phrase.
- $\mathcal{Q}_{choice}$ represents multiple choice questions, where a question and several options are generated, including one correct answer.
- $\mathcal{Q}_{bool}$ represents true or false questions, where a statement is generated, requiring the respondent to determine its truthfulness.
- $\mathcal{Q}_{complex}$ represents complex questions, involving multi-step reasoning or detailed explanations in the answer generation.

The generated section includes questions $\mathcal{Q} \in q_1, q_2, \ldots, q_n$ and corresponding answer pairs $\mathcal{A} \in \mathcal{A}_{fill}, \mathcal{A}_{choice}, \mathcal{A}_{bool}, \mathcal{A}_{complex}$. q_i represents the i-th word or phrase in the generated questions, and n is the length of the questions. The generated questions $\mathcal{Q}$ must be highly relevant to the input document $\mathcal{D}$ and the question type $\mathcal{Q}_{Type}$, and they should accurately reflect the content of the document. The answer pairs $\mathcal{A}$ represent the corresponding answers to the generated questions $\mathcal{Q}$. Depending on the question type $\mathcal{Q}_{Type}$, the form of the answers can be $\mathcal{A}_{fill}, \mathcal{A}_{choice}, \mathcal{A}_{bool}, \mathcal{A}_{complex}$, representing answers for fill-in-the-blank questions, multiple choice questions, true or false questions, and complex questions, respectively. For fill-in-the-blank questions, the answer is one or more words or phrases that need to be filled in the blank, $\mathcal{A}_{fill} = \{\alpha_1, \alpha_2, \ldots, \alpha_k\}$, where k is the length of the answer phrases. For multiple choice questions, the answer is the correct option selected from several options, $\mathcal{A}_{choice} = \alpha_{correct}$. For true or false questions, the answer is a Boolean value, $\mathcal{A}_{bool} \in \{True, False\}$. For complex questions, the answer may be a sequence of text containing detailed explanations or multi-step reasoning, $\mathcal{A}_{complex} = \{\alpha_1, \alpha_2, \ldots, \alpha_l\}$, where l is the length of the complex answer.

4   Method

Our method employs example and model selection along with APE for generating military equipment question-answer pairs. By choosing suitable examples and prompts, we enhance LLM's inference and adaptability. Initially, high-quality examples are selected based on a scoring strategy. We then test various LLMs to select the most suitable one, underscoring the value of thorough evaluation. Finally, APE is used to create and iteratively optimize prompts, leveraging LLM's reasoning capabilities.

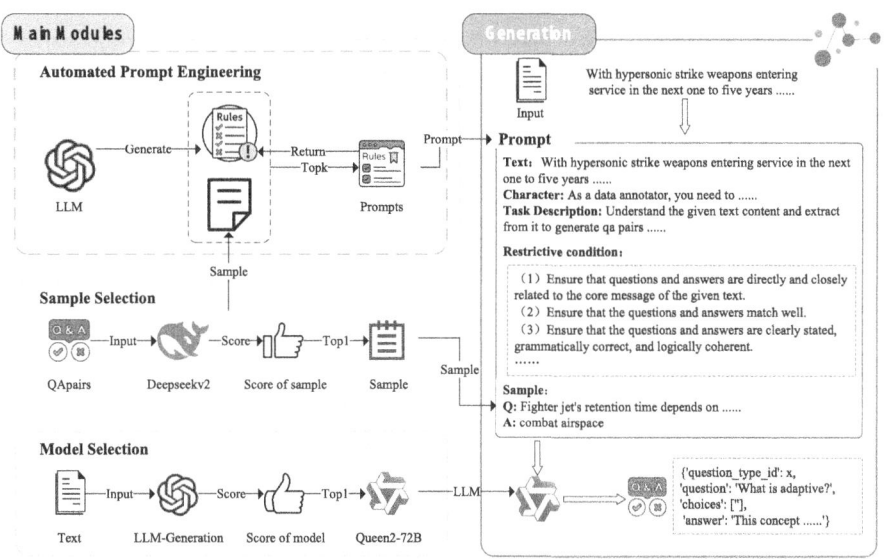

Fig. 2. Illustration of our method.

4.1   Example and Model Selection

Firstly, we select the highest-scoring example from each question type in the training set and choose the LLM that scores the highest in QAG for each question type.

Example Selection. In generating military equipment question-answer pairs, selecting appropriate samples to prompt LLMs enhances their reasoning and output format consistency. We employ a sample selection strategy based on question scoring to boost understanding and adaptability across different question types. The strategy focuses on choosing high-quality samples that accurately represent question characteristics and requirements, crucial for enhancing QAG precision in LLMs. This approach is particularly valuable for handling complex and varied question-answer pairs, enabling the model to effectively manage questions with high complexity.

This method involves selecting samples from 40 documents that comprehensively cover various aspects of military equipment such as technology, performance, application, support, and strategy. Each document contains 10 examples of question-answer pairs, distributed as 2 fill-in-the-blank, 3 multiple choice, 3 true or false, and 2 complex questions, ensuring a broad representation of topics. During the evaluation phase, the method assigns scores to 400 questions from these documents. To refine sample selection, the questions are categorized into four types, and the highest-scoring questions from each category are chosen as examples for prompting.

Below is a detailed description and calculation method of the scoring system, including how to score across different dimensions and how to integrate these scores to obtain the final results.

For each question-answer pair result, let its score be S_q.

Table 1. Various scoring criteria.

name		mean	range
S_1	Question relevance	Whether the generated questions are relevant to the text	0–100
S_3	QA matching degree	The matching degree between the questions and answers	0–100
S_2	Question fluency	The fluency of the question text	0–100
S_3	Answers fluency	The fluency of the answer text	0–100
S_4	Problem complexity	Judge the complexity level	0–100

a) for fill-in-the-blank questions, true or false questions, and multiple choice questions, the scoring method is:

$$S_q = 0.4 * S_1 + 0.4 * S_2 + 0.2 * S_3 \tag{1}$$

b) for complex questions, the scoring method is:

$$S_q = 0.2 * S_1 + 0.2 * S_2 + 0.1 * S_3 + 0.1 * S_4 + 0.4 * S_5 \tag{2}$$

For each type of question, let its score be TS_q. The score for each type of question is the average score of the question-answer pairs in that type. For fill-in-the-blank questions and complex questions, $K = 2$. In other cases, $K = 3$.

$$TS_q = \frac{1}{K} \sum_{k=1}^{K} S_{q_k} \tag{3}$$

For each article, let its score be S_d.

$$S_d = 0.5 * TS_{fill} + 0.2 * TS_{choice} + 0.2 * TS_{bool} + 0.1 * TS_{complex} \qquad (4)$$

The final score S of the submitted results is:

$$S = \frac{1}{10} \sum_{i=1}^{10} S_{d_i} \qquad (5)$$

Score the training set to generate ratings for forty articles, each with ten questions. Categorize all questions into four types based on question type and select the highest-scoring questions as examples in the prompts.

Model Selection. According to the aforementioned scoring, this method selects the highest-scoring LLM for the four question types to generate question-answer pairs. We tested three existing open-source LLM: **Qwen1.5-110B-chat**, **DeepSeek-V2-Chat**, and **Qwen2-72B-Instruct**.

4.2   Automated Prompt Engineering

LLMs exhibit robust inference capabilities, yet their performance in specific tasks hinges significantly on the quality of prompts, which are often most effective when crafted manually. To automate this process, we adopted APE for generating and selecting instructions automatically. APE conceptualizes instructions as "programs" and navigates through a set of candidate prompts initiated by LLMs to optimize a chosen scoring function. It produces various candidate instructions either directly through inference or via a recursive process anchored in semantic similarity. These instructions are then executed using the target model, and the most suitable ones are chosen based on their evaluation scores.

Initial Candidate Generation. Due to the infinitely large search space, finding the correct instructions can be extremely challenging, making traditional approaches inadequate. LLMs excel at generating various natural language texts. Therefore, APE leverages an LLM to propose a set of good candidate solutions to guide its search process. While random samples from LLMs are unlikely to produce desired question-answer pairs, we can ask the LLM to infer the most probable prompts with high scores given context examples.

Prompt Selection. APE initializes candidate prompts using an LLM and assesses their quality via zero-shot learning on another LLM. To enhance prompt selection, we use an iterative Monte Carlo search method that refines the instruction set through repeated iterations to optimize model performance. The subsequent sections explain the iterative search steps and their impact on instruction optimization.

Iterative Search. The iterative search begins by generating an initial set of prompt candidates U, consisting of m prompt sets $U = \{\rho_1, \rho_2, \ldots, \rho_m\}$, created by the LLM from input-output pairs $\mathcal{D}_{train}$. Each candidate ρ_i is assessed using the five scoring categories from the sample selection process. The top k instructions U_k are chosen for the next iteration based on these scores. The highest-scoring set U_k is retained, and new candidates U' are generated through resampling. This critical step allows the exploration of a broader solution space by creating semantically similar new candidates to existing high-scoring instructions. The iterative process continues until convergence, marked by insignificant score improvements or a set iteration limit. The final optimized instruction is the highest-scoring instruction at the iteration's end.

5   Experiment

5.1   Experimental Settings

Our experimental data is the dataset of CCKS 2024 Review Task 10. The data consists of 40 articles related to military equipment technology, effectiveness, utilization, security, and strategy. Each article will have a sample of 10 quiz pairs containing 2 fill-in-the-blanks, 3 single-choice questions, 3 judgment questions, and 2 complex quiz questions. Our experiments were conducted on Qwen2-72B-Instruct, Qwen1.5-110B-Chat, and Deepseek-v2-Chat for prompt learning and scoring. The scoring is shown in Table 1.

5.2   Experimental Results

Example Score. We scored the samples for each of the four question types, and the exact scores are shown in Table 2. We used the two highest-scoring samples as prompt samples.

Table 2. Sample scores for four question types.

Question Type	Min	Max	Average
$\mathcal{Q}_{fill}$	17.00	97.00	91.21
$\mathcal{Q}_{choice}$	17.00	97.00	89.68
$\mathcal{Q}_{bool}$	17.00	97.00	88.54
$\mathcal{Q}_{complex}$	62.00	97.00	87.96

Model Score. Based on the selected samples, we use the Few-shot cue to test the generation of different LLMs for the four question types. It can be seen that Qwen2-72B-Instruct performs better in multiple-choice and complex questions with 86.44% and 80.40%, respectively. Deepseek-v2-Chat and Qwen1.5-110B-Chat perform better in fill-in-the-blank and judgmental questions with 95.00% and 92.00%, respectively. The scoring is shown in Table 3.

Table 3. Select examples using few-shot prompts.

Model	Q_{fill}	Q_{choice}	Q_{bool}	$Q_{complex}$	score
Qwen2-72B-Instruct	94.24	86.44	91.70	80.40	85.25
Qwen1.5-110B-Chat	93.40	84.35	92.00	80.20	84.71
Deepseek-v2-Chat	95.00	80.50	90.32	77.35	82.34

Prompt Template Score. We used different types of prompt templates for scoring question generation and found that our method achieved the optimum in the three types of questions as well as in the total score, proving the effectiveness of our prompt optimization method. The scoring is shown in Table 4.

Table 4. Scoring results of different prompt templates.

Method	Q_{fill}	Q_{choice}	Q_{bool}	$Q_{complex}$	score
Zero-shot	94.24	86.44	91.70	80.40	85.25
Few-shot	94.60	95.00	93.80	80.43	87.43
CoT	95.12	95.20	80.48	93.40	87.47
Ours	95.90	95.40	95.20	85.42	90.42

5.3   Case Study

Samples of prompt templates are shown in Fig. 3, we took the fill-in-the-blank question as an example and compared four prompt templates. It can be seen that our prompt templates provide sample as well as output format constraints that outperform the rest of the templates. A comparison of the generation results of the four prompt templates on different types of problems is shown in Fig. 4. We compared the generation results of four prompt templates on four question types. Our results better met the evaluation requirements and had higher scores.

Zero-shot (Fill in the blanks)

prompt_template = '''
给出一段文本，请你理解给出的文本内容，从中抽取生成1个填空题，给出问题与对应的答案。
文本：{text}
注意输出格式，在问题中用下划线代替填空题的答案！，每一条结果按照示例格式进行返回，以下是两条示例：'''
example = '''
[
　{
　"question":"",
　"answer":""
　}
] '''

Few-shot (Fill in the blanks)

prompt_template = '''
给出一段文本，请你理解给出的文本内容，从中抽取生成1个填空题，给出问题与对应的答案。
文本：{text}
注意输出格式，在问题中用下划线代替填空题的答案！，每一条结果按照示例格式进行返回，以下是两条示例：'''
example = '''
[
　{
　"question":"导弹作为一型在复杂战场环境下使用的武器装备，其＿＿＿能力也非常重要。",
　"answer":"实战化"
　},
　{
　"question":"＿＿＿是导弹的顶层需求之一，通过完善导弹能力内涵，可以寻找到更准确的导弹装备发展方向，也可以作为导弹装备创新的切入点。",
　"answer":"导弹能力需求"
　}
] '''

CoT (Fill in the blanks)

prompt_template = '''
现在给你一段文本，你需要做的有：
1.理解给出的文本内容，识别出其中的关键信息，并思考这些信息之间的关系。
2.提出一个与关键信息相关的填空题，以考察解答者对文本的理解。
3.在问题中用下划线代替填空的答案，问题应清晰明确。
4.给出答案，说明为什么这是正确答案。
文本：{text}
注意输出格式，在问题中用下划线代替填空题的答案！，每一条结果按照示例格式进行返回，以下是两条示例：'''
example = '''
[
　{
　"question":"导弹作为一型在复杂战场环境下使用的武器装备，其＿＿＿能力也非常重要。",
　"answer":"实战化"
　},
　{
　"question":"＿＿＿是导弹的顶层需求之一，通过完善导弹能力内涵，可以寻找到更准确的导弹装备发展方向，也可以作为导弹装备创新的切入点。",
　"answer":"导弹能力需求"
　}
] '''

Ours (Fill in the blanks)

prompt_template = '''
###角色：{作为一个数据标注员，你的任务是从给出的文本中抽取填空题形式的问答对。}
###文本：{text}
###任务：{理解给出的文本内容，从中抽取生成1个填空题，给出问题与对应的答案。}
###限制条件，注意问答对内容需要满足：{
@保证问题与答案直接且紧密地关联于所给文本的核心信息。
@保证问题与答案的匹配程度很高。
@保证问题与答案表述清晰、语法正确、逻辑连贯。
}
###注意输出格式，在问题中用下划线代替填空题的答案！，每一条结果按照示例格式进行返回，以下是两条示例：
'''
example = '''
[
　{
　"question":"导弹作为一型在复杂战场环境下使用的武器装备，其＿＿＿能力也非常重要。",
　"answer":"实战化"
　},
　{
　"question":"＿＿＿是导弹的顶层需求之一，通过完善导弹能力内涵，可以寻找到更准确的导弹装备发展方向，也可以作为导弹装备创新的切入点。",
　"answer":"导弹能力需求"
　},
　{
　"question":"自适应基地的概念基于灵活分散的理念，提升作战＿＿＿能力。",
　"answer":"机动"
　}
] '''

Fig. 3. Examples of four prompt templates.

Fill in the Blanks

Zero-shot	Few-shot	CoT	Ours
question: 报告中提到, 空军的空中加油机机队现状无法满足印太司令部提出的＿＿＿概念的需求。 *answer:* 自适应基地	*question:* 美国海军战争学院发布的一份报告《超越美国空军加油机的自适应基地构想》提出, 空军的＿＿＿现状无法满足印太司令部提出的自适应基地概念的需求。 *answer:* 空中加油机机队	*question:* 空军的空中加油机机队现状无法满足＿＿＿提出的需求。 *answer:* 印太司令部	*question:* 报告认为, 由于受到后勤和作战方面的限制, 空军的空中加油机机队现状无法满足印太司令部提出的＿＿＿概念的需求。 *answer:* 自适应基地

Multiple Choice Question

Zero-shot	Few-shot	CoT	Ours
question: 根据知远导读中提及的报告, 美国空军的空中加油机机队现状为何无法满足印太司令部的自适应基地概念需求? *choices:* A. 因为机队规模过大, 难以灵活部署 B. 由于后勤和作战等方面的限制 C. 空中加油机的性能不足, 无法支持远程作战 D. 合作伙伴的空中加油机基地数量不足 *answer:* B	*question:* 印太司令部的军事战略中, 通过摆出战斗姿态, 将作战可靠的部队部署在前沿地带的目的是迫使敌人如何推进其利益? *choices:* A. 通过武力手段 B. 通过经济手段 C. 通过外交手段 D. 通过符合国际公认规则的良性手段 *answer:* D	*question:* 2019年《印太战略报告》阐明印太司令部的军事战略, 其中的核心思想是什么? *choices:* A. 通过增加军事预算, 提升军队整体实力 B. 通过前沿部署可靠作战部队, 迫使敌人以良性方式推进利益 C. 通过发展远程打击能力, 威慑潜在对手 D. 通过外交途径, 增强区域合作伙伴关系 *answer:* B	*question:* 印太司令部面临的主要限制因素, 影响其空中加油机机队满足自适应基地概念需求的原因有哪些?, *choices:* A. 缺乏空中加油机 B. 后勤限制和作战限制 C. 缺乏足够的基地 D. 空中加油机的老旧 *answer:* B

True or False Question

Zero-shot	Few-shot	CoT	Ours
question: 印太司令部提出的自适应基地概念的需求目前可以得到空军的空中加油机机队的完全满足。 *answer:* 错	*question:* 美空军的空中加油机机队现状无法满足印太司令部自适应基地概念的需求。 *answer:* 对	*question:* 印太司令部希望通过增加空中加油机中队数量与区域合作伙伴签订新基地协议来改善空中加油战备态势。 *answer:* 对	*question:* 空中加油机机队的现状能够满足印太司令部提出的自适应基地概念的需求。 *answer:* 错

Complex Question

Zero-shot	Few-shot	CoT	Ours
question: 美国印太司令部面临的空中加油机机队现状问题及解决方案是什么? *answer:* 印太司令部的空中加油机机队现状无法满足自适应基地概念的需求, 受到后勤和作战等方面的限制。解决方案包括在未来的全球部队管理计划中增加加油机中队的数量, 以及与遍及东亚、东南亚和大洋洲的区域合作伙伴签订新的空中加油机基地协议。	*question:* 为什么美国空军的空中加油机机队现状无法满足印太司令部的自适应基地概念需求? *answer:* 由于受到后勤和作战等方面的限制, 美国空军的空中加油机机队现状无法满足印太司令部的自适应基地概念需求。具体表现在空中加油机输油能力有限, 随着空中加油机运输距离的增加, 空中行动的总体燃料需求呈现指数级增长, 这影响了战斗机、轰炸机和侦察机的留空时间。在对抗性的反介入/区域拒止环境中, 受油飞机为了应对来自陆地、海上和空中的威胁而做出规避动作时会消耗更多燃油, 使空中加油支持的效率降低。	*question:* 根据美国海军战争学院报告, 印太司令部面临的加油机不足问题的两个替代方案是什么? *answer:* 印太司令部应通过以下两种方式解决加油机不足的问题; 第一, 寻求在未来的全球部队管理计划中增加加油机中队的数量; 第二, 美国与遍及东亚、东南亚和大洋洲的区域合作伙伴签订新的空中加油机基地协议。	*question:* 为什么美国空军的空中加油机机队无法满足印太司令部自适应基地概念的需求, 以及印太司令部应如何解决这一问题? *answer:* 美国空军的空中加油机机队受到后勤和作战等方面的限制, 无法满足印太司令部提出自适应基地概念的需求。印太司令部应通过两种替代方案解决这一问题: 首先, 寻求在未来的全球部队管理计划中增加空中加油机中队的数量, 以实现更强大的空中加油机战备态势。其次, 与遍及东亚、东南亚和大洋洲的区域合作伙伴签订新的空中加油机基地协议, 以确保更全面的作战范围, 这将有助于在面对中国在时间、空间和部队上的优势时, 提高进攻速度、扩大作战范围以及为机载武器系统提供更大的战术打击半径。

Fig. 4. Comparison of samples generated by different prompting methods.

6 Conclusion

This study introduces an innovative QAG method for the military equipment domain to overcome the challenges of data access and dataset construction costs. Leveraging LLMs' generative power, the method enhances QA accuracy and relevance via sample selection and automatic cue engineering. Initially, high-quality samples are identified using a QA scoring strategy for prompt generation. Then, several models are tested to find the most suitable for the task. Prompts are generated and refined through automatic prompt engineering, utilizing the models' reasoning abilities. The method proved effective, placing second with a score of 91.5% in CCKS 2024 Task 10, demonstrating a viable text generation solution for the military equipment domain.

Acknowledgments. This work was supported in part by the Science and Technology Innovation 2030-"New Generation of Artificial Intelligence" Major Project (No.2021ZD0111000), and Henan Provincial Science and Technology Research Project (No.232102211033).

References

1. Achiam, J., et al.: GPT-4 technical report. arXiv preprint arXiv:2303.08774 (2023)
2. Arora, S., et al.: Ask me anything: a simple strategy for prompting language models. In: The Eleventh International Conference on Learning Representations (2022)
3. Chang, Y., et al.: A survey on evaluation of large language models. ACM Trans. Intell. Syst. Technol. **15**(3), 1–45 (2024)
4. Duan, N., Tang, D., Chen, P., Zhou, M.: Question generation for question answering. In: Proceedings of the 2017 Conference on Empirical Methods in Natural Language Processing, pp. 866–874 (2017)
5. Gupta, D., Suleman, K., Adada, M., McNamara, A., Harris, J.: Improving neural question generation using world knowledge. arXiv preprint arXiv:1909.03716 (2019)
6. Hou, X., Zhu, C., Li, Y., Wang, P., Peng, X.: Question answering system based on military knowledge graph. In: International Conference on Electronic Information Engineering and Computer Communication (EIECC 2021), vol. 12172, pp. 33–39. SPIE (2022)
7. Mitra, C., Huang, B., Darrell, T., Herzig, R.: Compositional chain-of-thought prompting for large multimodal models. In: Proceedings of the IEEE/CVF Conference on Computer Vision and Pattern Recognition (CVPR), pp. 14420–14431 (2024)
8. Tan, H., Zhan, S., Lin, H., Zheng, H.T., Kin, W., et al.: QAEA-DR: a unified text augmentation framework for dense retrieval. arXiv preprint arXiv:2407.20207 (2024)
9. Touvron, H., et al.: LLaMA: open and efficient foundation language models. arXiv preprint arXiv:2302.13971 (2023)
10. Wei, J., et al.: Chain-of-thought prompting elicits reasoning in large language models. Adv. Neural. Inf. Process. Syst. **35**, 24824–24837 (2022)
11. Zhang, S., Bansal, M.: Addressing semantic drift in question generation for semi-supervised question answering. arXiv preprint arXiv:1909.06356 (2019)

12. Zhou, D., et al.: Least-to-most prompting enables complex reasoning in large language models. arXiv preprint arXiv:2205.10625 (2022)
13. Zhou, Y., Zhu, X.: A method for generating question-answer pairs based on Seq2Seq framework and text data processing. Comput. Digit. Eng. **50**(11), 2515–2520 (2022)

Improving SQL Generation with Schema Retrieval and Reaction Mechanism

Yang Shuangtao[⊠], Zhang Donghai, and Fu Bo

Lenovo Knowdee (Beijing) Intelligent Technology, Haidian District, Beijing, China
{yangst,zhangdh,fubo}@knowdee.com

Abstract. Text-to-SQL is one of the crucial tasks in natural language processing. We present an innovative approach to the text-to-SQL parsing challenge, leveraging retrieval-driven in-context learning to enhance the performance of large language model (LLMs) in translating natural language queries into SQL queries. We address the critical challenges of schema linking and SQL generation quality by incorporates a schema-similar retrieval module and a Reaction mechanism for error correction. Our schema-similar retrieval module identifies and integrates the most contextually relevant data from the database schema into the LLMs' input, significantly improving the model's reasoning capabilities. The Reaction mechanism introduces a novel verification process during SQL generation, ensuring syntactical correctness and query accuracy. Our method has been rigorously tested on the Archer, a complex bilingual text-to-SQL dataset known for its demanding reasoning requirements. The results demonstrate a remarkable improvement in query accuracy, achieving 42.56% on the Archer dataset, surpassing both Deepseek and GPT-4o baselines. This advancement not only streamlines the interaction between non-expert users and database access but also contributes to the broader field of natural language processing by pushing the boundaries of LLMs applicability in structured query generation.

Keywords: Text-to-SQL · Large Language Model · Schema Linking · Reaction

1 Introduction

Text-to-SQL parsing is a well-established task in the field of natural language processing. Its purpose is to convert natural language queries into SQL queries, bridging the gap between non-expert users and database access [1, 2]. Text-to-SQL primarily faces the following challenges: (1) Extracting the meaning of natural language utterances in a coded manner; (2) Translating the extracted meaning into an equivalent expression in another language; (3) Generating the corresponding SQL query [3].

To address these technical challenges, a large number of scholars have dedicated themselves to experiments from various perspectives such as representation learning, intermediate structures, decoding, model architectures, and training objectives. Early text-to-SQL approaches predominantly employed rule-based and template-based methods [4, 5]. Rule-based methods can effectively and stably handle simple queries, but their flexibility and scalability are limited due to the reliance on rules, templates, and expert knowledge. Currently, mainstream approaches are increasingly focusing on deep learning, particularly the optimization of encoders and decoders. IRNet [6] employs bidirectional LSTM [7] and self-attention mechanisms [8] to encode the question and table schema, respectively. RYANSQL [9] utilizes convolutional neural networks [11] with dense connections [10] for question/schema encoding. Meanwhile, with the advancements in pre-trained language models (PLMs), SQLova [12] first proposed leveraging pre-trained language models like BERT [13] as the base encoder. RATSQL [14], SADGA [15], and LGESQL [16] employ graph neural networks to encode the relational structure between the database schema and the given question. In terms of decoders, there are two categories of SQL generation methods: sketch-based and generation-based. Specifically, sketch-based methods [17] decompose the SQL generation process into sub-modules, each corresponding to a type of predicted slot to be filled. These sub-modules are later aggregated to generate the final SQL query. To enhance the performance of the generated SQL logical form, generation-based methods [16] typically use LSTM decoders to decode the SQL query into an abstract syntax tree in a depth-first traversal order. TaBERT [18] uses tabular data for pre-training, aiming at masked column prediction and cell value recovery to pre-train BERT. Grappa [19] synthesizes query-SQL pairs on tables and pre-trains BERT using masked language modeling objectives, as well as predicting whether columns appear in the SQL query and which SQL operations are triggered.

Large language models (LLMs) have emerged as a milestone for natural language processing and machine learning, with their vast number of parameters and pre-training data providing them with powerful zero-shot and few-shot learning and reasoning capabilities. LLMs excel in semantic encoding, feature extraction, and SQL generation, offering new insights for text-to-SQL [20].

Prompt engineering, often referred to as in-context learning, involves structuring instructions that LLMs can comprehend. From a developer's standpoint, it entails customizing the LLMs' output for specific tasks by crafting precise prompt words during interactions, given the autoregressive decoding nature of most LLMs, subsequent text is predicted based on all currently visible preceding text. This approach is crucial for optimizing performance [21].

In this paper, we present an effective method, based on retrieval-driven In-context Learning, to translate natural language queries into SQL. Previous studies [23] on text-to-SQL prompting using LLMs have been evaluated solely in a zero-shot setting. However, zero-shot prompting merely offers a lower bound on the potential capabilities of LLMs for most tasks. Our research demonstrates that our proposed method significantly outperforms the few-shot prompting approach.

Our method achieved a query accuracy of 42.56% on the Archer dataset, a challenging bilingual text-to-SQL dataset. Our contributions can be summarized as follows:

(1) Enhancing the performance of LLM-based text-to-SQL models through retrieval-driven.
(2) Addressing schema linking challenges through table creation statements and field information.
(3) Improving SQL generation quality by correcting large model errors through a Reaction mechanism.

2 Related Work

Sequence-to-sequence models [24] have shown considerable promise in tasks like text-to-SQL. The main idea is to jointly encode a natural language question and the database schema, then use a decoder to predict the corresponding SQL.

Encoder and decoder are two important optimization directions. Various methods have been focused on them. For example, bidirectional LSTM [25] and CNN [25] are used to represent the question and schema. Wang et al. [26] introduced a unified framework called RAT-SQL for encoding relational structures within database schemas and given questions. This framework employs relation-aware self-attention to integrate global reasoning over schema entities and question words with structured reasoning over predefined schema relations. Gan et al. [28] proposed an intermediate representation to connect natural language questions with SQL statements. Bogin et al. [29] introduced Global-GNN, an alternative approach to schema linking for the Spider dataset. This method applies global reasoning between question words and schema columns and tables. Binyuan and Ruiying [28] proposed a novel model based on schema dependency, designed to more effectively capture the complex interactions between questions and schemas to integrate schema dependency and SQL prediction simultaneously, adopting an adaptive multi-task approach.

Prompting techniques have also been applied to tasks like table understanding, reasoning, and table-to-text generation [27], with impressive results achieved using LLMs with just a few examples in the prompt. Gan [29] introduced an SQL intermediate representation known as Natural SQL (NatSQL), which retains the essential functionalities of SQL. NatSQL simplifies queries by eliminating operators and keywords like GROUP BY, HAVING, FROM, and JOIN ON. This reduction in complexity removes the need for nested subqueries and set operators, and it also makes schema linking more straightforward by minimizing the number of required schema items. Pourreza [32] investigated the effectiveness of decomposing tasks into smaller sub-tasks that breaking down the generation problem into sub-problems and incorporating the solutions of these sub-problems into large language models (LLMs) can significantly enhance their performance.

3 Methodology

We now describe our approach and its application to schema linking. First, we provide the definition of Text-to-SQL and the remainder of the section details our implementation.

3.1 Problem Definition

Given a query Q and a schema S = (D, C) for a relational database or knowledge base, we aim to generate the corresponding SQL to answer the given query by searching within the database. Here, the query $Q = q_1 \ldots q_{|n|}$ is a sequence of words, and the schema comprises columns $D = \{D_1, \ldots, D_{|m|}\}$ and tables $T = \{t_1, \ldots, t_{|n|}\}$. Each column name D_i contains words $c_1, \ldots, c_{|i|}$, and each table name t_j contains words $t_1, \ldots, t_{|j|}$. The desired program PP is represented as an abstract syntax tree TT in the context-free grammar of SQL. Some columns in the schema serve as primary keys, uniquely indexing the corresponding table, while others act as foreign keys, referencing a primary key column in a different table.

3.2 Schema-Similar Retrieval Module

With the continuous improvement of the capabilities of large pre-trained language models (LLMs), in-context learning (ICL) [32] has gradually become a new paradigm in the field of natural language processing. ICL enhances the context with a few task related examples or instructions to improve the prediction performance of language models. Exploring the performance of ICL to evaluate and infer the capabilities of LLMs has also become a new trend. The core of in-context learning lies in learning from task related analogy samples, so more relevant knowledge is more helpful in enhancing the reasoning capabilities of large models.

We introduce a schema-similar retrieval model for embedding non-structured and semi-structured input sequences and linking more relevant tables and data. To enable the retrieval model to obtain effective information from the schema, we establish dependencies between table creation instructions and column names, and reinforce the relationships between data columns in different tables through user queries. The retrieval method is illustrated in Fig. 1.

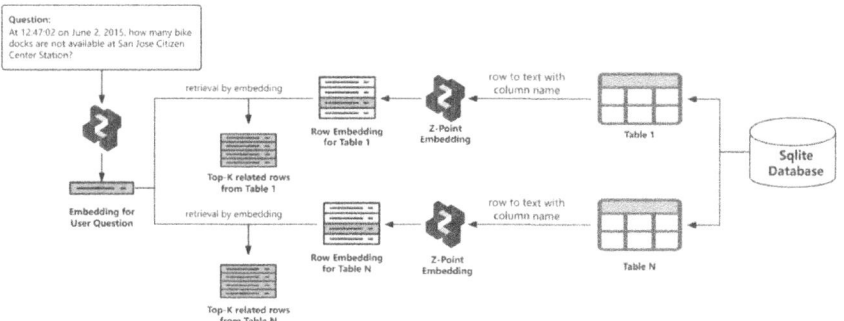

Fig. 1. Schema-Similar Retrieval Data Retrieval Process

Schema-similar retrieval is essentially an effective schema linking solution that addresses issues such as irrelevant information in In-Context Learning. It retrieves the K most relevant rows from various tables in the target database based on the user's question, integrating these rows as examples into the SQL generation instructions. This provides large models with the most relevant data inputs for generating SQL, thereby enhancing the accuracy of SQL generation.

3.3 SQL Generation Based on Reaction

The process of SQL generation by large models typically lacks a standardized verification procedure. To avoid the generation of invalid SQL, we propose the reaction mechanism, which requires large models to consider two questions during generation:

(1) Does the generated SQL contain syntax errors?
(2) Can the generated SQL retrieve the correct answer?

Unlike Chain of Thought (COT) [33], our proposed reaction tolerates errors in the generation process to obtain sufficient error information, guiding the generation of more meaningful SQL. The reaction process is illustrated in Fig. 2.

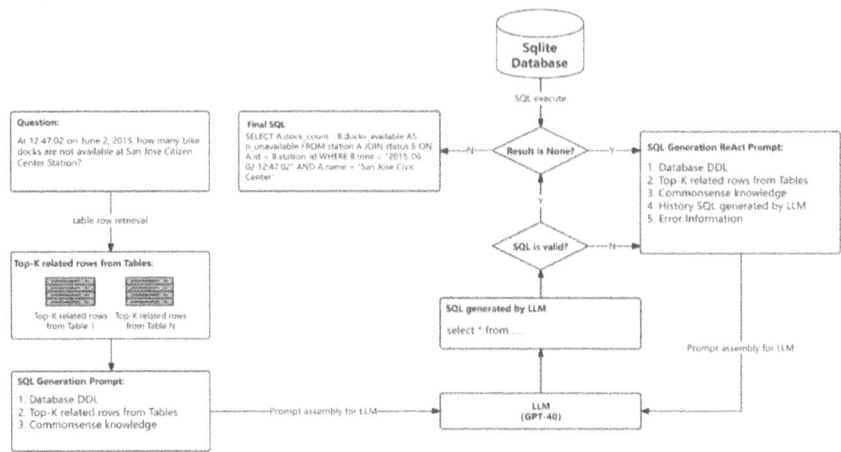

Fig. 2. SQL Generation Based on Reaction

Reaction utilizes the most relevant rows from each table retrieved by schema to construct SQL generation instructions, which are then used by the large model to generate SQL. The generated SQL is evaluated to determine if it is syntactically correct and can retrieve results from the database. If the SQL contains syntax errors or fails to retrieve results from the database, the error information is filled into the reaction prompt for regenerating SQL.

4   Experiments

4.1   Dataset

Archer[1] is a challenging bilingual text-to-SQL dataset focused on complex reasoning, encompassing arithmetic, commonsense, and hypothetical reasoning. It includes 1,042 English questions and 1,042 Chinese questions, along with 521 unique SQL queries, spanning 20 English databases across 20 domains.

Commonsense Reasoning Example

Query: Which 4-cylinder car needs the most fuel to drive 300 miles? List how many gallons it needs, and its make and model.

Commonsense Knowledge: Fuel used is calculated by divding distance driven by fuel consumption.

SQL:SELECT B. Make, B.Model, 1.0 * 300 / mpg AS n_gallon FROM cars_data A JOIN car_names B ON A.Id = B.MakeId WHERE cylinders = "4" ORDER BY mpg ASC LIMIT 1.

---
[1] https://sig4kg.github.io/archer-bench/.

Arithmetic Reasoning Example

Query: How much higher is the maximum power of a BMW car than the maximum power of a Fiat car?

SQL: SELECT MAX(horsepower) - (SELECT MAX (horsepower) FROM cars_data A JOIN car_names B ON A.id = B.makeid WHERE B.model = "fiat") AS diff FROM cars_data A JOIN car_names B ON A.id = B.makeid WHERE B.model = "bmw".

4.2 Baseline

We mainly use two LLMs as baseline model:

GPT-4o[2] accepts any combination of text, audio, image and video as input and generates any combination of text, audio and image. It represents a significant step towards more natural human-computer interaction.

Deepseek[3] has made comprehensive innovations to the self-attention mechanism in the transformer architecture, proposing the MLA (Multi-head Latent Attention) structure and contains 236B parameters, with 2.1B parameters activated per token, and supports a context length of up to 128K.

4.3 Results

We conducted preliminary experiments on Archer to test generalization and used execution accuracy to evaluate performance. The results as shown in Table 1.

Table 1. Execution accuracy (%) on the test set of Archer.

Model	English	Chinese
Deepseek	31.1	25.0
Deepseek + Z-point Embedding	48.0	36.7
GPT-4o	39.12	37.79
GPT-4o + Z-point Embedding	42.18	42.94

As shown in Table 1, our method, when compared to deepseek, demonstrates an average improvement of 14.4% in execution accuracy, and an average improvement of 4.1% on GPT-4o. Analysis reveals that schema-similar retrieval addresses the issue of schema linking, making the retrieved tables and data more relevant to user queries and thereby enhancing the model's reasoning capabilities. Additionally, our analysis of SQL execution results indicates the reaction mechanism significantly ensures the execution rate of SQL, with an average improvement of 2% in the accuracy of SQL execution.

Even the phased retrieval approach and reaction mechanism we proposed effectively address the issue of insufficient information during the SQL generation process, the 42%

---

[2] https://openai.com/index/hello-gpt-4o/.

[3] https://chat.deepseek.com/.

execution rate, subjectively, still falls short of industrial demands. We hypothesize that the primary reason is the lack of semantic information in the schema of certain tables, which leads to insufficient basis for retrieval and feedback, thereby failing to provide strong support for LLMs. Additionally, to ensure the generality of our solution, we did not fine-tune our model in domain-specific data. In industrial applications, training is typically based on domain-specific data, and we believe this can address the issue.

References

1. Zhong, V., Xiong, C., Socher, R.: Seq2SQL: generating structured queries from natural language using reinforcement learning. arXiv preprint arXiv:1709.00103 (2017)
2. Yu, T., et al.: Spider: a large scale human-labeled dataset for complex and crossdomain semantic parsing and text-to-SQL task. In: Proceedings of the 2018 Conference on Empirical Methods in Natural Language Processing, pp. 3911–3921, Brussels, Belgium. Association for Computational Linguistics (2018c)
3. Deng, N., Chen, Y., Zhang, Y.: Recent advances in Text-to-SQL: a survey of what we have and what we expect (2022). arXiv:2208.10099
4. Li, F., Jagadish, H.V.: Constructing an interactive natural language interface for relational databases. Proc. VLDB Endowment **8**(1), 73–84 (2014)
5. Mahmud, T., Azharul Hasan, K.M., Ahmed, M., Chak, T.H.C.: A rule based approach for NLP based query processing. In: 2015 2nd International Conference on Electrical Information and Communication Technologies (EICT), pp. 78–82. IEEE (2015)
6. Guo, J., et al.: Towards complex text-to-SQL in cross domain database with intermediate representation. Proc. ACL (2019)
7. Hochreiter, S., Schmidhuber, J.: Long short-term memory. Neural Comput. (1997)
8. Vaswani, A.: Attention is all you need. Proc. NeurIPS (2017)
9. Choi, D., Shin, M.C., Kim, E., Shin, D.R.: RYANSQL: recursively applying sketch-based slot fillings for complex text-to-SQL in cross-domain databases. Comput. Linguist. (2021)
10. Yoon, D., Lee, D., Lee, S.: Dynamic self-attention: computing attention over words dynamically for sentence embedding. ArXiv preprint (2018)
11. O'Shea, K., Nash, R.: An introduction to convolutional neural networks. ArXiv preprint (2015)
12. Hwang, W., Yim, J., Park, S., Seo, M.: A comprehensive exploration on WikiSQL with table-aware word contextualization. ArXiv preprint (2019)
13. Devlin, J., Chang, M.-W., Lee, K., Toutanova, K.: BERT: pre-training of deep bidirectional transformers for language understanding. Proc. AACL (2019)
14. Wang, B., Shin, R., Liu, X., Polozov, O., Richardson, M.: RAT-SQL: relation-aware schema encoding and linking for text-to-SQL parsers. Proc. ACL (2020)
15. Cai, R., Yuan, J., Xu, B., Hao, Z.: SADGA: structure aware dual graph aggregation network for Text-to-SQL. Proc. NeurIPS (2021)
16. Cao, R., Chen, L., Chen, Z., Zhao, Y., Zhu, S., Yu, K.: LGESQL: line graph enhanced text-to-SQL model with mixed local and non-local relations. Proc. ACL (2021)
17. Hui, B., et al.: Improving Text-to-SQL with schema dependency learning. ArXiv preprint (2021)
18. Yin, P., Neubig, G., Yih, W.-T., Riedel, S.: TaBERT: pretraining for joint understanding of textual and tabular data. In: Proceedings of the 58th Annual Meeting of the Association for Computational Linguistics, pp. 8413–8426, Online. Association for Computational Linguistics (2020)

19. Yu, T., et al.:. Grappa: grammar-augmented pre-training for table semantic parsing. In: 9th International Conference on Learning Representations, ICLR 2021, Virtual Event, Austria, May 3–7, 2021. OpenReview.net (2021)
20. Zhao, W.X., et al.: A survey of large language models. arXiv preprint arXiv:2303.18223 (2023)
21. Fu, Y., Bailis, P., Stoica, I., Zhang, H.: Break the sequential dependency of LLM inference using look ahead decoding. arXiv preprint arXiv:2402.02057 (2024)
22. Liu, A., Hu, X., Wen, L., Yu, P.S.: A comprehensive evaluation of ChatGPT's zero-shot Text-to-SQL capability. arXiv preprint arXiv:2303.13547 (2023a)
23. Sutskever, I., Vinyals, O., Le, Q.V.: Sequence to sequence learning with neural networks. In: Advances in Neural Information Processing Systems, vol. 27 (2014)
24. Graves, A., Graves, A.: Long short-term memory. Super-vised sequence labelling with recurrent neural networks, pp. 37–45 (2012)
25. Choi, D.H., Shin, M.C., Kim, E.G., Shin, D.R.: RyanSQL: recursively applying sketch-based slot fillings for complex Text-to-SQL in cross-domain databases. Comput. Linguist. **47**(2), 309–332 (2021)
26. Wang, B., Shin, R., Liu, X., Polozov, O., Richardson, M.: Rat SQL: relation-aware schema encoding and linking for Text-to-SQL parsers. arXiv preprint arXiv:1911.04942 (2019)
27. Gan, Y., et al.: Natural SQL: Making SQL easier to infer from natural language specifications. arXiv preprint arXiv:2109.05153 (2021)
28. Bogin, B., Gardner, M., Berant, J.: Global reasoning over database structures for Text to-SQL parsing. In: Proceedings of the 2019 Conference on Empirical Methods in Natural Language Processing and the 9th International Joint Conference on Natural Language Processing (EMNLP IJCNLP), pages 3657–3662 (2019b)
29. Hui, B., et al.: Improving Text-to-SQL with schema dependency learning. arXiv preprint arXiv:2103.04399 (2021)
30. Guo, Z., et al.: Few-shot table-to-text generation with prompt planning and knowledge memorization. arXiv preprint arXiv:2302.04415 (2023)
31. Pourreza, M., Rafiei, D: Din-SQL: decom-posed in-context learning of text-to-SQL with self-correction. In: Advances in Neural Information Processing Systems, vol. 36 (2024)
32. Xie, S.M., Raghunathan, A., Liang, P., Ma, T.: An explanation of In-Context learning as implicit Bayesian inference. In: ICLR (2022)
33. Wei, J., et al.: Chain-of-thought prompting elicits reasoning in large language models. In: NeurIPS (2022)

HIT-SCIR at CCKS-IJCKG2024: Enhancing Text-to-SQL with Multi-step Pipeline

Dingzirui Wang, Xuanliang Zhang, Keyan Xu, and Wanxiang Che(✉)

Harbin Institute of Technology, Harbin, China
{dzrwang,xuanliangzhang,kyxu,car}@ir.hit.edu.cn

Abstract. The text-to-SQL task translates natural language questions into SQL queries, simplifying database access. While large language models (LLMs) have shown strong performance, they often struggle with complex reasoning, such as commonsense and numerical reasoning, required for more challenging SQL generation. We propose a new pipeline that enhances SQL generation by incorporating advanced reasoning skills, alongside techniques like entity linking and self-correction. Tested on the Archer dataset, which requires more complex reasoning, our approach improves performance by 23.3% over the baseline, demonstrating its effectiveness in handling challenging queries.

Keywords: Text-to-SQL · Large Language Models

1 Introduction

The text-to-SQL task is crucial for reducing the difficulty of retrieving information from databases [4,12]. Considering the strong performance of LLMs without fine-tuning, methods based on LLMs have become the current mainstream in text-to-SQL tasks [11,15,24]. Specifically, the text-to-SQL task is defined as taking a user question and a related database as input and generating the corresponding SQL query.

Existing work shows that pipeline approaches can effectively enhance text-to-SQL performance [8,21,29]. For example, MAC-SQL [27] performs entity linking, question decomposition, and self-correction, while DART-SQL [19] rewrites the question before entity linking. However, these methods handle relatively simple SQL queries that are semantically consistent with user questions. In real-world applications, generating SQL often requires more complex reasoning, such as commonsense and numerical reasoning, which current methods struggle with [32].

To address these challenges, we propose a new method that enhances the complex reasoning capabilities when generating SQL queries. We also employ techniques like entity linking [18] and self-correction [3] to improve performance on tasks requiring complex reasoning.

B. Xu et al. (Eds.): CCKS-IJCKG 2024, CCIS 2229, pp. 496–504, 2025.
https://doi.org/10.1007/978-981-96-1809-5_41

To validate our approach, we conducted experiments on Archer [32], a dataset for text-to-SQL tasks demanding more complex reasoning ability than the previous works [17,30]. Our method improves performance by 23.3% over the baseline, demonstrating its effectiveness.

Our contributions are as follows:

– We propose a new text-to-SQL pipeline that significantly improves performance on Archer;
– Our method achieves a 23.3% improvement over the baseline on the Archer dataset, validating its effectiveness.

2 Methodology

In this section, we describe the pipeline we designed for enhancing the text-to-SQL task, particularly focusing on addressing the challenges of complex reasoning such as commonsense and numerical reasoning, which is shown in Fig. 1. Our approach involves a combination of advanced prompting techniques and multiple iterative self-improvement methods. The components of our pipeline are outlined below.

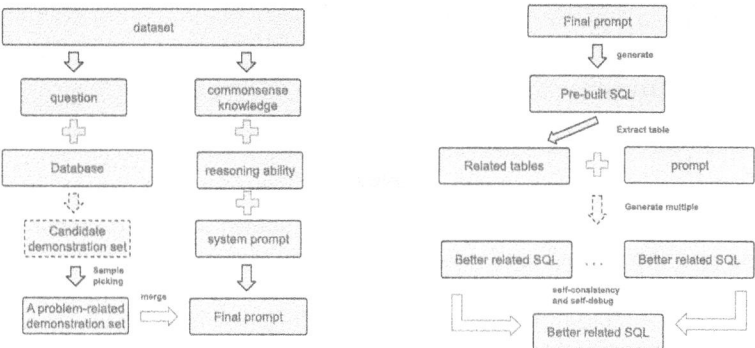

Fig. 1. The illustration of our method.

To improve the model's ability to generate complex SQL queries, we use a demonstration-based prompting strategy. The prompt format includes not only the user's question but also additional structured information that aids the model in solving complex queries.

We further enhance the effectiveness of the model by introducing a system-level prompt. This explicitly defines the task and constraints that the model should follow when generating SQL queries. The system prompt informs the model about its role as an SQL and DevOps expert and clarifies the conditions under which it should provide its response.

To further refine the generated SQL queries, we incorporate a self-debugging mechanism [3]. This involves multiple rounds of query generation, followed by self-assessment and correction. The model iteratively improves its output by detecting and correcting errors in the SQL, particularly those related to table joins, complex conditions, or logic inconsistencies. This multi-round process helps to enhance the robustness of the generated SQL queries.

Another key component of our method is the self-consistency approach [28]. In this approach, we generate multiple SQL queries for the same question and compare them to ensure consistency across different outputs. By selecting the most consistent query, we reduce the likelihood of generating incorrect or suboptimal SQL statements.

Finally, we introduce a link-table method to further enhance query accuracy. In this step, we first pre-generate SQL queries and extract relevant tables from them. Using this information, we optimize the prompt by focusing the model's attention on specific tables relevant to the query. This helps in reducing ambiguity and improving the precision of the generated SQL queries.

Our final pipeline integrates all the above components: demonstration-based prompting, system-level prompting, self-debugging, self-consistency, and link-table optimization. These techniques work together to enhance the model's ability to generate SQL queries requiring complex reasoning capabilities. We set hyperparameters based on the complexity of the task and adjust them dynamically during the iterative process.

3   Experiment

3.1   Experiment Settings

In our experiments, we use `Deepseek-Coder-33B` and `GPT-4o` as the primary models. `Deepseek-Coder-33B` is a high-performance language model designed for code generation and understanding tasks, with robust multimodal processing capabilities. This model, trained using advanced deep learning techniques, contains 33 billion parameters, enabling it to efficiently handle tasks such as code generation, debugging, code completion, and cross-language code translation. It excels across various programming languages, particularly in understanding and reasoning through complex syntactic structures, making it highly applicable in fields like automated software development and code optimization. `GPT-4o`, on the other hand, is an optimized version of the `GPT-4o` architecture, focusing on improving accuracy and efficiency in both code generation and natural language processing tasks. Compared to the base version of `GPT-4o`, `GPT-4o` enhances the model's adaptability to language-code multimodal inputs, showing significant advantages in complex language reasoning, context retention, and cross-domain knowledge application. It generates high-quality code and excels in code semantic understanding, error correction, and multi-task code processing, making it a powerful tool for solving problems across various domains. Both models play crucial roles in our experiments, providing solid support and data foundations for our research in code generation, code understanding, and multimodal tasks.

For the demonstration selection, we select 7 databases based on the BM25 similarity with the user question, selecting 5 questions from each database. For the self-debugging process, we set the maximum number of iterations to 3. For self-consistency, we set the temperature to 0.3 and generate 8 outputs for each input.

Table 1. The main experimental results of our method.

Method	Dev EX	Test EX
GPT-3.5 + CT-3 [32]	10.57	15.84
Deepseek-Coder-33B	11.54	–
+ Self-Debug	13.46	–
+ Schem-Linking	14.42	–
+ Self-Consistency	16.35	–
GPT-4o	34.62	39.12

3.2   Main Experiment

Our main experimental results are shown in Table 1. From the table, we can observe the following insights:

(i) Our proposed approach leads to a considerable performance improvement, yielding a 4.81% increase in performance over the baseline method (GPT-3.5 + CT-3), which demonstrates the effectiveness of our methodology. Specifically, the model incorporating self-debugging, schema-linking, and self-consistency consistently outperforms the base model across different experimental settings. These enhancements are particularly evident in the progression from 11.54% for the baseline Deepseek-Coder-33B to 16.35% when all improvements are combined.

(ii) The self-debugging mechanism introduces the most notable improvement among the individual techniques, raising performance from 11.54% to 13.46%. This significant increase suggests that the primary challenge lies in the complexity of the original questions, which often leads to the model producing SQL queries that cannot be executed. By iterative refining and debugging the generated SQL, the model can better address these difficulties, thus boosting its performance.

(iii) Schema-linking also plays an essential role in enhancing the model's capability, as it further improves performance to 14.42%. This indicates that the ability to link the input queries to relevant schema components effectively is crucial for solving complex SQL generation tasks. By explicitly modeling these relationships, the model is better equipped to produce accurate and executable SQL queries.

(iv) Self-consistency provides an additional performance boost, raising the performance to 16.35%. This technique, which leverages multiple sampled outputs to ensure consistency, proves highly effective for addressing ambiguity in the input queries. By combining several outputs and selecting the most consistent one, the model achieves more robust performance, reducing the likelihood of incorrect SQL generations.

(v) Furthermore, the results show that our method, when fully integrated, surpasses not only baseline methods but also more advanced models like GPT-4o, which achieves 34.62% and 39.12% on the development and test sets, respectively. Although GPT-4o shows superior overall performance, our methodology demonstrates significant incremental improvements in more specific aspects like self-debugging and schema-linking, contributing to performance enhancement even for models with fewer parameters.

In conclusion, our approach addresses the key challenges in SQL generation, particularly in handling execution failures, ambiguous schema mapping, and query consistency. These results underscore the potential of combining self-debugging, schema-linking, and self-consistency for improving model performance in complex tasks.

4   Related Work

Research on the text-to-SQL task has garnered significant attention due to its practical applications in enabling users to interact with databases using natural language. This task focuses on the automatic generation of SQL queries based on user questions, while also incorporating the structure of the underlying database schemas. The challenge lies in bridging the gap between the flexibility of human language and the rigid syntax of SQL, making this a complex problem at the intersection of natural language processing (NLP) and database management. As the field evolves, there has been a shift towards utilizing more sophisticated models, driven by advances in language understanding and representation.

In the early stages of this research, smaller models were commonly employed to tackle the text-to-SQL challenge. One prominent example of these early approaches is the use of encoder models, such as BERT [5]. BERT was primarily tasked with identifying relevant entities or keywords in a user's question that could be mapped to components of an SQL query. These models did not generate full SQL queries but instead facilitated the identification of important elements that would be used to form the queries. By leveraging the strong contextual understanding provided by models like BERT, these early approaches could extract key information from natural language input.

In addition to encoder models, another popular strategy in the early phases of text-to-SQL research was the application of encoder-decoder models. Models such as BART [14] were particularly effective in treating text-to-SQL as a natural language translation task. In this framework, the model's encoder would interpret the input question, while the decoder would generate a corresponding SQL query. This approach effectively reframed the problem as translating between two languages: one human and one machine. By leveraging both the contextual understanding from the encoder and the generative capacity of the decoder, these models could output more complete SQL queries, significantly advancing the field.

Several studies have contributed to refining the use of encoder-decoder models for text-to-SQL tasks. Researchers have explored methods to improve the models' understanding of both the question and the database schema. For example, studies like Guo et al. [9] and Bogin et al. [1] sought to incorporate better global reasoning about database structures, while Wang et al. [26] introduced more sophisticated techniques to represent the relationships between various components within a database. The goal of these works was to improve the precision and accuracy of the SQL queries generated by the models, especially in complex databases with many interrelated tables and fields.

More recent advancements have continued to push the boundaries of text-to-SQL systems by incorporating the latest pre-trained language models (PLMs) into their architectures. For instance, models like PICARD [23] and UniSAr [6] have adopted strategies to fine-tune large pre-trained models to better align with the specific needs of SQL generation tasks. These approaches have led to more robust and scalable solutions, as the models can now handle more varied user questions and larger, more complex databases. By continually refining the alignment between the natural language input and the database structure, these systems represent a significant step forward in the field.

In conclusion, the evolution of the text-to-SQL task has moved from the early use of simpler models like BERT and BART to more sophisticated, schema-aware systems that incorporate the latest in pre-trained language model technology. These advancements have enabled researchers to develop models that can not only identify relevant entities and keywords within user questions but also generate highly accurate and contextually appropriate SQL queries. As the field continues to evolve, further improvements are expected, particularly in terms of handling more complex database structures and diverse user input.

In recent years, methods based on LLMs have gained traction as the dominant approach for text-to-SQL tasks, thanks to their remarkable performance even without fine-tuning [11]. One approach is to fine-tune LLMs specifically for text-to-SQL tasks, either by leveraging text-to-SQL datasets directly [16,22,31] or by incorporating data from related tasks such as entity linking and knowledge generation [10,25]. Alternatively, many recent studies explore the use of few-shot inference with LLMs, which delivers superior performance without the need for fine-tuning. Techniques like in-context learning [2,7,21], self-consistency [13,20, 33], and multi-agent systems [27,29] have shown significant promise, allowing LLMs to solve text-to-SQL tasks more efficiently.

5 Conclusion

In this work, we address the challenge of improving the reasoning capabilities required for complex text-to-SQL tasks. While existing approaches prove effective for simple queries that align semantically with user inputs, they often struggle with real-world demands such as commonsense and numerical reasoning. To overcome these limitations, we propose a new text-to-SQL pipeline that incorporates advanced techniques like entity linking and self-correction to enhance performance on complex queries. Our experiments on the Archer dataset, which focuses on advanced reasoning capabilities, show a significant improvement of 23.3% over baseline methods. This demonstrates the effectiveness of our approach in handling more sophisticated SQL generation tasks. In conclusion, our contributions advance the text-to-SQL field, particularly in scenarios requiring complex reasoning. Future work may explore further optimizations to address even broader reasoning challenges across various domains.

References

1. Bogin, B., Gardner, M., Berant, J.: Global reasoning over database structures for text-to-SQL parsing. In: Inui, K., Jiang, J., Ng, V., Wan, X. (eds.) Proceedings of the 2019 Conference on Empirical Methods in Natural Language Processing and the 9th International Joint Conference on Natural Language Processing (EMNLP-IJCNLP), pp. 3659–3664. Association for Computational Linguistics, Hong Kong, China (2019). https://doi.org/10.18653/v1/D19-1378, https://aclanthology.org/D19-1378

2. Chang, S., Fosler-Lussier, E.: Selective demonstrations for cross-domain text-to-SQL. In: Bouamor, H., Pino, J., Bali, K. (eds.) Findings of the Association for Computational Linguistics: EMNLP 2023. pp. 14174–14189. Association for Computational Linguistics, Singapore (2023). https://doi.org/10.18653/v1/2023.findings-emnlp.944, https://aclanthology.org/2023.findings-emnlp.944

3. Chen, X., Lin, M., Schärli, N., Zhou, D.: Teaching large language models to self-debug. In: The Twelfth International Conference on Learning Representations (2024). https://openreview.net/forum?id=KuPixIqPiq

4. Deng, N., Chen, Y., Zhang, Y.: Recent advances in text-to-SQL: a survey of what we have and what we expect. In: Calzolari, N., et al. (eds.) Proceedings of the 29th International Conference on Computational Linguistics. International Committee on Computational Linguistics, Gyeongju, Republic of Korea, pp. 2166–2187 (O2022), https://aclanthology.org/2022.coling-1.190

5. Devlin, J., Chang, M.W., Lee, K., Toutanova, K.: BERT: pre-training of deep bidirectional transformers for language understanding. In: Burstein, J., Doran, C., Solorio, T. (eds.) Proceedings of the 2019 Conference of the North American Chapter of the Association for Computational Linguistics: Human Language Technologies, Volume 1 (Long and Short Papers), Minneapolis, Minnesota, pp. 4171–4186. Association for Computational Linguistics (2019). https://doi.org/10.18653/v1/N19-1423, https://aclanthology.org/N19-1423

6. Dou, L., et al.: Unisar: a unified structure-aware autoregressive language model for text-to-SQL. arXiv preprint arXiv:2203.07781 (2022). https://arxiv.org/abs/2203.07781

7. Gao, D., et al.: Text-to-SQL empowered by large language models: a benchmark evaluation. Proc. VLDB Endow. **17**(5), 1132–1145 (2024). https://doi.org/10.14778/3641204.3641221

8. Gu, Z., et al.: Interleaving pre-trained language models and large language models for zero-shot nl2sql generation (2023). https://arxiv.org/abs/2306.08891

9. Guo, J., et al.: Towards complex text-to-SQL in cross-domain database with intermediate representation. In: Korhonen, A., Traum, D., Màrquez, L. (eds.) Proceedings of the 57th Annual Meeting of the Association for Computational Linguistics, Florence, Italy, pp. 4524–4535. Association for Computational Linguistics (2019). https://doi.org/10.18653/v1/P19-1444, https://aclanthology.org/P19-1444

10. Hong, Z., Yuan, Z., Chen, H., Zhang, Q., Huang, F., Huang, X.: Knowledge-to-SQL: enhancing SQL generation with data expert LLM (2024). https://arxiv.org/abs/2402.11517

11. Hong, Z., et al.: Next-generation database interfaces: a survey of LLM-based text-to-SQL (2024). https://arxiv.org/abs/2406.08426

12. Kanburoğlu, A., Tek, F.: Text-to-SQL: a methodical review of challenges and models. Turkish J. Electr. Eng. Comput. Sci. **32**(3), 403–419 (2024). https://doi.org/10.55730/1300-0632.4077, publisher Copyright: TÜBİTAK

13. Lee, D., Park, C., Kim, J., Park, H.: MCS-SQL: leveraging multiple prompts and multiple-choice selection for text-to-SQL generation (2024). https://arxiv.org/abs/2405.07467

14. Lewis, M., et al.: BART: denoising sequence-to-sequence pre-training for natural language generation, translation, and comprehension. In: Jurafsky, D., Chai, J., Schluter, N., Tetreault, J. (eds.) Proceedings of the 58th Annual Meeting of the Association for Computational Linguistics, pp. 7871–7880. Association for Computational Linguistics (2020). https://doi.org/10.18653/v1/2020.acl-main.703, https://aclanthology.org/2020.acl-main.703

15. Li, B., Luo, Y., Chai, C., Li, G., Tang, N.: The dawn of natural language to SQL: are we fully ready? (2024). https://doi.org/10.14778/3681954.3682003, https://arxiv.org/abs/2406.01265

16. Li, H., et al.: Codes: towards building open-source language models for text-to-SQL. Proc. ACM Manag. Data **2**(3) (2024). https://doi.org/10.1145/3654930

17. Li, J., et al.: Can LLM already serve as a database interface? a BIg bench for large-scale database grounded text-to-SQLs. In: Thirty-seventh Conference on Neural Information Processing Systems Datasets and Benchmarks Track (2023). https://openreview.net/forum?id=dI4wzAE6uV

18. Liu, Q., Yang, D., Zhang, J., Guo, J., Zhou, B., Lou, J.G.: Awakening latent grounding from pretrained language models for semantic parsing. In: Zong, C., Xia, F., Li, W., Navigli, R. (eds.) Findings of the Association for Computational Linguistics: ACL-IJCNLP 2021, pp. 1174–1189. Association for Computational Linguistics (2021). https://doi.org/10.18653/v1/2021.findings-acl.100, https://aclanthology.org/2021.findings-acl.100

19. Mao, W., et al.: Enhancing text-to-SQL parsing through question rewriting and execution-guided refinement. In: Ku, L.W., Martins, A., Srikumar, V. (eds.) Findings of the Association for Computational Linguistics ACL 2024, Bangkok, Thailand and virtual meeting, pp. 2009–2024. Association for Computational Linguistics (2024). https://aclanthology.org/2024.findings-acl.120

20. Ni, A., et al.: Lever: learning to verify language-to-code generation with execution. In: Proceedings of the 40th International Conference on Machine Learning. ICML'23, JMLR.org (2023)

21. Pourreza, M., Rafiei, D.: DIN-SQL: decomposed in-context learning of text-to-SQL with self-correction. In: Thirty-seventh Conference on Neural Information Processing Systems (2023). https://openreview.net/forum?id=p53QDxSIc5

22. Pourreza, M., Rafiei, D.: DTS-SQL: Decomposed text-to-SQL with small large language models (2024). https://arxiv.org/abs/2402.01117

23. Scholak, T., Schucher, N., Bahdanau, D.: PICARD: Parsing incrementally for constrained auto-regressive decoding from language models. In: Moens, M.F., Huang, X., Specia, L., Yih, S.W.T. (eds.) Proceedings of the 2021 Conference on Empirical Methods in Natural Language Processing, Punta Cana, Dominican Republic, pp. 9895–9901. Association for Computational Linguistics (2021). https://doi.org/10.18653/v1/2021.emnlp-main.779, https://aclanthology.org/2021.emnlp-main.779

24. Shi, L., Tang, Z., Yang, Z.: A survey on employing large language models for text-to-SQL tasks (2024). https://arxiv.org/abs/2407.15186

25. Sun, Y., et al.: QDA-SQL: questions enhanced dialogue augmentation for multi-turn text-to-SQL (2024). https://arxiv.org/abs/2406.10593

26. Wang, B., Shin, R., Liu, X., Polozov, O., Richardson, M.: RAT-SQL: relation-aware schema encoding and linking for text-to-SQL parsers. In: Jurafsky, D., Chai, J., Schluter, N., Tetreault, J. (eds.) Proceedings of the 58th Annual Meeting of the Association for Computational Linguistics, pp. 7567–7578. Association for Computational Linguistics (2020). https://doi.org/10.18653/v1/2020.acl-main.677, https://aclanthology.org/2020.acl-main.677

27. Wang, B., et al.: MAC-SQL: a multi-agent collaborative framework for text-to-SQL (2024). https://arxiv.org/abs/2312.11242

28. Wang, X., et al.: Self-consistency improves chain of thought reasoning in language models. In: The Eleventh International Conference on Learning Representations (2023). https://openreview.net/forum?id=1PL1NIMMrw

29. Xie, Y., et al.: Decomposition for enhancing attention: Improving LLM-based text-to-SQL through workflow paradigm (2024). https://arxiv.org/abs/2402.10671

30. Yu, T., et al.: Spider: a large-scale human-labeled dataset for complex and cross-domain semantic parsing and text-to-SQL task. In: Riloff, E., Chiang, D., Hockenmaier, J., Tsujii, J. (eds.) Proceedings of the 2018 Conference on Empirical Methods in Natural Language Processing, Brussels, Belgium, pp. 3911–3921. Association for Computational Linguistics (2018). https://doi.org/10.18653/v1/D18-1425, https://aclanthology.org/D18-1425

31. Zhang, C., et al.: FinSQL: model-agnostic LLMS-based text-to-SQL framework for financial analysis. In: Companion of the 2024 International Conference on Management of Data, pp. 93–105. SIGMOD/PODS '24, New York, NY, USA. Association for Computing Machinery (2024). https://doi.org/10.1145/3626246.3653375

32. Zheng, D., Lapata, M., Pan, J.: Archer: a human-labeled text-to-SQL dataset with arithmetic, commonsense and hypothetical reasoning. In: Graham, Y., Purver, M. (eds.) Proceedings of the 18th Conference of the European Chapter of the Association for Computational Linguistics (Volume 1: Long Papers), St. Julian's, Malta, pp. 94–111. Association for Computational Linguistics (2024). https://aclanthology.org/2024.eacl-long.6

33. Zhong, R., Snell, C., Klein, D., Eisner, J.: Non-programmers can label programs indirectly via active examples: a case study with text-to-SQL. In: Bouamor, H., Pino, J., Bali, K. (eds.) Proceedings of the 2023 Conference on Empirical Methods in Natural Language Processing, Singapore, pp. 5126–5152. Association for Computational Linguistics (2023). https://doi.org/10.18653/v1/2023.emnlp-main.312, https://aclanthology.org/2023.emnlp-main.312

Author Index

© The Editor(s) (if applicable) and The Author(s), under exclusive license
to Springer Nature Singapore Pte Ltd. 2025
B. Xu et al. (Eds.): CCKS-IJCKG 2024, CCIS 2229, pp. 505–507, 2025.
https://doi.org/10.1007/978-981-96-1809-5

The manufacturer's authorised representative in the EU is Springer
Nature Customer Service Centre GmbH, Europaplatz 3, 69115 Heidelberg,
Germany. If you have any concerns regarding our products, please
contact ProductSafety@springernature.com

Printed and bound by CPI Group (UK) Ltd, Croydon, CR0 4YY
06/05/2026
02103601-0004